Waste-derived Biochar for Sustainable
Rural Revitalization

Waste-derived Biochar for Sustainable Rural Revitalization

Edited by
Yuqing Sun
Sun Yat-Sen University
China

Daniel C.W. Tsang
Hong Kong University of Science and Technology
China

WILEY

Published by John Wiley & Sons, Inc., Hoboken, New Jersey.

For general information on our other products and services or for technical support, please contact our Customer Care Department within the United States at (800) 762-2974, outside the United States at (317) 572-3993 or fax (317) 572-4002.

Wiley also publishes its books in a variety of electronic formats. Some content that appears in print may not be available in electronic formats. For more information about Wiley products, visit our web site at www.wiley.com.

Library of Congress Cataloging-in-Publication Data:

Names: Sun, Yuqing (Assistant professor) editor | Tsang, Daniel C. W. editor
Title: Waste-derived biochar for sustainable rural revitalization / edited by Yuqing Sun, Daniel C.W. Tsang.
Description: Hoboken, New Jersey: Wiley, [2026] | Includes bibliographical references and index.
Identifiers: LCCN 2025030995 | ISBN 9781394250271 hardback | ISBN 9781394250295 adobe pdf | ISBN 9781394250288 epub | ISBN 9781394250301
Subjects: LCSH: Biochar | Rural development | Sustainable agriculture
Classification: LCC TP248.B55 W36 2026
LC record available at https://lccn.loc.gov/2025030995

Cover Design: Wiley
Cover Image: © fotoVoyager/Getty Images

Contents

List of Contributors

Huijun Li
State Key Lab of Biocontrol
Guangdong Provincial Key Laboratory of Plant Stress Biology
School of Agriculture and Biotechnology
Shenzhen Campus of Sun Yat-sen University
Shenzhen
P. R. China

Jing Luo
School of Agriculture and Biotechnology
Sun Yat-sen University
Guangzhou
Guangdong
China

Leichang Cao
College of Chemistry and Molecular Sciences
Henan University
Kaifeng
China

Miami College, Henan University
Kaifeng
China

Qiang Chen
College of Environmental Science and Engineering
Hunan University
Changsha
Hunan
China

Kui Cheng
International Cooperation Joint Laboratory of Health in Cold Region
Black Soil Habitat of the Ministry of Education
Harbin
China

College of Engineering, Northeast Agricultural University
Harbin
China

Jiaqi Deng
College of Engineering, Huazhong Agricultural University
Wuhan
P. R. China

Pavani Dulanja Dissanayake
Department of Soil Science, Faculty of Agriculture
University of Peradeniya
Peradeniya
Sri Lanka

Le Fang
Research and Development Center for Watershed Environmental Eco-Engineering
Beijing Normal University
Zhuhai
P. R. China

State Key Laboratory of Wetland Conservation and Restoration
School of Environment
Beijing Normal University
Beijing
P. R. China

Key Laboratory of Coastal Water Environmental Management and Water Ecological Restoration of Guangdong Higher Education Institutes
Beijing Normal University
Zhuhai
P. R. China

Qianzhen Fang
College of Environmental Science and Engineering
Hunan University
Changsha
Hunan
China

Yipu Guo
Centre for Infrastructure Engineering and Safety, School of Civil and Environmental Engineering
The University of New South Wales
Kensington
NSW
Australia

Lanfang Han
Key Laboratory for City Cluster Environmental Safety and Green Development of the Ministry of Education
School of Ecology, Environment and Resources, Guangdong University of Technology
Guangzhou
China

Yingying Han
Research and Development Center for Watershed Environmental Eco-Engineering
Beijing Normal University
Zhuhai
P. R. China

State Key Laboratory of Wetland Conservation and Restoration
School of Environment
Beijing Normal University
Beijing
P. R. China

Key Laboratory of Coastal Water Environmental Management and Water Ecological Restoration of Guangdong Higher Education Institutes
Beijing Normal University
Zhuhai
P. R. China

Shih-Hsin Ho
State Key Laboratory of Urban Water Resource and Environment
School of Environment, Harbin Institute of Technology
Harbin
P. R. China

Haodong Hou
College of Chemistry and Molecular Sciences
Henan University
Kaifeng
China

Miami College, Henan University
Kaifeng
China

Fang Huang
School of Environment and Health, Jianghan University
Wuhan
P. R. China

Xiaochen Huang
State Key Lab of Biocontrol
Guangdong Provincial Key Laboratory of Plant Stress Biology
School of Agriculture and Biotechnology
Shenzhen Campus of Sun Yat-sen University
Shenzhen
P. R. China

Xiaoqian Jiang
School of Agriculture and Biotechnology, Sun Yat-sen University
Guangzhou
Guangdong
China

Fanying Kong
School of Water Conservancy and Civil Engineering, Northeast Agricultural University
Harbin
China

Jingyi Liang
Department of Civil and Environmental Engineering
The Hong Kong University of Science and Technology
Clear Water Bay
Hong Kong
China

Xin Liu
College of Environment and Ecology, Hunan Agricultural University
Changsha
China

Zhongwang Liu
School of Agriculture and Biotechnology, Sun Yat-sen University
Guangzhou
Guangdong
China

Zhuqing Liu
School of Water Conservancy and Civil Engineering
Northeast Agricultural University
Harbin
China

International Cooperation Joint Laboratory of Health in Cold Region

Black Soil Habitat of the Ministry of Education
Harbin
China

Yong Sik Ok
Korea Biochar Research Center
Department of Environmental Science and Ecological Engineering
Korea University
Seoul
Republic of Korea

International ESG Association (IESGA)
Seoul
Republic of Korea

Yutao Peng
School of Agriculture and Biotechnology, Sun Yat-sen University
Guangzhou
Guangdong
China

Fulin Qu
Centre for Infrastructure Engineering and Safety
School of Civil and Environmental Engineering, The University of New South Wales
Kensington
NSW
Australia

Jianhua Qu
School of Resources and Environment, Northeast Agricultural University
Harbin
China

Hong-Yu Ren
State Key Laboratory of Urban-Rural Water Resource and Environment, School of Environment
Harbin Institute of Technology
Harbin
China

Maheshika Senanayake
Korea Biochar Research Center, Department of Environmental Science and Ecological Engineering
Korea University
Seoul
Republic of Korea

Jiang Song
School of Water Conservancy and Civil Engineering
Northeast Agricultural University
Harbin
China

International Cooperation Joint Laboratory of Health in Cold Region Black Soil Habitat of the Ministry of Education
Harbin
China

Qingqing Song
State Key Laboratory of Urban-Rural Water Resource and Environment
School of Environment
Harbin Institute of Technology
Harbin
China

Xueting Song
State Key Laboratory of Urban Water Resource and Environment
School of Environment
Harbin Institute of Technology
Harbin
China

Ke Sun
State Key Laboratory of Water Simulation
School of Environment, Beijing Normal University
Beijing
China

Yuqing Sun
School of Agriculture and Biotechnology, Sun Yat-sen University
Guangzhou
Guangdong
China

Jinfang Tan
School of Agriculture and Biotechnology, Sun Yat-sen University
Guangzhou
Guangdong
China

Xiaofei Tan
College of Environmental Science and Engineering, Hunan University
Changsha
Hunan
China

Daniel C.W. Tsang
Department of Civil and Environmental Engineering
The Hong Kong University of Science and Technology
Clear Water Bay
Hong Kong
China

Jieni Wang
College of Chemistry and Molecular Sciences, Henan University
Kaifeng
China

Miami College
Henan University
Kaifeng
China

Yizhe Wang
Centre for Infrastructure Engineering and Safety
School of Civil and Environmental Engineering, The University of New South Wales
Kensington
NSW
Australia

Mi Wei
School of Agriculture and Biotechnology, Sun Yat-sen University
Guangzhou
Guangdong
China

Peng Xie
State Key Laboratory of Urban Water Resource and Environment
School of environment, Harbin Institute of Technology
Harbin
P. R. China

Yifan Xing
Research and Development Center for Watershed Environmental Eco-Engineering
Beijing Normal University
Zhuhai
P. R. China

State Key Laboratory of Wetland Conservation and Restoration
School of Environment
Beijing Normal University
Beijing
P. R. China

Key Laboratory of Coastal Water Environmental Management and Water Ecological Restoration of Guangdong Higher Education Institutes
Beijing Normal University
Zhuhai
P. R. China

Weijian Xu
Department of Civil and Environmental Engineering
The Hong Kong University of Science and Technology
Clear Water Bay
Hong Kong
China

Fan Yang
School of Water Conservancy and Civil Engineering, Northeast Agricultural University
Harbin
China

International Cooperation Joint Laboratory of Health in Cold Region Black Soil Habitat of the Ministry of Education
Harbin
China

Hailan Yang
College of Environmental Science and Engineering, Hunan University
Changsha
Hunan
China

Baojun Yi
College of Engineering, Huazhong Agricultural University
Wuhan
P. R. China

Su Yilin
Department of Civil and Environmental Engineering
The Hong Kong Polytechnic University
Kowloon
Hong Kong
China

Lingfeng Zeng
College of Environment and Ecology
Hunan Agricultural University
Changsha
China

Shuqin Zhang
College of Chemistry and Molecular Sciences, Henan University
Kaifeng
China

Miami College
Henan University
Kaifeng
China

Ying Zhang
School of Resources and Environment
Northeast Agricultural University
Harbin
China

Northeast Institute of Geography and Agroecology
Chinese Academy of Sciences
Changchun
China

Yuying Zhang
Department of Civil and Environmental Engineering
The Hong Kong University of Science and Technology
Clear Water Bay
Hong Kong
China

Ying Zhao
School of Water Conservancy & Civil Engineering
Northeast Agricultural University
Harbin
China

International Cooperation Joint Laboratory of Health in Cold Region Black Soil Habitat of the Ministry of Education
Harbin
China

Yaoyu Zhou
College of Environment and Ecology, Hunan Agricultural University
Changsha
China

Shishu Zhu
The Key Lab of Pollution Control and Ecosystem Restoration in Industry Clusters, Ministry of Education
School of Environment and Energy, South China University of Technology
Guangzhou
China

Xiefei Zhu
School of Advanced Energy, Sun Yat-sen University
Shenzhen
Guangdong
China

Preface

With advancements in agricultural productivity and rural living standards, the daily generation of waste biomass resources—such as crop residues, fruit tree prunings, animal manure, food waste, and domestic sludge—has risen significantly. However, the indiscriminate disposal and open burning of these wastes pose severe threats to rural ecosystems. The high-value utilization of waste biomass presents a viable pathway toward establishing zero-waste communities, fostering a circular economy, and supporting global pollution mitigation efforts. Furthermore, it contributes to achieving carbon emission peaks and carbon neutrality goals. Among emerging solutions, pyrolysis (300–700 °C under oxygen-limited conditions) converts waste biomass into a carbon-rich solid known as biochar.

This book synthesizes the latest research developments, state-of-the-art knowledge, and practical case studies while providing scientific insights, identifying current challenges, and outlining future directions for biochar technologies as a novel, cost-effective, and eco-friendly approach to sustainable rural revitalization. It systematically reviews application-oriented biochar production technologies using rural waste biomass, emphasizing the tailored selection of pyrolysis conditions and pre-/post-treatment methods. Additionally, the book explores diverse applications of biochar in rural development, including:

- Renewable biofuels
- Clean compost conditioners
- Organic seedling substrates
- Slow-release fertilizers and green pesticides
- Targeted plant disease suppression
- Farmland soil conditioning and amendment
- Adsorbents/catalysts for agricultural wastewater treatment
- Carbon sequestration in soils
- Low-carbon construction materials

This book also evaluates these technologies through Technical and Economic Analysis (TEA), Environmental, Social, and Governance (ESG) frameworks, and discusses potential environmental risks.

This work will serve as a valuable reference for engineers, scientists, researchers, and graduate students in waste recycling and management, sustainable rural development, environmental engineering, civil engineering, chemical engineering, and related fields.

1

Tailored Biochar Production from Waste Biomass Resource in Rural Areas

Yuqing Sun[1] *and Daniel C.W. Tsang*[2]

[1] *School of Agriculture and Biotechnology, Sun Yat-sen University, Guangzhou, Guangdong, China*
[2] *Department of Civil and Environmental Engineering, The Hong Kong University of Science and Technology, Clear Water Bay, Hong Kong, China*

1.1 Introduction

The conceptual foundation for biomass conversion into biochar was first proposed by Seifritz in 1993. Over a decade later, Marris (2006) expanded upon this idea through a comprehensive theoretical investigation, which subsequently stimulated widespread academic interest following the publication of his research. With progressive advancements in scientific knowledge and technological innovation, biochar-related methodologies have undergone continuous refinement and development.

Biochar represents a carbon-dense solid substance generated via thermochemical decomposition of biomass in oxygen-restricted environments. Recent scientific investigations have expanded the range of organic feedstocks suitable for biochar fabrication, encompassing not only conventional lignocellulosic biomass but also agricultural byproducts, food waste, livestock excreta, and residual sludge from wastewater treatment. The heterogeneous nature and complex treatment requirements of municipal and rural solid waste streams demand novel processing methodologies. Current research demonstrates that such waste materials exhibit considerable potential as precursor substances for biochar generation. This transformation process offers a dual benefit: it enables effective waste valorization while simultaneously providing an environmentally sustainable strategy for solid waste treatment.

Biochar exhibits a well-developed porous architecture, considerable surface area, and numerous stable surface functional moieties, endowing it with versatile

Waste-derived Biochar for Sustainable Rural Revitalization, First Edition. Edited by Yuqing Sun and Daniel C.W. Tsang.
Published 2026 by John Wiley & Sons, Inc.

applications across multiple environmental domains. These include but are not limited to soil enhancement and reclamation, aquatic contaminant mitigation, water resource regeneration, carbon capture and storage, as well as renewable energy applications. The material's effectiveness arises principally from its capacity to modulate carbon transformation processes, alleviate contamination by organic compounds and heavy metals, and modify microbial community dynamics and metabolic functions (Zhu et al. 2014). The remediation performance of biochar is fundamentally governed by its physicochemical characteristics, which are in turn dictated by production parameters and precursor materials. Consequently, systematic investigation of biochar fabrication methodologies and feedstock selection constitutes a critical research focus for optimizing its functional properties and environmental applications.

Contemporary advancements in biochar production technology have reached a sophisticated stage of development. Tailored synthesis approaches now enable the fabrication of high-performance biochar materials optimized for specific applications. This technological evolution is primarily motivated by the need to identify ideal preparation parameters that account for the heterogeneous nature of solid waste feedstocks, diverse application scenarios, and varying remediation objectives. With environmental pollution challenges growing increasingly intricate and multifaceted, there emerges a pressing requirement for biochar with superior functional characteristics. To meet these evolving demands, researchers are focusing on developing biochar-based composite materials through various modification strategies that augment conventional production techniques. Such engineered materials demonstrate enhanced capacity for pollutant sequestration and environmental remediation. In practical implementation, comprehensive assessment of biochar applications must consider multiple factors: the material's intrinsic properties and remediation performance, resource conversion efficiency, as well as the associated energy consumption and economic viability of production processes. This multidimensional evaluation framework ensures both optimal resource utilization and environmentally sustainable practices.

1.2 Overview of Rural Solid Wastes

In the context of accelerated economic growth in recent decades, environmental conservation has often been overlooked, resulting in numerous ecological challenges. These environmental disturbances not only compromise ecosystem equilibrium but also impose constraints on sustainable economic progress. This dual impact has consequently elevated environmental concerns to a prominent position in public discourse. Among various environmental pollutants, solid waste—classified as one of the "three major waste streams"—holds particular significance due to its direct relevance to public life. The management and treatment of solid residues have consistently remained a focal point in environmental discussions. Scientifically defined, solid waste encompasses materials discarded during industrial processes or daily activities, either having completely lost their initial utility or retaining only marginal functional value. Modern environmental classification systems have expanded this definition to incorporate materials failing to meet circular economy standards or violating environmental legislation (China Legal Publishing House 1995). From a spatial perspective, solid waste is typically differentiated between rural and urban sources, reflecting distinct generation patterns and composition characteristics across these settings.

1.2.1 Rural Solid Wastes

As a major agricultural country with extensive rural territories, China's agricultural and forestry production activities generate substantial quantities of organic residues. The processes of crop cultivation, harvesting, and post-harvest handling produce diverse biomass waste streams, encompassing crop straws, pruning residues, and various horticultural byproducts (Min et al. 2020). Recent technological advancements in cultivation techniques and irrigation systems have significantly boosted national grain productivity, with annual output reaching 669.5 million metric tons by 2020. This agricultural intensification has correspondingly increased the generation of associated biomass residues. Crop straws constitute a predominant fraction of rural solid waste, characterized by their high content of plant nutrients and organic components that make them valuable recyclable resources. However, several practical constraints hinder their effective utilization, including (1) massive production volumes with pronounced seasonal variation, (2) substantial spatial requirements for proper storage, and (3) limited environmental consciousness among rural populations regarding sustainable waste management practices. These factors collectively contribute to the ongoing challenges in developing efficient straw valorization systems.

With rising living standards and increased purchasing power, contemporary society has developed heightened expectations for dietary quality and nutritional intake. This socioeconomic transformation has precipitated a marked increase in consumption of animal-derived protein sources, including meat, eggs, and dairy products, thereby stimulating expansion in rural animal husbandry operations. While beneficial for food security, this agricultural intensification generates considerable environmental externalities. The proliferation of livestock production systems has led to accumulation of multiple waste streams, including unconsumed animal feed, pesticide container waste, and livestock manure—all of which constitute substantial components of rural solid waste inventory.

Agricultural plastic films serve as indispensable materials in modern farming practices, fulfilling multiple critical functions in crop production systems. These polymer-based coverings provide vital protection against adverse meteorological conditions, enable year-round cultivation of seasonal produce, and support the maintenance of high-value ornamental plants (Shen 2020). Beyond these primary applications, their implementation demonstrates additional agronomic benefits, including reduced herbicide requirements, decreased nutrient leaching, and improved irrigation efficiency. The widespread adoption of plastic mulching technology in rural agricultural operations has been accompanied by persistent environmental challenges. Although research into biodegradable alternatives has made considerable progress, economic constraints and market barriers continue to hinder their practical implementation (Yin et al. 2019). Consequently, conventional polyethylene films remain predominant in rural applications, characterized by inconsistent material quality, susceptibility to photodegradation, and prevalent single-use disposal patterns. These factors collectively contribute to significant plastic pollution in agricultural ecosystems. Compounding these environmental concerns, the residual packaging materials from agricultural inputs—particularly synthetic fertilizers and crop protection chemicals—represent another substantial component of rural solid waste streams. Effective mitigation of these issues will require integrated strategies encompassing both the development of environmentally benign alternatives and the implementation of comprehensive waste management systems.

The implementation of national policies emphasizing rural revitalization and agricultural modernization has stimulated substantial industrial development in countryside regions. This economic transformation has attracted numerous township enterprises to establish operations in rural areas, generating employment opportunities while simultaneously creating environmental pressures. Industrial activities, particularly in the textile sector, produce considerable quantities of non-biodegradable waste including synthetic fibers, elastomeric materials, and polymer residues. These materials often become mixed with other durable wastes like vulcanized rubber products and ceramic fragments during collection and transportation processes. The spatial distribution of waste management infrastructure in rural areas presents unique challenges. Low population density and inexpensive land values have led to the predominant siting of landfill facilities in peripheral or remote locations. Such geographical considerations increase the risk of environmental incidents due to logistical complications and potential operational deficiencies in waste-handling systems. Furthermore, domestic activities in rural communities contribute diverse waste streams encompassing organic residues, textile products, fecal matter, obsolete electronic devices, paper products, combustion byproducts, and construction debris. This heterogeneous waste composition necessitates the development of integrated management approaches that balance environmental protection with continued rural economic development.

1.2.2 Status of Rural Solid Waste Disposal

The solid waste management landscape in rural regions is predominantly characterized by agricultural residues and household waste streams. Contemporary socioeconomic transformations, including rural urbanization, technological progress, industrial restructuring, demographic changes, and evolving energy consumption profiles, have introduced significant temporal, spatial, and compositional variability in crop residue generation patterns (Li and Wang 2020). Current disposal practices often involve open-field burning or indiscriminate dumping by local residents, processes that emit substantial quantities of atmospheric pollutants (Zhang and Guo 2020). Statistical analyses reveal that China's annual theoretical biomass potential from crop residues attained 1.02 billion metric tons in 2017, with maize, wheat, and rice straw accounting for 42.2, 23.5, and 17.6% of the total biomass inventory, respectively (Shi 2018). Approximately 840 million tons of this agricultural residue is technically recoverable, finding applications in multiple value chains, including soil amendment through direct incorporation, animal husbandry feed supplements, bioenergy production (solid fuels and anaerobic digestion substrates), manufacture of artisanal products and composite materials, and pulp and paper industry feedstocks. While the current utilization rate exceeds 83%, significant opportunities remain for developing more sophisticated valorization pathways to enhance both the breadth and depth of agricultural residue applications. This requires innovative approaches to overcome existing technical and logistical constraints in rural waste management systems.

The management capacity for livestock manure—a predominant byproduct of rural animal husbandry—exhibits marked regional variability due to differences in economic development levels and arable land availability. Notably, areas characterized by high population density coupled with limited agricultural land resources encounter particularly acute challenges in assimilating substantial quantities of animal-derived waste materials. Paradoxically, these same regions typically demonstrate elevated consumption demands for animal protein products, including meat, eggs, and dairy commodities (Qiu et al. 2012). A promising strategy for addressing this spatial imbalance involves the valorization of agricultural waste through commercial organic fertilizer production. This technique, utilizing livestock manure as primary feedstock, enables effective nutrient cycling—particularly of nitrogen and phosphorus—while simultaneously reducing geographical inequities in waste-processing capabilities (Jiang et al. 2018). The efficacy of this solution is further enhanced by optimized logistics networks that facilitate the targeted distribution of processed fertilizers to nutrient-deficient areas, thereby completing the nutrient redistribution cycle.

The use of agricultural films has been empirically demonstrated to improve crop productivity significantly, while also serving as an incentive for farmers to engage more actively in agricultural practices. This dual effect not only enhances rural incomes but also contributes meaningfully to poverty reduction strategies. However, the accumulation of residual agricultural films in soil systems presents a growing environmental concern. Studies indicate that prolonged exposure to these films leads to progressive deterioration of soil structure and fertility. Despite their well-documented benefits, current methods for recycling and reprocessing agricultural films remain inconsistent and underdeveloped. Mechanized collection systems have been increasingly adopted in flat, arable regions, whereas topographically complex areas, such as mountainous terrains, still predominantly depend on manual recovery due to logistical challenges. Most agricultural films are composed of polyethylene, a synthetic polymer known for its high resistance to natural degradation. Given this persistent nature, converting waste agricultural films into reusable plastic granules offers a viable solution, combining economic value with environmental sustainability by reducing plastic pollution and promoting circular resource use.

The composition and quantity of household waste in rural areas are strongly influenced by the socioeconomic status of local inhabitants. In the past decade, the proper treatment of rural solid waste has garnered growing interest among researchers and policymakers. Unlike urban settings, waste management systems in rural communities have historically exhibited lower levels of organization and efficiency. Nevertheless, due to nationwide initiatives aimed at environmental remediation in these regions, substantial progress has been made in waste collection and processing infrastructure.

The management approaches for rural household waste largely align with urban municipal waste treatment systems, encompassing techniques such as aerobic composting, anaerobic digestion, thermal incineration, and engineered landfilling. Among these, aerobic composting stands as the most widely adopted method for organic waste valorization, utilizing microbial activity to degrade biodegradable components under thermoregulated conditions. This biological process evolves sequentially through four distinct phases—latency, mesophilic stabilization, thermophilic decomposition, and final maturation—ultimately yielding nutrient-rich humus suitable for soil amendment. Conversely, anaerobic digestion facilitates the metabolic breakdown of organic substrates by facultative and obligate anaerobes, resulting in the production of digestate and methane-rich biogas. Contemporary research has extensively examined this process across diverse waste streams, including lipid-rich sewer deposits, wastewater treatment sludge, and food waste (Dai et al. 2020; Hao et al. 2020; Luo et al. 2020). Thermal treatment through controlled incineration represents another prevalent strategy, capitalizing on the inherent calorific value of organic waste constituents. Modern incineration systems achieve substantial mass/volume reduction while enabling energy recovery through cogeneration (Liu 2020). Nevertheless, this method demands advanced emission control systems to mitigate secondary pollutants such as dioxins and reduced sulfur compounds, which pose significant atmospheric and public health concerns. Sanitary landfilling, while technologically mature, presents unique environmental challenges. Site selection necessitates rigorous geotechnical and hydrogeological assessments, as microbial degradation processes inevitably generate landfill gas emissions and nutrient-laden leachate. Consequently, modern landfill engineering mandates composite liner systems, leachate collection networks, and gas recovery infrastructure compliant with international environmental standards. The finite availability of suitable land further underscores the need for optimized waste diversion strategies to extend operational lifespans of disposal facilities.

1.2.3 Commonly Used Biochar Raw Material

Human settlements predominantly manifest in two distinct forms—rural communities and urban conglomerations—representing the principal interfaces between anthropogenic activities and natural ecosystems. The ecological integrity of these habitats remains vulnerable to irreversible degradation stemming from production systems and consumption patterns inherent to modern living. Such environmental deterioration ultimately compromises the fundamental life-support systems upon which human populations depend. Contemporary resource utilization patterns in both rural and urban contexts exhibit significant inefficiencies, characterized by substantial material misallocation. Implementing circular economy paradigms offers a viable pathway to address these challenges, particularly through systematic recovery and valorization of discarded materials within municipal and agricultural waste streams. This approach facilitates closed-loop material flows, enabling (1) equilibration between resource extraction and regeneration cycles, (2) enhanced ecological symbiosis through waste-to-resource conversion, and (3) concurrent realization of multiple UN Sustainable Development Goals (SDGs).

1.2.3.1 Agricultural and Forestry Waste

Lignocellulosic biomass residues from agricultural and silvicultural activities, including crop straws, coconut husks, tree barks, fruit peels, and bamboo, represent abundant sources of organic-rich materials. Due to their widespread availability and ease of collection, these renewable resources serve as ideal feedstocks for biochar synthesis through pyrolysis processes. Recent investigations by Jiang et al. (2020) demonstrated the potential of citrus-derived biochar in wastewater treatment applications, particularly in enhancing anaerobic co-digestion of municipal sludge and food waste. Their research revealed that citrus peel biochar effectively buffers acidic intermediates, facilitating the breakdown of short-chain fatty acids (e.g., acetic and propionic acids) while preventing digester acidification. Mechanistic studies further elucidated that the biochar's conductive properties mediated direct interspecies electron transfer, thereby accelerating methanogenic pathways and improving biogas yields. In a separate study, Luo et al. (2021) examined the phosphorus recovery potential of copper-modified biochar derived from walnut shells. Through comprehensive adsorption analyses incorporating kinetic and equilibrium studies, they determined that the phosphorus uptake mechanism followed Langmuir monolayer adsorption behavior, achieving a theoretical maximum capacity of 12.18 mg P/g adsorbent. Their findings suggested that the adsorption process was predominantly governed by surface complexation and limited by internal particle diffusion kinetics. This work provides valuable insights for designing functionalized biochars for nutrient recovery applications.

1.2.3.2 Animal Excrement

Animal-derived organic wastes, particularly livestock manure, are rich in essential plant nutrients including nitrogen and phosphorus compounds, traditionally serving as valuable soil amendments for agricultural production. However, the potential accumulation of heavy metal contaminants in these organic fertilizers may compromise soil health and pose risks to aquatic ecosystems through groundwater leaching. Thermochemical conversion of such waste materials into biochar presents a dual-benefit solution, enabling both resource recovery and environmental remediation. Recent studies have demonstrated the effectiveness of manure-derived biochars in wastewater treatment applications. Han et al. (2020) systematically investigated the uranium adsorption performance of pyrolyzed pig manure biochar subjected to various chemical modifications. Their comparative analysis revealed that sodium hydroxide-activated biochar achieved a fivefold enhancement in uranium uptake capacity compared to its unmodified counterpart. This significant improvement was attributed to the increased availability of oxygen-containing functional groups and enhanced surface properties following alkaline treatment. In a complementary study, Liu et al. (2021) conducted a comprehensive evaluation of bovine manure-derived biochars produced through different thermochemical pathways. Their research focused on phenol removal efficiency, with particular emphasis on the correlation between preparation methods and resultant biochar characteristics. The hydrothermal carbonization (HTC) approach yielded biochar with optimal properties, including (1) extensive microporous structure with high surface area, (2) abundant oxygenated surface functionalities, (3) advanced carbon matrix development, and (4) exceptional contaminant affinity. Notably, this synthesis method also demonstrated advantages in terms of operational simplicity and energy efficiency, suggesting its potential for scalable environmental applications.

1.2.3.3 Kitchen Waste

Food waste represents a heterogeneous organic matrix containing substantial proportions of animal- and plant-derived components alongside polysaccharide-based materials. Compositional analyses indicate this resource typically consists of 33%

animal-based residues, 50% plant matter, and 17% carbohydrate-rich fractions. The scientific community has increasingly recognized the value of converting these waste streams into functional biochar materials through thermochemical processes, with growing research attention directed toward their environmental applications. Recent investigations have demonstrated the versatility of food waste-derived biochars in pollution remediation. Lykoudi et al. (2020) produced highly porous biochar from spent coffee grounds using high-temperature pyrolysis (850 °C), which exhibited exceptional textural properties including extensive surface area and microporous architecture. These characteristics proved particularly effective in activating persulfate for the oxidative degradation of sulfamethoxazole, an emerging pharmaceutical contaminant. Complementary research by Yu et al. (2021) explored the adsorptive potential of chitinous waste-derived biochar, specifically using shrimp shell precursors. Their findings revealed two distinct adsorption mechanisms: (1) the macro-mesoporous network facilitated rapid contaminant uptake through enhanced mass transfer and (2) the graphitic domains provided strong adsorption sites for aromatic compounds through π–π interactions. Notably, the shrimp shell-based activated carbon demonstrated remarkable performance in pharmaceutical wastewater treatment applications, achieving removal efficiencies that suggest strong potential for industrial-scale implementation. The combination of hierarchical porosity and graphitized carbon structures in these waste-derived materials presents unique advantages for addressing complex pollution challenges.

1.2.3.4 Domestic Sewage Sludge

Wastewater treatment sludge represents a heterogeneous byproduct whose physicochemical characteristics are determined by influent pollutant profiles and operational parameters of treatment facilities. This complex matrix typically harbors multiple heavy metal species, pathogenic microorganisms, and recalcitrant organic compounds. Despite its hazardous components, the significant organic fraction present in sludge renders it suitable for valorization through thermochemical conversion processes. Recent advancements have demonstrated the efficacy of chemically modified sludge-derived biochars in wastewater remediation. Liu et al. (2019) developed an ammonia-treated biochar from municipal sludge, systematically optimizing modification parameters including ammonia concentration (15%, w/w), treatment duration (8 hours), and resultant pH (8). The optimized material achieved >65% removal efficiency for both phenolic compounds and cyanide species in coking wastewater at a dosage of 60 g/L within 80 minutes contact time. Alternative activation approaches have been explored by various research groups. Zhao et al. (2012) employed a combined methodology using peanut shell biomass and zinc chloride to produce sludge-based activated carbon, which exhibited exceptional adsorption capacity for triphenylmethane dyes. Their risk assessment confirmed that activation at 600 °C with 20% peanut shell additive effectively immobilized heavy metals, minimizing ecological hazards. In a separate innovation, Li et al. (2019) pioneered the use of titanium dioxide waste acid for sludge activation, followed by hydrogen peroxide (H_2O_2) modification. Application of this engineered biochar (12 g/L dosage, 50 minutes contact time) to abattoir effluent treatment yielded remarkable reductions exceeding 90% for both phosphorus species and organic load (COD), demonstrating exceptional performance without pH adjustment requirements.

1.3 Biochar Production Methods

Currently, the predominant techniques for converting rural solid wastes into biochar include pyrolysis, microwave-assisted pyrolysis, and hydrothermal carbonization. Among these, conventional pyrolysis involves the thermochemical decomposition of organic materials in an oxygen-deficient environment, generating three primary products: condensable vapors (bio-oil), non-condensable gaseous compounds, and a carbon-rich solid residue known as biochar. This approach offers notable advantages, such as comprehensive utilization of feedstock components, broad compatibility with various raw materials, and efficient conversion rates. The yield and composition of pyrolysis-derived products are highly dependent on the operational parameters and the specific technique applied. Based on variations in heating mechanisms and reaction conditions, pyrolysis methods can be categorized into slow pyrolysis, fast pyrolysis, microwave pyrolysis, flash pyrolysis, hydrothermal carbonization, and gasification, as summarized in Table 1.1.

Table 1.1 Different thermochemical techniques and biochar yields.

Techniques	Temperature (°C)	Residence time	Biochar yields (wt%)
Slow pyrolysis	150~700	min~d	15~89
Fast pyrolysis	300~1000	~s	12~37
Microwave pyrolysis	200~600	~30 min	25~50
Flash pyrolysis	300~1200	<10 s	27~46
Hydrothermal carbonization	180~300	30 min~16 h	36~72
Gasification	700~800	10~20 s	5~10

1.3.1 Slow Pyrolysis

Slow pyrolysis represents a fundamental thermochemical conversion method for processing biomass under oxygen-limited conditions, yielding three primary products: gaseous compounds, liquid bio-oil, and solid biochar. As a widely adopted technique for biochar production, this process is typically carried out at heating rates below 10 °C/min within a temperature range of 150–700 °C, with prolonged retention periods exceeding 1 hour at the target temperature. The underlying mechanisms are intricate, encompassing both physicochemical transformations, with process dynamics heavily dependent on feedstock properties, operational parameters, and reactor configurations. The efficiency and outcomes of slow pyrolysis are governed by multiple variables, which can be classified into two principal categories: (1) process-related factors, including pyrolysis temperature, heating rate, and residence time, and (2) feedstock-related factors, such as biomass type, intrinsic characteristics, and particle dimensions. Plant-derived biomass, a predominant source of biochar, is chiefly composed of hemicellulose, cellulose, and lignin, each exhibiting distinct thermal degradation behaviors as temperature escalates. Notably, lignin demonstrates greater thermal stability and slower carbonization kinetics compared to hemicellulose and cellulose. The pyrolysis process initiates with moisture evaporation at lower temperatures, followed by substantial devolatilization between 200 and 500 °C, which can be delineated into four sequential phases: (1) initial breakdown of hemicellulose, (2) complete decomposition of hemicellulose coupled with partial degradation of cellulose, (3) full cellulose decomposition alongside partial lignin breakdown, and (4) continued thermal cracking of residual components, leading to enhanced carbonization. Beyond 700 °C, mass loss stabilizes and further structural reorganization occurs. The biochar yield obtained from slow pyrolysis varies considerably (15–89%) depending on the feedstock, with a general inverse correlation observed between yield and pyrolysis temperature.

The elemental composition of biochar is predominantly determined by pyrolysis temperature, which also governs its physicochemical characteristics. Research demonstrates that elevated pyrolysis temperatures lead to an enrichment of carbon (C) content in biochar, accompanied by a decline in nitrogen (N), hydrogen (H), and oxygen (O) concentrations (Ahmad et al. 2012; Li et al. 2013). However, some studies suggest that N retention is more dependent on feedstock type than on thermal treatment conditions (Kim et al. 2013; Harvey et al. 2011). With increasing pyrolysis temperature, the progressive loss of H and O results in reduced molar ratios of H/C, O/C, and (O + N)/C. The H/C ratio, which reflects the abundance of aliphatic structures, serves as a reliable indicator of carbonization extent, as H is predominantly bound to organic constituents in the original biomass. Meanwhile, the O/C ratio provides insight into surface hydrophilicity, with lower values signifying a higher aromaticity due to the thermal decomposition of polar functional groups (Chen et al. 2011; Chun et al. 2004). Similarly, the (O + N)/C ratio, a measure of overall polarity, diminishes at higher temperatures, further confirming the depletion of oxygen- and nitrogen-containing moieties.

1.3.2 Fast Pyrolysis

The pyrolysis process can be categorized into two distinct regimes based on operational parameters. Slow pyrolysis is conducted under mild thermal conditions, featuring gradual heating rates and extended residence durations. Conversely, fast pyrolysis employs substantially higher heating rates with residence times often limited to mere seconds. Under rapid pyrolysis conditions, biomass undergoes instantaneous thermal degradation, generating both volatile pyrolysis vapors and solid biochar as secondary products. The condensable fraction of these vapors is subsequently liquefied through rapid cooling, yielding a viscous, dark brown liquid termed bio-oil. The yield of biochar is inversely correlated with both pyrolysis temperature and heating rate, as elevated thermal severity promotes enhanced devolatilization. At accelerated heating rates, biomass particles experience rapid thermal decomposition, and the resultant vapors are swiftly expelled from the reaction zone. This abbreviated exposure to pyrolytic conditions minimizes secondary char formation through vapor-phase carbon deposition. Experimental evidence demonstrates this phenomenon: Angin (2013) reported a 3–8% reduction in safflower seed biochar yield when heating rates increased from 10 to 50 °C/min. Analogous findings by Chen et al. (2016) showed poplar biochar yields declining from 34.83 to 31.95% at 400 °C under similar heating rate variations. The influence of heating rate on char yield has been further substantiated by Aguado et al. (2000), who observed a significant decrease in solid residue (from 38.8 to 26.4%) when heating rates were elevated from 5 to 40 °C/min. Notably, reactor configuration also impacts product distribution, as demonstrated by Antal et al. (1996), who achieved substantially higher biochar yields (41–62%) in high-pressure pyrolysis systems.

Fast pyrolysis induces rapid thermal decomposition of biomass, generating a product distribution dominated by condensable vapors (steam and tar), non-condensable gases, and solid char. Optimal liquid yields are achieved when employing low-ash biomass feedstocks under carefully controlled process conditions. Three critical operational parameters govern the efficiency of liquid production in fast pyrolysis systems: First, exceptionally rapid heating and heat transfer rates at the biomass particle interface require the use of finely comminuted feedstocks (typically <3 mm particle size) to compensate for the inherently poor thermal conductivity of lignocellulosic materials. Second, precise temperature regulation around 500 °C proves most favorable for maximizing liquid phase output across diverse biomass varieties. Third, strict control of vapor phase residence time (<2 s) is essential to prevent undesirable secondary cracking reactions of primary pyrolysis products. These carefully optimized parameters collectively promote efficient biomass conversion to liquid bio-oil while suppressing the formation of secondary byproducts. The process demonstrates particular sensitivity to feedstock particle size due to heat transfer limitations inherent in biomass materials.

Under optimal operating conditions, fast pyrolysis can achieve bio-oil yields exceeding 75 wt% (dry feed basis), while the co-produced char and gaseous fractions serve as potential heat sources to sustain the thermal process. Multiple parameters govern the liquid product distribution, notably feedstock characteristics (including ash content and lignocellulosic composition), reaction temperature profile, vapor phase residence duration, and the efficacy of solid-phase separation. The process sequence begins with feedstock preparation, where biomass undergoes drying to maintain moisture levels below 10%—a critical step for minimizing aqueous content in the final liquid product. Subsequent size reduction through mechanical grinding produces fine particulate matter (<3 mm), ensuring adequate heat transfer rates during the subsequent rapid thermochemical conversion. The pyrolysis stage is immediately followed by efficient char removal and rapid condensation of volatile components to maximize liquid recovery. This integrated approach optimizes both product quality and process energy efficiency.

The carbon content and specific surface area of biochar are significantly influenced by pyrolysis temperature due to the release of volatiles from biomass particles. Studies have demonstrated that elevated pyrolysis temperatures enhance these properties. For instance, when the pyrolysis temperature of rapeseed stem biochar was raised from 200 to 700 °C, its specific surface area exhibited a substantial increase from 1 to 45 m^2/g (Zhao et al. 2018). A comparable trend was observed in pine sawdust-derived biochar, where the carbon content increased from 70.68 to 78.75% as the temperature escalated from 550 to 750 °C (Peng et al. 2012). In contrast, the effect of heating rates on biochar properties during fast pyrolysis is more intricate. Research by Onay (2007) revealed that biochars generated at higher heating rates possess greater carbon content and specific surface area compared to those produced under slower heating conditions. This phenomenon can be attributed to variations in devolatilization efficiency, which subsequently alters the structural characteristics of biochar. However, conflicting observations exist in the literature. Chen et al. (2016) noted that the specific surface area of biochar initially increases but later declines with rising heating rates and carbon content. Conversely, Mohan et al. (2014) reported that excessively high heating rates lead to reduced specific surface area and pore volume, likely due to rapid surface depolymerization. These discrepancies suggest that while higher heating rates generally improve carbon retention, their impact on pore structure is nonlinear and dependent on specific pyrolysis conditions.

To maximize bio-oil production via fast pyrolysis, significant advancements have been made in reactor design (Qureshi et al. 2018). Commonly employed configurations include bubbling fluidized beds, circulating fluidized beds, ablation reactors, rotating cones, and auger or spiral reactors. Efficient separation of biochar from pyrolysis vapors is essential to reduce secondary cracking reactions, which can degrade oil quality. Studies indicate that fluidized bed reactors, rotating cones, and ablation reactors generate around 15% solid carbon byproduct (Bridgwater 2012), while spiral reactors exhibit higher biochar yields, reaching ~25% (Raclavská et al. 2015). Both biochar and pyrolysis gas are frequently repurposed as energy sources to sustain process heat in industrial settings. Fast pyrolysis-derived bio-oil offers distinct benefits, including ease of storage and transportation, along with its versatility as a feedstock for high-value chemicals. Recent research has focused on upgrading bio-oil properties, particularly for biofuel applications, broadening its potential utility. Furthermore, biorefineries provide a promising framework for process and product optimization in fast pyrolysis systems. Achieving this optimization requires the refinement of individual unit operations to improve overall system efficiency.

1.3.3 Microwave Pyrolysis

Microwave technology has found extensive application in food-processing operations including defrosting, microbial inactivation, thermal processing, dehydration, and pastry preparation. In recent years, its potential has been increasingly recognized in environmental remediation and chemical synthesis applications. When subjected to microwave irradiation, materials demonstrate three distinct response mechanisms: (1) reflective behavior exhibited by conductive materials, (2) energy absorption and subsequent thermal conversion by dielectric materials, and (3) transparent transmission characteristic of insulating materials. The fundamental principle of microwave heating involves electromagnetic wave penetration into materials, causing dipole moment reorientation at the molecular level and consequent volumetric heat generation through dielectric loss mechanisms. This heating modality demonstrates remarkable material-specific selectivity. A practical illustration can be observed in pyrolysis systems, where nitrogen carrier gas functions as a microwave-transparent medium, while metallic reactor components reflect electromagnetic waves to concentrate energy on microwave-absorbent feedstock materials. Comparative studies have demonstrated that microwave-assisted pyrolysis achieves significant reductions in process duration relative to conventional thermal methods, attributable to its superior energy transfer efficiency (Lam et al. 2019). Additional benefits include instantaneous response characteristics, enabling rapid initiation and termination of heating processes, which permits exceptional thermal regulation precision. This combination of features makes microwave heating particularly advantageous for controlled thermal processing applications.

Despite its advantages, microwave-assisted pyrolysis presents several technical constraints and operational challenges. A primary concern involves the inadequate dielectric response of certain feedstock materials to microwave irradiation. Research indicates that optimizing pyrolysis efficiency frequently necessitates either elevated power input or the incorporation of microwave susceptors like activated carbon or silicon carbide as additives (Martín et al. 2017; Tarves et al. 2017). The effectiveness of microwave heating is substantially influenced by field distribution irregularities, which are governed by multiple factors including cavity design, magnetron specifications, and mixing mechanisms (Binner et al. 2014; Chen et al. 2016). Furthermore, the intrinsic variability of biomass composition, coupled with differences in loading configuration and material-specific dielectric/thermal characteristics, introduces additional complexity to the pyrolysis process (Ciacci et al. 2010; Salvi et al. 2009). These non-uniform heating patterns generate localized temperature variations, creating distinct thermal zones where surface and core temperatures may diverge considerably (Zaini and Kamaruddin 2013). Such thermal gradients can significantly alter the decomposition pathways of agricultural residues during pyrolysis, ultimately influencing both the yield distribution and quality characteristics of the resultant gaseous, liquid, and solid products.

The microwave-assisted pyrolysis of agricultural residues typically progresses through three sequential phases: initial removal of unbound moisture, followed by primary thermochemical decomposition (encompassing water elimination, methyl group removal, and carboxyl group dissociation), and culminating in secondary transformation processes (including bio-oil repolymerization and molecular fragmentation) (Shafaghat et al. 2019). Research indicates that this technique exhibits superior heating efficiency relative to conventional pyrolysis approaches, while simultaneously generating enhanced quantities of economically valuable byproducts, particularly hydrogen-enriched syngas and furfural-concentrated bio-oils (Liew et al. 2019). Empirical studies by Zhao et al. (2011) revealed that rice straw subjected to microwave pyrolysis yielded significantly greater quantities of synthesis gas (25–50% increase) and biochar compared to traditional thermal decomposition methods. Complementary research by Afolabi et al. (2017) established that microwave pyrolysis substantially improves both the energy content and structural integrity of biochar through particle densification mechanisms. This versatile conversion technology has been successfully applied to diverse agricultural byproducts, with documented cases including coniferous wood dust (Ren et al. 2012), birch biomass (Huang et al. 2017), deciduous wood (Arshanitsa et al. 2016), cereal straws (Satpathy et al. 2014), and rice plant residues (Huang et al. 2012), all of which can be effectively transformed into high-quality solid biochar through microwave-mediated pyrolysis processes.

While microwave pyrolysis of agricultural wastes has been extensively studied, there is still a notable need for further investigation and process optimization to enable efficient continuous operation. In such systems, the conveyance of biomass feedstock and the removal of solid biochar can be accomplished through mechanical handling equipment, including screw augers, conveyor belts, or gravity-fed mechanisms. Additionally, the gaseous and liquid byproducts, namely syngas and bio-oil, can be collected and processed for bioenergy generation (Zhao et al. 2012; Zhou et al. 2018). For large-scale processing of agricultural residues, a multimode microwave cavity equipped with distributed low-power magnetrons (e.g., 1000 W) has been proposed as an effective configuration (Zhang et al. 2015). A critical requirement for successful microwave-assisted pyrolysis is the maintenance of an oxygen-free atmosphere to ensure proper carbonization. This can be established through several approaches, such as inert gas purging (typically N_2 or Ar), vacuum extraction, or a self-sustaining syngas retention mechanism. Although nitrogen purging is widely adopted, it may introduce operational inefficiencies and increase overall costs. Alternatively, the self-purging technique utilizes internally generated syngas to displace oxygen, whereas vacuum systems actively remove air from the reaction chamber (Domínguez et al. 2008). Both of these methods offer economically viable solutions for transforming agricultural waste into high-value biofuels via microwave pyrolysis (Kong et al. 2019).

1.3.4 Flash Pyrolysis

Flash pyrolysis represents a technologically advanced variant of fast pyrolysis, characterized by ultra-rapid heating rates and precise thermal control. In this process, biomass is subjected to extreme heating, often exceeding 1000 °C/s, achieving target temperatures between 300 and 1200 °C within a remarkably short duration of under 10 seconds (Demirbaş and Arin 2002). The process efficiency is governed by complex interactions among heat and mass transfer mechanisms, reaction kinetics, and biomass phase transitions, all of which collectively determine the final product distribution. Optimizing operational parameters, such as maintaining high heating rates, controlled peak temperatures, and minimal vapor residence times, enhances liquid bio-oil production while suppressing undesirable solid char formation. However, scaling this technology for industrial use presents significant engineering challenges, particularly in reactor design, where achieving consistent ultra-fast heating remains a critical hurdle. Furthermore, the practical application of flash pyrolysis is constrained by bio-oil quality concerns. High ash content in the feedstock can degrade oil purity, while reactive carbon species within the oil may promote polymerization, leading to increased viscosity and instability. Overcoming these limitations is essential for improving reactor performance and facilitating the large-scale adoption of flash pyrolysis for sustainable biofuel production.

Flash pyrolysis is distinguished from conventional fast pyrolysis by its exceptionally short vapor residence time (<10 seconds) and substantially higher heating rates, resulting primarily in bio-oil production. Under high-temperature conditions (300–1200 °C), biomass undergoes rapid depolymerization and thermal decomposition, yielding a bio-oil with viscosity characteristics comparable to diesel fuel. The predominant variants of this process include flash hydropyrolysis and vacuum flash pyrolysis, each employing distinct operational parameters to optimize product recovery (Sri Shalini et al. 2021). Flash hydropyrolysis is typically performed at 350–600 °C with heating rates of 10–300 °C/s and residence times exceeding 15 seconds, under elevated pressures (5–20 MPa) to facilitate condensable product separation. In contrast, vacuum flash pyrolysis operates at lower pressures (0.01–0.02 MPa) and temperatures of 300–600 °C, but with markedly reduced heating rates (0.1–1.0 °C/s) and residence times (0.001–1.0 seconds), enhancing rapid volatilization of light fractions (Tripathi et al. 2016). These differences in process conditions significantly influence product distribution and bio-oil quality, necessitating careful optimization for industrial scalability.

1.3.5 Hydrothermal Carbonization

Hydrothermal carbonization (HTC) represents an innovative thermochemical approach that enables direct processing of high-moisture biomass without energy-intensive drying pretreatment. This technique is particularly advantageous for converting diverse organic feedstocks, such as sewage sludge and algal biomass, into a carbon-rich solid product termed hydrochar. The resulting hydrochar has attracted considerable research attention due to its versatile applications, including but not limited to environmental remediation (wastewater purification and soil amendment) and renewable energy production. Furthermore, HTC technology offers a sustainable solution for managing various rural and urban waste streams, simultaneously mitigating solid waste disposal challenges while creating value-added carbon materials.

Municipal and agricultural waste streams represent abundant renewable feedstocks that can be converted into valuable solid, liquid, and gaseous products via various conversion pathways, including biological, physicochemical, and thermochemical processes. Despite this potential, several inherent characteristics limit their utilization efficiency, such as compositional heterogeneity, low calorific value, poor compaction properties, elevated water content, and contamination issues. Among these constraints, the substantial moisture content in biomass feedstocks presents particular challenges for pyrolysis applications, as it requires considerable energy input for water removal prior to thermal decomposition. This limitation has driven growing attention toward hydrothermal conversion technologies, which demonstrate superior economic feasibility compared to traditional thermal treatment approaches for wet biomass processing.

The HTC process typically operates within a moderate temperature range of 180–300 °C under self-generated pressure conditions. During this thermochemical conversion, feedstock undergoes significant chemical transformation through dehydration and decarboxylation mechanisms, leading to substantial reduction in oxygen and hydrogen content (Funke and Ziegler 2010). Reaction intensity and biomass carbonization extent are predominantly influenced by key process variables, particularly temperature and retention time (Qian et al. 2013). The resulting carbonaceous solid product, hydrochar, demonstrates distinct physicochemical advantages over untreated biomass. Characterized by enhanced hydrophobicity and friability, it can be readily separated from the aqueous phase. Furthermore, hydrochar possesses significantly improved fuel characteristics, including elevated mass and energy densities, superior dehydration capacity, and enhanced combustibility (Saidur et al. 2011). These properties, coupled with its multifunctional applications in carbon capture, agricultural amendment, renewable energy generation, and water purification, make hydrochar a versatile material whose performance is dictated by both feedstock selection and processing parameters (Saidur et al. 2011). The liquid fraction obtained from HTC requires additional processing steps, such as extraction and fractionation, for bio-oil production. Studies indicate this aqueous phase exhibits favorable biodegradability owing to its rich organic composition (Passos and Ferrer 2015). In summary, hydrothermal treatment yields a high-value, carbon-enriched material (hydrochar) with substantially reduced heteroatom content, demonstrating the effectiveness of this conversion process (Funke and Ziegler 2010).

Recent scientific literature has demonstrated growing interest in hydrochar's versatile applications, especially for developing functional carbon materials and addressing challenges in energy storage and environmental remediation. Research output on hydrochar synthesis and utilization has shown exponential growth since 2009. The HTC process operates under subcritical water conditions, where water's physicochemical properties undergo significant alterations. At temperatures below the critical point (374 °C), water exhibits reduced dielectric constants, weakened hydrogen bonding, and enhanced ionic product, facilitating its autoionization into H_3O^+ and OH^- species (Marcus 1999; Savage 1999). This unique environment provides inherent catalytic activity, eliminating the requirement for external acid catalysts in organic transformations (Ruiz et al. 2013; Savage 1999). Notably, the water present in biomass serves dual roles as both solvent and reaction medium during HTC (Libra et al. 2011). The HTC process yields three distinct product fractions: a carbon-rich solid phase, an aqueous liquid containing soluble organics, and minor gaseous products (mainly CO_2). Product distribution and characteristics are strongly dependent on both feedstock properties and operational parameters. The solid fraction, distinguished by its uniform composition and pronounced hydrophobic nature, represents the principal output and can be efficiently isolated from the reaction mixture. Importantly, the hydrothermal processing conditions critically influence the final hydrochar's physical and chemical attributes. A comprehensive understanding of the governing reaction mechanisms and key process variables is therefore imperative for optimizing hydrochar properties and expanding its potential applications.

The HTC process is profoundly influenced by temperature, which governs the dominant reaction pathways in the subcritical regime. Below the critical point, ionic reactions prevail, while supercritical conditions favor radical-mediated mechanisms (Savage 1999). Within the HTC operating range, elevated temperatures induce two key effects: (1) reduction of water's viscosity, facilitating improved penetration into biomass matrices and accelerating decomposition (Funke and Ziegler 2010; Möller et al. 2011), and (2) enhancement of concurrent hydrolysis, dehydration, and decarboxylation reactions. A distinctive feature of hydrothermal systems is their ability to dissolve organic components at sustained elevated temperatures, enabling secondary char formation through polymerization pathways (Knežević et al. 2010). This temperature-dependent transformation is quantitatively reflected in the elemental composition of hydrochar, as demonstrated by systematic shifts in van Krevelen diagrams. Experimental studies have consistently shown that increasing HTC temperature from 230 to 250 °C leads to progressive reduction of both O/C and H/C ratios, indicative of advanced carbonization through condensation reactions (Sevilla and Fuertes 2009). This thermal behavior has been validated across diverse biomass feedstocks. Parshetti et al. (2013) documented analogous trends in lignocellulosic materials, observing continuous decreases in heteroatom ratios with temperature elevation during palm fruit bunch conversion. Comparable carbonization patterns have been established for starchy biomass, confirming the universal nature of these temperature-dependent transformations (Hwang et al. 2012; Kim et al. 2014).

Residence time represents a fundamental process parameter that significantly affects hydrochar formation by modulating polymerization kinetics. While its influence on solid product yield is secondary to temperature, this variable plays a critical role in determining the structural characteristics of the final material. Initial stages of hydrothermal treatment with short durations typically yield higher solid hydrate content, which progressively decreases with extended processing periods. The temporal dimension of HTC is particularly important for secondary hydrochar formation, as prolonged reaction durations promote the recombination of dissolved organic fragments into polyaromatic structures (He et al. 2013;

Kang et al. 2012). This phenomenon exhibits feedstock-dependent behavior: lignocellulosic materials demonstrate pronounced sensitivity to residence time due to their complex polymerization requirements, while undissolved monomers respond more significantly to thermal conditions. Experimental investigations have revealed distinct morphological evolution patterns associated with residence time variations. Gao et al. (2013) systematically examined water hyacinth conversion at 240 °C across different durations (30 minutes to 24 hours), observing progressive surface modifications. Shorter processing times produced hydrochar with distinctive surface fissures and channels, while microsphere formation required minimum 6-hour treatment. Extended durations (>24 hours) resulted in microsphere coalescence, demonstrating the time-dependent nature of carbon surface development. The mechanistic role of residence time involves dual regulation of both depolymerization and repolymerization processes. By controlling the duration of monomer hydrolysis and subsequent condensation reactions at fixed temperatures, this parameter ultimately determines the hierarchical architecture of hydrochar, particularly during post-microsphere structural organization.

1.4 Biochar Modification Methods

For practical implementation, biochar often requires tailored modification to enhance its functional properties. The principal modification strategies can be categorized into physical and chemical methods. Physical approaches primarily utilize activating agents such as steam or gases (e.g., CO_2 or air) to alter porosity and surface characteristics. In contrast, chemical modification—more widely adopted due to its versatility—encompasses treatments with acidic or alkaline solutions, oxidizing agents, and metal-catalyzed graphitization processes.

1.4.1 Physical Modification

Physical activation of biochar involves a two-stage thermochemical treatment process. In the initial carbonization phase, biomass feedstock is thermally decomposed at 400–850 °C under oxygen-limited conditions to yield primary biochar. The second stage employs oxidizing gases (CO_2, steam, air, or combinations thereof) at elevated temperatures (600–1200 °C) to enhance the char's porosity. Pyrolytic tar decomposition often generates amorphous carbon deposits that may obstruct pore channels and diminish surface area. The gas activation process serves to selectively remove these carbonaceous residues while simultaneously developing an interconnected pore network through controlled oxidation. Pore formation mechanisms are governed by both volatile matter evolution and the specific reactivity of the activating gas medium.

Steam and CO_2 activation represent the predominant gaseous media employed in biomass thermochemical conversion and biochar physical activation processes. During biomass pyrolysis, these reactive gases participate in dual-phase reaction mechanisms that significantly influence product distribution. In the homogeneous phase, volatile organic compounds interact with steam or CO_2 at elevated temperatures, effectively suppressing tar formation while enhancing syngas production. Concurrently, heterogeneous reactions occur at the biomass–gas interface, modifying both the pyrolysis environment and the surface chemistry of resulting biochar. These synergistic effects also contribute to improved bio-oil quality during fast pyrolysis conditions (Lee et al. 2017). For biochar activation specifically, steam or CO_2 can be introduced either as an integrated step following pyrolysis or as a separate post-treatment process. The endothermic nature of steam-carbon reactions offers distinct advantages over air activation, including superior process control and enhanced suitability for gasification applications. This stems from steam's ability to (1) effectively remove volatiles, (2) decompose residual tars, (3) generate new micropores, and (4) expand existing pore structures. Such modifications lead to substantial improvements in specific surface area, microporous surface area, and pore volume distribution. Optimal activation typically occurs under superheated steam conditions (800–900 °C) with residence times ranging from 30 minutes to 3 hours (Demiral et al. 2011; Girgis et al. 2011). The steam activation mechanism involves three principal steps: water chemisorption on carbon surfaces, subsequent carbon gasification, and water conversion through the water-gas shift reaction. Mesopore formation is particularly dependent on this reaction pathway and associated with carbon consumption (Rajapaksha et al. 2015). Furthermore, steam activation addresses incomplete pyrolysis by eliminating residual organic compounds (including acids, aldehydes, and ketones) while simultaneously modifying surface properties—reducing hydrophobicity and increasing polarity through oxygen-containing functional group formation.

The efficacy of steam activation is predominantly governed by three critical operational parameters: activation temperature (T), steam-to-carbon mass ratio (S/C), and residence time (t). Elevated activation temperatures and prolonged treatment durations generally promote the development of pore structures and surface area, as these conditions facilitate the progressive gasification of surface carbon atoms. This phenomenon becomes particularly pronounced under high-temperature regimes. Notably, steam activation at elevated temperatures has been shown to counteract the adverse effects of low-temperature pyrolysis on biochar's textural properties, significantly enhancing both porosity and surface area (Kołtowski et al. 2016). For example, under optimized conditions, steam activation can increase the specific surface area of biochar by approximately sixfold compared to low-temperature treatment, with exceptional cases demonstrating up to a 76-fold improvement relative to non-activated biochar. However, it should be noted that suboptimal activation temperatures may induce pore collapse, resulting in diminished surface area—a phenomenon well-documented in previous studies (Rajapaksha et al. 2015).

While elevated steam temperatures (≤800 °C) generally enhance biochar's physicochemical properties, excessive activation durations (45–60 minutes) can compromise process efficiency. Extended exposure to high temperatures accelerates gasification kinetics, leading to over-activation characterized by two concurrent phenomena: (1) excessive macropore formation and (2) structural degradation of carbon matrices. This transformation results in micropore coalescence into larger meso- and macropores, ultimately diminishing the relative micropore volume fraction while increasing ash content and reducing specific surface area. Interestingly, augmenting vapor flow rates at lower temperatures shows limited efficacy in improving pore size distribution. This observation can be explained by the dilution theory (Arriagada et al. 1997; Martín-Gullón et al. 1996), where gasification byproducts (CO and H_2) dilute the internal vapor concentration, creating a vapor concentration gradient that promotes preferential external gasification. Consequently, this leads to particle mass loss and non-uniform pore enlargement. Contrastingly, at elevated temperatures, increased steam flow has been shown to positively correlate with mesopore development (Rodríguez-Reinoso et al. 1995). This temperature-dependent behavior underscores the importance of optimizing both thermal conditions and vapor flow rates to achieve desired pore architecture.

Gas-phase activation represents an effective approach for modifying biochar's textural properties through controlled reactions between carbon surfaces and oxidizing gases. Among the commonly utilized activating agents (CO_2, N_2, NH_2, O_2, and their mixtures), CO_2 demonstrates particular advantages due to its operational simplicity and environmental compatibility, reacting stoichiometrically with carbon atoms in the biochar matrix. The degree of carbon activation is governed by multiple operational parameters, including thermal conditions, exposure duration, gas flux, and reactor configuration. Experimental studies have demonstrated the efficacy of CO_2 activation in producing biochars with tailored porosity. Zhang et al. (2004) successfully developed porous carbons from diverse lignocellulosic feedstocks, achieving specific surface areas of 400–1000 m^2/g with micropore fractions representing 38–66% of total porosity. These findings align with the general observation that elevated activation temperatures promote pore development, as evidenced by Guo et al. (2009) in their systematic investigation of coconut shell-derived carbons. Their work revealed that thermal energy input not only enhances pore formation but also facilitates mesopore expansion, while prolonged activation durations must be carefully optimized to balance micropore generation against structural degradation. The temperature-dependent nature of pore evolution was further corroborated by Taer et al. (2010) in their examination of rubber wood-based activated carbons. Their results showed a positive correlation between activation temperature (700–1000 °C) and key performance metrics, including specific surface area (282–684 m^2/g), electrical conductivity (0.0075–0.0687 S/cm), and electrochemical capacitance. This improvement was attributed to progressive micropore development, though temperatures beyond 900 °C were found to induce pore network distortion.

Steam activation produces a more heterogeneous pore size distribution compared to CO_2 activation, creating a balanced network of both micropores and mesopores. This structural characteristic proves particularly beneficial for adsorption applications, as mesopores and macropores serve dual functions: (1) enabling the uptake of bulky molecular species and (2) providing transport pathways that enhance accessibility to adsorption sites located within micropores. However, for small molecule adsorption, the predominance of micropores in activated biochar becomes the determining factor for adsorption efficiency. Comparative research by Kołtowski et al. (2017) demonstrated this structure–performance relationship through systematic evaluation of CO_2, steam, and microwave activation methods. Interestingly, while steam activation yielded biochar with greater total surface area, CO_2-activated samples exhibited superior performance in removing 16 priority PAHs from aqueous solutions. This apparent paradox was resolved by examining the pore size distribution—CO_2 activation preferentially develops microporous structures, with micropores constituting a significantly larger proportion of the total pore volume, thereby optimizing the material for small molecule capture.

1.4.2 Chemical Modification

In the chemical activation process, acids, alkalis, and oxidants are frequently employed to enhance the properties of biochar. Common chemical agents used include $ZnCl_2$, H_3PO_4, H_2SO_4, KCNS, HNO_3, H_2O_2, $KMnO_4$, $(NH_4)_2S_2O_8$, NaOH, KOH, and K_2CO_3. Typically, the final step in the preparation of activated biochar involves washing to remove excess inorganic residues.

1.4.2.1 Acid Modification

Acid treatment serves dual purposes in biochar modification: surface purification and functionalization through the introduction of acidic groups. Typical acid reagents employed include mineral acids (HCl, H_2SO_4, HNO_3, H_3PO_4) and organic acids ($H_2C_2O_4$, $C_6H_8O_7$). Experimental evidence demonstrates that HCl treatment (1 mol/L) of reed-based biochar achieves substantial ash removal (29.5–11.8%), consequently exposing additional hydrophobic sites for pentachlorophenol adsorption (Peng et al. 2016). HNO_3 modification (2 mol/L) has been shown to effectively graft oxygen-containing functional groups (carboxyl, lactone, phenol, carbonyl) onto bamboo-derived biochar surfaces (Li et al. 2014). The impact of acid treatment on textural properties exhibits significant acid-dependent variation. HCl treatment (1 mol/L) enhances reed biochar's surface area by 50% (58.75–88.35 m^2/g), while H_3PO_4 modification reduces both surface area (522.5–517.1 m^2/g) and pore volume (1.2–0.65 mL/g) in rice straw biochar (Taha et al. 2014; Peng et al. 2016). Notably, H_3PO_4 concentration emerges as a critical factor governing the ultimate adsorption performance of acid-treated biochars.

Organic acids, including oxalic acid ($H_2C_2O_4$) and citric acid ($C_6H_8O_7$), can effectively functionalize biochar surfaces through esterification reactions, leading to the incorporation of carboxyl groups. However, while these weak acids enhance the concentration of oxygen-containing functional groups, they simultaneously reduce the specific surface area, as demonstrated by the decrease from 1.57 to 0.69 m^2/g ($H_2C_2O_4$) and 1.21 m^2/g ($C_6H_8O_7$) (Peng et al. 2016). Conversely, synergistic treatment with 30% sulfuric acid (H_2SO_4) and $H_2C_2O_4$ markedly enhances porosity, elevating the surface area from 2.31 to 571.00 m^2/g (Vithanage et al. 2015). This pronounced discrepancy arises from the distinct corrosive properties of the acids, with strong mineral acids like H_2SO_4 exhibiting more aggressive pore-forming capabilities compared to their organic counterparts.

1.4.2.2 Alkali Modification

Alkali modification serves as an effective approach for optimizing biochar's textural and surface chemical properties, primarily targeting surface area enhancement and oxygen functional group incorporation. Among the alkaline reagents employed, KOH and NaOH are most widely utilized due to their distinct activation mechanisms. Research by Jin et al. (2014) demonstrated that KOH treatment could elevate the surface area of biochar from 14.4 to 49.1 m^2/g while simultaneously introducing oxygen-containing functional groups, resulting in improved As(V) adsorption capacity. However, contrasting results were observed by Sun et al. (2015), where KOH modification reduced the surface area of hydrothermal carbonized wheat straw biochar from 4.4 to 0.69 m^2/g, highlighting the critical influence of feedstock characteristics and carbonization methods on modification outcomes. Comparative studies reveal that NaOH, being less corrosive and more economical than KOH, can achieve superior surface area development in certain biomass precursors. Tan et al. (2008) reported exceptional surface area enhancement (2885 m^2/g) for coconut-derived biochar using NaOH, surpassing the performance of KOH modification (1940 m^2/g). Nevertheless, the effectiveness varies across materials, as demonstrated by Fan et al. (2010), who observed minimal surface area changes in NaOH-treated bamboo biochars. The modification efficiency is further governed by process parameters, particularly the alkali-to-biochar ratio. Shen and Zhang (2019) identified a 3:1 mass ratio as optimal for KOH modification, emphasizing the importance of precise stoichiometric control. These findings collectively indicate that the success of alkali modification is contingent upon a complex interplay of factors including feedstock selection, carbonization conditions, and modification protocol specifics.

1.4.2.3 Oxidation Modification

Oxidative treatment represents an effective strategy for enhancing the surface functionality of biochar through the introduction of oxygen-containing groups. Experimental investigations have demonstrated that H_2O_2 modification of hydrothermally carbonized peanut shell biochar substantially augments surface oxygen moieties, with carboxyl groups showing particular prominence, leading to improved Pb(II) adsorption performance (Xue et al. 2012). This oxidative process also induces significant surface acidity, as evidenced by Huff and Lee (2016), who observed a pH reduction from 7.16 to 5.66 in pine-derived biochar following H_2O_2 treatment. Nevertheless, excessive oxidant concentration (>10% H_2O_2) was found to diminish methylene blue adsorption capacity, underscoring the importance of optimizing modification intensity relative to target contaminant characteristics. Alternative oxidants such as potassium permanganate ($KMnO_4$) similarly enhance surface oxygenation while concurrently improving textural properties. Wang et al. (2015) documented a twofold increase in specific surface area (101–205 m^2/g) for $KMnO_4$-modified pecan shell biochar, corresponding with enhanced heavy metal (Pb(II), Cu(II), Cd(II)) uptake capacity. Ozone oxidation presents another viable approach, generating diverse surface oxides including phenolic, hydroxyl, quinone, carboxyl, and ether functionalities (Gómez-Serrano et al. 2002). The efficacy of ozonation is temperature-dependent, as thermal decomposition above 100 °C reduces ozone reactivity, thereby influencing both the quantity and nature of formed surface groups.

1.4.2.4 Catalytic Graphitization

Catalytic graphitization represents a specialized chemical modification approach for transforming biomass-derived carbon structures. Biomass pyrolysis typically yields hard carbon composed of disordered graphitic crystallites embedded in an amorphous carbon matrix, which persists even at extreme temperatures exceeding 2500 °C (Thambiliyagodage et al. 2018; Thompson et al. 2015). This structurally imperfect carbon material often demonstrates limited electrical conductivity despite its porous nature, necessitating structural optimization for enhanced performance. Two principal methodologies exist for producing graphitic carbon from biomass: conventional high-temperature pyrolysis and catalytic graphitization. While thermal treatment at ≥2500 °C can improve graphitization degree, such energy-intensive processes conflict with SDGs. In contrast, catalytic graphitization employs metal catalysts to lower the transformation energy barrier, enabling graphitic carbon production at significantly reduced temperatures (~1000 °C). This approach preserves biomass-derived morphological features while simultaneously improving both graphitic ordering and porous architecture. The underlying mechanisms of catalytic graphitization involve complex physicochemical processes. Current understanding identifies two dominant pathways: (1) dissolution–reprecipitation, where catalysts disrupt carbon–carbon bonds, solubilize amorphous carbon, and subsequently reprecipitate graphitic crystallites (Fuertes and Alvarez 2004); and (2) carbide transformation, involving intermediate metal carbide formation followed by decomposition into graphitic carbon (Fuertes and Alvarez 2004). Transition metals (Fe, Co, Ni) and their oxides (Cr_2O_3, MnO_2) serve as effective catalysts, often exhibiting dual functionality as both graphitization promoters and pore-forming agents. Despite these advances, the fundamental mechanisms governing biomass graphitization remain incompletely understood, presenting significant opportunities for further research.

The physicochemical characteristics and structural features of biochar are determined by multiple variables, such as precursor biomass selection, graphitization catalyst species and loading, processing methodology, and treatment sequence. In a seminal study by Sevilla et al. (2007), wood sawdust impregnated with either $Fe(NO_3)_3$ or $Ni(NO_3)_2$ was carbonized at 900–1000 °C, yielding carbon materials with enhanced graphitic ordering. Ye et al. (2020) demonstrated that K_2FeO_4-mediated graphitization of sesame stalk biochar followed by nitrogen doping resulted in PGBF-N, an effective persulfate-activating catalyst. Investigations by Liu et al. (2012) revealed that $Fe(NO_3)_3$-catalyzed pyrolysis of coconut shells between 500 and 1000 °C generated three-dimensional graphitic carbon with hierarchical porosity, where elevated temperatures correlated with improved graphitic crystallinity. Similar findings were reported by Ru et al. (2016) during microalgae pyrolysis under N_2 atmosphere (10 °C/min heating rate), where 900 °C treatment produced materials containing abundant graphitic microdomains and well-developed layered porosity. Notably, temperature-dependent structural evolution showed progressive micropore coalescence into meso/macropores at higher temperatures (1100 °C). X-ray diffraction (XRD) and Raman spectroscopic analyses substantiated that thermal treatment enhanced carbon ordering, as evidenced by intensified (002) diffraction peaks and reduced ID/IG ratios. All synthesized materials exhibited lamellar pore architectures, with both Brunauer–Emmett–Teller (BET) surface area and total pore volume showing positive correlations with pyrolysis temperature. These observations underscore the pivotal role of thermal conditions in governing both textural and crystalline properties. Optimal pore development, particularly micropore formation, occurs within specific temperature windows. Yu et al. (2020) observed that shrimp shell-derived carbons developed increasingly hierarchical porosity and graphitic character at higher carbonization temperatures, accompanied by nitrogen speciation transitions from pyrrolic to pyridinic and eventually graphitic configurations. Contrastingly, Thompson et al. (2015) reported that cork sawdust-derived biochars maintained consistent graphitic quality despite decreasing iron catalyst concentrations, while exhibiting progressive surface area enhancement. Their proposed mechanism involves Fe_3C-mediated dissolution–reprecipitation processes that simultaneously etch pore networks and catalyze graphite nanotube growth. Alternative synthesis approaches have been explored, including Wu et al.'s (2016) demonstration of iron acetylacetonate-impregnated cotton pyrolysis yielding graphitic carbon via iron nanoparticle-catalyzed vapor–liquid–solid growth. Wang et al. (2008) developed a microwave-assisted pyrolysis strategy for converting iron/polymer/biomass composites into nanostructured graphitic carbons that preserve the intricate morphological features of biological precursors while achieving high porosity and crystallinity. This collective evidence highlights that pyrolysis temperature serves as a master variable controlling the competing processes of pore development and graphitic ordering during biochar formation, with catalyst selection and processing methods providing additional tuning parameters for material optimization.

Current literature reveals a limited number of investigations dedicated to the synthesis of biochar via catalytic graphitization of biomass feedstocks. This research gap constrains a thorough understanding of the structural characteristics and functional properties of graphitized biochar. Of particular interest are biochar materials possessing both enhanced graphitic ordering and hierarchical porosity, as these features could substantially expand their potential applications in various fields.

1.5 Conclusion

The thermochemical conversion of agricultural and urban solid wastes into biochar presents a viable approach for eco-friendly waste valorization. To optimize the physicochemical characteristics of biochar and improve its remediation performance for sustainable utilization, selection of appropriate activation techniques must be tailored to intended application scenarios. Chemical activation methods are particularly effective for modifying surface characteristics, resulting in substantial improvements in specific surface area and pore structure. These engineered biochars demonstrate enhanced adsorption capabilities for diverse contaminants, including both organic compounds and heavy metals, while simultaneously exhibiting improved catalytic performance. As a result, chemically activated biochar emerges as a promising functional material for advancing environmental sustainability initiatives.

Biomass is acknowledged as a versatile raw material due to its structural diversity, property variability, and widespread availability. Recent years have witnessed growing research interest in developing precisely engineered biochar with controlled characteristics derived from different biomass feedstocks. However, the field currently lacks comprehensive theoretical frameworks and standardized methodologies to guide the targeted synthesis of biochar with predetermined properties. To address these limitations, several strategic approaches merit consideration:

1. Strategic selection of biomass feedstocks: The fundamental characteristics of biochar—including its porous architecture, graphitic domains, surface area, and heteroatom incorporation—are intrinsically linked to the physicochemical properties of the precursor biomass. Therefore, careful identification and utilization of biomass with appropriate inherent structures and elemental compositions is paramount for producing biochar with tailored functionalities. To enable large-scale manufacturing, a comprehensive classification and analysis of biomass resource characteristics must be conducted, with particular emphasis on establishing quantitative correlations between feedstock categories and the resultant biochar properties. Such methodology development will ultimately allow for predictable control over biochar's structural and chemical attributes during the synthesis process.
2. Establishing an integrated production framework: Current research has predominantly focused on biochar synthesis from singular biomass feedstocks, leaving the systematic optimization of mixed biomass conversion relatively unexplored. This knowledge gap highlights the necessity to develop unified processing protocols capable of accommodating diverse biomass combinations. Such an integrated approach would provide crucial methodological guidance for controlling biochar properties when utilizing heterogeneous feedstock sources.
3. Development of a predictive framework for biochar synthesis: Current research must systematically evaluate how biomass characteristics and processing parameters—particularly activation methods—collectively influence biochar's structural and functional properties. A critical knowledge gap exists in understanding the synergistic effects among these variables during thermochemical conversion. By constructing a quantitative structure–property–application relationship model, researchers can establish predictive correlations between feedstock attributes, processing conditions, and end-use performance. Such a framework would facilitate the customized production of biochar at industrial scales, ensuring optimal material properties for site-specific environmental applications.

References

Afolabi, O.O.D., Sohail, M., and Thomas, C.L.P. (2017). Characterization of solid fuel chars recovered from microwave hydrothermal carbonization of human biowaste. *Energy* 134: 74–89.

Aguado, R., Olazar, M., San José, M.J. et al. (2000). Pyrolysis of sawdust in a conical spouted bed reactor. Yield and product composition. *Industrial & Engineering Chemistry Research* 39 (6): 1925–1933.

Ahmad, M., Lee, S.S., Dou, X. et al. (2012). Effects of pyrolysis temperature on soybean stover- and peanut shell-derived biochar properties and TCE adsorption in water. *Bioresource Technology* 118: 536–544.

Angin, D. (2013). Effect of pyrolysis temperature and heating rate on biochar obtained from pyrolysis of safflower seed press cake. *Bioresource Technology* 128: 593–597.

Antal, M.J., Croiset, E., Dai, X. et al. (1996). High-yield biomass charcoal. *Energy & Fuels* 10 (3): 652–658.

Arriagada, R., García, R., Molina-Sabio, M. et al. (1997). Effect of steam activation on the porosity and chemical nature of activated carbons from *Eucalyptus globulus* and peach stones. *Microporous Materials* 8 (3–4): 123–130.

Arshanitsa, A., Dizhbite, T., Bikovens, O. et al. (2016). Effects of microwave treatment on the chemical structure of lignocarbohydrate matrix of softwood and hardwood. *Energy and Fuels* 30 (1): 457–464.

Binner, E., Mediero-Munoyerro, M., Huddle, T. et al. (2014). Factors affecting the microwave coking of coals and the implications on microwave cavity design. *Fuel Processing Technology* 125: 8–17.

Bridgwater, A.V. (2012). Review of fast pyrolysis of biomass and product upgrading. *Biomass and Bioenergy* 38: 68–94.

Chen, D., Li, Y., Cen, K. et al. (2016a). Pyrolysis polygeneration of poplar wood: effect of heating rate and pyrolysis temperature. *Bioresource Technology* 218: 780–788.

Chen, J., Pitchai, K., Birla, S. et al. (2016b). Modeling heat and mass transport during microwave heating of frozen food rotating on a turntable. *Food and Bioproducts Processing* 99: 116–127.

Chen, X., Chen, G., Chen, L. et al. (2011). Adsorption of copper and zinc by biochars produced from pyrolysis of hardwood and corn straw in aqueous solution. *Bioresource Technology* 102 (19): 8877–8884.

China Legal Publishing House. (1995). *Law of the People's Republic of China on the Prevention and Control of Environmental Pollution by Solid Wastes.*

Chun, Y., Sheng, G., Chiou, C.T. et al. (2004). Compositions and sorptive properties of crop residue-derived chars. *Environmental Science & Technology* 38 (17): 4649–4655.

Ciacci, T., Galgano, A., and Di Blasi, C. (2010). Numerical simulation of the electromagnetic field and the heat and mass transfer processes during microwave-induced pyrolysis of a wood block. *Chemical Engineering Science* 65 (14): 4117–4133.

Dai, J., Niu, C., Pan, Y. et al. (2020). Performance of anaerobic membrane bioreactors for the co-digestion of sewage sludge and food waste. *Environmental Science* 41 (8): 3740–3747.

Demiral, H., Demiral, I., Karabacakoĝlu, B. et al. (2011). Production of activated carbon from olive bagasse by physical activation. *Chemical Engineering Research and Design* 89 (2): 206–213.

Demirbaş, A. and Arin, G. (2002). An overview of biomass pyrolysis. *Energy Sources* 24 (5): 471–482.

Domínguez, A., Fernández, Y., Fidalgo, B. et al. (2008). Bio-syngas production with low concentrations of CO_2 and CH_4 from microwave-induced pyrolysis of wet and dried sewage sludge. *Chemosphere* 70 (3): 397–403.

Fan, Y., Wang, B., Yuan, S. et al. (2010). Adsorptive removal of chloramphenicol from wastewater by NaOH modified bamboo charcoal. *Bioresource Technology* 101 (19): 7661–7664.

Fuertes, A.B. and Alvarez, S. (2004). Graphitic mesoporous carbons synthesised through mesostructured silica templates. *Carbon* 42 (15): 3049–3055.

Funke, A. and Ziegler, F. (2010). Hydrothermal carbonization of biomass: a summary and discussion of chemical mechanisms for process engineering. *Biofuels, Bioproducts and Biorefining* 4: 160–177.

Gao, Y., Wang, X., Wang, J. et al. (2013). Effect of residence time on chemical and structural properties of hydrochar obtained by hydrothermal carbonization of water hyacinth. *Energy* 58: 376–383.

Girgis, B.S., Soliman, A.M., and Fathy, N.A. (2011). Development of micro-mesoporous carbons from several seed hulls under varying conditions of activation. *Microporous and Mesoporous Materials* 142 (2–3): 518–525.

Gómez-Serrano, V., Álvarez, P.M., Jaramillo, J. et al. (2002). Formation of oxygen complexes by ozonation of carbonaceous materials prepared from cherry stones I. Thermal effects. *Carbon* 40 (4): 513–522.

Guo, S., Peng, J., Li, W. et al. (2009). Effect of CO_2 activation on porous structures of coconut shell-based activated carbons. *Applied Surface Science* 255 (20): 8443–8449.

Han, L., Zhan, E., Yang, Y. et al. (2020). Highly efficient U(VI) removal by chemically modified hyrochar and pyrochar derived from animal manure. *Journal of Cleaner Production* 264: 121542.

Hao, J., de Los Reyes III, F.L., and He, X. (2020). Fat, oil, and grease (FOG) deposits yield higher methane than FOG in anaerobic co-digestion with waste activated sludge. *Journal of Environmental Management* 268: 110708.

Harvey, O.R., Herbert, B.E., Rhue, R.D. et al. (2011). Metal interactions at the biochar-water interface: energetics and structure-sorption relationships elucidated by flow adsorption microcalorimetry. *Environmental Science & Technology* 45 (13): 5550–5556.

He, C., Giannis, A., and Wang, J.Y. (2013). Conversion of sewage sludge to clean solid fuel using hydrothermal carbonization: hydrochar fuel characteristics and combustion behavior. *Applied Energy* 111: 257–266.

Huang, Y.F., Chen, W.R., Chiueh, P.T. et al. (2012). Microwave torrefaction of rice straw and pennisetum. *Bioresource Technology* 123: 1–7.

Huang, Y.F., Sung, H.T., Chiueh, P.T. et al. (2017). Microwave torrefaction of sewage sludge and leucaena. *Journal of the Taiwan Institute of Chemical Engineers* 70: 236–243.

Huff, M.D. and Lee, J.W. (2016). Biochar-surface oxygenation with hydrogen peroxide. *Journal of Environmental Management* 165: 17–21.

Hwang, I.H., Aoyama, H., Matsuto, T. et al. (2012). Recovery of solid fuel from municipal solid waste by hydrothermal treatment using subcritical water. *Waste Management* 32 (3): 410–416.

Jiang, Q., Chen, Y., Yu, S. et al. (2020). Effects of citrus peel biochar on anaerobic co-digestion of food waste and sewage sludge and its direct interspecies electron transfer pathway study. *Chemical Engineering Journal* 398: 125643.

Jiang, Q., Wang, R., and Sun, W. (2018). Potential evaluation and countermeasures on livestock manure resource utilization-based on perspective of commercial organic fertilizer utilization. *Journal of Huazhong Agricultural University (Social Sciences Edition)* 4 (30–37): 166–167.

Jin, H., Capareda, S., Chang, Z. et al. (2014). Biochar pyrolytically produced from municipal solid wastes for aqueous As(V) removal: adsorption property and its improvement with KOH activation. *Bioresource Technology* 169: 622–629.

Kang, S., Li, X., Fan, J. et al. (2012). Characterization of hydrochars produced by hydrothermal carbonization of lignin, cellulose, d-xylose, and wood meal. *Industrial and Engineering Chemistry Research* 51 (26): 9023–9031.

Kim, D., Lee, K., and Park, K.Y. (2014). Hydrothermal carbonization of anaerobically digested sludge for solid fuel production and energy recovery. *Fuel* 130: 120–125.

Kim, W.K., Shim, T., Kim, Y.S. et al. (2013). Characterization of cadmium removal from aqueous solution by biochar produced from a giant miscanthus at different pyrolytic temperatures. *Bioresource Technology* 138: 266–270.

Knežević, D., Van Swaaij, W., and Kersten, S. (2010). Hydrothermal conversion of biomass. II. conversion of wood, pyrolysis oil, and glucose in hot compressed water. *Industrial and Engineering Chemistry Research* 49 (1): 104–112.

Kołtowski, M., Hilber, I., Bucheli, T.D. et al. (2016). Effect of steam activated biochar application to industrially contaminated soils on bioavailability of polycyclic aromatic hydrocarbons and ecotoxicity of soils. *Science of the Total Environment* 556–557: 1023–1031.

Kołtowski, M., Hilber, I., Bucheli, T.D. et al. (2017). Activated biochars reduce the exposure of polycyclic aromatic hydrocarbons in industrially contaminated soils. *Chemical Engineering Journal* 310: 33–40.

Kong, S., Lam, S.S., Yek, P.N.Y. et al. (2019). Self-purging microwave pyrolysis: an innovative approach to convert oil palm shell into carbon-rich biochar for methylene blue adsorption. *Journal of Chemical Technology and Biotechnology* 94 (5): 1397–1405.

Lam, S., Wan Mahari, W.A., Ok, Y.S. et al. (2019). Microwave vacuum pyrolysis of waste plastic and used cooking oil for simultaneous waste reduction and sustainable energy conversion: recovery of cleaner liquid fuel and techno-economic analysis. *Renewable and Sustainable Energy Reviews* 115: 109359.

Lee, J., Yang, X., Song, H. et al. (2017). Effects of carbon dioxide on pyrolysis of peat. *Energy* 120: 929–936.

Li, H., Zhang, T., Lv, L. et al. (2019). Modification of sludge activated carbon made by activation of waste acid from titanium dioxide production and applications. *Modern Chemical Industry* 39 (4): 131–136.

Li, X., Shen, Q., Zhang, D. et al. (2013). Functional groups determine biochar properties (pH and EC) as studied by two-dimensional ^{13}C NMR correlation spectroscopy. *PLoS One* 8 (6): e65949.

Li, Y. and Wang, Q. (2020). Study on potential of straw resource nutrient return to field and application technology in China. *Soil and Fertilizer Sciences in China* 1: 119–126.

Li, Y., Shao, J., Wang, X. et al. (2014). Characterization of modified biochars derived from bamboo pyrolysis and their utilization for target component (furfural) adsorption. *Energy and Fuels* 28 (8): 5119–5127.

Libra, J.A., Ro, K.S., Kammann, C. et al. (2011). Hydrothermal carbonization of biomass residuals: a comparative review of the chemistry, processes and applications of wet and dry pyrolysis. *Biofuels* 2 (1): 71–106.

Liew, R.K., Chai, C., Yek, P.N.Y. et al. (2019). Innovative production of highly porous carbon for industrial effluent remediation via microwave vacuum pyrolysis plus sodium-potassium hydroxide mixture activation. *Journal of Cleaner Production* 208: 1436–1445.

Liu, S. (2020). Research of present situation of solid waste disposal in the city and the development path. *Heilongjiang Science* 11 (16): 144–146.

Liu, Q., Gu, J., Zhang, W. et al. (2012). Biomorphic porous graphitic carbon for electromagnetic interference shielding. *Journal of Materials Chemistry* 22 (39): 21183–21188.

Liu, Y., Li, X., Jing, Z. et al. (2019). Application of low-temperature aqueous ammonia-modified sludge-based activated carbon to coking wastewater treatment. *Industrial Water Treatment* 39 (1): 25–28.

Liu, Z., Wang, Z., Tang, S. et al. (2021). Fabrication, characterization and sorption properties of activated biochar from livestock manure via three different approaches. *Resources, Conservation and Recycling* 168: 105254.

Luo, J., Huang, X., Guo, W. et al. (2020). Novel strategy to stimulate the food wastes anaerobic fermentation performance by eggshell wastes conditioning and the underlying mechanisms. *Chemical Engineering Journal* 398: 125560.

Luo, Y., Xie, K., Feng, Y. et al. (2021). Preparation of lanthanum modified walnut shell biochar and adsorption of phosphate from aqueous solutions. *Chemical Industry and Engineering Progress* 40 (2): 1121–1129.

Lykoudi, A., Frontistis, Z., Vakros, J. et al. (2020). Degradation of sulfamethoxazole with persulfate using spent coffee grounds biochar as activator. *Journal of Environmental Management* 271: 111022.

Marcus, Y. (1999). On transport properties of hot liquid and supercritical water and their relationship to the hydrogen bonding. *Fluid Phase Equilibria* 164 (1): 131–142.

Marris, E. (2006). Putting the carbon back: black is the new green. *Nature* 442 (7103): 624–626.

Martín, M.T., Sanz, A.B., Nozal, L. et al. (2017). Microwave-assisted pyrolysis of Mediterranean forest biomass waste: bioproduct characterization. *Journal of Analytical and Applied Pyrolysis* 127: 278–285.

Martín-Gullón, I., Asensio, M., Font, R. et al. (1996). Steam-activated carbons from a bituminous coal in a continuous multistage fluidized bed pilot plant. *Carbon* 34 (12): 1515–1520.

Min, C., An, D., Wang, Y. et al. (2020). Progress of rural solid waste resource utilization in China. *Journal of Agricultural Resources and Environment* 37 (2): 151–160.

Mohan, D., Sarswat, A., Ok, Y.S. et al. (2014). Organic and inorganic contaminants removal from water with biochar, a renewable, low cost and sustainable absorbent – A critical review. *Bioresource Technology* 160: 191–202.

Möller, M., Nilges, P., Harnisch, F. et al. (2011). Subcritical water as reaction environment: fundamental of hydrothermal biomass transformation. *ChemSusChem* 4 (5): 566–579.

Onay, O. (2007). Influence of pyrolysis temperature and heating rate on the production of bio-oil and char from safflower seed by pyrolysis, using a well-swept fixed-bed reactor. *Fuel Processing Technology* 88 (5): 523–531.

Parshetti, G.K., Kent Hoekman, S., and Balasubramanian, R. (2013). Chemical, structural and combustion characteristics of carbonaceous products obtained by hydrothermal carbonization of palm empty fruit bunches. *Bioresource Technology* 135: 683–689.

Passos, F. and Ferrer, I. (2015). Influence of hydrothermal pretreatment on microalgal biomass anaerobic digestion and bioenergy production. *Water Research* 68: 364–373.

Peng, F., He, P.W., Luo, Y., et al. (2012). Adsorption of phosphate by biomass char deriving from fast pyrolysis of biomass waste. *CLEAN – Soil, Air, Water* 40 (5): 493–498.

Peng, P., Lang, Y.H., and Wang, X.M. (2016). Adsorption behavior and mechanism of pentachlorophenol on reed biochars: pH effect, pyrolysis temperature, hydrochloric acid treatment and isotherms. *Ecological Engineering* 90: 225–233.

Qian, X., Qian, Q., Quek, A. et al. (2013). Hydrothermal carbonization of macroalgae and the effects of experimental parameters on the properties of hydrochar. *ACS Sustainable Chemistry & Engineering* 1 (9): 1092–1101.

Qiu, H., Yan, J., Cai, Y. et al. (2012). Analysis of pollution emission and control strategies for specialized livestock and poultry farming in China – An empirical study based on a survey in five provinces. *Journal of Agrotechnical Economics* 5: 29–35.

Qureshi, K.M., Kay Lup, A.N., Khan, S. et al. (2018). A technical review on semi-continuous and continuous pyrolysis process of biomass to bio-oil. *Journal of Analytical and Applied Pyrolysis* 131: 52–75.

Raclavská, H., Corsaro, A., Juchelková, D. et al. (2015). Effect of temperature on the enrichment and volatility of 18 elements during pyrolysis of biomass, coal, and tires. *Fuel Processing Technology* 131: 330–337.

Rajapaksha, A.U., Vithanage, M., Ahmad, M. et al. (2015). Enhanced sulfamethazine removal by steam-activated invasive plant-derived biochar. *Journal of Hazardous Materials* 290: 43–50.

Ren, S., Lei, H., Wang, L. et al. (2012). Microwave torrefaction of Douglas fir sawdust pellets. *Energy and Fuels* 26 (9): 5936–5943.

Rodríguez-Reinoso, F., Molina-Sabio, M., and González, M.T. (1995). The use of steam and CO_2 as activating agents in the preparation of activated carbons. *Carbon* 33 (1): 15–23.

Ru, H., Bai, N., Xiang, K. et al. (2016). Porous carbons derived from microalgae with enhanced electrochemical performance for lithium-ion batteries. *Electrochimica Acta* 194: 10–16.

Ruiz, H.A., Rodríguez-Jasso, R.M., Fernandes, B.D. et al. (2013). Hydrothermal processing, as an alternative for upgrading agriculture residues and marine biomass according to the biorefinery concept: a review. *Renewable and Sustainable Energy Reviews*. 21: 35–51.

Saidur, R., Abdelaziz, E.A., Demirbas, A. et al. (2011). A review on biomass as a fuel for boilers. *Renewable and Sustainable Energy Reviews* 15 (5): 2262–2289.

Salvi, D., Ortego, J., Arauz, C. et al. (2009). Experimental study of the effect of dielectric and physical properties on temperature distribution in fluids during continuous flow microwave heating. *Journal of Food Engineering* 93 (2): 149–157.

Satpathy, S.K., Tabil, L.G., Meda, V. et al. (2014). Torrefaction of wheat and barley straw after microwave heating. *Fuel* 124: 269–278.

Savage, P.E. (1999). Organic chemical reactions in supercritical water. *Chemical Reviews* 99 (2–3): 603–621.

Seifritz, W. (1993). Should we store carbon in charcoal? *International Journal of Hydrogen Energy* 18 (5): 405–407.

Sevilla, M. and Fuertes, A.B. (2009). The production of carbon materials by hydrothermal carbonization of cellulose. *Carbon* 47 (9): 2281–2289.

Sevilla, M., Sanchís, C., Valdés-Soh, T., et al. (2007). Synthesis of graphitic carbon nanostructures from sawdust and their application as electrocatalyst supports. *Journal of Physical Chemistry C* 111 (27): 9749–9756.

Shafaghat, H., Lee, H.W., Tsang, Y.F. et al. (2019). *In-situ* and *ex-situ* catalytic pyrolysis/co-pyrolysis of empty fruit bunches using mesostructured aluminosilicate catalysts. *Chemical Engineering Journal* 336: 330–338.

Shen, H. (2020). Effect of mulch on crop yield. *Agricultural Development & Equipments* 7: 75–77.

Shen, Y. and Zhang, N. (2019). Facile synthesis of porous carbons from silica-rich rice husk char for volatile organic compounds (VOCs) sorption. *Bioresource Technology* 282: 294–300.

Shi, Z. (2018). Current situation of utilization of straw resources in China and relevant countermeasures and suggestions. *World Environment* 5: 16–18.

Sri Shalini, S., Palanivelu, K., Ramachandran, A. et al. (2021). Biochar from biomass waste as a renewable carbon material for climate change mitigation in reducing greenhouse gas emissions – A review. *Biomass Conversion and Biorefinery* 11 (5): 2247–2267.

Sun, K., Tang, J., Gong, Y. et al. (2015). Characterization of potassium hydroxide (KOH) modified hydrochars from different feedstocks for enhanced removal of heavy metals from water. *Environmental Science and Pollution Research* 22: 16640–16651.

Taer, E., Deraman, M., Talib, I.A. et al. (2010). Physical, electrochemical and supercapacitive properties of activated carbon pellets from pre-carbonized rubber wood sawdust by CO_2 activation. *Current Applied Physics* 10 (4): 1071–1075.

Taha, S.M., Amer, M.E., Elmarsafy, A.E. et al. (2014). Adsorption of 15 different pesticides on untreated and phosphoric acid treated biochar and charcoal from water. *Journal of Environmental Chemical Engineering* 2 (4): 2013–2025.

Tan, I.A.W., Ahmad, A.L., Hameed B.H. et al. (2008). Adsorption of basic dye on high-surface-area activated carbon prepared from coconut husk: equilibrium, kinetic and thermodynamic studies. *Journal of Hazardous Materials* 154 (1–3): 337–346.

Tarves, P.C., Serapiglia, M.J., Mullen, C.A. et al. (2017). Effects of hot water extraction pretreatment on pyrolysis of shrub willow. *Biomass and Bioenergy* 107: 299–304.

Thambiliyagodage, C., Ulrich, S., Araujo, P.T. et al. (2018). Catalytic graphitization in nanocast carbon monoliths by iron, cobalt and nickel nanoparticles. *Carbon* 134: 452–463.

Thompson, E., Danks, A.E., Bourgeois, L. et al. (2015). Iron-catalyzed graphitization of biomass. *Green Chemistry* 17 (1): 551–556.

Tripathi, M., Sahu J.N., and Ganesan P. (2016). Effect of process parameters on production of biochar from biomass waste through pyrolysis: a review. *Renewable and Sustainable Energy Reviews* 55: 467–481.

Vithanage, M., Rajapaksha, A.U., Zhang, M. et al. (2015). Acid-activated biochar increased sulfamethazine retention in soils. *Environmental Science and Pollution Research* 22 (3): 2175–2186.

Wang, C., Ma, D., and Bao, X. (2008). Transformation of biomass into porous graphitic carbon nanostructures by microwave irradiation. *Journal of Physical Chemistry C* 112 (45): 17596–17602.

Wang, H., Zhang, Z., Sun, R. et al. (2015). HPV infection and anemia status stratify the survival of early T2 laryngeal squamous cell carcinoma. *Journal of Voice* 29 (3): 356–362.

Wu, F., Huang, R., Mu, D. et al. (2016). Controlled synthesis of graphitic carbon-encapsulated α-Fe_2O_3 nanocomposite via low-temperature catalytic graphitization of biomass and its lithium storage property. *Electrochimica Acta* 187: 508–516.

Xue, Y., Gao, B., Yao, Y. et al. (2012). Hydrogen peroxide modification enhances the ability of biochar (hydrochar) produced from hydrothermal carbonization of peanut hull to remove aqueous heavy metals: batch and column tests. *Chemical Engineering Journal* 200–202: 673–680.

Ye, S., Zeng, G., Tan, X. et al. (2020). Nitrogen-doped biochar fiber with graphitization from *Boehmeria nivea* for promoted peroxymonosulfate activation and non-radical degradation pathways with enhancing electron transfer. *Applied Catalysis B: Environmental* 269: 118850.

Yin, X., Zhang, Y., Yang, Y. et al. (2019). An overview of solid waste pollution in rural areas of China. *Low Carbon World* 9 (10): 14–16.

Yu, J., Feng, H., Tang, L. et al. (2021). Insight into the key factors in fast adsorption of organic pollutants by hierarchical porous biochar. *Journal of Hazardous Materials* 403: 123610.

Yu, J., Tang, L., Pang, Y. et al. (2020). Hierarchical porous biochar from shrimp shell for persulfate activation: a two-electron transfer path and key impact factors. *Applied Catalysis B: Environmental* 260: 118160.

Zaini, M. and Kamaruddin, M. (2013). Critical issues in microwave-assisted activated carbon preparation. *Journal of Analytical and Applied Pyrolysis* 101: 238–241.

Zhang, L. and Guo, Q. (2020). Air pollutant emission mechanism, spatio-temporal patterns and emission reduction strategies of resource-oriented utilization of crop straw: a literature review. *Journal of Southwest University (Natural Science Edition)* 42 (7): 143–153.

Zhang, S., Dong, Q., Zhang, L. et al. (2015). High quality syngas production from microwave pyrolysis of rice husk with char-supported metallic catalysts. *Bioresource Technology* 191: 17–23.

Zhang, T., Walawender, W., Fan, L.T. et al. (2004). Preparation of activated carbon from forest and agricultural residues through CO_2 activation. *Chemical Engineering Journal* 105 (1–2): 53–59.

Zhao, B., O'Connor, D., Zhang, J. et al. (2018). Effect of pyrolysis temperature, heating rate, and residence time on rapeseed stem derived biochar. *Journal of Cleaner Production* 174: 977–987.

Zhao, J., Zhou, S., Chen, A. et al. (2012a). Morphological analysis and ecological risk assessment on heavy metals in the activated carbon prepared from sewage sludge and peanut hull. *Journal of Agro-Environment Science* 31 (11): 2284–2289.

Zhao, X., Wang, M., Liu, H. et al. (2012b). A microwave reactor for characterization of pyrolyzed biomass. *Bioresource Technology* 104: 673–678.

Zhao, X., Zhang, J., Song, Z. et al. (2011). Microwave pyrolysis of straw bale and energy balance analysis. *Journal of Analytical and Applied Pyrolysis* 92 (1): 43–49.

Zhou, J., Liu, S., Zhou, N. et al. (2018). Development and application of a continuous fast microwave pyrolysis system for sewage sludge utilization. *Bioresource Technology* 256: 295–301.

Zhu, J., Sun, C., Yu, H. et al. (2014). *Guangdong Agricultural Sciences* 41 (03): 65–69, 73.

2

Waste-derived Biochar as Renewable Bio-fuels

Xiefei Zhu[1], *Yuqing Sun*[2], *and Daniel C.W. Tsang*[3]

[1] *School of Advanced Energy, Sun Yat-sen University, Shenzhen, Guangdong, China*
[2] *School of Agriculture and Biotechnology, Sun Yat-sen University, Guangzhou, Guangdong, China*
[3] *Department of Civil and Environmental Engineering, The Hong Kong University of Science and Technology, Clear Water Bay, Hong Kong, China*

2.1 Feedstocks Type of Waste-derived Biochar Fuels

Agricultural waste, as one of the largest sources of pollution in rural areas, is huge in quantity, poor in quality, and low in beneficial components. The overall waste grade and utilization efficiency are not high, which seriously restricts the sustainable development of rural areas (Sudalai et al. 2024; Wu et al. 2022). Although the whole world has been committed to the resource utilization of agricultural waste and some positive results have been achieved, there is still a large space for development. Typical agricultural waste mainly includes crop waste and livestock manure. How to deal with and utilize the exponentially increasing organic waste is now a real problem facing the sustainable development of rural areas. Tapping into advanced agricultural waste utilization technology to truly realize the transformation of agricultural waste into "treasure" is of great significance in eliminating environmental pollution, improving the rural ecological environment, promoting the sustainable development of agriculture, guaranteeing food security, and improving the ecological environment (Shahbaz et al. 2021; Ghodake et al. 2021; Jeníček et al. 2023).

Agricultural production waste and rural household rubbish, as the main components of rural waste, are dispersed, complex, and diverse. They are difficult to collect, treat, and utilize in a resourceful manner, seriously threatening the

Waste-derived Biochar for Sustainable Rural Revitalization, First Edition. Edited by Yuqing Sun and Daniel C.W. Tsang.
Published 2026 by John Wiley & Sons, Inc.

quality of the ecological environment in rural areas. Typical agricultural waste mainly includes residues from farmland and orchards (such as crop stalks, fruit tree branches, grain husks, fruit shells, leaves, etc.), animal and poultry manure, and residues from the processing of agricultural byproducts. According to statistics, about 900 million tonnes of plant straw and 3.8 billion tonnes of livestock and poultry manure annually are produced in China, and these wastes contain abundant biomass energy (Mengqi et al. 2021; Alengebawy et al. 2023; Wang et al. 2024). However, for a long time, most of this biomass energy has not been effectively utilized as resources but has been casually discarded or incinerated. This not only causes a waste of resources but also may cause environmental problems such as fires and pollution of air and water. Therefore, how to convert these rural wastes into valuable resources has become an urgent problem at present. Most of the rural biomass energy use is based on direct combustion, which has a low thermal efficiency (less than 10%). Although direct combustion treatment can increase the energy conversion potential of agricultural wastes, the emission of large amounts of soot and residual ash is deteriorating the living environment and damaging human health. The use of advanced biomass energy conversion technology can increase thermal efficiency by 35–40%, further saving resources, improving the living environment of farmers, and raising living standards. For example, the development of biomass carbonization, rural biogas and other energy projects, and eco-agriculture models can effectively promote the virtuous ecological cycle, reduce the destruction of forest resources and soil erosion, and protect biodiversity. In particular, biomass carbonization technology has great potential for application because of its wide adaptability to a wide range of feedstocks, which can convert most rural biomass into high-value biochar for use as fuels, adsorbents, and soil conditioners (Qin et al. 2022; Amalina et al. 2022; Seow et al. 2022).

2.2 Preparation Technologies of Waste-derived Biochar Fuels

Biochar is a black, fluffy, lightweight, carbon-rich solid material made by heating biomass (e.g., agricultural waste, wood, or organic materials) to high temperatures (typically between 300 and 1000 °C) under anoxic conditions through thermochemical conversion techniques. This material is mainly composed of carbon, hydrogen, oxygen, nitrogen, and other elements. The carbon content is usually more than 70%, showing excellent combustion performance and high heating value (HHV). In recent years, the preparation and application technologies of biochar have been widely investigated and applied, which are of great economic and environmental significance (Wen et al. 2023; Zhu et al.

2023). In this part, the technology of biochar preparation from biomass will be introduced in detail.

Biochar preparation technology, as an important way of biomass energy development and utilization, has received extensive attention at home and abroad. Typical biochar preparation technologies include slow pyrolysis, fast pyrolysis, gasification, microwave pyrolysis, and hydrothermal carbonization. These production technologies differ in terms of reaction conditions, equipment requirements, and production costs. Suitable processes and conditions need to be selected according to the characteristics of feedstocks and product requirements. Typical biochar preparation conditions are summarized in Table 2.1. Meanwhile, the key factors affecting the properties of biochar in the biochar preparation process are also pointed out, such as pyrolysis temperature, heating rate, holding time, and reaction atmosphere. It is generally believed that higher heating rates and pyrolysis temperatures are not favorable to the yield of solid product biochar.

2.2.1 Typical Technologies for Biochar Preparation

Pyrolysis, as a typical biomass thermal conversion technology, occurs mainly in the absence of oxygen and at high temperatures (about 300–1000 °C), and the main pyrolysis products generated are biochar, bio-oil, and syngas (Li et al. 2023). Pyrolysis is generally divided into slow pyrolysis and fast pyrolysis according to the heating rate. The biochar from fast pyrolysis is produced as a byproduct, generally at a yield of less than 30%. Slow pyrolysis is one of the main methods for producing biochar and is commonly used to make biochar from biomass. When the pyrolysis temperature is below 300 °C, this thermochemical process is known as torrefaction and the product can also be considered as a special type of biochar (Chen et al. 2022a; Cao et al. 2023).

Gasification is a technology that uses different biomass feedstocks to produce syngas, and the product gas can be used for energy supply and chemical synthesis (Li et al. 2024; Wu et al. 2024). The main process of gasification is to introduce biomass with a certain gasification agent such as oxygen, air, or water vapor into a gasifier, so that the feedstock undergoes a redox reaction at high temperatures (about 800–1000 °C). At the same time, a small amount of biochar is produced during the gasification process, with a yield typically below 25%. The gasifier not only facilitates the generation of syngas components such as hydrogen and CO but also the formation of the pore structure of the biochar. The carbon content and yield of biochar are closely related to the equivalence ratio of gasification parameters, feedstock type, gasifier, and pressure. Gasification is suitable for biomass feedstocks containing high ash and volatile matter (VM) (Rubinsin et al. 2024; Tezer et al. 2022).

Microwave pyrolysis for the production of biochar is an advanced technology that has attracted much attention in recent years (Li et al. 2022; Suresh et al. 2021;

Table 2.1 Biochar production technology.

Heating method	Temperature (°C)	Heating rate	Residence time	Atmospheres	Major product	Biochar yield (%)
Slow pyrolysis	300–600	< 10 °C/min	Hours to days	Oxygen free or limited	Biochar	35–55
Fast pyrolysis	400–600	1000 °C/min	2 s	Oxygen free or limited	Bio-oil	< 20
Flash pyrolysis	400–700	1000 °C/s	2 s	Oxygen free or limited	Bio-oil	< 30
Vacuum pyrolysis	400–600	< 50 °C/min	Minutes to hours	Oxygen free	Bio-oil	< 30
Microwave pyrolysis	400–600	Up to 330 °C/s	Several minutes	Oxygen free or limited	Bio-oil	< 40
Gasification	800–1600	1000 °C/min	Several seconds	—	Syngas	< 25
Torrefaction	200–300	< 50 °C/min	Minutes to hours	Oxygen free or limited	Torrefied biochar	60–80
Hydrothermal carbonization	150–300	< 10 °C/min	Minutes to hours	—	Biochar	> 50

Source: Wang et al. 2020; Yaashikaa et al. 2020; Zhou et al. 2021; Rathore et al. 2021; Li et al. 2022; Baath et al. 2022.

Ge et al. 2021). Specifically, the biomass feedstock is placed in a microwave reactor and then undergoes a thermal decomposition reaction under the action of microwaves, which in turn produces biochar. The heat source of microwave heating is generated inside the biomass, which is more conducive to the cracking of the feedstock and the release of the products. Usually, microwave heating is more advantageous than conventional heating in biomass pyrolysis. Microwave pyrolysis has the advantages of fast reaction rate, high yield, and low energy consumption, but further optimization is needed to improve the productivity and quality of biochar.

Hydrothermal carbonization, as an advanced technology for thermochemical treatment of biomass in subcritical or supercritical water heated to 200–300 °C in a sealed system, is also an effective way to produce biochar (Zhuang et al. 2022; Zeng et al. 2023). Hydrothermal carbonization does not require drying pretreatment of the feedstock and is suitable for processing biomass and organic matter with high moisture content; typical feedstocks include animal manure and sewage sludge. In hydrothermal carbonization, the reactions do not take place in a certain order but rather interact with each other. Similar to pyrolysis, the solid products, liquid products, and gaseous products can be obtained through hydrothermal carbonization. Generally, solid products show a higher carbon content. The carbonization process is controlled and is suitable for the preparation of homogeneous carbon materials.

2.2.2 Key Influencing Factors for Biochar Preparation

Induction temperature is one of the key factors affecting the properties and yield of biochar. Usually, as the induction temperature increases, the carbon content of biochar increases, while the content of hydrogen, oxygen, nitrogen, and other elements decreases (Janu et al. 2021; Tripathi et al. 2016). Meanwhile, induction temperature also affects the physical properties of biochar such as porosity and specific surface area. The rate of temperature increase refers to how fast or slow the feedstock is heated during the induction process. Rapid thermal decomposition is usually carried out at relatively high temperatures. The intermediate products of rapid thermal decomposition of biomass have a very short residence time in the reaction system, which mainly produces bio-oil with a low yield of biochar. Whereas, slow thermal decomposition exhibits a high biochar yield.

The carbonization pressure is the pressure inside the reactor during the biochar production process. Changes in pressure affect the thermal decomposition reaction of biomass and the properties of the products. Generally, the yield and carbon content of biochar increase as the pressure increases. Reaction residence time is the duration of retention of the biomass in the reactor. The length of residence time affects the yield and properties of biochar. Shorter residence time favors bio-oil production, while longer residence time facilitates biochar production. In addition, suitable catalysts may promote the thermal decomposition reaction of biomass and improve the yield and quality of biochar. Commonly used catalysts include alkali metals, alkaline earth metals, and transition metals. The type and amount of catalyst affect the properties and yield of biochar.

2.3 Preparation of Biochar Fuel by Carbonization

Biomass carbonization, as the main method of biochar production, is the process of converting biomass into its main target product under high temperature, anaerobic, or low oxygen conditions. The main conditions for biomass carbonization include slow heating rates (below 10 °C/min), long reaction times (ranging from hours to days), and low reaction temperatures (300–600 °C) (Kong et al. 2022; Wang et al. 2021). This process involves pyrolysis, condensation, and carbonization of organic matter in the biomass feedstock at high temperatures to produce biochar and the byproducts bio-oil and syngas. Specifically, pyrolysis refers to the process of decomposition of biomass at high temperatures to produce gases, liquids, and solid products, and is the main reaction of biomass carbonization. Condensation is the further reaction of gas and liquid products at high temperatures to produce more stable products. Carbonization refers to the further chemical reaction of solid products at high temperatures to produce a more stable biochar.

The preparation of biochar consists of pretreatment, carbonization, and post-treatment stages. Raw material pretreatment is the first step of biomass carbonization, which mainly includes the collection, removal of impurities, crushing, drying, and sieving of biomass raw materials. The collection of biomass feedstock should ensure the cleanliness and dryness of the feedstock in order to reduce the impact of impurities and moisture on the products during the carbonization process. It is worth noting that rural waste biomass generally consists of a wide range of biomasses, resulting in complex feedstocks and diverse impurities, which increases the difficulty of utilizing rural waste biomass. During biomass collection process, the purity of the same type of waste biomass needs to be maintained in order to ensure the singularity and cleanliness of the feedstock. For biochar, the process of removing impurities from the biomass feedstock is also essential. For example, the ash in the biomass is removed to obtain purer feedstocks, so as to ensure the quality of the carbonization product and improve its utilization value.

Due to the large size of biomass, which is not conducive to the heat transfer process, crushing is generally required. After crushing, the biomass particles become smaller and the surface area increases, which is favorable to the carbonization process, and the biochar with ideal particle size can also be obtained. In addition, sieving is also an important means of biomass pretreatment, generally following the crushing step. The purpose of sieving the crushed biomass is to obtain biomass feedstock with uniform particle size, and the sieved biomass feedstock is more conducive to the carbonization process and product quality control. Meanwhile, high moisture, as one of the main characteristics of biomass, also affects the yield and quality of biomass carbonization products. Therefore, the biomass needs to be dried to reduce its moisture content before carbonization.

The biomass carbonization process can be generally divided into three stages: drying stage, volatile release stage, and carbonization stage. The first stage is mainly the evaporation of water by heat, and the internal structure of biomass is almost unchanged; the second stage of biomass macromolecules undergo bond breaking and rearrangement, and the organic matter is gradually volatilized, and at the same time part of the organic matter undergoes anoxic combustion; and the third stage is generally the slow formation of biochar. Carbonization is the core step of pyrolysis and the key process of biomass conversion to biochar. The carbonization reaction is usually carried out under high temperature, anaerobic, or low oxygen conditions to ensure that the biomass feedstock can be fully pyrolyzed and carbonized. Parameters such as temperature, pressure, and residence time of the carbonization reaction have a significant impact on the quality and yield of the product (Emenike et al. 2022).

The posttreatment stage is mainly product separation. It is the process of separating the gas, liquid, and solid products after the carbonization reaction. Gas products mainly include carbon monoxide, carbon dioxide, hydrogen, and other combustible gases, which can be used as energy. Liquid products mainly include bio-oil, tar, etc., which can be used as chemical raw materials or fuels. The solid product is biochar, which is the main target product of biomass carbonization. In addition, the product biochar can be further refined to improve its quality and performance, mainly including crushing, screening, washing, drying, and other steps, in order to remove impurities and uncarbonized organic matter in the biochar and improve its purity and stability.

2.4 Types and Characteristics of Carbonation Reactors

As the core equipment of biomass carbonization, the type and characteristics of the reactor have an important impact on the whole process as well as the efficiency and quality of biomass carbonization products. According to different sources of heat energy, biomass pyrolysis carbonization equipment can be divided into two types: internal combustion type and external combustion heating type (Cai et al. 2021; Gao et al. 2022; Luo et al. 2022). The internal combustion type mainly adopts the way of self-ignition of raw materials in the carbonization furnace to heat the materials directly, while the external combustion heating type mainly relies on the external heat source to provide energy. According to the continuous mode of operation, biomass carbonization equipment can be divided into two categories: fixed bed and moving bed (Zhu et al. 2022b). Fixed-bed equipment mainly includes kiln and kettle equipment, which are intermittent equipment, while moving bed can realize continuous pyrolysis and carbonization of biomass. Moving bed has the advantages of continuous production, high productivity, convenient process control, and relatively stable product quality, which is more promising.

2.4.1 Types of Carbonation Reactors

The fixed-bed reactor is one of the earliest reactor types used for biomass carbonization. Its structure is simple, easy to operate and maintain, but it suffers from low heat transfer efficiency and uneven carbonization (Raza et al. 2021; Fonseca and Soares Dias 2021; Jiang et al. 2022). Fixed-bed reactors are usually suitable for small-scale biomass carbonization production. Typical fixed-bed reactors include traditional charcoal kilns, open thermal cracking charcoal kilns, and fixed-bed pyrolysis furnaces (Garcia-Nunez et al. 2017). Fixed bed is mainly used to control the oxygen content in the reactor, so that the biomass in the reactor undergoes partial oxidation and combustion to generate heat, and finally obtain biochar products. However, the kiln furnace has certain limitations in terms of pyrolysis efficiency and product quality, such as the pyrolysis temperature is not easy to control, and the impurity content in the product is high. Fixed-bed pyrolysis furnace is a more advanced biomass carbonization equipment, its structure is relatively complex, but the pyrolysis efficiency and product quality are higher. Fixed-bed pyrolysis furnace can be further divided into externally heated, internally combusted, and recirculating gas-heated types according to the different heat transfer methods. These different types of fixed-bed pyrolysis furnaces have their own advantages and disadvantages in terms of pyrolysis efficiency, product quality, and operational flexibility.

Fluidized bed is an efficient biomass carbonization reactor type. It mainly makes the biomass raw material form a fluidized state in the reactor through the flow of gas, so as to achieve efficient heat transfer and carbonization reaction (Cui et al. 2023; Kaminsky 2021). In the fluidized state, biomass particles and gases can be in full contact with each other, thus facilitating thermochemical decomposition of biomass in a high temperature environment. Then, feedstocks (e.g., organic materials such as wood, straw, algae, etc.) are converted into more valuable products (e.g., bio-oil, syngas, etc.). Fluidized bed is mainly composed of air-distributing plate, reaction chamber, heating system, and product collection system. Biomass raw materials are fed into the carbonization furnace through the feeding system. Under high temperature conditions, the organic components in the raw materials decompose, the non-carbon elements are discharged in the form of gas, and the remaining carbon elements form biochar. The products that have completed the decomposition reaction enter the cooling system and are cooled down to room temperature by natural or forced cooling, and finally exhausted through the outlet system. The air-distributing plate is mainly used to evenly distribute the gas entering the reaction chamber, so that the biomass can be well fluidized in the reaction chamber. The reaction chamber is usually made of high-temperature and corrosion-resistant materials, such as alloy steel or ceramic materials, in order to withstand the high-temperature pyrolysis reaction environment. Its shape is usually cylindrical, and its height and diameter are designed according to factors such as treatment capacity and reaction time. The heating system mainly provides the required heat for the pyrolysis reaction, and the common heating methods are external heating and internal heating. External heating is mainly done by setting up a heating furnace or heating jacket outside the reaction chamber, and using heat radiation and heat conduction to transfer the heat to the reaction chamber. Internal heating is to set up heating elements directly in the reaction chamber, such as electric heating rods or gas burners. Generally speaking, the internal heating method can make the heat distribution more uniform, but it requires higher sealing and safety of the equipment. In addition, the feeding system is also very important in fluidized beds, which is mainly used to transport the feedstocks to the reaction chamber. This part includes storage silos, screw feeders, or pneumatic conveying devices. For example, a screw feeder can precisely control the feed rate of biomass and ensure the stability of the reaction. Pneumatic conveying devices mainly use airflow to transport biomass particles to the reaction chamber, which has the advantage of a cleaner conveying process and reduces the risk of clogging. The product collection system is mainly used to collect the liquid bio-oil, syngas, and solid biochar products from pyrolysis. When the gases from pyrolysis pass through the condenser, the condensable components (e.g., bio-oil vapors) are cooled and condensed into a liquid that flows into the collection tank. The gaseous product can be piped to a subsequent treatment unit, such as a gas purification unit or a burner. The biochar is then transported to the solid product storage silo after gas–solid separation by means of a screw conveyor or other equipment. Fluidized bed reactor has the advantages of high carbonization efficiency and good product quality, but the equipment is complicated and difficult to operate. Continuous biomass carbonization furnace has the advantages of high efficiency and energy saving, environmental protection and low carbon, high degree of automation, strong adaptability, and so on.

Moving bed is a new type of biomass carbonization reactor using mechanical rolling propulsion. It has wide feedstock adaptability (Gao et al. 2020; Zhang et al. 2020). The reactor can handle mixed feedstocks of varying shapes, such as municipal waste, and is limited only by the maximum size of the feedstock.

The constant rotation of the drums results in a slow relative displacement of the materials and the inner walls of the drums, as well as a thorough mixing of the materials. This enables the materials to be alternately heated by heat conduction, heat convection, and heat radiation, which improves the rate of heat and mass transfer. Compared with the fixed bed, at the same final pyrolysis temperature, it shows faster pyrolysis rate, larger capacity, and more uniform pyrolysis carbon with smaller particle size, which is favorable for subsequent use. The feedstock is continuously moved within the reactor for efficient carbonization and product separation. Moving bed reactors have the advantages of high carbonization efficiency, good product quality, and easy operation, but the equipment is complex and costly.

Mechanical contact reactors are designed to transfer energy in the form of heat transfer to the biomass by direct or indirect contact between the biomass and a searing hot reactor surface, causing it to heat up rapidly to the temperature required for thermal cracking. This type of reactor includes rotating cone reactors and ablative reactors (Schnabel et al. 2022; Luo et al. 2017). Rotating cone reactors are based on the principle of centrifugal force to separate the non-condensable gases from the pyrolytic char after the reaction. The biomass is fed into the bottom of the rotating cone of reactor together with an excess of the heat carrier quartz sand. As the mixture of biomass and quartz sand spirals up the wall of the high-temperature cone, the biomass undergoes thermal cracking to produce volatiles and pyrolytic char. In the ablative reactor, heat is transferred to the biomass by direct contact with the biomass particles through the hot reactor surface, resulting in a rapid heating of the biomass for rapid thermal cracking.

In conclusion, different types of biomass carbonization reactors show obvious differences in terms of pyrolysis performance, mainly including carbonization efficiency, product quality, equipment cost, and operation difficulty. Therefore, the selection of biomass carbonization reactors is generally based on specific production requirements and process conditions.

2.4.2 Performance of Carbonization Reactors

As the core of biomass carbonization technology, the performance of the reactor directly affects the conversion efficiency of biomass energy. Therefore, a comprehensive and in-depth evaluation of the performance of the biomass carbonization reactor is of great significance in promoting the development of biomass energy. The performance evaluation of biomass carbonization reactors is a complex process that requires comprehensive consideration of several factors, mainly including pyrolysis efficiency, mass transfer performance, energy consumption, product quality, stability, and operability.

Pyrolysis efficiency is one of the most important indicators of biomass carbonization reactor performance. It reflects the ability of the reactor to convert biomass feedstock into carbonaceous products under given conditions. An efficient biomass carbonization reactor should exhibit high pyrolysis efficiency and be able to convert biomass feedstock into high-quality carbonaceous products in a relatively short period of time and minimize the generation of byproducts. The efficiency of pyrolysis directly affects the conversion efficiency of biomass energy and product quality.

Mass transfer efficiency is another important index to evaluate the performance of biomass carbonization reactor. Good mass transfer can ensure that biomass feedstock, oxygen, and other substances are evenly distributed in the reaction system. At the same time, the reaction products are discharged in time to avoid their accumulation and inhibition of the reaction. In the process of biomass carbonization, mass transfer performance directly affects the pyrolysis efficiency and product quality. In order to evaluate the mass transfer performance, it is necessary to pay attention to the parameters such as the material transfer rate in the reaction system, the mixing effect, and the homogeneity of the reaction system. These parameters can be studied by experimental measurement and numerical simulation. Through experimental measurement, parameters such as concentration distribution and flow rate of substances in the reaction process can be obtained directly. Numerical simulation can simulate the substance transfer and mixing process in the reaction system and predict the mass transfer performance under different conditions.

Energy consumption is also an important indicator for evaluating the economics of biomass carbonization reactors. It reflects the energy consumption required by the reactor in the biochar production process. An efficient biomass carbonization reactor should have low energy consumption and be able to ensure product quality while reducing production costs.

Product quality is another important indicator of biomass carbonization reactor performance. It reflects the quality and stability of the carbonaceous products produced by the reactor. An efficient biomass carbonization reactor should be able to produce high-quality carbon products, such as high fixed carbon (FC) content, low ash, low VM, and so on.

Stability and operability are two other important indicators for evaluating the performance of a biomass carbonization reactor. Stability implies that the biomass carbonization reactor is able to maintain stable performance over a long period of time, with fewer breakdowns and fluctuations. Operability relates to the difficulty of controlling the equipment, the cost of maintenance, and the ability to adapt to different reaction conditions. In order to evaluate stability and operability, attention needs to be paid to parameters such as reactor operating time, failure rate, maintenance cost, and operational flexibility. Through experimental measurements, key data such as reactor runtime and failure rate can be directly obtained to further evaluate its stability and operability.

2.4.3 Application Prospects and Challenges of Carbonization Reactors

As an important biomass energy conversion equipment, the biomass carbonization reactor has a broad application prospect and great development potential in the fields of energy, chemical industry, and agriculture. Especially in the field of energy, biomass carbonization reactor can be used as an important renewable energy conversion equipment, contributing to the alleviation of energy crisis and the reduction of greenhouse gas emissions. Carbonization reactor is the core equipment of biomass carbonization technology, and its type and performance have an important impact on the whole pyrolysis process. However, in practical applications, biomass carbonization reactors also face some challenges and problems. For example, the types and qualities of biomass feedstocks vary greatly, which has a certain impact on the performance and product quality of biomass carbonization reactors. The pyrolysis efficiency and product quality of the biomass carbonization process still need to be improved and the stability and operability of biomass carbonization reactors need to be further strengthened.

In order to address these challenges and problems, the research and development of biomass carbonization reactors need to be further enhanced. The performance and stability of biomass carbonation reactors can be improved by optimizing and improving their structure and operating conditions. The pyrolysis efficiency and product quality of the biomass carbonation can be further improved by optimizing the material transfer and mixing process in the reaction system. By strengthening the development of new technologies, the wide application of biomass carbonization reactors in energy, chemical, and agricultural fields can be further promoted.

2.5 Fuel Properties of Biochar

As a renewable energy source, biochar has received widespread attention in the energy field in recent years. It has the advantages of HHV, low ash content, and low pollutant emissions, and is considered to be an important alternative to traditional fossil fuels. Biochar generally has a rich pore structure and a large specific surface area and a high number of surface oxygen-containing active groups, which makes biochar exhibit unique fuel properties and a wide range of application prospects. The physicochemical properties of biochar, including density, pore structure, ash content, FC content, volatile content, and heating value, all have a certain influence on the combustion process (Lubwama et al. 2020). The density of biochar is generally low, which is favorable to the gas diffusion and combustion reaction during the combustion process. The rich pore structure of biochar increases its specific surface area, which is conducive to the diffusion of oxygen and the contact of reactants during the combustion process. The lower ash content of biochar helps to reduce the emission of pollutants during the combustion process. Biochar has a high FC content, which determines its heating value in the combustion process. Biochar has moderate volatile content, which is conducive to its ignition and stable combustion in the combustion process. Biochar with HHV shows desirable fuel properties.

Proximate and ultimate analyses of biochar are important techniques for analyzing the composition of biochar. The fuel properties of biochar are mainly closely related to its composition, and the heating value is a direct indicator of the energy potential of biochar. The moisture, VM, FC, and ash content of biochar can be measured by proximate analysis. The quality of biochar is related to its FC content, VM, ash, and HHV. The higher the FC content, the lower the ash and VM, the better the quality of biochar. Ash content mainly affects the ash characteristics and heat generation of biochar after combustion. Generally speaking, high ash content has a negative effect on heat generation. The heating value of the fuel is a key indicator for evaluating the performance of biochar fuels, and the heat generation of solid and liquid fuels is generally determined by oxygen bomb calorimetry. In addition, in order to facilitate the estimation of the heating value of fuels based on the composition of biochar, the use of the results of ultimate and proximate analyses is also a simple and convenient way to estimate the heating value of fuels.

Proximate analyses of typical biochar are summarized in Table 2.2. With the increase of temperature, the ash content and FC of biochar showed a gradual increase, while the VM of biochar gradually decreased. Generally speaking, the volatile fraction of biochar is basically released before 600 °C. For different biochar, its maximum heating value corresponds to different induction temperatures.

The ultimate analysis of typical biochar is summarized in Table 2.3. With the increase of temperature, the carbon element of biochar increases gradually, and the elements of hydrogen (H), oxygen (O), nitrogen (N) and sulfur (S) all decrease

Table 2.2 Proximate analysis and HHV of biochar.

Biochar	Pyrolysis temperature (°C)	Ash (wt%)	VM (wt%)	FC (wt%)	HHV (MJ/kg)	References
Corn straw	550	19.48	15.15	65.37	26.20	Qian et al. (2020)
Soybean straw	500	25.08	19.96	54.96	20.95	Chen et al. (2022b)
Cassava peel	500	5.00	47.09	47.91	24.18	Egbosiuba (2022)
Pinewood	500	2.87	22.77	74.36	31.42	Li et al. (2016)
Cow manure	350	24.2	53.5	23.2	20.90	Cantrell et al. (2012)
Rich husk	500	30.30	7.44	60.10	22.02	Vieira et al. (2020)

Table 2.3 Ultimate analysis of biochar.

Biochar	Pyrolysis temperature (°C)	C (wt%)	H (wt%)	O (wt%)	N (wt%)	S (wt%)	References
Rich husk	350	51.91	3.17	12.62	0.72	0.03	Qian et al. (2020)
Cotton straw	450	66.70	2.50	10.80	1.48	0.28	Malik and Laura (2025)
Orange peel	450	39.70	4.26	53.12	1.23	0.21	Malik and Laura (2025)
Eucalyptus wood	550	78.13	2.68	16.82	—	—	Chaves Fernandes et al. (2020)
Chicken manure	400	22.01	2.62	28.34	2.25	—	Cuixia et al. (2020)
Corncob	400	54.36	4.32	39.02	0.63	—	Adekanye et al. (2022)

Source: Adapted from Qian et al. 2020.

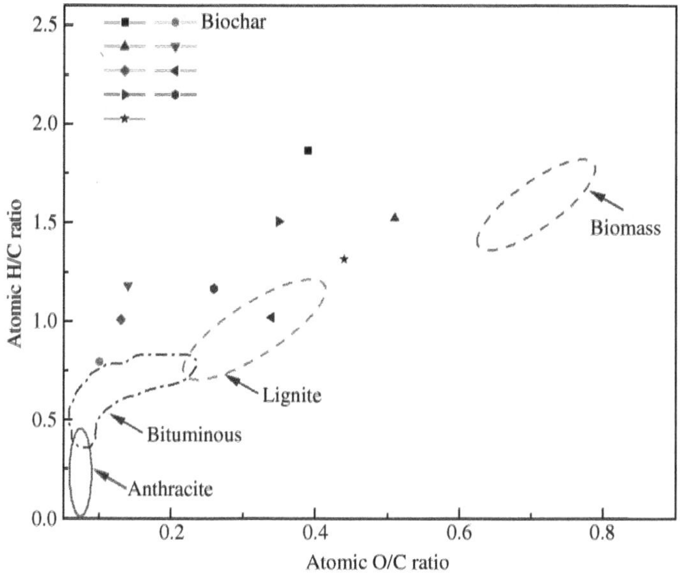

Figure 2.1 The van Krevelen diagram of various fuels.

gradually. Meanwhile, the elemental compositions of biochar originating from different types of biomass can be significantly different, such as eucalyptus wood biochar displaying higher carbon content. Even with the same induction temperature, biochar can produce significant differences in elemental composition. Based on the elemental composition of the feedstocks, the van Krevelen diagram of typical fuels is used to evaluate the fuel properties of different biochar very clearly (Zhu et al. 2024) (Figure 2.1). From the van Krevelen diagram, it is clear that high-quality biochar fuels can be produced by regulating the appropriate carbon–hydrogen and carbon–oxygen ratios.

2.6 Combustion of Biochar

2.6.1 Combustion Process of Biochar

The combustion of biochar is a complex chemical reaction and physical process, which mainly includes the stages of preheating, drying, formation and release of volatile components, combustion of volatile components, formation of coke, and combustion of coke. In these stages, the intensity of coke combustion determines the intensity of the whole fuel combustion process.

Biochar needs to be preheated and dried before combustion to remove moisture and volatile substances from it. The temperature at this stage is low, which mainly affects the burning speed and combustion stability of biochar. As the temperature increases, the volatile substances in the biochar begin to decompose and release, and their combustion has an important impact on the overall combustion process of biochar. Generally, the volatile content of biochar prepared at high temperatures is low. The released volatiles are combusted in the presence of sufficient oxygen to produce carbon dioxide, water, and combustion heat. With the combustion of VM, the remaining solid portion of biochar is gradually converted into coke. The formation of coke marks the high-temperature stage of the biochar combustion process. The coke continues to burn in oxygen, producing carbon dioxide and ash.

The combustion products of biochar consist mainly of carbon dioxide, water, and a small amount of ash. Carbon dioxide is the main combustion product during biochar combustion. In the presence of sufficient oxygen, biochar is completely oxidized to produce carbon dioxide and water. The emission of carbon dioxide is closely related to the carbon content of biochar and the combustion conditions. Water is a byproduct of the combustion process and is mainly produced by the reaction of hydrogen in biochar with oxygen. The amount of water produced depends on the hydrogen content of the biochar and the combustion temperature. Ash is the product of inorganic matter contained in biochar that remains at high temperature. The content and composition of the ash depend on the raw material and preparation conditions of the biochar (Shanmugam et al. 2022).

2.6.2 Influencing Factors of Biochar Combustion

The combustion characteristics of biochar are affected by several factors, including the raw material of biochar, moisture content, carbonization temperature, particle size, reaction temperature, and oxygen concentration. The raw material of biochar has an important influence on its combustion characteristics. Different kinds of biomass feedstocks have different chemical compositions and physical structures, resulting in different reaction rates and combustion products during the combustion process of biochar. The moisture content of biochar affects its combustion speed and combustion stability. Biochar with high moisture content requires more energy to evaporate the moisture during the combustion process, resulting in lower combustion speed and unstable combustion. The induction temperature is an important factor affecting the combustion characteristics of biochar. Higher temperatures lead to more complete carbonization of volatile and organic substances in the biomass, thus increasing the quality and combustion temperature of the biochar. As the carbonization temperature increases, the heat generation of biochar increases, the FC and ash content increases, and the volatile content decreases.

The particle size of biochar has an important effect on its combustion speed. Biochar particles with smaller particle size have larger specific surface area and shorter diffusion path, which is conducive to the full contact and reaction between oxygen and fuel, thus increasing the combustion speed. The reaction temperature is an important factor affecting the combustion rate of biochar. As the reaction temperature increases, the combustion rate of biochar increases. Under high-temperature conditions, the chemical reaction rate is accelerated, which is conducive to the rapid combustion of biochar. Oxygen concentration is also a key factor affecting the combustion process of biochar. The higher the oxygen concentration, the faster the combustion speed of biochar. With sufficient oxygen, biochar can completely oxidize to produce carbon dioxide and water, releasing more heat energy.

2.6.3 Combustion Technologies for Biochar

Combustion technologies for biochar include direct combustion, gasification combustion, and combined combustion (Safarian 2023; Ayanwusi et al. 2024; Barbanera et al. 2018). Direct combustion is the simplest biochar combustion method. The biochar is fed directly into the boiler or burner, mixed with air and burned. This method is suitable for small- or medium-sized biochar combustion systems. Gasification combustion involves reacting biochar with a gasifying agent (e.g., air, water vapor, or carbon dioxide) at high temperatures to produce a combustible gas (e.g., carbon monoxide, hydrogen, and methane). The combustible gases are then fed into a burner for combustion. This method increases the combustion efficiency and heating value of biochar and reduces pollutant emissions. Combined combustion is the combustion of biochar with conventional fossil fuels such as coal, oil, or natural gas. This method can make full use of the combustion energy of biochar while reducing the reliance on traditional fossil fuels. The combustion reactor of biochar is the key equipment to achieve efficient combustion of biochar, and its design and operation have a significant impact on the combustion characteristics and environmental impact of biochar.

2.7 Combustion Characteristics of Biochar

The combustion process of biochar can usually be divided into three stages: the preheating ignition stage, the volatile fraction combustion stage, and the coke combustion stage. In the preheating stage, the biomass is heated, the water gradually evaporates, and when the temperature reaches 160 °C, it starts to release volatile components, which mainly include low molecular hydrocarbons and other combustible gases. In the volatile combustion stage, these combustible

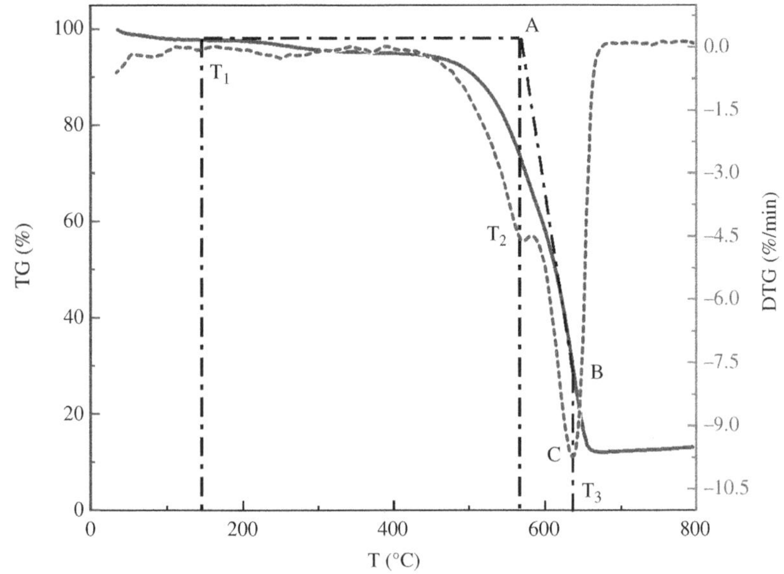

Figure 2.2 TG-DTG curve.

gases start to burn at high temperatures, releasing a large amount of heat energy. In the coke combustion stage, the remaining red-hot coke comes into contact with oxygen and undergoes a combustion reaction.

The combustion characteristic parameters of biochar mainly include ignition temperature, ignition stability, burnout temperature, the comprehensive combustion index (CCI), and the combustion stability index (CSI) (He et al. 2022; Su et al. 2021). Among them, the ignition temperature, the burnout temperature, the CCI, and the CSI were adopted to evaluate the combustion characteristics of biochar. Generally, the combustion characteristic parameters of biochar can be determined by the Thermogravimetric Analysis-Derivative Thermal Gravity (TG-DTG) curve (Figure 2.2). The ignition temperature is generally determined according to the TG-DTG tangent method (Koçer and Özçimen 2021; Yi et al. 2012). A plumb line is drawn through the peak of the DTG curve at point C to intersect with the TG curve at point B, and a tangent line is drawn through point B to intersect with the parallel line at the beginning of the TG curve at the point C. The temperature corresponding to the intersection point A is the ignition temperature of the fuel, and the intersection point of the first peak and the TG curve is taken to determine the ignition temperature in the case of multiple peaks. The burnout temperature is generally taken as the corresponding temperature when the conversion rate is 99%. CCI and CSI are shown in Eqs. (2.1) and (2.2), respectively.

$$CCI = \frac{(dw/dt)_{max} (dw/dt)_{mean}}{T_i^2 T_b} \quad (2.1)$$

$$CSI = 8.5875 \times 10^7 \times \frac{(dw/dt)_{mean}}{T_i T_{max}} \quad (2.2)$$

where $(dw/dt)_{max}$ and $(dw/dt)_{mean}$ represent the maximum mass loss rate and the average mass loss rate (% °C^{-1}), respectively; T_i, T_b, and T_{max} represent the ignition temperature, the burnout temperature, and the temperature with the maximum mass loss rate, respectively.

The combustion characteristics of different types of biochar show significant differences. For example, the initial combustion temperature and the burnout temperature of wood waste biochar are higher than those of food waste digestate biochar (Zhu et al. 2022a). Meanwhile, the CCI of wood waste biochar is significantly higher than that of food waste digestate biochar. By adjusting the ratio of feedstock, it was found that the combustion performance of biochar could be optimized.

The comparison of combustion characteristics of biochar and coal is summarized in Table 2.4. At the same induction temperature, cotton stalk biochar showed the smallest ignition temperature and burnout temperature, followed by wood chip biochar, while bamboo chip biochar had the largest ignition temperature and burnout temperature. With the increase of induction temperature, the ignition and burnout temperatures of cotton stalk biochar, bamboo chip biochar, and wood chip biochar increased, the CCI values decreased, and the combustion characteristics deteriorated. Further, the combustion interval moves to the high temperature region. The maximum combustion rate and average combustion rate show a decreasing trend, but there are slight differences between different types of biochar. Moreover, the volatile fraction component of biochar decreases with increasing induction temperature, which decreases the reactivity of biochar.

At present, charcoal and barbecue charcoal are the two most used fuels. They are mainly used in metallurgical industry, catering industry, and domestic fuel. The quality standards of charcoal and barbecue charcoal are summarized in Tables 2.5 and 2.6, respectively (Shaowu et al. 2015). Combustion performance is an important indicator for evaluating the quality of fuel charcoal, but there are no standards on the combustion performance of fuel charcoal. Charcoal prepared in a traditional kiln for rural households and barbecue charcoal prepared by machines were chosen as controls to evaluate the fuel properties of biochar. Table 2.4 summarizes that the ignition temperature, burnout temperature, and CCI of biochar were better than those of charcoal, except that the ignition temperature of bamboo biochar induced at 800 °C was slightly higher than that of charcoal and the CCI value of wood biochar induced at 800 °C was slightly smaller than

Table 2.4 Comparison of combustion characteristics of biochar and coal.

Biochar	Pyrolysis temperature (°C)	$(dw/dt)_{max}$ (% min^{-1})	$(dw/dt)_{mean}$ (% min^{-1})	T_i (°C)	T_b (°C)	CCI (10^{-7}%^{2}°C^{-5})	References
Cotton stalks	500	6.021	4.002	336.2	496.1	4.297	Shaowu et al. (2015)
Grape pomace	450	—	—	299	499	10.1	Kluska and Ramotowski (2024)
Oatmeal shell	450	—	—	327	514	7.08	Kluska and Ramotowski (2024)
Ficus virens	600	0.85	0.25	368	456	0.253	Tong et al. (2020)
Charcoal	—	6.055	3.799	487.2	663.3	1.461	Shaowu et al. (2015)
Barbecue charcoal	—	6.111	3.854	361.7	528.9	3.405	Shaowu et al. (2015)

Table 2.5 Charcoal quality indexes.

Types	M	A	FC	Mixed sundries
Hard broadleaf charcoal	≤ 10	≤ 3.5	≥ 75	≤ 3
Broadleaf charcoal	≤ 12	≤ 5.0	≥ 65	≤ 5
Pine charcoal	≤ 12	≤ 3.0	≥ 65	≤ 5

Table 2.6 Quality indexes of barbecue charcoal and barbecue charcoal briquettes.

Biochar	M (%)	A (%)	FC (%)	Heap density
Barbecue charcoal	≤ 8	≤ 8	≥ 75	≥ 130
Barbecue charcoal briquettes	≤ 8	≤ 18	≥ 60	—

that of charcoal. Cotton stalk biochar induced at 500 and 600 °C was better than that of barbecue charcoal in all the combustion characteristics. The ignition and burnout temperatures of cotton stalk biochar induced at 700 °C were lower than that of barbecue charcoal, and the CCI value was slightly smaller than that of barbecue charcoal. The ignition temperature and burnout temperature of wood and bamboo biochar were higher than those of barbecue charcoal, while the CCI values were lower than those of barbecue charcoal.

Taking into account the composition and combustion characteristics of biochar, biochar prepared under specific conditions can be used as fuel charcoal. Typically, although the ash content of cotton stalk biochar does not meet the requirements for barbecue charcoal, it is much lower than that of barbecue charcoal briquettes. And the excellent combustion characteristics of cotton stalk biochar make it a good raw material for barbecue charcoal production. Generally, in order to improve the combustion performance of barbecue charcoal, it is necessary to add combustion aids, which are mostly non-combustible components. The excellent combustion characteristics of cotton stalk biochar allow it to be used with little or no additives, so cotton straw can be used to make barbecue charcoal briquettes. Considering the biochar yield, the optimum induction temperature for cotton stalks to make barbecue biochar is 500 °C.

In addition, by comparing Tables 2.2 and 2.5, it can be found that the ash content of biochar is generally higher than that of charcoal. Only the ash content of white pine and pinus sylvestris were able to meet the quality standards of hard broadleaf charcoal. All biochar prepared from agricultural waste contained higher ash fractions. Barbecue charcoal can have a higher ash content than charcoal, while barbecue charcoal briquettes can have less FC content, which allows more biochar to be used as barbecue charcoal. Corncob biochar and bamboo willow biochar can easily meet the barbecue charcoal standards. Although other biochar show high ash content, the ash content of biochar can be reduced by deashing treatment during the production process, thus increasing the FC content and meeting the quality standards for barbecue charcoal and charcoal.

2.8 Combustion Kinetics of Biochar

As a sustainable energy source, biochar has unique fuel properties. With the continuous development of biomass energy technology, the study of fuel kinetics of biochar has also made remarkable progress. The fuel kinetic properties of biochar mainly include its combustion reaction kinetics, pyrolysis kinetics, and diffusion kinetics (Tong et al. 2020; McMurchie et al. 2024; Peterson and Brown 2020). On the one hand, parameters such as combustion reaction rate, pyrolysis rate, and diffusion rate of biochar can be investigated more accurately through advanced experimental techniques and analytical methods. On the other hand, through methods such as numerical simulation, the combustion process and pyrolysis process of biochar can be investigated more deeply.

2.8.1 Combustion Reaction Kinetics of Biochar

In order to reveal the combustion process of biochar in depth, combustion reaction kinetics models were developed (Vasudev et al. 2021). Combustion reaction kinetics is an efficient method to study the behavior of chemical reactions of fuels during combustion. Combustion reaction kinetics models are usually based on the Arrhenius equation and chemical reaction kinetics (Sharma et al. 2023; Yang et al. 2022). The characteristic parameters of the combustion reaction kinetics of biochar mainly include reaction rate, activation energy, and pre-exponential factor (Wang et al. 2014).

The combustion reaction rate of biochar is affected by a variety of factors, including temperature, oxygen concentration, particle size, and so on (Chen et al. 2021; Jia et al. 2023). Generally, the combustion reaction rate increases with the increase of temperature. At the same time, the increase of oxygen concentration also increases the combustion reaction rate. The effect of particle size on the combustion reaction rate is more complicated. Typically, smaller particle size is favorable to oxygen diffusion and combustion reaction. Activation energy is an important parameter to describe the sensitivity of chemical reaction rate to temperature. The activation energy of combustion reaction of biochar is usually high, which means that the combustion reaction requires higher energy to initiate. However, the activation energy of the combustion reaction of biochar can be reduced by methods such as additives and pretreatment, thus improving its combustion performance. The pre-exponential factor is another important parameter that describes the relationship between chemical reaction rate and temperature. The pre-exponential factor of combustion reaction of biochar is usually large, which means that the combustion reaction rate increases rapidly at higher temperatures. Therefore, in practical applications, the combustion temperature needs to be reasonably controlled to ensure the stable combustion of biochar.

2.8.2 Pyrolysis Kinetics of Biochar

The combustion of biochar mainly includes pyrolysis process and combustion process. Pyrolysis kinetics is the key to study the decomposition process before combustion (Zhang et al. 2023). The pyrolysis kinetics of biochar mainly includes pyrolysis temperature, pyrolysis rate, pyrolysis products, etc., and also has similar characteristics with the combustion reaction, such as the reaction rate, activation energy, and the pre-exponential factor. The pyrolysis kinetics of biochar can be investigated by the same research method as that of combustion kinetics. The pyrolysis temperature of biochar is usually high, and the pyrolysis rate of biochar increases and the pyrolysis products change as the temperature increases. At lower temperatures, VM and tars are mainly produced, which can be used as a direct fuel for combustion. At higher temperatures, gas and solid char are mainly produced. The pyrolysis rate of biochar is affected by a number of factors, including the type of raw material, pyrolysis temperature, and pyrolysis time. Generally speaking, the pyrolysis rate increases with the increase of temperature and time. At the same time, the difference in the type of feedstock also leads to the difference in the pyrolysis rate.

Pyrolysis kinetic models are usually established based on thermogravimetric analysis and pyrolysis product analysis methods. These models can describe the changing rules of pyrolysis rate and pyrolysis products of biochar at different temperatures and times. Through model calculation, parameters such as pyrolysis temperature range and peak pyrolysis rate of biochar can be obtained to provide theoretical support for the optimization of pyrolysis process of biochar.

2.8.3 Diffusion Kinetics of Biochar

Diffusion kinetics focuses on the phenomenon of diffusion of substances in the reactor during the combustion of biochar (Chang et al. 2021; Zhu et al. 2017). The diffusion of oxygen and combustion products has an important influence on the combustion rate and combustion efficiency. Diffusion kinetic models are usually established based on Fick's law of diffusion and the theory of diffusion in porous media. These models can describe the changing law of diffusion rate of oxygen and combustion products in biochar particles. Through model calculations, the diffusion coefficient, pore structure, and other parameters of biochar can be obtained, which can help optimize the combustion process of biochar.

Oxygen diffusion is one of the key steps in the biochar combustion process. The rate of oxygen diffusion is affected by a variety of factors, including biochar particle size, pore structure, and oxygen concentration (Baath et al. 2022). Smaller particle size and rich pore structure are favorable for oxygen diffusion and combustion reaction. At the same time, higher oxygen concentration will also increase the diffusion rate of oxygen and combustion reaction rate. The diffusion of combustion products has an important impact on the combustion efficiency and pollutant emission of biochar. The combustion products include carbon dioxide, water vapor, nitrogen oxides, and so on. The diffusion rates of these products are affected by a variety of factors, including temperature, pressure, and particle size.

2.9 Emission Characteristics of Biochar Combustion

2.9.1 Emission Characteristics of Biochar During Combustion Processes

The combustion process of biochar is similar to the combustion of coal or wood, but with different emission characteristics. The main emissions include particulate matter (PM), sulfur dioxide (SO_2), nitrogen oxides (NO_x), carbon monoxide (CO), and carbon dioxide (CO_2) (Zhu et al. 2024; Dunnigan et al. 2018a; Dunnigan et al. 2018b). The types and quantities of pollutants released during the combustion of biochar are essential for assessing its environmental impacts and guiding its production and use.

PM from biochar combustion consists mainly of organic and inorganic carbon. Its emission concentration is affected by factors such as the type of biochar, combustion conditions (e.g., temperature, oxygen concentration), and the type of combustion equipment. Typically, the pyrolysis temperature can significantly affect the emission of PM2.5 from biochar combustion (Itoh et al. 2020). As the pyrolysis temperature increases, the emission of PM2.5 from biochar combustion gradually decreases. The high temperature and sufficient oxygen help to reduce the generation of PM. In addition, the ash content in biochar affects the particulate emissions. Biochar has a low sulfur content and therefore SO_2 emissions from combustion are usually lower. However, if the biochar is derived from sulfur-rich biomass material, SO_2 emissions may increase. In addition, the oxygen concentration and temperature of the combustion process can affect SO_2 production and emissions. NO_x produced during biochar combustion is mainly derived from the reaction of nitrogen in the fuel with oxygen in the air. Combustion temperature, oxygen concentration, and nitrogen content of biochar are key factors affecting NO_x emissions. High temperature and sufficient oxygen will promote NO_x production, and the higher the nitrogen content in biochar, the higher the NO_x emission.

CO is a product of incomplete combustion of biochar. During combustion, if the oxygen supply is insufficient or the combustion temperature is too low, more CO will be produced. CO emissions not only affect the air quality, but also pose a hazard to human health. CO_2 is one of the main products of biochar combustion. Although the combustion of biochar reduces net CO_2 emissions relative to fossil fuels because the carbon in biochar is derived from atmospheric CO_2. The combustion of biochar still releases a significant amount of CO_2. Therefore, biochar combustion also has important implications for global climate change.

2.9.2 Factors Affecting the Emission of Biochar Combustion

The emission characteristics of biochar are affected by a number of factors, including the type and properties of the biochar, combustion conditions, combustion equipment, and posttreatment technologies (Feng and Wu 2018; Huang et al. 2022). Different types of biochar show different properties such as elemental composition, heating value, and pore structure, which affect the chemical reaction and emission generation during the combustion process. Combustion conditions are important factors that affect the emission characteristics of biochar. Changes in conditions such as combustion temperature, oxygen concentration, and combustion rate directly affect the rate of chemical reactions and the amounts of emissions generated during the combustion process. The type and structure of combustion equipment also affect the emission characteristics of biochar. Different combustion equipment has different combustion efficiency and emission control performance, so choosing the right combustion equipment is important for reducing emissions (Mau and Gross 2018; Wang et al. 2019).

In order to reduce emissions from biochar combustion, effective strategies are important steps toward increasing the energy potential of biochar. For example, increasing the combustion temperature and increasing the oxygen concentration can promote complete combustion of the fuel, thereby reducing emissions of CO and unburned hydrocarbons. Selection of efficient combustion equipment can significantly improve combustion efficiency and emission control performance. For example, the use of advanced burners, furnace structures, and flue gas-cleaning devices can reduce emission concentrations and emissions. Further, adding appropriate inhibitors or catalysts to the biochar can inhibit the generation of harmful substances or promote their conversion into harmless substances. For example, the addition of a calcium- or magnesium-based desulfurizer can reduce SO_2 emissions, the addition of a nitrogen oxide reduction catalyst can reduce NO_x emissions. Posttreatment technologies such as flue gas desulfurization, denitrification, and dedusting can further reduce emissions from biochar combustion processes. These technologies can be chosen to be used individually or in combination according to the actual situation in order to achieve the best emission reduction effect.

Biochar, as a renewable resource, has a wide range of applications and potential. However, the emission characteristics of its combustion process have an important impact on the environment. In the future, with the continuous advancement of technology and increased environmental awareness, the emission characteristics of biochar will receive more attention to achieve the win–win goal of its sustainable use and environmental protection.

2.10 Migration of Alkali Metals During Biochar Combustion

Alkali metal content in biomass is generally high, and in the combustion process, alkali metal elements in biomass can cause a series of problems, such as agglomeration, deposition, corrosion, etc., thus affecting the combustion efficiency and the service life of the equipment (Wang et al. 2018; Wu et al. 2021). Therefore, it is of great significance to study the migration patterns of alkali metals and their influence mechanisms during the combustion process of biochar, in order to improve the utilization efficiency of biomass energy and solve the related problems.

Biomass consists mainly of cellulose, hemicellulose, lignin, and ash. The component content of biomass can vary between different species, growth periods, and growth conditions. Ash in biomass includes the ash inherent in the feedstocks itself and the ash brought in during processing and treatment, in which inorganic elements such as aluminum, calcium, iron, potassium, magnesium, phosphorus, sulfur, silicon are the main components of ash (Yousaf et al. 2017). The alkali metal elements calcium and potassium are high in biomass. These elements are easy to volatilize and migrate during the combustion process of biochar, and cause problems such as agglomeration of bed material, staining and corrosion of heated surfaces, slagging, and deposition, which have become one of the main factors hindering the extensive use of biomass fuels (Niu et al. 2016). Therefore, it is of great significance to understand the migration pattern of alkali metals in biochar combustion process to optimize the combustion process and improve the combustion efficiency and energy efficiency of biochar.

2.10.1 Alkali Metal Migration Pattern in Biochar Combustion Process

In order to gain an in-depth understanding of the migration pattern of alkali metals during the combustion of biochar, a large number of experimental studies have been conducted. These studies usually use tube furnace combustion test, thermogravimetric analysis, X-ray diffraction (XRD), and scanning electron microscopy (SEM) to simulate and analyze the migration process of alkali metals in the biochar combustion process (Wang et al. 2022). The tube furnace combustion test is a commonly used method to simulate the biochar combustion process. By controlling the temperature and atmosphere conditions inside the tube furnace, the alkali metal migration process of biochar under different combustion stages can be simulated. Combined with chemical heat balance software such as FactSage, the precipitation and transformation patterns of alkali metals during the combustion process can be further analyzed. Thermogravimetric analysis is a method to study the thermal stability and chemical reaction process of a sample by measuring its mass change during the heating process. Through thermogravimetric analysis, the effects of different atmospheres, heating rates, and catalysts on the combustion of biochar during the combustion process can be investigated. The catalytic effect of alkali metals on the combustion process of biochar can also be analyzed. XRD analysis is a method used to determine the crystal structure and chemical composition of a substance. XRD analysis can be used to understand the types of alkali metals in the ash fraction of biochar after combustion. This is important for studying the migration paths and morphological transformations of alkali metals during combustion. SEM is a method used to observe the surface morphology and microstructure of samples. SEM observation can visualize the morphological characteristics of the ash and the distribution pattern of alkali metal elements in the biochar after combustion. This is of great significance to study the migration and transformation process of alkali metals in the ash.

Alkali metal elements are mainly present in organic form in biochar during the preheating and volatile release phases of biochar ignition. As the temperature rises, some of the organic alkali metals are released into the gas phase along with the VM. These alkali metal elements mainly exist in the form of gaseous compounds, such as potassium chloride and potassium sulfate. With further combustion of the VM, these gaseous compounds may be converted into more stable inorganic salts or oxides. The migration of alkali metals during this process is mainly influenced by factors such as temperature, biochar type, and combustion conditions. During the coke combustion stage, the remaining FC and inorganic alkali metals in the volatile fraction continue to burn. At this time, the alkali metal elements in the coke mainly exist in solid form, and the migration of alkali metals mainly occurs within the coke and at the interface between the coke and the gas phase. These alkali metal elements may react with other elements in the coke to form more complex compounds. The alkali metals in the coke may be gradually released into the gas phase as the coke burns, or they may be converted to inorganic potassium or sodium during coke oxidation. At the end of biochar combustion, the remaining ash still contains a certain amount of alkali metal elements. These alkali metal elements exist mainly in the form of inorganic salts such as potassium chloride and potassium sulfate. The alkali metal elements in the ash may migrate further into the gas phase or combine with other substances to form more stable compounds by means of further chemical reactions or physical processes, such as melting and evaporation.

2.10.2 Effect of Alkali Metal Migration on Combustion Characteristics of Biochar

In the biochar combustion process, the factors affecting alkali metal migration mainly include temperature, biomass type, combustion conditions, and ash properties. Temperature is the key factor affecting alkali metal migration. With the increase of temperature, alkali metal elements in biochar are more easily transferred from the solid phase to the gas phase. Especially at high temperatures, the evaporation rate of alkali metals increases significantly. There are differences in the content and forms of alkali metals in different species of biomass, which will affect the migration pathways of alkali metals. For example, alkali metals in herbaceous plants are mainly present in the water-soluble form, whereas others may be present as inorganic salts. These differences can lead to differences in the migration pathways and transformations of alkali metals during combustion. Combustion conditions such as oxygen concentration and combustion rate may also affect the migration of alkali metals. For example, oxygen-enriched conditions may favor oxidation and evaporation of alkali metals, whereas low-oxygen conditions may promote the combination of alkali metals with other elements in the biomass to form more stable compounds. In addition, the structure and operating parameters of the combustion equipment may also have an effect on the migration of alkali metals. The content and morphology of alkali metal elements in the ash resulting from biochar combustion are closely related to the properties and composition of the ash. For example, minerals such as silicates in the ash may react chemically with the alkali metal elements to form more stable compounds, while chlorine in the ash may promote the evaporation and migration of alkali metals.

The migration of alkali metal elements during the combustion process of biochar has an important impact on the combustion characteristics. On the one hand, the migration of alkali metal elements will change the combustion characteristics of biochar, such as combustion rate and ignition point. On the other hand, the migration of alkali metal elements will also affect the emission of pollutants and the environmental impact of the combustion process of biochar. Alkali metal elements as catalysts can promote the combustion rate and ignition point advancement of biochar. This is because alkali metal elements can reduce the activation energy of combustion of biochar, thus accelerating the combustion process. However, when the content of alkali metal elements is too high, it may lead to problems such as unstable combustion process or slagging. Alkali metal elements may promote the conversion and emission of nitrogen and sulfur in biochar. At the same time, alkali metal elements may combine with other elements to form more complex compounds, such as potassium chloride and potassium sulfate, which may produce secondary pollution during the combustion process.

The migration of alkali metals in the combustion process of biochar is a complex process, which is affected by a variety of factors. Through in-depth study of the influencing factors and migration rules of alkali metals, it can provide strong theoretical support and technical guarantee for the efficient use of biomass energy and environmental protection. In the future, with the continuous development and improvement of biomass energy technology, it is believed that the problem of alkali metal migration in the combustion process of biochar can be better solved to promote the wide application and sustainable development of biomass energy.

2.10.3 Control Strategies for Alkali Metal Migration

In order to reduce the problems of alkali metal staining and corrosion, slagging and deposition on the heating surface during biochar combustion, effective control strategies are developed. These strategies mainly include changing the type of biomass, optimizing combustion conditions, and adding catalysts or inhibitors.

The migration and release of alkali metals can be reduced by choosing biochar with a lower alkali metal content or a more stable form as a fuel. Meanwhile, the content of alkali metals in biochar can also be further reduced by biomass pretreatment such as washing and deashing. The migration process of alkali metals can be influenced by optimizing the combustion conditions such as controlling the oxygen concentration and adjusting the combustion rate. For example, appropriately lowering the oxygen concentration may reduce the oxidation and evaporation rates of alkali metals, while increasing the combustion rate may promote the combination of alkali metals with other elements in the biomass to form more stable compounds. The addition of appropriate catalysts or inhibitors can affect the alkali metal migration and transformation processes. For example, some metal oxides such as copper oxide can act as catalysts to promote combustion rate and ignition temperature advancement of biochar, while certain silicates can act as inhibitors to inhibit the evaporation and migration of alkali metals. By choosing the appropriate inhibitors or catalysts, the migration of alkali metal elements and their impact on the environment can be effectively reduced.

2.11 Prospects and Challenges of Biochar Energy

With the growth of global energy demand and increased awareness of environmental protection, biochar will receive more and more attention as a clean and efficient fuel. As a renewable energy source, the combustion process of biochar has a low environmental impact. Compared with fossil fuels, the combustion of biochar produces lower greenhouse gas emissions, which contributes to the mitigation of global climate change. In addition, the preparation of biochar mainly utilizes waste resources such as agricultural waste and municipal waste, thereby recycling resources and reducing environmental pollution. As a high-quality fuel, biochar has a wide range of application prospects. Biochar can be used in power generation, heating, industrial production, and transport. In the field of power generation, biochar can be used as fuel in biomass power plants to replace traditional fossil fuels. In the field of heating, biochar can be used in home heating, industrial heating, and other fields to provide clean and efficient heat energy. In the field of industrial production, biochar can be used as reducing agent, catalyst, and adsorbent in metal smelting, chemical production, and wastewater treatment. However, biochar preparation and combustion technologies still face some challenges, such as the diversity of feedstock sources, cost control of the preparation process, and control of pollutant emissions during combustion. In the future, the investigation and development of biochar preparation and combustion technologies need to be further strengthened to promote the wide application of biochar in the energy field.

References

Adekanye, T., Dada, O., and Kolapo, J. (2022). Pyrolysis of maize cob at different temperatures for biochar production: proximate, ultimate and spectroscopic characterisation. *Research in Agricultural Engineering* 68: 27–34.

Alengebawy, A., Ran, Y., Ghimire, N. et al. (2023). Rice straw for energy and value-added products in China: a review. *Environmental Chemistry Letters* 21: 2729–2760.

Amalina, F., Razak, A.S.A., Krishnan, S. et al. (2022). Biochar production techniques utilizing biomass waste-derived materials and environmental applications—a review. *Journal of Hazardous Materials Advances* 7: 100134.

Ayanwusi, O.R., Abdulkareem, S.A., Micheal, K.T. et al. (2024). Influence of varying aluminum-based combustion compartment sizes on groundnut shell biochar properties. *Biofuels* 16: 1–12.

Baath, Y.S., Nikrityuk, P.A., and Gupta, R. (2022). Experimental and numerical verifications of biochar gasification kinetics using TGA. *Renewable Energy* 185: 717–733.

Barbanera, M., Cotana, F., and Di Matteo, U. (2018). Co-combustion performance and kinetic study of solid digestate with gasification biochar. *Renewable Energy* 121: 597–605.

Cai, W., Luo, Z., Zhou, J. et al. (2021). A review on the selection of raw materials and reactors for biomass fast pyrolysis in China. *Fuel Processing Technology* 221: 106919.

Cantrell, K.B., Hunt, P.G., Uchimiya, M. et al. (2012). Impact of pyrolysis temperature and manure source on physicochemical characteristics of biochar. *Bioresource Technology* 107: 419–428.

Cao, X., Luo, Q., Song, F. et al. (2023). Effects of oxidative torrefaction on the physicochemical properties and pyrolysis products of hemicellulose in bamboo processing residues. *Industrial Crops and Products* 191: 115986.

Chang, G., Wang, G., Li, Y. et al. (2021). Synergistic interactions between biochar reacted with steam and CO_2 originating from a diffusion reaction state and intrinsic ash. *Fuel Processing Technology* 215: 106754.

Chaves Fernandes, B.C., Ferreira Mendes, K., Dias Júnior, A.F. et al. (2020). Impact of pyrolysis temperature on the properties of eucalyptus wood-derived biochar. *Materials* 13: 5841.

Chen, D., Cen, K., Chen, F. et al. (2022a). Solar pyrolysis of cotton stalks: combined effects of torrefaction pretreatment and HZSM-5 zeolite on the bio-fuels upgradation. *Energy Conversion and Management* 261: 115640.

Chen, J., Ding, L., Wang, P. et al. (2022b). The estimation of the higher heating value of biochar by data-driven modeling. *Journal of Renewable Materials.* 10: 1555–1574.

Chen, R., Sheng, Q., Dai, X. et al. (2021). Upgrading of sewage sludge by low temperature pyrolysis: biochar fuel properties and combustion behavior. *Fuel* 300: 121007.

Cui, Y., Zhang, Y., Cui, L. et al. (2023). Microwave-assisted fluidized bed reactor pyrolysis of polypropylene plastic for pyrolysis gas production towards a sustainable development. *Applied Energy* 342: 121099.

Cuixia, Y., Yingming, X., Lin, W. et al. (2020). Effect of different pyrolysis temperatures on physico-chemical characteristics and lead(II) removal of biochar derived from chicken manure. *RSC Advances* 10: 3667–3674.

Dunnigan, L., Ashman, P.J., Zhang, X. et al. (2018a). Production of biochar from rice husk: particulate emissions from the combustion of raw pyrolysis volatiles. *Journal of Cleaner Production* 172: 1639–1645.

Dunnigan, L., Morton, B.J., Ashman, P.J. et al. (2018b). Emission characteristics of a pyrolysis-combustion system for the co-production of biochar and bioenergy from agricultural wastes. *Waste Management* 77: 59–66.

Egbosiuba, T.C. (2022). Biochar and bio-oil fuel properties from nickel nanoparticles assisted pyrolysis of cassava peel. *Heliyon* 8: e10114.

Emenike, E.C., Iwuozor, K.O., Agbana, S.A. et al. (2022). Efficient recycling of disposable face masks via co-carbonization with waste biomass: a pathway to a cleaner environment. *Cleaner Environmental Systems* 6: 100094.
Feng, C. and Wu, H. (2018). Synergy on particulate matter emission during the combustion of bio-oil/biochar slurry (bioslurry). *Fuel* 214: 546–553.
Fonseca, F.G. and Soares Dias, A.P. (2021). Almond shells: catalytic fixed-bed pyrolysis and volatilization kinetics. *Renewable Energy* 180: 1380–1390.
Gao, A., Wang, Y., Lin, G. et al. (2022). Volatile-char interactions during biomass pyrolysis: reactor design toward product control. *Renewable Energy* 185: 1–7.
Gao, S., Hao, S., Huang, Z. et al. (2020). Synthesis of high-entropy alloy nanoparticles on supports by the fast moving bed pyrolysis. *Nature Communications* 11: 2016.
Garcia-Nunez, J.A., Pelaez-Samaniego, M.R., Garcia-Perez, M.E. et al. (2017). Historical developments of pyrolysis reactors: a review. *Energy Fuels* 31: 5751–5775.
Ge, S., Yek, P.N.Y., Cheng, Y.W. et al. (2021). Progress in microwave pyrolysis conversion of agricultural waste to value-added biofuels: a batch to continuous approach. *Renewable and Sustainable Energy Reviews* 135: 110148.
Ghodake, G.S., Shinde, S.K., Kadam, A.A. et al. (2021). Review on biomass feedstocks, pyrolysis mechanism and physicochemical properties of biochar: state-of-the-art framework to speed up vision of circular bioeconomy. *Journal of Cleaner Production* 297: 126645.
He, M., Zhu, X., Dutta, S. et al. (2022). Catalytic co-hydrothermal carbonization of food waste digestate and yard waste for energy application and nutrient recovery. *Bioresource Technology* 344: 126395.
Huang, H., Liu, J., Liu, H. et al. (2022). Turning the co-combustion synergy of textile dyeing sludge and waste biochar into emission-to-bottom slag pollution controls toward a circular economy. *Renewable Energy* 194: 760–777.
Itoh, T., Fujiwara, N., Iwabuchi, K. et al. (2020). Effects of pyrolysis temperature and feedstock type on particulate matter emission characteristics during biochar combustion. *Fuel Processing Technology* 204: 106408.
Janu, R., Mrlik, V., Ribitsch, D. et al. (2021). Biochar surface functional groups as affected by biomass feedstock, biochar composition and pyrolysis temperature. *Carbon Resources Conversion* 4: 36–46.
Jeníček, L., Tunklová, B., Malaťák, J. et al. (2023). The impact of nutshell biochar on the environment as an alternative fuel or as a soil amendment. *Materials (Basel)* 16: 2074.
Jia, W., Guo, Y., Guo, F. et al. (2023). Co-combustion of carbon-rich fraction from coal gasification fine slag and biochar: gas emission, ash sintering, heavy metals evolutions and environmental risk evaluation. *Chemical Engineering Journal* 471: 144312.
Jiang, W., Li, D., Chen, S. et al. (2022). Design and operation of a fixed-bed pyrolysis-gasification-combustion pilot plant for rural solid waste disposal. *Bioresource Technology* 362: 127799.

Kaminsky, W. (2021). Chemical recycling of plastics by fluidized bed pyrolysis. *Fuel Communications* 8: 100023.

Kluska, J. and Ramotowski, J. (2024). Combustion characteristics of biochar from food processing waste. *Archives of Thermodynamics* 45–51.

Koçer, A.T. and Özçimen, D. (2021). Determination of combustion characteristics and kinetic parameters of *Ulva lactuca* and its biochar. *Biomass Conversion and Biorefinery* 14: 5913–5922.

Kong, G., Wang, K., Zhang, X. et al. (2022). Torrefaction/carbonization-enhanced gasification-steam reforming of biomass for promoting hydrogen-enriched syngas production and tar elimination over gasification biochars. *Bioresource Technology* 363: 127960.

Li, A., Liu, H.-L., Wang, H. et al. (2016). Effects of temperature and heating rate on the characteristics of molded bio-char. *Bioresources* 11: 16.

Li, N., Pan, Y., Yan, Z. et al. (2024). Cornstalk steam gasification for syngas in a two-stage electromagnetic induction reactor. *Chemical Engineering Journal* 488: 150948.

Li, X., Peng, B., Liu, Q. et al. (2022). Microwave pyrolysis coupled with conventional pre-pyrolysis of the stalk for syngas and biochar. *Bioresource Technology* 348: 126745.

Li, Y., Gupta, R., Zhang, Q. et al. (2023). Review of biochar production via crop residue pyrolysis: development and perspectives. *Bioresource Technology* 369: 128423.

Lubwama, M., Yiga, V.A., Muhairwe, F. et al. (2020). Physical and combustion properties of agricultural residue bio-char bio-composite briquettes as sustainable domestic energy sources. *Renewable Energy* 148: 1002–1016.

Luo, G., Chandler, D.S., Anjos, L.C.A. et al. (2017). Pyrolysis of whole wood chips and rods in a novel ablative reactor. *Fuel* 194: 229–238.

Luo, H. Wang, X. Liu, X. et al. (2022). A review on CFD simulation of biomass pyrolysis in fluidized bed reactors with emphasis on particle-scale models. *Journal of Analytical and Applied Pyrolysis* 162: 105433.

Malik, S. and Laura, J.S. (2025). Evaluating biochar properties derived from various agro-residues via pyrolysis: a comparative study. *Oriental Journal of Chemistry* 41: 120–126.

Mau, V. and Gross, A. (2018). Energy conversion and gas emissions from production and combustion of poultry-litter-derived hydrochar and biochar. *Applied Energy* 213 510–519.

McMurchie, E.J., Olatunji, J., Kwong, K.Y. et al. (2024). The kinetics and mechanism of the combustion of the carbon in biochar from oak, as studied in an electrically heated, fluidised bed of sand. *Fuel* 378: 132677.

Mengqi, Z., Shi, A., Ajmal, M. et al. (2021). Comprehensive review on agricultural waste utilization and high-temperature fermentation and composting. *Biomass Conversion and Biorefinery* 13: 5445–5468.

Niu, Y., Tan, H., and Hui. S. (2016) Ash-related issues during biomass combustion: alkali-induced slagging, silicate melt-induced slagging (ash fusion), agglomeration, corrosion, ash utilization, and related countermeasures. *Progress in Energy and Combustion Science* 52: 1–61.

Peterson, C.A. and Brown, R.C. (2020). Oxidation kinetics of biochar from woody and herbaceous biomass. *Chemical Engineering Journal* 401: 126043.

Qian, C., Li, Q., Zhang, Z. et al. (2020). Prediction of higher heating values of biochar from proximate and ultimate analysis. *Fuel* 265: 116925.

Qin, F., Zhang, C., Zeng, G. et al. (2022). Lignocellulosic biomass carbonization for biochar production and characterization of biochar reactivity. *Renewable and Sustainable Energy Reviews* 157: 112056.

Rathore, N.S., Pawar, A., and Panwar, N.L. (2021). Kinetic analysis and thermal degradation study on wheat straw and its biochar from vacuum pyrolysis under non-isothermal condition. *Biomass Conversion and Biorefinery* 13: 7547–7559.

Raza, M., Inayat, A., Ahmed, A. et al. (2021). Progress of the pyrolyzer reactors and advanced technologies for biomass pyrolysis processing. *Sustainability* 13: 11061.

Rubinsin, N.J., Karim, N.A., Timmiati, S.N. et al. (2024). An overview of the enhanced biomass gasification for hydrogen production. *International Journal of Hydrogen Energy* 49: 1139–1164.

Safarian, S. (2023). Climate impact comparison of biomass combustion and pyrolysis with different applications for biochar based on LCA. *Energies* 16: 5541.

Schnabel, K., Brück, F., Pohl, S. et al. (2022). Development and test of a rotating drum reactor for the simultaneous hydration and carbonation of dry biomass bottom ash. *Waste and Biomass Valorization* 13: 4319–4330.

Seow, Y.X., Tan, Y.H., Mubarak, N.M. et al. (2022). A review on biochar production from different biomass wastes by recent carbonization technologies and its sustainable applications. *Journal of Environmental Chemical Engineering* 10: 107017.

Shahbaz, M., AlNouss, A., Ghiat, I. et al. (2021). A comprehensive review of biomass based thermochemical conversion technologies integrated with CO_2 capture and utilisation within BECCS networks. *Resources, Conservation and Recycling* 173: 105734.

Shanmugam, V., Sreenivasan, S.N., Mensah, R.A. et al. (2022). A review on combustion and mechanical behaviour of pyrolysis biochar. *Materials Today Communications* 31: 103629.

Shaowu, X., Shouyu, Z., Qiaomei, W. et al. (2015). Study on biomass carbonization for fuel charcoal preparation. *Acta Energiae Solaris Sinica* 36: 7.

Sharma, A., Kumar, A.A., Mohanty, B. et al. (2023). Synthesis, characterization and thermo-kinetic analysis of sawdust biochar for adsorbent and combustion drives. *Bioresource Technology Reports* 22: 101485.

Su, H., Zhou, X., Zheng, R. et al. (2021). Hydrothermal carbonization of food waste after oil extraction pre-treatment: study on hydrochar fuel characteristics, combustion behavior, and removal behavior of sodium and potassium. *Science of the Total Environment* 754: 142192.

Sudalai, S., Prabakaran, S., Varalakksmi, V. et al. (2024). A review on oilcake biomass waste into biofuels: current conversion techniques, sustainable applications, and challenges: waste to energy approach (WtE). *Energy Conversion and Management* 314: 118724.

Suresh, A., Alagusundaram, A., Kumar, P.S. et al. (2021). Microwave pyrolysis of coal, biomass and plastic waste: a review. *Environmental Chemistry Letters* 19 3609–3629.

Tezer, Ö., Karabağ, N., Öngen, A. et al. (2022). Biomass gasification for sustainable energy production: a review. *International Journal of Hydrogen Energy* 47: 15419–15433.

Tong, W., Cai, Z., Liu, Q. et al. (2020). Evaluation of biochar combustion reactivity under pyrolysis temperature: microstructure characterization, kinetics and thermodynamics. *Journal of the Energy Institute* 93: 1914–1923.

Tripathi, M., Sahu, J.N., and Ganesan, P. (2016). Effect of process parameters on production of biochar from biomass waste through pyrolysis: a review. *Renewable and Sustainable Energy Reviews* 55: 467–481.

Vasudev, V., Ku, X., and Lin, J. (2021). Combustion behavior of algal biochars obtained at different pyrolysis heating rates. *ACS Omega* 6: 19144–19152.

Vieira, F.R., Romero Luna, C.M., Arce, G.L.A.F. et al. (2020). Optimization of slow pyrolysis process parameters using a fixed bed reactor for biochar yield from rice husk. *Biomass and Bioenergy* 132: 105412.

Wang, D., Jiang, P., Zhang, H. et al. (2020). Biochar production and applications in agro and forestry systems: a review. *Science of the Total Environment* 723: 137775.

Wang, Q., Han, K., Qi, J. et al. (2018). Investigation of potassium transformation characteristics and the influence of additives during biochar briquette combustion. *Fuel* 222: 407–415.

Wang, T., Li, Y., Zhi, D. et al. (2019). Assessment of combustion and emission behavior of corn straw biochar briquette fuels under different temperatures. *Journal of Environmental Management* 250: 109399.

Wang, W., Lemaire, R., Bensakhria, A. et al. (2022). Review on the catalytic effects of alkali and alkaline earth metals (AAEMs) including sodium, potassium, calcium and magnesium on the pyrolysis of lignocellulosic biomass and on the co-pyrolysis of coal with biomass. *Journal of Analytical and Applied Pyrolysis* 163: 105479.

Wang, X., Hu, Z., Deng, S. et al. (2014). Kinetics investigation on the combustion of biochar in O_2/CO_2 atmosphere. *Environmental Progress & Sustainable Energy* 34: 923–932.

Wang, X., Liu, Q., Yang, L. et al. (2024). Perceptions of biomass energy sustainability in policy scenarios of China. *Heliyon* 10: e37180.

Wang, Y.-H., Bayatpour, S., Qian, X. et al. (2021). Activated carbon fibers via reductive carbonization of cellulosic biomass for adsorption of nonpolar volatile organic compounds. *Colloids and Surfaces A: Physicochemical and Engineering Aspects* 612: 125908.

Wen, C., Liu, T., Wang, D. et al. (2023). Biochar as the effective adsorbent to combustion gaseous pollutants: preparation, activation, functionalization and the adsorption mechanisms. *Progress in Energy and Combustion Science* 99: 101098.

Wu, L., Yan, Z., Xie, J. et al. (2024). Enhancing the catalytic H_2 production performance of magnetic Ni-Fe_2O_3-C catalyst in biomass steam gasification using electromagnetic induction heating. *Bioresource Technology* 402: 130844.

Wu, W., Yan, B., Zhong, L. et al. (2021). Combustion ash addition promotes the production of K-enriched biochar and K release characteristics. *Journal of Cleaner Production* 311: 127557.

Wu, Y., Wang, H., Li, H. et al. (2022). Applications of catalysts in thermochemical conversion of biomass (pyrolysis, hydrothermal liquefaction and gasification): a critical review. *Renewable Energy* 196: 462–481.

Yaashikaa, P.R., Kumar, P.S., Varjani, S. et al. (2020). A critical review on the biochar production techniques, characterization, stability and applications for circular bioeconomy. *Biotechnology Reports* 28: e00570.

Yang, G., Liu, Y., Gao, L. et al. (2022). Investigation of the synergistic effect and kinetic behavior of anthracite and biochar during co-combustion process in pure oxygen atmosphere. *Journal of the Energy Institute* 101: 1–18.

Yi, Q., Qi, F., Cheng, G. et al. (2012). Thermogravimetric analysis of co-combustion of biomass and biochar. *Journal of Thermal Analysis and Calorimetry* 112: 1475–1479.

Yousaf, B., Liu, G., Abbas, Q. et al. (2017). Systematic investigation on combustion characteristics and emission-reduction mechanism of potentially toxic elements in biomass- and biochar-coal co-combustion systems. *Applied Energy* 208: 142–157.

Zeng, M., Ge, Z., Ma, Y. et al. (2023). Hydrothermal carbonization coupled with gasification for collaborative disposal of kitchen waste and yard waste. *Energy Conversion and Management* 283: 116864.

Zhang, J., Wang, G., Xu, S. et al. (2020). Upgrading of biomass fast pyrolysis oil over a moving bed of coal char. *Carbon Resources Conversion* 3: 130–139.

Zhang, X., Li, Y., Zhang, X. et al. (2023). Co-combustion of municipal solid waste and hydrochars under non-isothermal conditions: thermal behaviors, gaseous emissions and kinetic analyses by TGA–FTIR. *Energy* 265: 126373.

Zhou, Y., Qin, S., Verma, S. et al. (2021). Production and beneficial impact of biochar for environmental application: a comprehensive review. *Bioresource Technology* 337: 125451.

Zhu, G., Wen, C., Liu, T. et al. (2024). Combustion and co-combustion of biochar: combustion performance and pollutant emissions. *Applied Energy* 376: 124292.

Zhu, M., Zhang, Z., Zhang, Y. et al. (2017). An experimental investigation into the ignition and combustion characteristics of single droplets of biochar water slurry fuels in air. *Applied Energy* 185: 2160–2167.

Zhu, X., He, M., Xu, Z. et al. (2022a). Combined acid pretreatment and co-hydrothermal carbonization to enhance energy recovery from food waste digestate. *Energy Conversion and Management* 266: 115855.

Zhu, X., Labianca, C., He, M. et al. (2022b). Life-cycle assessment of pyrolysis processes for sustainable production of biochar from agro-residues. *Bioresource Technology* 360: 127601.

Zhu, X., Sun, M., Zhu, X. et al. (2023). Molten salt shielded pyrolysis of biomass waste: development of hierarchical biochar, salt recovery, CO_2 adsorption. *Fuel* 334: 126565.

Zhuang, X., Liu, J., Zhang, Q. et al. (2022). A review on the utilization of industrial biowaste via hydrothermal carbonization. *Renewable and Sustainable Energy Reviews* 154: 111877.

3

Agricultural Waste-derived Biochar for Energy Recovery from Waste Fermentation

Hong-Yu Ren[1], Qingqing Song[1], Fanying Kong[2], and Xueting Song[1]*

[1] *State Key Laboratory of Urban-Rural Water Resource and Environment, School of Environment, Harbin Institute of Technology, Harbin, China*
[2] *School of Water Conservancy and Civil Engineering, Northeast Agricultural University, Harbin, China*

*Corresponding author: rhy@hit.edu.cn; rhongyu@126.com

3.1 Introduction

As the global economy continues to grow rapidly, agricultural production has also benefited, with significant increases in productivity and capacity (Aziz et al. 2023; Kumar and Walia 2024). However, this economic boom has been accompanied by a sharp increase in the amount of agricultural waste, posing unprecedented challenges to waste management systems (Wani et al. 2024). Looking back over the past few decades, the boom in agricultural activities has led directly to a surge in the global production of waste from the agricultural sector, with annual growth rates climbing to billions of tons per year and accumulating to tens of billions of tons per day (Chandrappa and Das 2024). This trend not only highlights the urgency of agricultural waste management but also the need to develop efficient and sustainable waste treatment and resource utilization technologies. Agricultural wastes are organic and inorganic materials discarded during the agricultural production process, which can be raw materials, intermediates, or agricultural products. Agricultural wastes can be categorized according to different criteria, for example, according to biodegradability they can be classified as biodegradable and non-biodegradable; according to physical state they can be classified as solid and liquid; according to toxicity they can be classified as hazardous and non-hazardous; and according to cultivation activities they can be classified as livestock activities, animal wastes, aquaculture, and horticultural wastes. Figure 3.1 shows the main classifications and examples of agricultural waste.

Waste-derived Biochar for Sustainable Rural Revitalization, First Edition. Edited by Yuqing Sun and Daniel C.W. Tsang.
Published 2026 by John Wiley & Sons, Inc.

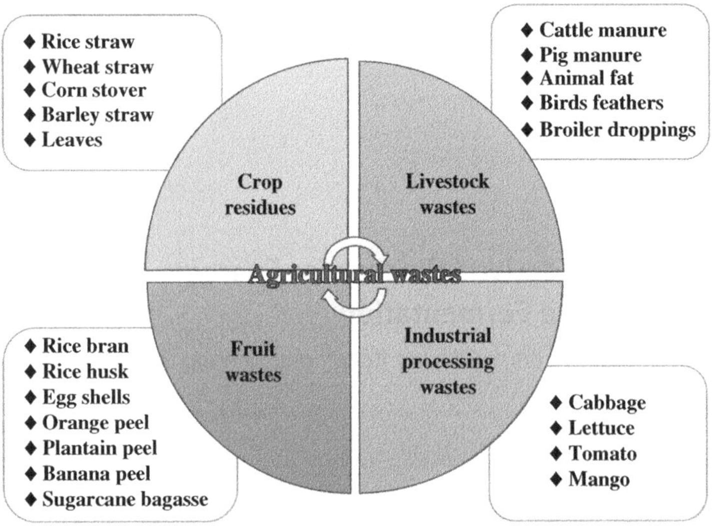

Figure 3.1 Classifications of agricultural waste.

Overall, the agricultural waste system broadly encompasses two major categories: crop residues and livestock manure. Specifically, crop residues, such as cornstalks, wheat stubble, and rice hulls, account for a significant portion of the agricultural production system and constitute an important component of agricultural by-products. On the other hand, livestock, and poultry manure, which originates from biological excretory activities during livestock production and covers a wide range of types such as pig manure, cow manure, and chicken manure, is another large group of agricultural waste that should not be ignored. In addition, agricultural wastes include agricultural processing wastes and residential processing wastes, such as wastes from the meat-processing industry, human waste, and other wastes from daily life. These wastes, like other biomasses, consist of complex molecular structures including cellulose, hemicellulose, lignin, and some extractives. However, these wastes can pose serious threat to human society, health, and the environment if they are not properly disposed of and managed, e.g., through unsuitable and unsustainable incineration. If properly disposed of and managed, agricultural wastes can be recycled, value-added, and conserved, and their potential value can be tapped, thus avoiding a large amount of unused and wasted resources to a large extent. Because of the extensive scale, sustainability, renewability, and environmental friendliness of agricultural waste, exploring and implementing efficient and sustainable utilization strategies is of far-reaching strategic significance for promoting resource recycling, reducing environmental loads, and realizing the green transformation of agricultural production.

3.2 Biochar Preparation from Agricultural Waste

A key strategy for an eco-friendly, sustainable, and cost-effective approach to agricultural waste management is to convert it into forms that have practical applications (Koul et al. 2022). Converting agricultural waste into biochar is a cost-effective, environmentally friendly, and sustainable approach. Biochar is a solid carbonaceous material with a porous structure, amorphous and rich in carbon, with characteristics similar to charcoal (Amonette and Joseph 2012). Its characteristics are similar to those of charcoal, but it is derived from a wider range of biomass feedstocks through a process of pyrolytic carbonization. One of the characteristics of biochar is its porous nature, which gives it excellent adsorption properties, enabling it to effectively fix nutrients in the soil and adsorb heavy metals and organic pollutants. In addition, biochar is highly aromatic, high in carbon content, rich in functional groups, highly stable, specific in physical and chemical properties, and has a wide range of raw material sources, which makes it play an important role in soil remediation, improving soil fertility and stability, carbon sequestration and emission reduction, and water purification (Awogbemi and Von Kallon 2023). Converting agricultural waste into biochar is an easy-to-implement and cost-effective method, which not only effectively solves the problem of agricultural waste treatment, but also reduces the direct pollution of the environment caused by the accumulation of waste, and realizes the resourceful utilization of waste and energy recovery through the preparation of biochar.

To convert agricultural waste into biochar, the first step is the selection of agricultural waste. When selecting agricultural waste, its accessibility, treatment cost, and impact on the environment should be considered, such as common straw, rice husk, animal, and poultry manure. These wastes are rich in a large amount of biomass, which is a high-quality raw material for biochar extraction (Tripathi et al. 2016). Secondly, raw materials need to be pre-treated in the extraction process of biochar, for example, straw needs to be crushed and dried to improve the efficiency of carbonization, and the carbonization process needs to work in a low-oxygen or anaerobic environment, and the biomass will be pyrolyzed and carbonized through high temperatures (usually 400–600 °C), and the lignin, cellulose, and other organic matter in the biomass in the process will be pyrolyzed to generate biochar; finally, there is the technology of converting agricultural waste into biochar, and the specific methods can be divided into the following three types, i.e., traditional, modern, and innovative methods; the conventional approach involves excavating an earth pit to bury lignocellulosic materials like agricultural waste, subsequently igniting them, and allowing them to undergo combustion in an oxygen-deprived environment (Awogbemi and Von Kallon 2023). This crude method was later changed to the use of refractory brick pits, clay burners, brick kilns, and iron stills as reactors instead of earth pits to convert agricultural wastes into biochar by heating.

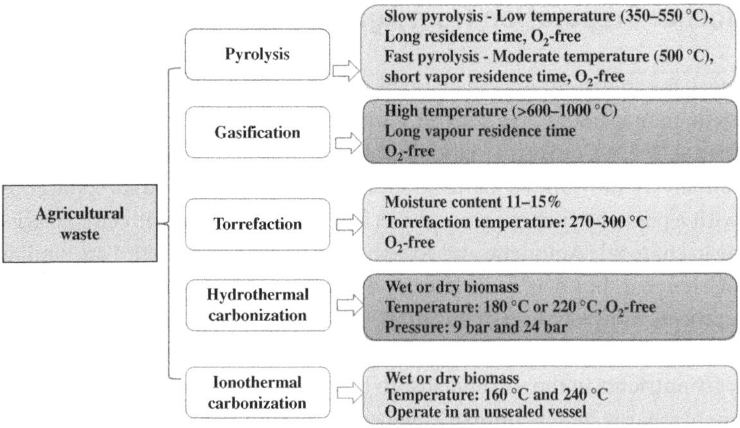

Figure 3.2 Schematic illustration of various biochar production methods.

While traditional methods may be crude and laborious, they offer the advantage of being relatively inexpensive, needing no specialized sophisticated equipment, and being easily adaptable for local implementation. In addition to the traditional methods, several modern technologies have been developed in which the original heat sources, heating chambers, and heat applications have been replaced by advanced methods, such as replacing localized, contaminated heat sources with modern, clean sources of energy, replacing flat furnaces and refractory brick pits. In traditional technology, furnaces and ovens are employed, where the heating medium is furnished with instruments for measuring temperature, heating rate, and atmospheric pressure, significantly enhancing the conversion efficiency. Advancements in technology have led to the emergence of novel biochar production techniques, including electrically modified biochar, magnetic biochar, and plasma-based methods, which address the limitations and drawbacks of older, traditional conversion methods that rely on electronic, chemical, and other processes for biochar creation and modification. Continuous improvement of these technologies is needed to shorten reaction times, reduce energy consumption, and improve the quality of biochar (Sakhiya et al. 2020) (Figure 3.2).

3.3 Energy Recovery from Waste Biomass Fermentation Based on Biochar Treatment

The discarding of waste biomass and the depletion of energy supplies have made renewable energy from waste biomass one of the top global concerns (Khedulkar et al. 2023). Preparation of agricultural waste into biochar can be applied as a high value-added material in fermentation engineering of waste biomass to produce renewable energy and mitigate the energy crisis. In addition, the process increases the value of agricultural waste utilization and improves the agriculture-based circular economy (Song et al. 2022).

In the background of the development of green circular economy, using microbial fermentation process to convert waste biomass, such as agricultural waste, residual sludge, and kitchen waste, into renewable energy sources such as methane, hydrogen, acid, and bio-fertilizers is a sustainable method of organic waste treatment (Zhang et al. 2023). Biochar prepared from agricultural waste has the advantages of low cost, diversified raw materials, and green environment. Biochar is uniquely superior in terms of its potential to increase the fermentation capacity of biomass, due to its rich surface functional groups, high adsorption capacity, and immobilization effects (Lehmann et al. 2021). As summarized in Table 3.1, it is now widely practiced in various biomass fermentation processes.

The mechanism by which biochar promotes the fermentation capacity of waste biomass is closely related to the properties of biochar (Yang et al. 2021). Biochar materials can be used as additives in fermentation systems to improve the efficiency and stability of the fermentation system due to their high thermal stability, high specific surface area, porous size, and multifunctional groups (Shao et al. 2024). The alkalinity of biochar neutralizes the volatile fatty acids produced by the fermentation process and increases the abundance and activity of the fermenting microflora, thereby increasing fermentation gas production. The porous structure of biochar and the rich functional group structure of biochar make biochar have the characteristics of adsorption and exchange of ions, which can adsorb the excessive ammonia nitrogen produced by the fermentation system and reduce the inhibitory effect of ammonia nitrogen on the fermentation system (Shao et al. 2024). Therefore, increasing the number of functional groups on the surface of biochar through the preparation of modified biochar is a practical way to improve the performance of biochar (Wang et al. 2019). The high surface area and porous structure of biochar can also change the attachment state between fermentation microorganisms and substrates, providing more ideal growth habitats and attachment sites for microorganisms, thus improving the activity of microbial communities in anaerobic fermentation systems and accelerating the hydrolysis process of substrates (He et al. 2024). The nitrogen and carbon contained in the biochar can provide the necessary nutrients for the growth of microorganisms and promote the growth and reproduction of microbial flora, thus increasing the gas production rate of the whole fermentation system (Wang et al. 2018).

Biochar has good electron transfer conductivity, provides abundant active sites and unsaturated electron pairs through its own unsaturated carbon and specific functional groups, mediates electron transfer between microorganisms, establishes a stable electron transfer chain, promotes the interspecies relationship between

Table 3.1 Application of biochar prepared from agricultural waste in waste biomass fermentation.

Product	Materials for biochar preparation	Fermentation substrate	Effect of production capacity	References
Hydrogen	Residue cornstalk left	Corn stalk hydrolysate	The biochar enabled a maximal fermentation bio-H_2 yield of 154.3 mL H_2/g cornstalk	Zhao et al. (2021b)
	Pinewood	Food waste	Cumulative H_2 production and maximum H_2 productivity increased by 107 and 54%, respectively	Sugiarto et al. (2021)
	The fallen leaves	Bagasse	Hydrogen production was further increased by 158.1% with the addition of biochar	Huang et al. (2022)
	Rice straw	Corn stalk	Biochar enabled a maximum bio-H_2 yield of 286.1 mL H_2/g substrate	Zhao et al. (2021a)
Methane	Rice husk	Corn stover and chicken manure	Volumetric CH_4 production rate improved 27.8–96.4%	Yu et al. (2021)
	Corn straw	Corn stalk	The cumulative methane production of corn stalk by anaerobic fermentation exhibited the highest value of 164.54 mL/g-VS	Pan et al. (2022)
	Corn stover	Food waste	The addition of 10 g/L HNO_3-modified biochar increased CH_4 production by 90%	Gao et al. (2021)
Others	Wheat straw	Defatted *Schizochytrium* sp. biomass	Improved erythritol yield of 223.2 mg/gds	Liu et al. (2021)
	Wheat straw	Peanut press cake and sesame meal	An enhanced erythritol yield of 207.3 mg/gds was achieved	Liu et al. (2020)
	Corncob	Food waste	The highest caproate yield of 0.17 g-COD/g-VS was obtained	Tang et al. (2024)

microorganisms, and promotes the anaerobic fermentation process (Wu et al. 2022). In addition, it has been shown that the addition of biochar can also increase the activity of hydrolytic enzymes in the fermentation system and accelerate the hydrolysis process of the fermentation substrate, further improving the efficiency of the fermentation system (Mohtaram et al. 2024). Thus, biochar as an additive in fermentation systems has a large development potential and a broad application prospect.

3.4 A Case for Biochar in Energy Recovery

Shandong Tairan Biotech is a circular economy enterprise founded in 2011, and its business includes biochar, distributed energy cogeneration, biomass liquid fertilizer production, and biomass pyrolysis (Shanghai Zhigeng Technology Service Co., LTD 2024). The core technology in biochar production includes an optimized pyrolysis process and structural manipulation, which can process agricultural wastes such as straw and wood chips at high temperatures and convert them into solid materials with a carbon content of more than 70%. The product has very low solubility and carboxyl and aromatic structure, and is suitable for soil improvement in agriculture, environmental pollutant adsorption, and industrial high-performance materials. The company has developed a 330,000 ton/year biomass pyrolysis project, forming an industrial system integrating biochar-based fertilizer, biomass power generation, bioactive and efficient organic liquid fertilizer, and organic agricultural tourism park.

3.5 Conclusion

Due to the limited energy reserves, the waste of resources from waste biomass and agricultural waste has attracted great attention, so the preparation of agricultural waste into biochar as an additive to promote the fermentation system to generate renewable energy is of great practical significance to the green economic cycle of agriculture. Studies have shown that biochar prepared from agricultural waste can be a promising strategy to effectively improve the efficiency of energy recovery from methane, hydrogen, organic acids, and alcohols in waste fermentation. Although biochar as a fermentation system additive can significantly improve the operational stability and energy yield of fermentation systems, there are still some limitations and challenges associated with this technology. In order to achieve practical applications, new and inexpensive biochar needs to be developed to meet commercial needs. In addition, laboratory research should be combined with commercial pilot projects to progressively commercialize biochar-promoted fermentation systems prepared from agricultural wastes.

References

Amonette, J.E. and Joseph, S. (2012). *Biochar for Environmental Management*, 65–84. Routledge.

Awogbemi, O. and Von Kallon, D.V. (2023). Progress in agricultural waste derived biochar as adsorbents for wastewater treatment. *Applied Surface Science Advances* 18: 100518.

Aziz, G., Sarwar, S., Waheed, R. et al. (2023). *The Significance of Renewable Energy, Globalization, and Agriculture on Sustainable Economic Growth and Green Environment: Metaphorically, A Two-Sided Blade*. Wiley Online Library.

Chandrappa, R. and Das, D.B. (2024). *Solid Waste Management: Principles and Practice*, 47–87. Springer.

Gao, B., Wang, Y., Huang, L. et al. (2021). Study on the performance of HNO-modified biochar for enhanced medium temperature anaerobic digestion of food waste. *Waste Management* 135: 338–346.

He, Z.-W., Li, A.-H., Tang, C.-C. et al. (2024). Biochar regulates anaerobic digestion: insights to the roles of pore size. *Chemical Engineering Journal* 480: 148219.

Huang, J.R., Chen, X., Hu, B.B. et al. (2022). Bioaugmentation combined with biochar to enhance thermophilic hydrogen production from sugarcane bagasse. *Bioresource Technology* 348: 126790.

Khedulkar, A.P., Pandit, B., Dang, V.D. et al. (2023). Agricultural waste to real worth biochar as a sustainable material for supercapacitor. *Science of the Total Environment* 869: 161441.

Koul, B., Yakoob, M., and Shah, M.P. (2022). Agricultural waste management strategies for environmental sustainability. *Environmental Research* 206: 112285.

Kumar, K.P. and Walia, P. (2024). Speed breeding in cereal crops: accelerating genetic improvement for rapid agricultural advancement. *Journal of Experimental Agriculture International* 46 (6): 465–477.

Lehmann, J., Cowie, A., Masiello, C.A. et al. (2021). Biochar in climate change mitigation. *Nature Geoscience* 14 (12): 883–892.

Liu, X., Wang, Z., Yan, Y. et al. (2020). Novel strategy of incorporating biochar in solid-state fermentation for enhancing erythritol production by forming "microzones". *Bioresource Technology* 306: 123141.

Liu, X.Y., Dong, X.J., Chen, S.Y. et al. (2021). Enhancing erythritol production by wheat straw biochar-incorporated solid-state fermentation of agricultural wastes using defatted *Schizochytrium* sp. biomass as supplementary feedstock. *Industrial Crops and Products* 170: 113703.

Mohtaram, S., Sina Mohtaram, M., Sabbaghi, S. et al. (2024). Enhancement strategies in CO_2 conversion and management of biochar supported photocatalyst for effective generation of renewable and sustainable solar energy. *Energy Conversion and Management* 300: 117987.

Pan, X., Zhang, Y., He, C. et al. (2022). Enhancement of anaerobic fermentation with corn straw by pig bone-derived biochar. *Science of the Total Environment* 829: 154326.

Sakhiya, A.K., Anand, A., and Kaushal, P. (2020). Production, activation, and applications of biochar in recent times. *Biochar* 2: 253–285.

Shao, Y.F., Song, J.Z., and Sun, Y.L. (2024). Advances in biochar's effect on anaerobic fermentation systems. *Polish Journal of Environmental Studies* 33 (3): 3009–3018.

Shanghai Zhigeng Technology Service Co., LTD. (2024). Biochar for black gold: analysis of market dynamics behind 30-fold growth. https://www.thepaper.cn/newsDetail_forward_27810186

Song, B., Cao, X.W., Gao, W.R. et al. (2022). Preparation of nano-biochar from conventional biorefineries for high-value applications. *Renewable & Sustainable Energy Reviews* 157: 112057.

Sugiarto, Y., Sunyoto, N.M.S., Zhu, M. et al. (2021). Effect of biochar in enhancing hydrogen production by mesophilic anaerobic digestion of food wastes: the role of minerals. *International Journal of Hydrogen Energy* 46 (5): 3695–3703.

Tang, J., Hu, Z., Yang, H. et al. (2024). Enhancement of caproate production from food waste using biochar produced from agricultural wastes: a microbial perspective. *Industrial Crops and Products* 221: 119401.

Tripathi, M., Sahu, J.N., and Ganesan, P. (2016). Effect of process parameters on production of biochar from biomass waste through pyrolysis: a review. *Renewable and Sustainable Energy Reviews* 55: 467–481.

Wang, G., Li, Q., Gao, X. et al. (2018). Synergetic promotion of syntrophic methane production from anaerobic digestion of complex organic wastes by biochar: performance and associated mechanisms. *Bioresource Technology* 250: 812–820.

Wang, X., Li, C., Li, Z. et al. (2019). Effect of pyrolysis temperature on characteristics, chemical speciation and risk evaluation of heavy metals in biochar derived from textile dyeing sludge. *Ecotoxicology and Environmental Safety* 168: 45–52.

Wani, N.R., Rather, R.A., Farooq, A. et al. (2024). New insights in food security and environmental sustainability through waste food management. *Environmental Science and Pollution Research* 31 (12): 17835–17857.

Wu, F., Xie, J., Xin, X. et al. (2022). Effect of activated carbon/graphite on enhancing anaerobic digestion of waste activated sludge. *Frontiers in Microbiology* 13: 999647.

Yang, S., Wen, Q.X., and Chen, Z.Q. (2021). Effect of KH_2PO-modified biochar on immobilization of Cr, Cu, Pb, Zn and as during anaerobic digestion of swine manure. *Bioresource Technology* 339: 125570.

Yu, Q., Sun, C., Liu, R. et al. (2021). Anaerobic co-digestion of corn stover and chicken manure using continuous stirred tank reactor: the effect of biochar addition and urea pretreatment. *Bioresource Technology* 319: 124197.

Zhang, L., Tsui, T.H., Wah Tong, Y. et al. (2023). Biochar applications in microbial fermentation processes for producing non-methane products: current status and future prospects. *Bioresource Technology* 386: 129478.
Zhao, L., Wang, Z., Ren, H.Y. et al. (2021a). Residue cornstalk derived biochar promotes direct bio-hydrogen production from anaerobic fermentation of cornstalk. *Bioresource Technology* 320: 124338.
Zhao, L., Wu, K.K., Chen, C. et al. (2021b). Role of residue cornstalk derived biochar for the enhanced bio-hydrogen production via simultaneous saccharification and fermentation of cornstalk. *Bioresource Technology* 330: 125006.

4

Waste-derived Biochar as Organic Seeding Substrate

Yutao Peng[1]*, Yuqing Sun*[1]*, and Daniel C.W. Tsang*[2]

[1] *School of Agriculture and Biotechnology, Sun Yat-sen University, Shenzhen, Guangdong, China*
[2] *Department of Civil and Environmental Engineering, The Hong Kong University of Science and Technology, Clear Water Bay, Hong Kong, China*

4.1 Peat-based Growing Media Substituted by Biochar

The recent studies have highlighted the potential of biochar as an alternative or partial substitute for conventional peat and perlite growing media in seedling cultivation (Kaudal et al. 2016). For example, Rozas et al. (2023) demonstrated that peat-based media mixed with compost and biochar could support biological activity and maintain seedling growth comparable to pure peat, suggesting biochar as a sustainable alternative to reduce peat dependency. Meanwhile, Tian et al. (2012) investigated green waste-derived biochar and found that it not only improved plant growth and biomass in ornamental plants but also contributed to the reduction of green waste, suggesting its potential as a partial replacement for peat. Further, Blok et al. (2017) demonstrated that biochar derived from nutrient-poor feedstocks could replace up to 20% of peat in potting mixes without compromising plant growth, while biochars from nutrient-rich feedstocks showed excessive salinity and alkalinity, requiring neutralization with acidic peat.

The replacement ratio of biochar is in the range of 5–50%. Chrysargyris et al. (2019) examined the effects of commercial biochars mixed with peat on cabbage seedlings and observed that low concentrations (5–10%) improved plant growth, increasing nitrogen, potassium, and phosphorus content in leaves. Rathnayake et al. (2021) tested biochar from sawmill residue as a partial peat replacement and found that a 25–50% biochar inclusion ratio promoted seed germination and plant growth, outperforming peat-only and biochar-only treatments. Glaser and

Asomah (2022) found that biochar-containing media, especially Sonnenerde, increased cereal yields by 30–40% compared to peat-based media, while fungal activity was slightly higher in biochar mixtures. However, replacing seedling substrates with higher biochar concentrations (exceeding 25%) does not necessarily benefit seedling growth. For example, Dorais et al. (2016) observed that high biochar concentrations (50%) negatively impacted tomato growth and did not improve disease resistance, suggesting that excessive biochar can have adverse effects. Further, Sarauer and Coleman (2018) assessed biochar as a peat amendment for Douglas-fir regeneration in the northwestern United States but found that high biochar concentrations (25–50%) negatively impacted seedling growth, primarily due to changes in pH and phosphorus availability.

The replacement of peat by biochar significantly changed the physical and chemical properties of the original growing media. Prasad et al. (2019) highlighted that biochar's nutrient content and pH depend on the feedstock and particle size, with finer fractions having higher nutrient availability. Additionally, Chrysargyris et al. (2020) observed that wood-based biochars increased media pH and nutrient levels, improving seedling performance, particularly at lower biochar concentrations, and noted species-specific responses to biochar incorporation. Kuo et al. (2019) found that sorghum distillery residue biochar (SDRBC) improved the pH, fertility index, and seedling performance of growth media, particularly when carbonized at lower temperatures. Massa et al. (2019) highlighted biochar's role in improving plant biomass and potassium content in tomato cultivation, though it did not significantly boost fruit yield or quality under nutrient-limited conditions. Nemati et al. (2015) evaluated various biochars in organic substrates and found that biochar increased pH and cation exchange capacity (CEC), reducing nutrient leaching by 11%, and showed promise as a replacement for perlite and partial peat moss substitution, although variability in biochar properties poses challenges for its consistent use. Banitalebi et al. (2019) evaluated agricultural waste-derived biochars mixed with various organic and inorganic substrates and found that biochar-based mixtures enhanced water retention and buffering capacity compared to peat-perlite, with some mixtures showing strong potential for plant growth. Vandecasteele et al. (2024) compared conventional mineral wool and peat-based media with peat-reduced and peat-free organic blends, with or without biochar amendments. They found that organic media with biochar showed improved carbon content, C : N ratio, and CEC, although biochar did not enhance carbon stability.

In past decade, biochar mixed with peat were well explored and used for planting ornamental plant seedings. Specifically, Steiner and Harttung (2014) demonstrated that biochar, when incorporated into peat mixtures, can support sunflower growth similarly to peat or perlite, with its favorable pH and structural stability, though optimization in feedstock selection and particle size is needed for tailored formulations. Altaf et al. (2021) tested biochar combined with peat moss and leaf compost in flower production, finding that it led to superior plant growth, including larger flowers and better transpiration rates. Lastly, Álvarez et al. (2017) demonstrated that biochar–vermicompost mixtures supported ornamental plant growth at levels comparable to commercial peat-based media.

It's widely acknowledged that substitution of peat-based growing media with biochar could promote the growth of many plant seedings, including vegetable and tree seedings. For instance, Méndez et al. (2017) showed that sewage sludge-derived biochar, when mixed with peat, significantly increased lettuce biomass and microbial activity, despite no improvement in hydrophysical properties. Similarly, Nieto et al. (2016) evaluated biochar from pruning waste, finding that at appropriate concentrations (50–75%), biochar improved air space volume and water-holding capacity, resulting in enhanced growth of lettuce, a key crop in soilless vegetable production. Moreover, Iii et al. (2018) explored the use of sugarcane bagasse biochar (SBB) in soilless media for green bean seedlings, finding that a mix of SBB and pine biochar enhanced seedling growth, with certain combinations proving most effective. Also, Vandecasteele et al. (2023) found that blending wood-based biochar with peat did not significantly affect nutrient dynamics in soilless strawberry cultivation but improved yields with reduced fertigation. Lastly, Köster et al. (2020) examined the impact of willow biochar on forest seedling growth, showing that biochar improved growth and nutrient concentrations, particularly for birch seedlings, without significantly reducing water retention.

At present, the field of crop breeding in which biochar replaced peat substrates can also be developed in the research direction of biochar types, pyrolysis temperature differences, and pyrolysis methods of biochar. For example, the study by Nobile et al. (2020) explored how the physical and chemical properties of biochar influence plant biomass production in greenhouse conditions. Six different biochars were incorporated into growing media for basil, lettuce, and pansy. The results showed that biochars with low density and high porosity promoted biomass production in basil and lettuce, while nutrient concentration had minimal impact on growth. Additionally, higher biochar pH negatively affected lettuce and pansy biomass, and increased electrical conductivity reduced basil biomass. These findings emphasize the importance of selecting biochar based on its properties for horticultural applications. Kaudal et al. (2018) showed that microwave-pyrolyzed biochar from biosolids improved plant growth and nutrient efficiency while reducing nutrient leaching.

4.2 Coir-based Growing Media Substituted by Biochar

Besides the peat media, coir was also a kind of popular growing media source. For example, Fornes and Belda (2018) evaluated three types of char—biochar from forest waste (FWB), biochar from olive mill waste (OMB), and hydrochar from forest waste (FWH)—as component substitution of coir growth media. Their study found that FWB exhibited optimal properties, such as balanced air/water ratios, low salinity, and minimal CO_2 emissions, which supported good plant growth at up to 25% inclusion. In contrast, OMB showed high salinity, while FWH led to excessive water retention, poor aeration, and increased CO_2 emissions, requiring further refinement before use. Banitalebi et al. (2021) also explored biochar-based mixtures by incorporating biochars derived from agricultural residues like sawdust, wheat straw, rice hulls, and palm bunches into organic and inorganic substrates. Their findings revealed that biochar–coir mixtures enhanced water retention, buffering capacities, and plant growth, with mixtures containing biochar from wheat straw and rice hulls showing promise as substitutes for traditional media like cocopeat-perlite. Similarly, Kim et al. (2017) incorporated rice hull-derived biochar into a coir-based medium for kale cultivation, demonstrating significant improvements in nutrient retention, water-holding capacity, and plant growth, with a 150% increase in shoot dry weight at 5% biochar inclusion. The leachate analysis confirmed reduced nutrient losses, further supporting biochar's role in enhancing media fertility. In addition, Martins et al. (2023) developed peat-free substrates by combining coir with municipal compost and acacia biochar, showing that mixes with 10% biochar and 12% compost supported lettuce seedling growth comparable to conventional substrates, while significantly reducing peat use and incorporating locally produced renewable resources. Finally, Fornes et al. (2017) evaluated biochar from forest and olive mill wastes in coir-based media for tomato cultivation, finding that FWB impaired nutrient retention, while hydrochar caused poor aeration and root anoxia. However, OMB proved to be more tolerant, especially for certain tomato cultivars.

Collectively, these studies highlighted the potential of biochar-based substrates as coir alternatives to traditional coir growth media, though challenges such as high salinity, poor aeration, and the need for further optimization remain. Future work should optimize the effects of biochar properties, which are affected by materials and pyrolysis conditions, on the physicochemical properties of coconut-based growing media.

4.3 Growing Media of Compost, Humic Acid, and Activated Carbon

To date, except the peat and coir, there are other carbon-based materials mixed with biochar for growing media production and utilization. Firstly, Di Lonardo et al. (2013) showed that biochar, in concentrations of 0.5–1.5 g/dm^3, could adsorb ethylene, prolonging its atmospheric reduction for 14–21 days, and enhancing shoot elongation, root biomass, and root number in *Populus alba* tissue culture. These benefits were independent of genotype, and biochar's efficacy was comparable to sole activated charcoal without any adverse effects. Similarly, Abdipour et al. (2019) highlighted the significant improvements in green basil's shoot and root biomass, nutrient uptake, and vegetative growth when applying cow manure biochar (3%) in combination with humic acid (7.5 g/L) in the culture medium, offering a sustainable strategy to optimize basil production quality and yield. Ferlito et al. (2020) assessed the use of conifer wood biochar as a substitute for black peat in citrus nurseries. The study found that mixtures containing 25% biochar (with black peat and compost) yielded the best seedling performance, although root growth was limited to shallow layers due to substrate compaction. Lastly, Radin et al. (2018) explored the integration of oil palm empty fruit bunch (EFB) biochar with compost in polybag media for oil palm seedling growth. Their results indicated that biochar and compost amendments improved the chemical properties of the growth medium, such as pH, C : N ratio, and CEC, while significantly promoting root growth and improving the shoot-to-root ratio, particularly at higher biochar and compost levels.

In general, this chapter concluded that biochar source growing media has a good effect in promoting plant growth and improving plant quality. Going forward, further research could explore optimizing the source, concentration, and combination of biochar with other organic amendments to meet the needs of specific crop seedlings and evaluate its long-term performance as a soilless cultivation substrate. The potential of biochar to reduce reliance on peat and synthetic fertilizers provides an exciting opportunity for sustainable agricultural practices.

References

Abdipour, M., Hosseinifarahi, M., and Najafian, S. (2019). Effects of humic acid and cow manure biochar (CMB) in culture medium on growth and mineral concentrations of basil plant. *International Journal of Horticultural Science and Technology* 6 (1): 27–38.

Altaf, K., Younis, A., Ramzan, Y. et al. (2021). Effect of composition of agricultural wastes and biochar as a growing media on the growth of potted Stock (*Matthiola incana*) and Geranium (*Pelargonium spp*). *Journal of Plant Nutrition* 44 (7): 919–930.

Álvarez, J.M., Pasian, C., Lal, R. et al. (2017). Vermicompost and biochar as substitutes of growing media in ornamental-plant production. *Journal of Applied Horticulture* 19 (3): 205–214.

Banitalebi, G., Mosaddeghi, M.R., and Shariatmadari, H. (2019). Feasibility of agricultural residues and their biochars for plant growing media: physical and hydraulic properties. *Waste Management* 87: 577–589.

Banitalebi, G., Mosaddeghi, M.R., and Shariatmadari, H. (2021). Evaluation of physico-chemical properties of biochar-based mixtures for soilless growth media. *Journal of Material Cycles and Waste Management* 23 (3): 950–964.

Blok, C., Van der Salm, C., Hofland-Zijlstra, J. et al. (2017). Biochar for horticultural rooting media improvement: evaluation of biochar from gasification and slow pyrolysis. *Agronomy* 7 (1): 6.

Chrysargyris, A., Prasad, M., Kavanagh, A. et al. (2019). Biochar type and ratio as a peat additive/partial peat replacement in growing media for cabbage seedling production. *Agronomy* 9 (11): 693.

Chrysargyris, A., Prasad, M., Kavanagh, A. et al. (2020). Biochar type, ratio, and nutrient levels in growing media affects seedling production and plant performance. *Agronomy* 10 (9): 1421.

Di Lonardo, S., Vaccari, F.P., Baronti, S. et al. (2013). Biochar successfully replaces activated charcoal for in vitro culture of two white poplar clones reducing ethylene concentration. *Plant Growth Regulation* 69: 43–50.

Dorais, M., Martinez, C., Diop, M. et al. (2016). Assessing the potential of biochar as a growing media component for potted plants. *Acta Horticulturae* 1137: 19–26.

Ferlito, F., Torrisi, B., Allegra, M. et al. (2020). Evaluation of conifer wood biochar as growing media component for citrus nursery. *Applied Sciences* 10 (5): 1618. https://www.mdpi.com/2076-3417/10/5/1618

Fornes, F. and Belda, R.M. (2018). Biochar versus hydrochar as growth media constituents for ornamental plant cultivation. *Scientia Agricola* 75: 304–312.

Fornes, F., Belda, R.M., Fernández de Córdova, P. et al. (2017). Assessment of biochar and hydrochar as minor to major constituents of growing media for containerized tomato production. *Journal of the Science of Food and Agriculture* 97 (11): 3675–3684.

Glaser, B. and Asomah, A.A.A. (2022). Plant growth and chemical properties of commercial biochar-versus peat-based growing media. *Horticulturae* 8 (4): 339.

Iii, C.L.W., White Jr, P.M., Gu, M. et al. (2018). Sugarcane and pine biochar as amendments for greenhouse growing media for the production of bean (*Phaseolus vulgaris* L.) seedlings. *Journal of Agricultural Science* 10 (4): 58.

Kaudal, B.B., Aponte, C., and Brodie, G. (2018). Biochar from biosolids microwaved-pyrolysis: characteristics and potential for use as growing media amendment. *Journal of Analytical and Applied Pyrolysis* 130: 181–189.

Kaudal, B.B., Chen, D., Madhavan, D.B. et al. (2016). An examination of physical and chemical properties of urban biochar for use as growing media substrate. *Biomass and Bioenergy* 84: 49–58.

Kim, H.S., Kim, K.R., Yang, J.E. et al. (2017). Amelioration of horticultural growing media properties through rice hull biochar incorporation. *Waste and Biomass Valorization* 8: 483–492.

Köster, E., Pumpanen, J., Palviainen, M. et al. (2020). Effect of biochar amendment on the properties of growing media and growth of containerized Norway spruce, Scots pine, and silver birch seedlings. *Canadian Journal of Forest Research* 51 (1): 31–40. https://doi.org/10.1139/cjfr-2019-0399.

Kuo, W.R., Yeh, N.C., Fujimoto, N. et al. (2019). Physicochemical properties and fertility index of culture media containing distillery residue biochar and their applications to plug seedling. *Journal of the Faculty of Agriculture Kyushu University* 64 (1):127–135.

Martins, T.C., Machado, R.M.A., Alves-Pereira, I. et al. (2023). Coir-based growing media with municipal compost and biochar and their impacts on growth and some quality parameters in lettuce seedlings. *Horticulturae* 9 (1): 105.

Massa, D., Bonetti, A., Cacini, S. et al. (2019). Soilless tomato grown under nutritional stress increases green biomass but not yield or quality in presence of biochar as growing medium. *Horticulture, Environment, and Biotechnology* 60: 871–881.

Méndez, A., Cárdenas-Aguiar, E., Paz-Ferreiro, J. et al. (2017). The effect of sewage sludge biochar on peat-based growing media. *Biological Agriculture & Horticulture* 33 (1): 40–51.

Méndez, A., Paz-Ferreiro, J., Gil, E. et al. (2015). The effect of paper sludge and biochar addition on brown peat and coir based growing media properties. *Scientia Horticulturae* 193: 225–230.

Nemati, M.R., Simard, F., Fortin, J.-P. et al. (2015). Potential use of biochar in growing media. *Vadose Zone Journal* 14 (6): vzj2014–2006.

Nieto, A., Gasco, G., Paz-Ferreiro, J. et al. (2016). The effect of pruning waste and biochar addition on brown peat based growing media properties. *Scientia Horticulturae* 199: 142–148.

Nobile, C., Denier, J., and Houben, D. (2020). Linking biochar properties to biomass of basil, lettuce and pansy cultivated in growing media. *Scientia Horticulturae* 261: 109001.

Prasad, M., Chrysargyris, A., McDaniel, N. et al. (2019). Plant nutrient availability and pH of biochars and their fractions, with the possible use as a component in a growing media. *Agronomy* 10 (1): 10.

Radin, R., Abu Bakar, R., Ishak, C.F. et al. (2018). Biochar-compost mixture as amendment for improvement of polybag-growing media and oil palm seedlings at main nursery stage. *International Journal of Recycling of Organic Waste in Agriculture* 7: 11–23.

Rathnayake, D., Creber, H., Van Poucke, R. et al. (2021). Biochar from sawmill residues: characterization and evaluation for its potential use in the horticultural growing media. *Biochar* 3: 201–212.

Rozas, A., Aponte, H., Maldonado, C. et al. (2023). Evaluation of compost and biochar as partial substitutes of peat in growing media and their influence in microbial counts, enzyme activity and *Lactuca sativa* L. seedling growth. *Horticulturae* 9 (2): 168.

Sarauer, J.L. and Coleman, M.D. (2018). Biochar as a growing media component for containerized production of Douglas-fir. *Canadian Journal of Forest Research* 48 (5): 581–588.

Steiner, C. and Harttung, T. (2014). Biochar as a growing media additive and peat substitute. *Solid Earth* 5 (2): 995–999.

Tian, Y., Sun, X., Li, S. et al. (2012). Biochar made from green waste as peat substitute in growth media for *Calathea rotundifola cv. Fasciata*. *Scientia Horticulturae* 143: 15–18.

Vandecasteele, B., Hofkens, M., De Zaeytijd, J. et al. (2023). Towards environmentally sustainable growing media for strawberry cultivation: effect of biochar and fertigation on circular use of nutrients. *Agricultural Water Management* 284: 108361. https://doi.org/10.1016/j.agwat.2023.108361.

Vandecasteele, B., Similon, L., Moelants, J. et al. (2024). End-of-life stage of renewable growing media with biochar versus spent peat or mineral wool. *Nutrient Cycling in Agroecosystems* 128 (3): 447–461.

5

Waste-derived Biochar as Slow-release Fertilizers

Xiaoqian Jiang, Yuqing Sun, and Jing Luo

School of Agriculture and Biotechnology, Sun Yat-sen University, Guangzhou, Guangdong, China

5.1 Introduction

Fertilizers play an indispensable role in enhancing soil fertility, promoting crop growth, and increasing agricultural yields. The scientific application of chemical fertilizers significantly improves the efficiency of agricultural production, making them a critical component of modern agriculture. By supplying 17 essential nutrients, including nitrogen, phosphorus, and potassium, fertilizers directly contribute to the healthy growth and high yield of crops. This, in turn, ensures food security and supports sustainable agricultural development.

The growth of the global population has led to an increasing demand for food production, resulting in a yearly rise in the usage of chemical fertilizers. Over the past few decades, the application of these fertilizers in agricultural practices has significantly boosted food production, accounting for ~50% of the additional crop yields (Chen et al. 2018c). However, conventional fertilizers, due to their high solubility, poor thermal stability, and small molecular weight, often result in substantial nutrient loss through surface runoff, denitrification, leaching, and volatilization. This leads to low fertilizer efficiency, declining soil fertility, and environmental pollution (Yang et al. 2018). Additionally, the excessive and unreasonable use of chemical fertilizers exacerbates soil quality degradation and poses severe environmental challenges (Ahmad et al. 2015). Due to various factors, the utilization rate of traditional fertilizers remains limited, with estimates suggesting that 40–70% of nitrogen fertilizer, 80–90% of phosphorus fertilizer, and 50–70% of potassium fertilizer are lost to the environment

Waste-derived Biochar for Sustainable Rural Revitalization, First Edition. Edited by Yuqing Sun and Daniel C.W. Tsang.
Published 2026 by John Wiley & Sons, Inc.

(Mir et al. 2018). Consequently, there is an urgent need to develop precision fertilizers with slow-release mechanisms.

In the past decade, extensive research has been conducted on various slow/controlled release fertilizers, which can enhance nutrient retention time and improve fertilizer efficiency by reducing nutrient solubility and controlling their release, thus providing appropriate nutrients for plant growth (Eghbali Babadi et al. 2015). Recently, the incorporation of biochar into slow-release fertilizers (SRFs) has emerged as a promising trend. Biochar effectively reduces nutrient leaching and enhances nutrient content (Shi et al. 2020). Given the escalating severity of global climate and environmental issues, the use of chemical fertilizers in traditional agriculture presents numerous challenges to soil and environmental health. This has spurred interest in finding safer and more efficient alternatives, such as developing new green fertilizers and utilizing SRFs. The application of biochar-based fertilizers has garnered significant global attention, making the improvement of fertilizer utilization and reduction of environmental pollution critical issues for the sustainable development of agriculture.

5.2 Research Progress of SRF

SRF is designed to release nutrients gradually, aligning nutrient availability with crop demand. This synchronization enhances the efficiency of chemical fertilizer utilization, reduces the required amount of fertilizer, and represents an essential strategy for achieving reductions in fertilizer usage while maintaining efficacy (Wei et al. 2018). Currently, there are various types of SRFs, each characterized by distinct mechanisms and advantages/disadvantages. However, only a few types are suitable for industrial production and practical field application (Zhao et al. 2024). Based on their release control principles, SRFs can be categorized into several main types: physically coated SRFs, chemically synthesized SRFs, stable fertilizers, and compound SRFs, including those incorporating biochar.

Physical barrier coated SRFs primarily involve coating urea particles with either natural or synthetic materials on their surface to decelerate the nitrogen release rate, thereby achieving a slow-release effect. These coating materials encompass both inorganic and organic varieties. Inorganic coating agents primarily consist of certain minerals, such as white wax, asphalt, clay, kaolin, and calcium magnesium phosphate. While these materials are cost-effective, they are prone to shell cracking, which compromises their slow-release performance (Mustafa et al. 2022). Organic coating materials comprise natural polymers and synthetic polymers. Among natural polymers, starch can be utilized to fabricate water-absorbing gel substances, while chitosan is employed to create coating layers that release carbon and nitrogen during degradation, thereby enhancing soil quality. However, chitosan coating alone does not significantly delay urea release and necessitates combination with magnesium sulfate for optimal effectiveness (Adlim et al. 2019). Additionally, the hydrophilic nature of chitosan poses a developmental limitation. Lignin has the capacity to adsorb and encapsulate nutrients; upon decomposition, it forms humus, although most lignin requires modification before application. Synthetic polymer materials constitute a diverse category within coating agents, with modified polyurethanes and aliphatic polyesters being prominent examples. Traditional polyurethane exhibits poor biodegradability, whereas aliphatic polyester, featuring hydrolyzable ester bonds, includes polylactic acid and polycaprolactone among its variants. Nonetheless, the production costs associated with aliphatic polyesters are relatively high, rendering them more suitable for application in high-yield and high-quality crop cultivation (Zhao et al. 2024).

Chemically synthesized SRFs are nitrogen-containing polymers formed through a series of chemical reactions using urea as the primary raw material and aldehydes as additives. These fertilizers exhibit slightly soluble properties in water, allowing them to remain in the soil for extended periods without being leached away by soil moisture (Li et al. 2022). The nutrient release mechanism involves the gradual decomposition of the polymer structure by soil microorganisms and enzymes, eventually breaking down into simple nitrogen ions that are slowly absorbed and utilized by plants. SRFs primarily consist of inorganic nitrides such as magnesium ammonium phosphate and organic nitrogen compounds like urea-formaldehyde and oxamide fertilizers. Inorganic nitrides have very low solubility in water, enabling them to store nitrogen in the soil after application. This stored nitrogen gradually undergoes nitrification and decomposes slowly under microbial action. On the other hand, organic nitrogen compound fertilizers, such as urea-formaldehyde, possess high nitrogen content but have an excessively long release period, often spanning 2–3 years. This prolonged release period can hinder plant absorption and limit their practical application (Smolander et al. 2022).

Stable fertilizers are long-acting, SRFs designed to extend nutrient availability over time. They achieve this by incorporating urease inhibitors or nitrification inhibitors during the production process. Urease inhibitors delay the hydrolysis of urea to ammonium nitrogen (NH_4^+—N) by inhibiting soil urease activity, thereby prolonging the efficacy of urea-based fertilizers. Similarly, nitrification inhibitors suppress the conversion of NH_4^+—N to nitrate nitrogen (NO_3^-—N) by inhibiting the activity of nitrifying and nitrosobacteria, which extends the effectiveness of NH_4^+—N fertilizers and reduces the formation, leaching, and subsequent denitrification loss of NO_3^-—N (Zhang et al. 2021a). The inclusion of these inhibitors significantly enhances crop yields and nitrogen use efficiency while markedly reducing environmental pollution. However, the effectiveness of stable fertilizers is influenced by several factors, including the type and dosage of inhibitors, the physical and chemical properties of the fertilizer itself, soil and environmental conditions, as well as field management practices. Consequently, stable fertilizers exhibit varying degrees of efficacy on different soils and crops across diverse regions (Zhang et al. 2021a).

Currently, each variety of SRF faces specific challenges that necessitate resolution. One approach involves combining different types of fertilizers to develop composite slow-release formulations with enhanced performance characteristics, including improved nutrient release profiles and utilization rates. For instance, incorporating inhibitors into coated fertilizers can mitigate the rapid decomposition issue associated with urea post-application, thereby stabilizing nutrient availability and reducing environmental impact. Recent investigations have focused on the integration of exogenous materials into SRFs to further optimize their functionality. By leveraging biochar's adsorption properties, nutrient release from fertilizers can be slowed. Biochar degradation provides nutrients to crops and enhances soil fertility through increased soil organic matter (SOM) (Waqas et al. 2021). A notable example is the incorporation of biochar, which has garnered significant interest due to its unique properties (Wang et al. 2022). Biochar, a carbonaceous material produced via the thermochemical conversion of agricultural residues (e.g., crop residues, rice hulls, peanut shells, and sugarcane bagasse) under oxygen-depleted and high-temperature conditions (Du et al. 2024), serves as an effective slow-release medium. Its adoption stems from advantageous attributes such as a large surface area per unit mass, low bulk density, heightened stability, and benefits related to environmental sustainability, cost-effectiveness, and ease of application (Tang et al. 2025). The presence of both organic constituents (like fulvic acid and humic substances) and inorganic nutrients (including salts of nitrogen, phosphorus, and potassium) within biochar renders it a dual-purpose agent. These components can directly nourish plants and microorganisms while also contributing to soil health enhancement through mechanisms like increased SOM content (Ding et al. 2016), thus highlighting biochar's multifaceted role in sustainable agriculture practices.

Biochar, functioning as a SRF, enhances soil fertility and carbon sequestration capacity. Its high specific surface area, porosity, and presence of various functional groups make it an excellent substrate for producing slow-release nutrient fertilizers (Dong et al. 2021). Application of biochar in soil has been shown to promote nutrient supply and crop yield, improve soil texture and properties, reduce greenhouse gas emissions, immobilize heavy metals and organic pollutants, and minimize nutrient loss (Dong et al. 2021; Karthik et al. 2021). Co-composting, co-pyrolysis, or co-application of biochar with synthetic fertilizers can facilitate biomass utilization and contribute to environmental protection (Karthik et al. 2021). Biochar is derived from organic wastes generated in daily life, including agricultural and forestry residues, domestic garbage, and industrial wastes. It can be produced through pyrolysis, hydrothermal carbonization, gasification, and roasting. Due to its porosity and adsorption capacity, biochar is effectively used as a SRF carrier, promoting the growth of rhizosphere bacteria and thereby enhancing crop growth (Ajeng et al. 2020). Compared to traditional fertilizers, biochar-based SRFs exhibit lower nutrient release rates, extend the efficacy of plant uptake, and reduce environmental nutrient loss (Wang et al. 2022).

The incorporation of biochar, derived from agricultural and forestry residues, into soil has been shown to effectively mitigate soil acidity and enhance the plant growth milieu. This practice also augments soil fertility, enriches the carbon reservoir, and bolsters nutrient retention capacities. Nonetheless, pristine biochar exhibits limited nutrient content; thus, the enrichment of biochar with essential nutrients emerges as a pivotal strategy for augmenting its fertilizing potential. Consequently, employing biochar as a matrix for nutrient delivery, in conjunction with conventional fertilizers, facilitates the formulation of biochar-based composite fertilizers. In comparison to traditional fertilization methods, these biochar-fortified fertilizers demonstrate a reduced rate of nutrient release. This characteristic fosters heightened efficiency in plant nutrient assimilation, minimizes non-target environmental nutrient dispersion, and mitigates ecological contamination associated with agricultural practices. Ultimately, this approach harnesses the dual benefits of sustainable waste management and enhanced crop productivity through the judicious utilization of biomass resources.

5.3 The Preparation Technologies of Biochar-based SRFs

The raw biochar obtained through direct thermochemical treatment of biomass feedstock generally lacks essential nutrients required for effective fertilization. Therefore, it is imperative to modify biochar to enhance its fertilizer efficiency. The inherent properties of biochar provide a solid foundation for the synthesis of SRFs based on this material. Numerous methods have been developed to produce effective biochar-based SRFs, including impregnation (An et al. 2020b), encapsulation (Sim et al. 2021), granulation (Zheng et al. 2017), and co-pyrolysis (An et al. 2020a).

5.3.1 *In Situ* Pyrolysis

In certain instances, biomass raw materials are rich in nutrients and can be directly converted into biochar-based fertilizers. Biochar serves as a SRF by providing nutrients directly to the soil (Ding et al. 2016). The depletion of non-renewable resources such as phosphate rock and silicate minerals necessitates the recovery of nutrients from waste. Consequently, nutrient-rich biomass waste has significant potential for the production of biochar-based fertilizers (Wang et al. 2022). Biochar is typically produced through pyrolysis. Nutrient-rich raw materials, including manure, poultry litter, sewage sludge, bone meal, and biologically fermented materials, can be directly pyrolyzed to produce biochar-based SRFs (Jiao et al. 2022; Luyima et al. 2020; Kim et al. 2018). Studies have demonstrated the preparation of potassium-rich biochar from banana peduncles, mushroom compost, and seaweed (Wang et al. 2022). Co-pyrolysis or plasma treatment of potassium-rich banana peduncles and phosphorus-rich waste sludge has been utilized for biochar production. Compared to plasma treatment, pyrolysis yields biochar with higher overall carbon content, while plasma treatment results in higher nutrient content (Karim et al. 2019). The properties of biochar, including pore structure, surface area, functional groups, and Zeta potential, are influenced by factors such as the type of biomass feedstock, heating rate, residence time, and pyrolysis temperature (Sun et al. 2014). For wood-derived biochar, 650 °C is the effective temperature for forming the biochar structure, while 750 °C is critical for developing a porous structure. Additionally, extending the pyrolysis time beyond 60 min can increase pore volume without affecting the yield (He et al. 2021).

5.3.2 Co-pyrolysis

The co-pyrolysis of biomass with phosphorus sources presents an alternative methodology for producing biochar-based SRFs. This process immobilizes phosphorus within the biochar matrix through mechanisms of adsorption and chemical bonding. Biochar-based SRFs synthesized via co-pyrolysis typically exhibit significant advantages, including ease of synthesis, high phosphorus loading capacity, and superior slow-release characteristics (An et al. 2021). For instance, Zhao et al. (2014a) employed various phosphorus sources—H_3PO_4, phosphate rock, and superphosphate—in the co-pyrolysis of wheat straw. Their findings revealed a reduction in carbon loss, decreased microbial mineralization, and enhanced biochar stability during pyrolysis. Similarly, other researchers have utilized the co-pyrolysis of biomass (such as pine sawdust or switchgrass) with fertilizers (like superphosphate or bone meal) to produce SRFs, thereby enabling gradual phosphorus release (Zhao et al. 2016). Additionally, co-pyrolysis technology has been applied to the mixture of sludge and walnut shells to create biochar aimed at removing pollutants such as phosphate and ammonia nitrogen from water (Mohamed et al. 2023). Lignin, selected as the primary component of biomass, was co-pyrolyzed with sewage sludge, which facilitated the decomposition of the mixture, lowered the reaction temperature, and augmented the biochar's adsorption capacity (Dai et al. 2022). Overall, co-pyrolysis of biomass with diverse materials is a promising technique for enhancing both the yield and quality of biochar, making it suitable for producing high-quality biochar with versatile applications (Dafalla et al. 2024).

5.3.3 Impregnation

The nutrient content of biochar can be effectively enhanced through soaking in nutrient solutions due to its inherently low initial nutrient levels. Its porous structure and extensive surface area facilitate the storage of nutrients, thereby improving plant nutrient utilization. The impregnation method, characterized by its simplicity and cost-effectiveness, has been utilized to prepare SRFs. This method involves the impregnation of biochar with ammonium salts (An et al. 2021), phosphates (Xu et al. 2014), and nitrates (Hale et al. 2013), which helps reduce nutrient loss and accelerates crop growth. For instance, oil palm kernel shell biochar, when impregnated with NH_4NO_3 and KH_2PO_4, exhibits high nutrient retention (Dominguez et al. 2020). Studies have demonstrated that mixing biochar with nutrient-rich salt solutions (containing P, K, Ca, and N) for 3–4 hours allows for nutrient adsorption, followed by filtration and drying. Infrared spectroscopy reveals surface groups on the biochar that bind with nutrients. Thermogravimetric analysis (TGA) confirms the stability of the biochar, showing significant improvements in salt index, water retention, and nutrient release after impregnation (Khan et al. 2021). Additionally, corncob biochar impregnated with macronutrients (N, Ca, P, K, and Mg) and micronutrients (Na, Zn, and Fe) has shown enhanced water retention and prolonged nutrient release (Lateef et al. 2019). Biochar can also function as a soil conditioner or fertilizer by concentrating and adsorbing nutrients from various wastewater sources, such as urine, anaerobic digestion effluent, landfill leachate, and livestock wastewater (Takaya et al. 2016). However, the original biochar's limited nutrient adsorption capacity necessitates modifications. To address this, materials such as MgO, K_2O, bentonite, and coal gangue have been developed to enhance biochar's adsorption capacity (Maghsoodi et al. 2020).

5.3.4 Encapsulation

Encapsulation is a widely used method for the production of slow-release biochar fertilizers. This technique involves coating inorganic or organic substrates with fertilizers to reduce nutrient loss and improve nutrient use efficiency. Encapsulation serves as an important method for producing SRFs by using coating materials as diffusion barriers to control the release of nutrients. Examples include biochar-biodegradable polymer-coated urea, biochar-aqueous copolymer-coated urea, and biochar-coated urea (Wang et al. 2022). The biochar is co-composted with raw materials such as poultry manure and green waste to form an organic coating rich in nitrogen and potassium nutrients on the surface of the biochar and within its inner pores. A SRF was synthesized from rapeseed meal biochar and sodium alginate through a graft copolymerization process, significantly improving the water retention, nitrogen content, and slow-release performance of the graft copolymer. Additionally, a granular biochar SRF prepared with modified corn starch adhesive as a coating material reduces nitrogen leaching loss and runoff at the sowing and tillering stages, providing more nutrients for rice at the heading and maturity stages (Dong et al. 2020). The fertilizer efficiency of biochar-coated fertilizers varies and is influenced by the properties of the biochar, the choice of coating materials, and the coating process. Common coating materials include organic polymers and inorganic minerals. Biochar SRF coatings often utilize natural biodegradable materials, including waste palm oil, starch, chitosan, and polyvinyl alcohol mixtures.

5.3.5 Granulation

Granulation is a widely employed technique in fertilizer production. Biochar, intended for soil application, presents notable challenges in storage and transportation due to its brittle nature, low density, and minute particle size. These characteristics contribute to significant biochar loss through wind erosion or surface runoff during heavy rainfall events. Additionally, fine biochar particles pose health risks, causing irritation to human eyes, skin, and respiratory systems (Chen et al. 2018a). In light of these concerns, granulation emerges as an effective strategy to mitigate biochar losses and reduce the costs associated with soil application treatments. It has been successfully implemented on a large scale in agricultural practices. The granulation of biochar and nutrient mixtures enhances the efficiency of nutrient delivery by increasing diffusion resistance. Biochar-based SRFs may consist of either of carbonized, nutrient-enriched, pelletized biomass or pelletized biochar blended with fertilizers. If the initial biomass lacks sufficient nutrients, the pelletized biochar can be further impregnated with additional nutrients. For instance, poultry litter and switchgrass have been combined, pelletized, and subsequently pyrolyzed at varying temperatures to produce biochar pellets (Lu et al. 2014). Alternatively, the granulation process involves thorough mixing of uniformly sized biochar and fertilizer granules, followed by disc granulation or mechanical extrusion. This procedure commonly incorporates a binder; for example, switchgrass-derived biochar mixed with liquid fertilizer and lignin can be granulated using an appropriate granulator (Kim et al. 2014).

5.3.6 The Integrated Approach

Innovative methods have been developed for the production of biochar-based SRFs. For instance, biochar derived from rice husk can be initially mixed with phosphoric acid. This mixture is then reacted with ammonia gas, leading to the formation of ammonium phosphate within the pores of the biochar. Notably, the release rate of effective nitrogen (NH_4^+—N and NO_3^-—N) from this granular fertilizer is slower compared to that of pure ammonium phosphate. Pot experiments have demonstrated that the incorporation of granular biochar-ammonium phosphate enhances nitrogen content and promotes plant growth (El Sharkawi et al. 2018). This approach offers a novel method for creating non-coated SRFs, contributing to more sustainable agricultural practices.

5.4 Mechanism of Slow Release of Biochar

Biochar-based SRFs can gradually release nutrients, thereby providing a prolonged supply of essential elements. This mechanism helps to avoid leaching losses, enhance fertilizer efficiency, and reduce environmental pollution (Wang et al. 2024). The primary factors influencing the slow-release performance of carbon-based fertilizers include the specific surface area, porosity, and surface charge of biochar. These characteristics are predominantly influenced by the type of biomass raw materials used. A larger surface area, higher porosity, favorable surface charge, and an abundance of surface functional groups collectively contribute to a reduced nutrient release rate and improved nutrient utilization efficiency (Jia et al. 2021).

Biochar adsorption is closely related to both the substances being adsorbed (fertilizers, pollutants, and pesticides) and the properties of biochar. Possible adsorption mechanisms include ion exchange, surface complexation, electrostatic attraction, physical adsorption, and surface precipitation (Tan et al. 2015). For organic substances, these mechanisms involve hydrogen bonding, pore filling, electrostatic interactions, hydrophobic interactions, and surface complexation. Most adsorption mechanisms depend on the characteristics of biochar, such as surface area, porous structure, surface groups, and mineral composition. The adsorption effect enhances the nutrient retention on biochar and facilitates the preparation of biochar-based SRFs.

Biochar is an excellent carrier for nitrogen, phosphorus, potassium, and other nutrients due to its porous structure, which provides a large loading capacity and strong adsorption capabilities for nitrate, ammonium, phosphate, and other elements. Studies have shown that biochar surfaces prepared under CO_2 or $H_2O(g)$ contain more K adsorption groups, enhancing their ability to adsorb N and P. The recovery of phosphate from water by biochar is governed by precipitation and adsorption mechanisms on Mg particles (Ye et al. 2019). For instance, the adsorption of phosphorus by MgO-modified magnetic biochar primarily occurs through precipitation, electrostatic attraction, ligand exchange, and complexation with surface groups and metals (Li et al. 2016). Similarly, the adsorption of P on cow-dung biochar mainly involves surface precipitation (Chen et al. 2018b).

Research on the nutrient adsorption of biochar/montmorillonite composites has revealed that NH_4^+ adsorption is predominantly achieved via surface adsorption and ion exchange, whereas PO_4^{3-} adsorption is driven by electrostatic attraction or ionic bonding between PO_4^{3-} and metal ions (Chen et al. 2017). Additionally, phosphate recovery from synthetic urine using a goethite/biochar composite indicated that phosphate adsorption occurs through Fe—OH bonds and the formation of internal sphere complexes (Fe—O—P) (Zhang et al. 2021b). Overall, these findings highlight the potential of biochar and its composites in nutrient recovery and environmental remediation.

5.5 The Specific Applications of Biochar SRF

Biochar is an excellent carrier of nitrogen, phosphorus, potassium, and other nutrients due to its strong loading capacity through its porous structure and its robust adsorption capabilities for nitrate, ammonium, and phosphate. Inorganic fertilizers such as urea provide nitrogen, potassium dihydrogen phosphate supplies phosphorus, and potassium chloride offers potash. Alternatively, two or more nutrient elements can be compounded to prepare a biochar-based fertilizer (Phares et al. 2022). The combination of these inorganic fertilizers ensures that the biochar is enriched with various specific nutrient elements, which are then slowly released into the soil. This gradual release improves the utilization efficiency of the fertilizer. Zhao et al. (2016) utilized two commonly used phosphate fertilizers—superphosphate and bone meal—to prepare a biochar-based fertilizer. They determined the release kinetics of phosphorus in the biochar fertilizer using a leaching test. The release rate constants were found to be 0.0012–0.0024 and 0.89–0.91, respectively, which are significantly lower than the rates of 0.012 and 1.79 observed for the fertilizer itself. This indicates that biochar has potential as a SRF. According to Yang et al. (2014), biochar can produce electrostatic interactions, complexation, and mineralization with N, P, K, and other nutrients in fertilizers, thereby effectively retaining these nutrients. When a biochar-based fertilizer is applied to soil, N, P, K, and other nutrients can be retained for extended periods and gradually released in a stable manner, allowing crops to fully absorb and utilize them (Sim et al. 2021). Studies have demonstrated that the coupled application of biochar and nitrogen fertilizer can significantly promote the growth and yield of maize under fertility stress conditions. This suggests that the combined use of biochar and nitrogen fertilizer synergistically improves soil fertility and the plant growth environment (Li et al. 2024). Zhou et al. (2024) screened appropriate application amounts of nitrogen fertilizer and biochar, finding significant effects on tomato plant height, stem diameter, and leaf area. These findings indicate that the coupled application plays a crucial role in improving tomato yield and quality. Some studies have explored the effects of combining biochar with potassium fertilizer versus their individual applications on sweet potato performance. Results showed that the combined application had a greater impact on sweet potato growth and yield, optimizing soil physical and chemical properties (Adekiya et al. 2022).

Biochar-based organic SRFs primarily consist of biochar compounded with various organic fertilizers, such as decomposed manure and biogas slurry. Previous studies have demonstrated that organic fertilizers serve as natural soil conditioners and provide essential nutrients for plants, including nitrogen, potassium, phosphorus, calcium, magnesium, and sodium, thereby significantly enhancing crop yields (Chen et al. 2022). However, challenges associated with organic fertilizers include rapid nutrient release, low nutrient utilization efficiency, and potential environmental risks due to insufficiently decomposed materials (Bashir et al. 2020). Biochar offers solutions by immobilizing pollutants, slowing nutrient release, and improving soil properties, thus complementing the benefits of organic fertilizers (Pandit et al. 2020). In a study conducted by Pandit et al. (2020), pot experiments revealed that the application of biochar-based organic fertilizer increased crop mass more than twofold compared to direct applications of nitrogen, phosphorus, and potassium fertilizers. Additionally, the soil content of magnesium (Mg^{2+}), calcium (Ca^{2+}), and potassium (K^{+}) ions showed significant increases. Available nitrogen levels rose from 5.1 to 8.7 mg/kg, and available phosphorus levels increased from 34.8 to 105.1 mg/kg. This synergistic effect of combining biochar and organic fertilizer not only enriches the nutrient profile but also enhances the performance of organic fertilizers, leading to improved soil properties, promoted crop growth, and increased nutrient use efficiency.

Biochar is inherently rich in carbon and other essential nutrients, thereby increasing the content of soil organic carbon and various nutrients. This enhancement improves the soil's organic matter content, boosts soil enzyme activity, and enhances the adsorption of nitrogen, phosphorus, potassium, and other elements. Consequently, nutrient leaching is reduced, and the utilization rate of fertilizer nutrients is improved (Rehrah et al. 2016). Additionally, biochar possesses an abundant pore structure, a large specific surface area, and strong adsorption capacity. When applied to soil, it adjusts soil pH and the soil's capacity to adsorb and release cations (Zhao et al. 2014b). These properties collectively improve soil physical and chemical properties, enhance soil fertility, and increase crop yields. Biochar's pore characteristics also enhance water retention, stabilize soil aggregates, and improve water-holding capacity (Kavitha et al. 2018). Furthermore, the inorganic minerals and organic matter in biochar provide essential nutrients for plant growth. Biochar also influences other soil properties such as tensile strength, expansion/contraction, and cracking density (Zong et al. 2014). A key feature of biochar is its stability, characterized by a very low decomposition rate in soil.

5.6 Summary and Outlook

SRFs possess several advantages, including high utilization rates, low consumption, minimal environmental pollution, and convenient application. The research directions in this field are extensive. A key area of focus is the development of biochar-based SRFs, which offer excellent slow-release performance, affordability, degradability, and environmental friendliness. By integrating biochar with other types of SRFs, we can mitigate the limitations inherent in any single type of product. Tailoring biochar-based SRFs to specific ecological environments and soil characteristics enhances their efficacy across diverse agricultural settings. Future research should aim to diversify, practicalize, and industrialize these advanced fertilizer formulations.

Biochar, characterized by its diverse pore structure, large specific surface area, abundant oxygen-containing functional groups, strong adsorption capacity, high biodegradation resistance, and robust antioxidant properties, has emerged as a versatile material with applications in water purification, soil improvement, and clean energy production. The sustainability of agricultural practices necessitates the effective management of crop residues. Utilizing crop waste as a raw material for biochar production presents an opportunity due to its abundance and cost-effectiveness. Consequently, biochar derived from crop waste holds significant potential for broad application across multiple fields.

Biochar SRF holds significant potential for agricultural applications. However, it encounters several challenges, including low nutrient content, difficulty in controlling the release rate, and an unstable market supply. Addressing these issues requires precise regulation of the nutrient release rate to align with the crop growth cycle, reduction of production costs, enhancement of market competitiveness, and rigorous long-term effects, and safety evaluations to ensure the sustainability and safety of biochar SRFs. Solutions to these problems are essential. As research advances and technology progresses, it is anticipated that biochar SRF will become increasingly influential in agricultural production.

References

Adekiya, A.O., Adebiyi, O.V., Ibaba, A.L. et al. (2022). Effects of wood biochar and potassium fertilizer on soil properties, growth and yield of sweet potato. *Heliyon* 8 (11): e11728.

Adlim, M., Zarlaida, F., Rahmayani, R.F.I. et al. (2019). Nutrient release properties of a urea-magnesium-natural rubber composite coated with chitosan. *Environmental Technology & Innovation* 16: 100442.

Ahmad, M.T., Asghar, H.N., Saleem, M. et al. (2015). Synergistic effect of rhizobia and biochar on growth and physiology of maize. *Agronomy Journal* 107 (6): 2327–2334.

Ajeng, A., Abdullah, R., Ling, T.C. et al. (2020). Bioformulation of biochar as a potential inoculant carrier for sustainable agriculture. *Environmental Technology & Innovation* 20: 101168.

An, X., Qu, Z., Liu, X. et al. (2021). A new class of biochar-based slow-release phosphorus fertilizers with high water retention based on integrated co-pyrolysis and co-polymerization. *Chemosphere* 285: 131481.

An, X., Wu, Z., Yu, J. et al. (2020a). Copyrolysis of biomass, bentonite, and nutrients as a new strategy for the synthesis of improved biochar-based slow-release fertilizers. *ACS Sustainable Chemistry & Engineering* 8 (8): 3181–3190.

An, X., Wu, Z., Yu, J. et al. (2020b). High-efficiency reclaiming phosphate from an aqueous solution by bentonite modified biochars: a slow release fertilizer with a precise rate regulation. *ACS Sustainable Chemistry & Engineering* 8 (15): 6090–6099.

Bashir, A., Rizwan, M., ur Rehman, M.Z. et al. (2020). Application of co-composted farm manure and biochar increased the wheat growth and decreased cadmium accumulation in plants under different water regimes. *Chemosphere* 246: 125809.

Chen, L., Chen, Q., Rao, P. et al. (2018a). Formulating and optimizing a novel biochar-based fertilizer for simultaneous slow-release of nitrogen and immobilization of cadmium. *Sustainability* 10 (8): 2740.

Chen, L., Chen, X., Zhou, C. et al. (2017). Environmental-friendly montmorillonite-biochar composites: facile production and tunable adsorption-release of ammonium and phosphate. *Journal of Cleaner Production* 156: 648–659.

Chen, L., Li, X., Peng, Y. et al. (2022). Co-application of biochar and organic fertilizer promotes the yield and quality of red pitaya (*Hylocereus polyrhizus*) by improving soil properties. *Chemosphere* 294: 133619.

Chen, Q., Qin, J., Sun, P. et al. (2018b). Cow dung-derived engineered biochar for reclaiming phosphate from aqueous solution and its validation as slow-release fertilizer in soil-crop system. *Journal of Cleaner Production* 172: 2009–2018.

Chen, S., Yang, M., Ba, C. et al. (2018c). Preparation and characterization of slow-release fertilizer encapsulated by biochar-based waterborne copolymers. *Science of the Total Environment* 615: 431–437

Dafalla, M., Inayat, A., Jamil, F. et al. (2024). Prospective of biochar material production and process optimization using co-pyrolysis approach-A mini-review. *Journal of Physics: Conference Series* 2751 (1): 012024.

Dai, Q., Xiang, W., Liu, Q. et al. (2022). Co-pyrolysis biochar derived from sewage sludge and lignin: synergetic effect and adsorption properties. *Journal of Environmental Chemical Engineering* 10 (3): 107898.

Ding, Y., Liu, Y., Liu, S. et al. (2016). Biochar to improve soil fertility. A review. *Agronomy for Sustainable Development* 36 (2): 36.

Dominguez, E.L., Uttran, A., Loh, S.K. et al. (2020). Characterisation of industrially produced oil palm kernel shell biochar and its potential as slow release nitrogen-phosphate fertilizer and carbon sink. *Materials Today: Proceedings* 31: 221–227.

Dong, D., Li, J., Ying, S. et al. (2021). Mitigation of methane emission in a rice paddy field amended with biochar-based slow-release fertilizer. *Science of the Total Environment* 792: 148460.

Dong, D., Wang, C., Van Zwieten, L. et al. (2020). An effective biochar-based slow-release fertilizer for reducing nitrogen loss in paddy fields. *Journal of Soils and Sediments* 20 (8): 3027–3040.

Du, H., Jiang, J., and Qi, Y. (2024). Research progress on application effects of agricultural waste biochar. *Journal of Cold-Arid Agricultural Sciences* 3 (03): 203–207.

Eghbali Babadi, F., Yunus, R., Rashid, S.A. et al. (2015). New coating formulation for the slow release of urea using a mixture of gypsum and dolomitic limestone. *Particuology* 23: 62–67.

El Sharkawi, H.M., Tojo, S., Chosa, T. et al. (2018). Biochar-ammonium phosphate as an uncoated-slow release fertilizer in sandy soil. *Biomass & Bioenergy* 117: 154–160.

Hale, S.E., Alling V., Martinsen V. et al. (2013). The sorption and desorption of phosphate-P, ammonium-N and nitrate-N in cacao shell and corn cob biochars. *Chemosphere* 91 (11): 1612–1619.

He, M., Xu, Z., Sun, Y. et al. (2021). Critical impacts of pyrolysis conditions and activation methods on application-oriented production of wood waste-derived biochar. *Bioresource Technology* 341: 125811.

Jia, Y., Hu, Z., Ba, Y. et al. (2021). Application of biochar-coated urea controlled loss of fertilizer nitrogen and increased nitrogen use efficiency. *Chemical and Biological Technologies in Agriculture* 8 (1): 3.

Jiao, M., Liu, T., Chen, X. et al. (2022). Biochar for clean composting and organic fertilizer production. In: *Biochar in Agriculture for Achieving Sustainable Development Goals*, 105–116. Academic Press.

Karim, A., Kumar, M., Mohapatra, S. et al. (2019). Nutrient rich rich biomass and effluent sludge wastes co-utilization for production of biochar fertilizer through different thermal treatments. *Journal of Cleaner Production* 228: 570–579.

Karthik, V., Kumar, P.S., Vo, D.N. et al. (2021). Hydrothermal production of algal biochar for environmental and fertilizer applications: a review. *Environmental Chemistry Letters* 19 (2): 1025–1042.

Kavitha, B., Reddy, P.V.L., Kim, B. et al. (2018). Benefits and limitations of biochar amendment in agricultural soils: a review. *Journal of Environmental Management* 227: 146–154.

Khan, H.A., Naqvi, S.R., Mehran, M.T. et al. (2021). A performance evaluation study of nano-biochar as a potential slow-release nano-fertilizer from wheat straw residue for sustainable agriculture. *Chemosphere* 285: 131382.

Kim, J.A., Vijayaraghavan K., Reddy, D.H. et al. (2018). A phosphorus-enriched biochar fertilizer from bio-fermentation waste: a potential alternative source for phosphorus fertilizers. *Journal of Cleaner Production* 196: 163–171.

Kim, P., Hensley, D., and Labbé, N. (2014). Nutrient release from switchgrass-derived biochar pellets embedded with fertilizers. *Geoderma* 232–234: 341–351.

Lateef, A., Nazir, R., Jamil, N. et al. (2019). Synthesis and characterization of environmental friendly corncob biochar based nano-composite – a potential slow release nano-fertilizer for sustainable agriculture. *Environmental Nanotechnology, Monitoring & Management* 11: 100212.

Li, F. (2024). The influence of the synergistic application of biochar and nitrogen fertilizer on corn growth and yield under conditions of nutrient stress. *Agricultural Development & Equipments* 07: 151–153.

Li, J., Liu, Y., Liu, J. et al. (2022). A novel synthetic slow release fertilizer with low energy production for efficient nutrient management. *Science of the Total Environment* 831: 154844.

Li, R., Wang, J.J., Zhou, B. et al. (2016). Recovery of phosphate from aqueous solution by magnesium oxide decorated magnetic biochar and its potential as phosphate-based fertilizer substitute. *Bioresource Technology* 215: 209–214.

Lu, X., Li, Z., Zhang, Y. et al. (2014). Characterization of human manure-derived biochar and energy-balance analysis of slow pyrolysis process. *Waste Management* 34 (9): 1619–1626.

Luyima, D., Lee, J.H., Sung, J. et al. (2020). Co-pyrolysed animal manure and bone meal-based urea hydrogen peroxide (UHP) fertilisers are an effective technique of combating ammonia emissions. *Journal of Material Cycles and Waste Management* 22 (6): 1887–1898.

Maghsoodi, M.R., Najafi, N., Reyhanitabar, A. et al. (2020). Hydroxyapatite nanorods, hydrochar, biochar, and zeolite for controlled-release urea fertilizers. *Geoderma* 379: 114644.

Mir, S.A., Shah, M.A., Mir, M.M. et al. (2018). New horizons of nanotechnology in agriculture and food processing industry. In: *Food Science and Nutrition: Breakthroughs in Research and Practice* 196–217. IGI Global.

Mohamed, B.A., Ruan, R., Bilal, M. et al. (2023). Co-pyrolysis of sewage sludge and biomass for stabilizing heavy metals and reducing biochar toxicity: a review. *Environmental Chemistry Letters* 21 (2): 1231–1250.

Mustafa, A., Athar, F., Khan, I. et al. (2022). Improving crop productivity and nitrogen use efficiency using sulfur and zinc-coated urea: a review. *Frontiers in Plant Science* 13: 942384.

Pandit, N.R., Schmidt, H.P., Mulder, J. et al. (2020). Nutrient effect of various composting methods with and without biochar on soil fertility and maize growth. *Archives of Agronomy and Soil Science* 66 (2): 250–265.

Phares, C.A. and Akaba, S. (2022). Co-application of compost or inorganic NPK fertilizer with biochar influences soil quality, grain yield and net income of rice. *Journal of Integrative Agriculture* 21 (12): 3600–3610.

Rehrah D., Bansode R.R., Hassan O. et al. (2016). Physico-chemical characterization of biochars from solid municipal waste for use in soil amendment. *Journal of Analytical and Applied Pyrolysis* 118: 42–53.

Shi, W., Ju, Y., Bian, R. et al. (2020). Biochar bound urea boosts plant growth and reduces nitrogen leaching. *Science of the Total Environment* 701: 134424.

Sim, D.H.H., Tan, I.A.W., Lim, L.L.P. et al. (2021). Encapsulated biochar-based sustained release fertilizer for precision agriculture: a review. *Journal of Cleaner Production* 303: 127018.

Smolander, A., Martikainen, P.J., and Henttonen, H.M. (2022). Half-a-century effects of a slow-release nitrogen fertilizer, ureaformaldehyde, on stand growth and soil processes in a Scots pine stand. *Forest Ecology and Management* 519: 120320.

Sun, Y., Gao, B., Yao, Y. et al. (2014). Effects of feedstock type, production method, and pyrolysis temperature on biochar and hydrochar properties. *Chemical Engineering Journal* 240: 574–578.

Takaya, C.A., Fletcher, L.A., Singh, S. et al. (2016). Phosphate and ammonium sorption capacity of biochar and hydrochar from different wastes. *Chemosphere* 145: 518–527.

Tan, X., Liu, Y., Zeng, G. et al. (2015). Application of biochar for the removal of pollutants from aqueous solutions. *Chemosphere* 125: 70–85.

Tang, L., Chen, S., Wand, N. et al. (2025). Comparative study of adsorption and slow-release performance and mechanism of phosphoric acid and ammonia-modified biochars on 2,4-dichlorophenoxyacetic acid. *Separation and Purification Technology* 358: 130470.

Wang, C., Luo, D., Zhang, X. et al. (2022). Biochar-based slow-release of fertilizers for sustainable agriculture: a mini review. *Environmental Science and Ecotechnology* 10: 100167.

Wang, H., Zhang, Z., Xu, P. et al. (2024). Research update on preparation and modification of biochar from crop residues. *Guangdong Chemical Industry* 51 (07): 95–97.

Waqas, M., Asam, Z., Rehan, M. et al. (2021). Development of biomass-derived biochar for agronomic and environmental remediation applications. *Biomass Conversion and Biorefinery* 11 (2): 339–361.

Wei, H.Y., Chen, Z.F., Xing, Z.P. et al. (2018). Effects of slow or controlled release fertilizer types and fertilization modes on yield and quality of rice. *Journal of Integrative Agriculture* 17 (10): 2222–2234.

Xu, G., Sun, J., Shao, H. et al. (2014). Biochar had effects on phosphorus sorption and desorption in three soils with differing acidity. *Ecological Engineering* 62: 54–60.

Yang, F., Tian, J., Meersmans, J. et al. (2018). Functional soil organic matter fractions in response to long-term fertilization in upland and paddy systems in South China. *Catena* 162: 270–277.

Yang, J., Li, H., Zhang, D. et al. (2014). Limited role of biochars in nitrogen fixation through nitrate adsorption. *Science of the Total Environment* 592: 758–765.

Ye, Z., Zhang, L., Huang, Q. et al. (2019). Development of a carbon-based slow release fertilizer treated by bio-oil coating and study on its feedback effect on farmland application. *Journal of Cleaner Production* 239: 118085.

Zhang, L., Wang, L., Fang, N. et al. (2021a). Effect of stabilized fertilizer in different regions of China and the suitable application rate. *Journal of Plant Nutrition and Fertilizers* 27 (2): 215–230.

Zhang, X., Gang, D.D., Sun, P. et al. (2021b). Goethite dispersed corn straw-derived biochar for phosphate recovery from synthetic urine and its potential as a slow-release fertilizer. *Chemosphere* 262: 127861.

Zhao, L., Cao, X., and Zhang, W. (2014a). Phosphorus-assisted biomass thermal conversion: reducing carbon loss and improving biochar stability. *PLos One* 9 (12): e115373.

Zhao, L., Cao, X., Zhang, W. et al. (2016). Copyrolysis of biomass with phosphate fertilizers to improve biochar carbon retention, slow nutrient release, and stabilize heavy metals in soil. *ACS Sustainable Chemistry & Engineering* 4 (3): 1630–1636.

Zhao, X., Wang, S., and Xing, G. (2014b). Nitrification, acidification, and nitrogen leaching from subtropical cropland soils as affected by rice straw-based biochar: laboratory incubation and column leaching studies. *Journal of Soils and Sediments* 14 (3): 471–482.

Zhao, X.E.S., Yuan, J., Wu, B. et al. (2024). Mechanism and research progress of nutrient slow-release in slow-release fertilizer. *Eco-Industry Science & Phosphorus Fluorine Engineering* 39 (07): 42–48.

Zheng, J., Han, J., Liu, Z. et al. (2017). Biochar compound fertilizer increases nitrogen productivity and economic benefits but decreases carbon emission of maize production. *Agriculture Ecosystems & Environment* 241: 70–78.

Zhou, Q., Wu, F., Wang, Z. et al. (2024). Effects of nitrogen fertilizer and biochar application rate interaction on growth and root-knot nematode disease of greenhouse tomatoes. *Crops* 40 (05): 212–219.

Zong, Y., Chen, D., and Lu, S. (2014). Impact of biochars on swell-shrinkage behavior, mechanical strength, and surface cracking of clayey soil. *Journal of Plant Nutrition and Soil Science* 177 (6): 920–926.

6

Waste-derived Biochar as Slow-release Pesticides

Xiaoqian Jiang, Yuqing Sun, and Jing Luo

School of Agriculture and Biotechnology, Sun Yat-sen University, Guangzhou, Guangdong, China

6.1 Introduction

Pesticides play a crucial role in controlling pests, weeds, improving crop yield and quality, and mitigating biological disasters. However, excessive pesticide use can result in harmful residues in the environment or crops, posing serious threats to human life and health. With the progression of agricultural modernization, there has been a significant increase in pesticide demand. Concurrently, improper application and low effective utilization rates have led to accumulating pesticide residues and difficult-to-metabolize compounds in the environment, exacerbating environmental issues and presenting severe challenges to ecological balance and human health (Yang et al. 2023). According to reports, global annual pesticide usage is estimated at 4.6 million tons (Zhao et al. 2018a). However, due to factors such as application techniques, pesticide properties, environmental conditions (e.g., wind speed, humidity, and temperature), and processes like leaching, evaporation, deposition, and degradation via photolysis, hydrolysis, and microbial activity, ~90% of chemical pesticides do not reach their intended targets within a specific timeframe (Ghormade et al. 2011). This widespread use has severely impacted soil, water, and air quality. Irrational pesticide application has resulted in significant harm to human health, including teratogenicity, carcinogenicity, mutagenicity, blood toxicity, neurotoxicity, endocrine disruption, and reproductive effects (Rani et al. 2021). Pesticide pollution adversely affects soil quality and crop production. When pesticide residue accumulation exceeds soil's

Waste-derived Biochar for Sustainable Rural Revitalization, First Edition. Edited by Yuqing Sun and Daniel C.W. Tsang.
Published 2026 by John Wiley & Sons, Inc.

self-purification capacity, it leads to severe soil degradation and quality decline (Liu et al. 2018), jeopardizing ecosystems, nutrient cycles, enzyme activities, soil biological populations, and biodiversity (Yu et al. 2019). Soil's poor adsorption of certain pesticides facilitates their leaching and migration into nearby water bodies and groundwater, increasing the risk of water contamination. In response to these challenges, recent national policies such as the 13th Five-year Plan advocate for a "double reduction of pesticides and fertilizers," aiming to promote soil ecological restoration, stable crop yields, cost savings, efficiency improvements, and enhanced product quality (Wang et al. 2024). Consequently, optimizing pesticide efficiency while ensuring high crop yields has become a critical issue. Given the rising demands for plant protection and resource efficiency, enhancing pesticide efficacy, minimizing soil migration and loss, improving bioavailability, and reducing environmental disturbances have become paramount topics of concern.

Environmental pollution caused by the excessive use of pesticides has garnered increasing attention. Given the current problems stemming from the unscientific and unreasonable application of pesticides, it is imperative to develop and implement novel pesticides that are highly efficient, environmentally friendly, safe, and cost-effective. There are two primary strategies to enhance pesticide efficacy while reducing pesticide pollution: firstly, the development of new eco-friendly pesticides with high specificity and toxicity toward target organisms, thereby minimizing pesticide usage and phytotoxic effects; secondly, the advancement of slow/controlled release formulations of pesticides, which extend pesticide action times through adsorption and embedding mechanisms, reducing losses due to decomposition and leaching, and improving overall utilization efficiency. Developing new pesticides is challenging, requiring substantial funding and prolonged research cycles. Conversely, the development of slow/controlled release formulations presents relatively lower difficulty and costs (Zhao et al. 2018a). Consequently, the application of pesticide slow-release technology emerges as an effective strategy to mitigate issues associated with excessive pesticide usage.

In recent years, biochar has many advantages, such as high specific surface area (SSA), large porosity, good pore structure, and a large number of oxygen-containing functional groups, which provide favorable conditions for the loading of pesticides. In addition, biochar has a wide range of raw materials, cheap and easy to obtain, and has a certain economic value while protecting the environment. At the same time, the application of biochar in slow-release pesticides is also conducive to soil improvement and environmental remediation, and biochar slow-release pesticides have attracted more and more attention (Khalid et al. 2020).

6.2 Research Progress of Biochar-based Slow-release Pesticides

New slow-release pesticides should exhibit advantages such as low toxicity, reduced environmental pollution, safety and reliability, convenience of operation, affordability, and high utilization rates of pesticides (Liu et al. 2015). Based on the rules, characteristics, and environmental conditions of pests and diseases, slow and controlled release pesticides can be dispensed continuously and steadily at specified dosages and intervals by enhancing processing methods, thereby achieving the most cost-effective, secure, and efficient control of pests and diseases. Numerous materials are utilized as carriers to develop sustained and controlled release formulations, including biochar, porous SiO_2, clay minerals, gels, chitosan, polymers, and others. For different types of pesticides, carrier materials can be applied individually or in combination to fabricate either open systems or controlled membrane systems to enhance their efficiency (Singh et al. 2020). Notably, open systems do not chemically interact with the active components of pesticides, allowing for complete release of the pesticides. Biochar stands out due to its high SSA, extensive porosity, well-defined pore structure comprising micropores, mesopores, and macropores, and numerous oxygen-containing functional groups, all of which facilitate optimal pesticide loading. Furthermore, biochar is abundantly available and cost-effective, making it an excellent carrier material for slow-release pesticides. Research on biochar as a sustained release carrier is growing. Studies have shown that when biochar and bentonite are incorporated into cyromazine, imidacloprid, and isoproturon pesticide formulations using sodium alginate as the controlled-release medium, the encapsulation rate exceeds that of formulations containing only sodium alginate, and the release rate is lower (Garrido-Herrera et al. 2006). Additionally, some researchers have employed rice straw biochar produced at a relatively low temperature (350 °C) both as a soil conditioner and as a carrier for the herbicides, acetochlor and 2,4-D. Their findings indicate that this biochar reduces herbicide leaching without compromising efficacy and exhibits a slow-release effect on the biochar/herbicide mixture (Lü et al. 2012).

Biochar is a stable carbon-rich solid material derived from the thermochemical conversion of biomass under relatively low temperatures (≤700 °C) in completely anaerobic or partially anoxic conditions (Tang et al. 2025). Primarily composed of carbon, hydrogen, oxygen, nitrogen, and other elements, biochar also contains potassium, sodium, magnesium, and other metal cations. Its surface is rich in oxygen-containing functional groups such as —OH and —COOH (Cao et al.

2019), and it possesses a well-developed pore structure with high pH value and cation exchange capacity. Biochar exhibits characteristics such as a large SSA, high porosity, and strong physical and chemical stability, coupled with excellent surface reactivity (Liu et al. 2018). These properties make biochar suitable for various application functions (Tan et al. 2017). Biochar demonstrates a significant adsorption capacity for pesticides. Studies have shown that biochar can adsorb pesticides through mechanisms such as affinity, pore filling, and specific adsorption. Additionally, biochar enhances the adsorption of pesticides in soil, thereby reducing the likelihood of pesticide loss (Yu et al. 2011). In recent years, biochar has proven effective in soil improvement and crop growth promotion, gradually revealing its potential as a green pesticide (Yu et al. 2011).

Biochar production can utilize a diverse array of feedstocks, encompassing virtually all carbon-rich agricultural and forestry residues. This includes, but is not limited to, crop residues, sawdust, animal manure, and kitchen waste (Waqas et al. 2018). As agricultural productivity advances and the agricultural economy expands rapidly, there is a concomitant surge in the generation of agricultural by-products. Globally, millions of tons of such waste are produced annually (Duque-Acevedo et al. 2020), presenting a substantial opportunity for the procurement of cost-effective raw materials for biochar synthesis. Regrettably, a significant portion of these agricultural residues is either incinerated directly or discarded. Harnessing agricultural waste for biochar production emerges as a strategic approach to resource optimization, aligning with principles of low-carbon, eco-friendly, efficient, circular, and sustainable development paradigms (Bose et al. 2023).

The elemental content, pore structure, and distribution of functional groups in biochar vary according to the source of the biomass and the pyrolysis conditions employed. Biochar's abundant pore structure, large SSA, and highly abundant oxygen-containing functional groups on its surface endow it with significant potential for soil remediation, soil fertility improvement, and as a carrier for slow-release pesticides or fertilizers (Tang et al. 2025). In Europe, biochar is recognized for its potential in sustainable agricultural development and climate change mitigation. Similarly, in Asian countries, such as Japan, China, and South Korea, biochar is also extensively utilized for soil improvement and pollution control (Wang et al. 2022).

Currently, the application of biochar in mitigating environmental pollution caused by pesticides has garnered significant attention. Research indicates that applying biochar to agricultural soils adjacent to lakes or rivers can effectively reduce water pollution resulting from pesticide runoff. Furthermore, the incorporation of biochar into soil not only enhances the adsorption of pesticide residues but also influences the degradation of these chemicals. Numerous studies have documented the effects of biochar on pesticide adsorption and desorption (Table 6.1) (Wang et al. 2022). Biochar

Table 6.1 Adsorption and desorption rates of different pesticides on different biochars.

Biochar feedstock	Pesticides	Adsorption rate (%)	Desorption rate (%)
Wood pellets	Topramezone	20–35	—
Macadamia shell	Aminocyclozine	18–25	—
Conch	Topramezone	13	—
Hardwood	Aminoazine	50	—
Hardwood	Topramezone	40	—
Beech wood	Imidazolinone	5	3
Hardwood shavings	2,4-D	55	85
Pig manure	Carbaryl	30	27
Corn straw	Atrazine	6	30

Source: Wang et al. 2022/with permission of Elsevier.

derived from various raw materials exhibits different adsorption (6–55%) and desorption capacities (3–85%) for diverse pesticides. These findings confirm that biochar holds promise for pesticide adsorption and controlled release. However, its current adsorption capacity for pesticides remains low. Therefore, improving the adsorption capacity of biochar represents a key area of ongoing research.

6.3 The Mechanisms of Biochar Loading and Slow Release of Pesticide

6.3.1 Physical and Chemical Characteristics of Biochar

Biochar is a stable, porous, and carbon-rich material produced by pyrolyzing biomass under oxygen-limited conditions (Awasthi 2022). The range of pyrolysis temperatures extends from 350 to 1500 °C, with fast pyrolysis occurring at 400–600 °C, slow pyrolysis at 350–800 °C, and gasification at 700–1500 °C (Tomczyk et al. 2020). Biochar exhibits properties such as high elemental content, large SSA, complex pore structures, electrostatic attraction, and ion exchange capabilities. These characteristics enable its direct application as an adsorbent, modifier, catalyst, and catalyst carrier (Liu et al. 2015). The physical and chemical properties of biochar, which largely determine its strong adsorption capacity for pesticides in soil environments, depend significantly on the raw materials used (such as sawdust, straw, rice husk, sewage sludge, etc.) and the pyrolysis conditions (including temperature, heating rate, and residence time) (Yavari et al. 2015).

The physical and chemical properties of biochar are significantly influenced by the varying sources of raw materials and the specific conditions under which it is prepared. Biochar can be derived from a diverse array of feedstocks, including agricultural residues, food waste, animal manure, and algae (O'Connor et al. 2021). The pyrolysis process, when applied to these differing raw materials, results in biochar exhibiting distinct characteristics such as variations in ash content, the presence of surface oxygen-containing functional groups, carbon-to-hydrogen ratios, porosity, and SSA (Song et al. 2019). These variations, in turn, impact the adsorptive capacity of biochar toward pesticides. For instance, the compositional differences in hemicellulose, cellulose, and lignin among raw materials lead to discrepancies in the resulting biochar's physicochemical attributes after pyrolysis (Jevrosimov et al. 2021). Furthermore, the structural breakdown of algae during pyrolysis is nearly complete (Yu et al. 2017), contrasting with the maintenance of structure in other biomasses. Notably, macroalgae-derived biochar features a more extensive porous structure compared to that from microalgae, attributed to the higher carbohydrate content inherent in macroalgae (Lee et al. 2020). The interaction among diverse biomass types during co-pyrolysis also modifies the pore architecture of the resultant biochar. A case in point is the observation of waste tire-derived carbon surfaces being coated with decomposition products from pine bark, leading to pore obstruction and a subsequent reduction in both the SSA and overall porosity of the biochar (Wang et al. 2020). Mandal et al. (2017) conducted a study examining the adsorption behaviors of various biochars—specifically those derived from eucalyptus bark, corncob, bamboo chips, rice husk, and rice straw—toward atrazine and imidacloprid. Among these, rice straw biochar demonstrated superior adsorption rates for atrazine (ranging from 37.5 to 70.7%) and imidacloprid (from 39.9 to 77.8%), highlighting the potential variability in performance based on feedstock origin. This comprehensive analysis underscores the importance of considering raw material selection and pyrolysis conditions when designing biochar for specific applications, particularly in the realm of environmental remediation where adsorptive properties are crucial.

The pyrolysis temperature significantly influences the physicochemical properties of biochar, including its surface area, porosity, ash content, and pH value. The adsorption capacity of biochar for pesticides depends on these physical and chemical attributes. Zhang et al. (2018) prepared six pig manure biochars at varying pyrolysis temperatures ranging from 200 to 700 °C and observed that both the SSA and ash content increased with rising pyrolysis temperatures, while the total pore volume peaked at 500 °C. Similarly, Sbizzaro et al. (2021) investigated bamboo stalk biochar and found that the surface acidity decreased as the pyrolysis temperature increased. Additionally, pyrolysis temperature alters the chemical properties of biochar, such as surface polarity and functional groups. The H/C and O/C ratios serve as indicators of polarity; a decrease in these ratios suggests a gradual reduction in biochar polarity. Consequently, an increase in hydrophobicity corresponds to a decrease in hydrophilicity, enhancing the potential adsorption performance for pesticides. Mayakaduwa et al. (2017) prepared rice husk biochar at pyrolysis temperatures of 300, 500, and 700 °C and noted that increasing the pyrolysis temperature resulted in higher carbon content, lower oxygen and hydrogen contents, greater aromaticity, and reduced polarity. Specifically, raising the pyrolysis temperature from 300 to 700 °C more than tripled the adsorption capacity. Higher pyrolysis temperatures yield biochar with more developed pore structures and diverse surface functional groups (Gao et al. 2019).

6.3.2 The Adsorption and Slow-release Mechanisms of Biochar-based Pesticides

Biochar, as a novel slow-release pesticide carrier, exhibits its adsorption and slow-release mechanisms primarily through its unique physical and chemical properties, along with interactions with soil microorganisms. These characteristics collectively enhance the loading capacity of biochar for pesticides, ensuring their stable release and prolonged efficacy in soil. The adsorption capability of biochar is fundamental to its application in slow release, representing one of the key processes influencing the environmental fate of pesticides. This process involves the retention of pesticide molecules on the solid surface via various forces, including bonding, hydrogen bonding, ionic bonding, covalent bonding, and Van der Waals interactions between charged sites on soil colloids (Wu et al. 2019). Such interactions dictate the accumulation and migration patterns of pesticides within the environment, impacting both environmental organisms and food safety. Empirical studies have substantiated that incorporating biochar into soil significantly enhances the adsorption of herbicides such as atrazine, nicosulfuron, and diuron (Liu et al. 2018).

Biochar has different mechanisms to bind pesticides in the soil (Twagirayezu et al. 2024). The adsorption process of biochar to pesticides can be divided into physical adsorption and chemical adsorption. Pore filling effects such as pore volume, pore size, and SSA are the factors affecting physical adsorption (Cheng et al. 2022). Most pesticides have strong affinity to biochar. Chemisorption includes intermolecular π–π conjugation effect and hydrogen bonding force between pesticide and biochar. In this interaction process, the oxygen atoms in the oxygen-containing groups of pesticides are highly electronegative and easy to form hydrogen bonds with the functional groups on the surface of biochar. Pesticide molecules can also act as hydrogen bond donors and acceptors at the same time, which is beneficial to bond formation through hydrogen interaction with various functional groups (—OH, N—H, C—O, C=O, and C—N). On the other hand, the π–π electron donor–acceptor interaction between the π-electron-rich graphene surface of biochar and the positively charged organic matter lacking π electrons is also the main factor affecting the chemisorption (Mandal et al. 2016). The adsorption capacity of biochar for hydrophilic/hydrophobic organic compounds depends on the surface hydrophilic groups, the carbonized and non-carbonized parts of itself, the bulk properties of the biochar, and the porous surface (Mohan et al. 2014). Other mechanisms affecting biochar-pesticide adsorption behavior include polymer matrix adsorption, surface coverage, multilayer adsorption, and condensation in capillary pores (Khalid et al. 2020). The sorption characteristics of biochar in soil can change over time due to weathering processes, interactions of organic matter with soil minerals, and oxidation by microorganisms in the soil (Xie et al. 2019). Therefore, the adsorption relationship between biochar and pesticides depends on biochar characteristics (aromaticity, carbon content, porosity, surface area, surface functional groups, and electrostatic attraction) (Li et al. 2017), pesticide characteristics (hydrophilic/hydrophobic), and soil characteristics (organic matter content and pH) (Dong et al. 2023).

The ambient temperature significantly influences the biochar adsorption capacity of pesticides. Specifically, when the adsorption process is endothermic, the adsorption capacity escalates with rising environmental temperatures. According to thermodynamic experiments conducted by Zhang et al. (2022), an increase in environmental temperature leads to the emergence of numerous oxygen-containing functional groups—namely, carboxyl (—COOH), aldehyde (—CHO), and hydroxyl (—OH) groups—on the surface of almond shell biochar. These functionalities render the biochar surface negatively charged, creating multiple cation exchange sites within the aromatic ring structure. Consequently, this enhances the cation exchange capacity of biochar, thereby augmenting its adsorption capability for atrazine. Conversely, in exothermic adsorption processes, the trend is reversed. In a separate study, Zhao et al. (2018b) successfully synthesized biochars (BC, ZnBC, and PBC) from wheat straw employing diverse pretreatment methodologies, including the absence or presence of zinc chloride or phosphoric acid addition. Their research revealed a marked enhancement in atrazine adsorption onto ZnBC and PBC upon lowering the temperature.

The porous structure and large SSA of biochar play pivotal roles in its application as a sustained release carrier. This unique characteristic enables biochar to retain and immobilize pesticide molecules, thereby decelerating their dissemination through soil matrices. Studies indicate that the pore architecture and surface characteristics of biochar exert a substantial influence on the release dynamics of pesticides. Specifically, micropores and mesopores within the pore structure function to constrain the liberation of pesticide molecules, whereas mesopores and macropores facilitate the gradual diffusion and subsequent release of these molecules in the soil environment. For instance, research on biochar loaded with abamectin has demonstrated that the pore morphology of biochar effectively modulates the release profile of abamectin, allowing for a rapid initial concentration followed by a sustained, prolonged release, enhancing the duration of pesticidal efficacy (Huang et al. 2018). Furthermore, the functional moieties and ionic charges present on biochar surfaces can engage in chemical adsorption or ion-exchange interactions with pesticide molecules, further decelerating their release rates. The pH and charge attributes of biochar can modify the speciation and transport behavior of pesticides within soil, influencing their slow-release characteristics. Notably, phosphoric acid-modified rape straw biochar exhibited a marked enhancement in the adsorption capacities for carbendazim and pyrazosulfuron-ethyl, with increases of 249-fold and 105-fold compared to untreated materials (Zhong et al. 2017), respectively, highlighting the significance of surface modification strategies. Additionally, biochar can indirectly modulate the release and transformation processes of pesticides by forming associations or complexes with soil constituents, both organic and inorganic. Investigations have revealed that the release kinetics of glyphosate when loaded onto biochar are influenced by surface functionalities and soil organic matter, leading to the formation of stable glyphosate-biochar-soil organic matter complexes. This complexation slows down glyphosate release and improves its bioavailability to target organisms (Rallet et al. 2022). Moreover, microbial metabolic activities can contribute to nutrient cycling, potentially participating in pesticide degradation pathways and altering soil conditions, thereby impacting pesticide bioaccessibility and release patterns (Schommer et al. 2023). For example, the presence of biochar has been shown to stimulate soil microbial communities, accelerating both the release and degradation of imidacloprid, an insecticide (Lu et al. 2024). In summary, the slow-release mechanism of biochar as a pesticide carrier is multifaceted, encompassing aspects of physical structure manipulation, chemical property adjustments, and microbial activity modulation. Collectively, these elements underscore biochar's potential as an effective platform for controlled release pesticide delivery systems.

6.4 Determinants Influencing Biochar Loading Efficiency and Pesticidal Release Capacity

6.4.1 Surface Area and Porosity

The adsorption capacity of biochar with a large surface area and high porosity is typically high. Biochar forms a porous structure during biomass pyrolysis due to water loss during dehydration (Bagreev et al. 2001). Conversely, biochar with low porosity is less effective at adsorbing pesticide molecules (Twagirayezu et al. 2024). The surface area of biochar is the most critical determinant of its adsorption capacity, while temperature also plays a significant role in its production.

Increasing the pyrolysis temperature from 500 to 900 °C has been reported to enhance the surface area (from 25.4 to 67.6 m^2/g) and pore volume (from 0.056 to 0.099 cm^3/g) (Chen et al. 2014). However, in certain instances, biochar produced at temperatures exceeding 550 °C exhibits reduced surface area and porosity. This reduction may be attributed to the formation of more macropores than micropores at higher temperatures or blockage by tar, leading to a smaller surface area (Li et al. 2017).

The composition of biochar feedstock significantly influences its performance characteristics. For instance, plant-derived biochar, such as oak, corn stover, and pine needles, typically exhibits a much higher surface area (ranging from 112 to 642 m^2/g) compared to manure (3.32–20.5 m^2/g) and biosolid biochar (50.9–94.2 m^2/g) (Li et al. 2017). Biomass with high lignin content, like bamboo and coconut shell, tends to produce biochar with a macroporous structure, whereas biomass rich in cellulose, such as fruit shells, primarily yields microporous biochar (Abhishek et al. 2022). Additionally, biochar produced without an activation step generally has a smaller surface area and fewer pores (Kim et al. 2012). Consequently, an activation process is essential to enhance both the surface area and porosity of the biochar.

6.4.2 Biochar pH

The pH value plays a crucial role in the adsorption of pesticides by biochar. It influences the chemical properties of the pesticides, the surface charge of the biochar, and the types of ionizable adsorbates present. Consequently, this impacts the adsorption of ionic pesticides onto the biochar's surface. Depending on their chemical structure and the surrounding pH, pesticides can carry either a positive or negative charge. For instance, if the biochar surface is positively charged, it may attract negatively charged pesticides, and vice versa (Zhang et al. 2013).

The surface area, porosity, and pH of biochar are influenced by the raw materials and pyrolysis temperature used during its production. Due to the diversity in raw material quality, the surface functionality of biochar is closely related to its pH value (Zhou et al. 2019). An increase in pH is mainly associated with high mineral ash content in the biochar. While the pH of any given feedstock will rise with temperature, the pH of biochar can either increase or decrease depending on the feedstock used and soil characteristics. For example, the pH of mineral-poor biochar (oak) decreased from 4.9 to 4.7, while the pH of mineral-rich corn stover biochar increased from 6.7 to 8.1 (Twagirayezu et al. 2024). The formation of acidic carboxyl groups upon carbon oxidation generally leads to a decrease in pH; however, an increase in biochar pH may result from the dissolution of alkaline minerals. A study on rice husk biochar showed that glyphosate adsorption increased from 75 to 85% between pH 3 and 5, decreased from 75 to 65% between pH 6 and 8, and then significantly dropped to 55% at pH 9 (Herath et al. 2017). Variations in pH affect pesticide solubility and thus their ability to adsorb onto biochar. Lee et al. (2018) concluded that raising the pH value to 7.7 or 7.9 would lead to the deprotonation of acidic carboxyl groups, a reduction in the adsorption affinity of biochar for neutral simazine due to increased negative charge density on the biochar surface, and a decrease in adsorption capacity due to enhanced electrostatic/hydrophobic repulsion. Zhang et al. (2022) found that as the pH value of the solution increases, the dissociation reaction of atrazine in water occurs, leading to proton loss, carboxyl hydrogen bond breakage, and transformation of the organic polymer into an anion with improved hydrophilicity. As a result, the adsorption capacity of biochar for atrazine also decreases. Tulun et al. (2021) studied the pH adsorption mechanism between walnut shell biochar and chlorpyrifos, finding that the zero point charge of the biochar was 8.35. Their results showed that although the pH value of the solution decreased with increasing H^+ concentration, the electrostatic interaction became more favorable on the negatively charged biochar surface.

6.4.3 Surface Functional Groups

The adsorption capacity of biochar is influenced by its surface functional groups, including amides, hydroxyls, amines, and lactones (Usman et al. 2019). The raw materials used and the pyrolysis temperature significantly affect these functional groups (El-Naggar et al. 2019). As the pyrolysis temperature increases, the atomic ratios of O/C, N/C, and H/C decrease. Higher temperatures result in a reduction of these surface functional groups on biochar.

Pyrolysis at varying temperatures yields biochars with distinct structural compositions. Specifically, pyrolysis of raw materials at 100–200 °C does not produce significant changes in Fourier Transform Infrared Spectroscopy (FTIR) spectra. However, dehydration of cellulose, lignin, and hemicellulose occurs at 300 °C, as indicated by the spectral range of 3500–3200 cm^{-1}. At 400 °C, various products from the decomposition of carboxylic acids and phenolic compounds in lignin and cellulose are observed, presenting multiple peaks between 1600 and 700 cm^{-1}. According to Keiluweit et al. (2010), the intensity of aromatic condensation, which typically occurs at temperatures higher than 500 °C, may be lower in the range of 1650–1500 cm^{-1} compared to the range of 885–752 cm^{-1}.

6.4.4 Carbon Content and Aromatic Structure

Both carbon content and aromatic structure significantly influence biochar's capacity to adsorb pesticides (Keiluweit et al. 2010). The H/C ratio serves as an indicator of the degree of aromaticity, which in turn impacts the pore size and surface area of biochar. Generally, the carbon content of biochar increases with temperature when the pyrolysis temperature exceeds 500 °C (Liu et al. 2017). Specifically, biochar produced at 600 °C exhibits higher carbon content compared to that produced at 400 °C. This trend is more pronounced in pine sawdust biochar, which has carbon contents of 92.1 and 78.6% at these temperatures, respectively, compared to corn residue biochar, which has carbon contents of 65.5 and 63.9% (Heitkötter and Marschner 2015), respectively. Due to its high degree of carbonization and aromatization, biochar possesses excellent anti-decay stability, a long half-life, and is resistant to decomposition and mineralization (Liu et al. 2018). Consequently, in the long term, biochar holds promise as a green alternative for controlling soil pollution through its pesticide adsorption capabilities.

6.4.5 Biochar Feedstock

The type of raw material used to produce biochar significantly affects its chemical composition and the level of pesticide adsorption. The feedstock influences the physical structure of the biochar, including its surface area and porosity. Raw materials high in lignocellulose content can enhance the porosity of biochar, thereby improving its ability to adsorb pesticides (Liu et al. 2018). The mineral content of the feedstock impacts the ash content in the biochar. Higher ash content may alter the soil pH, affecting pesticide behavior (Yadav et al. 2023). A positive relationship exists between biochar dose and pesticide adsorption rate, attributed to increased adsorption surface area and more available active sites. However, as the dosage of biochar increases, adsorbent accumulation on active sites may reduce their availability. Jacob et al. (2020) demonstrated that bagasse biochar effectively removed chlorpyrifos in a study, with adsorption capacity continuously increasing after adding 0.25 g of biochar but slightly decreasing after adding 0.5 g of biochar. Additionally, the raw materials used to produce biochar can influence the soil's microbial community, affecting pesticide degradation or the persistence of slow-release formulations (Deshoux et al. 2023).

6.4.6 Types of Pesticides

The adsorption–desorption behavior of pesticides with biochar is significantly influenced by pesticide properties such as persistence, solubility, and volatility. Among these properties, water solubility stands out as a crucial factor governing their affinity for biochar. Water-soluble pesticides tend to have weaker physicochemical interactions with biochar, making them more prone to leaching through soil profiles and contaminating groundwater resources. Conversely, persistent pesticides exhibit prolonged environmental stability, attributed to their extended interaction with biosolids, highlighting the dynamic nature of these interactions over time. Moreover, the evolving composition of biochar further complicates the adsorption–desorption dynamics, as its hydrophilicity and hydrophobicity characteristics play pivotal roles in modulating pesticide retention or release.

6.5 Modification of Biochar for Sustained Release of Pesticide

Due to the limited surface functional groups, pore volume, and SSA of biochar obtained from biomass pyrolysis, its adsorption capacity is weak, and it has a limited range of adsorbable substances. This limitation restricts its broad application. To address these issues, biochar needs to be modified to optimize its pore structure and surface properties. Such modifications alter the physical and chemical characteristics of biochar. Different modification methods and preparation techniques have varying effects on the drug loading and pesticide release capacities of biochar (Schommer et al. 2023).

Biochar pore structure regulation encompasses various methods, including physical activation, chemical activation, ball milling, coating, heteroatom doping, and the template method (Leng et al. 2021). The first five techniques are primarily effective for producing microporous biochars, whereas the template method is advantageous for generating mesoporous or macroporous biochars. Notably, the template method allows for the adjustment of pore shapes. For instance, when corn straw is activated with NH_3 at 800 °C for 3 hour, the resulting biochar exhibits a SSA of 418.7 m^2/g and a pore volume of 0.277 cm^3/g, approximately sevenfold higher than that of the original biochar (Chen et al. 2018). Additionally, NH_3 activation increases the number of nitrogen-containing functional groups within the biochar. Chu et al. (2018) utilized H_3PO_4 to activate pine sawdust, yielding biochar with a substantial SSA of 1627 m^2/g and a highly abundant porous structure. This modification significantly enhances the adsorption capacity of the biochar for bisphenol A and carbamazepine by an order of magnitude. Moreover, phosphoric acid-activated biochar can serve as a phosphorus-rich fertilizer. Guo et al. (2016) employed reed as a raw material and activated it with ammonium phosphate to produce biochar (BC-APs) characterized by developed micropores and abundant amino groups. Their study demonstrated that biochar modified with $(NH_4)_2HPO_4$ attained an SSA of 495.7 m^2/g, showing excellent adsorption capabilities for three heavy metal ions: Ni(II), Cu(II), and Cd(II). Naghdi et al. (2017) used pine biochar as a starting material and optimized milling parameters using a planetary ball mill via central composite experimental design and response surface methodology. They successfully produced nano-biochar with a surface area of 47.25 m^2/g and a total pore volume of 0.034 cm^3/g, achieving this in a short time and with low energy consumption. In contrast, the untreated nanobiochar had a surface area and total pore volume of 3.12 m^2/g and 0.005 cm^3/g, respectively. Furthermore, high-SSA biochar (634 m^2/g) with a large mesopore ratio (70–80%) was prepared using crab shell as the carbon precursor, natural calcium carbonate as the in-situ template, and hydrochloric acid as the etchant at 900 °C (Xi et al. 2020).

Surface modification primarily aims to alter the physical and chemical properties of a biochar's surface through the application of oxidants, reductants, polymers, clay minerals, and surfactants under specific conditions. This process enhances the material's ability to adsorb target substances. For instance, Jin et al. (2018) chemically modified biochar derived from wheat straw using HNO_3, resulting in an oxidized biochar with a U(VI) loading capacity 40 times greater than that of the untreated biochar. Similarly, Yang and Jiang (2014) prepared amino-modified biochar via nitrification and reduction treatments, achieving an adsorption capacity for Cu^{2+} fivefold higher than that of the original biochar. Additionally, composites of chitosan-modified biochar (CS/BC) have been utilized as effective adsorbents for removing both inorganic and organic pollutants from wastewater (Kaczorowska and Bożejewicz 2024). Ren et al. (2020) employed rice straw as a precursor to produce biomass carbon, which they subsequently modified with KOH and cetyltrimethylammonium bromide (CTAB). This modified biomass carbon (MBC) exhibited a loading capacity for 2,4-dichlorophenol reaching up to 85.18 mg/g.

6.6 Summary and Outlook

Biochar is a reliable and sustainable material. When combined with pesticides, biochar and its modified materials can serve as effective green pesticides. Additionally, biochar possesses the capability to adsorb pesticides, heavy metals, antibiotics, and inorganic pollutants from environmental matrices. Its physical and chemical properties, such as a large SSA, rich surface functional groups, and multi-pore distribution, help reduce the environmental risk associated with pesticides. Furthermore, these properties enhance the utilization efficiency of pesticides. Although the preparation of biochar materials is straightforward, certain modification methods or composite materials are necessary to improve the physical and chemical properties of the adsorbents. The emergence of advanced technologies will further augment the properties of biochar, thereby enhancing its adsorption capacity for pesticides. The progression of science and technology will also facilitate the development of more efficient and economical techniques for the large-scale synthetic production of biochar.

Currently, research on biochar for pesticide applications is predominantly focused on indoor and field experiments, indicating a significant gap before large-scale implementation. In practical applications, factors such as soil temperature, humidity, pH, soil microorganisms, and plant cultivation practices significantly influence the efficacy of slow-release pesticides. Using biochar to prepare slow-release pesticides necessitates high concentrations of active ingredients, particularly for acidic pesticides. Direct use of biochar as a pesticide carrier can lead to pesticide decomposition during application, thereby reducing efficacy. Furthermore, during the preparation process, attention must be given to optimizing the loading of biochar with pesticides; insufficient loading can result in wasted raw materials and potential environmental contamination. Hence, developing biochar with high loading capacity that aligns with the characteristics of both fertilizers and pesticides is crucial in the preparation of slow-release formulations. For the same crop and disease, different types of biochar can have contrasting effects: some may enhance plant growth but favor disease development, whereas others may suppress plant growth and mitigate disease incidence. Thus, selecting appropriate biochar types tailored to specific crops and diseases is essential for effective pest management.

With the actual development and popularization of biochar, biochar may be widely used in agricultural production as a slow-release pesticide in the future. It brings remarkable economic and environmental benefits to modern agriculture.

References

Abhishek, K., Shrivastava, A., Vimal, V. et al. (2022). Biochar application for greenhouse gas mitigation, contaminants immobilization and soil fertility enhancement: a state-of-the-art review. *Science of the Total Environment* 853: 158652.

Awasthi, M.K. (2022). Engineered biochar: a multifunctional material for energy and environment. *Environmental Pollution* 298: 118831.

Bagreev, A., Bandosz, T.J., and Locke, D.C. (2001). Pore structure and surface chemistry of adsorbents obtained by pyrolysis of sewage sludge-derived fertilizer. *Carbon* 39 (13): 1971–1979.

Bose, S., Senthil, K.P., Rangasamy, G. et al. (2023). A review on the applicability of adsorption techniques for remediation of recalcitrant pesticides. *Chemosphere* 313: 137481.

Cao, Y., Xiao, W., Shen, G. et al. (2019). Carbonization and ball milling on the enhancement of Pb(II) adsorption by wheat straw: competitive effects of ion exchange and precipitation. *Bioresource Technology* 273: 70–76.

Chen, T., Zhang, Y., Wang, H. et al. (2014). Influence of pyrolysis temperature on characteristics and heavy metal adsorptive performance of biochar derived from municipal sewage sludge. *Bioresource Technology* 164: 47–54.

Chen, W., Li, K., Xia, M. et al. (2018). Influence of NH_3 concentration on biomass nitrogen-enriched pyrolysis. *Bioresource Technology* 263: 350–357.

Cheng, Y., Wang, B., Shen, J. et al. (2022). Preparation of novel N-doped biochar and its high adsorption capacity for atrazine based on π–π electron donor-acceptor interaction. *Journal of Hazardous Materials* 432: 128757.

Chu, G., Zhao, J., Huang, Y. et al. (2018). Phosphoric acid pretreatment enhances the specific surface areas of biochars by generation of micropores. *Environmental Pollution* 240: 1–9.

Deshoux, M., Sadet-Bourgeteau, S., Gentil, S. et al. (2023). Effects of biochar on soil microbial communities: a meta-analysis. *Science of the Total Environment* 902: 166079.

Dong, X., Chu, Y., Tong, Z. et al. (2023). Mechanism and functionalization of biochar for pesticide adsorption: research progress. *Chinese Agricultural Science Bulletin* 39 (13): 117–124.

Duque-Acevedo, M., Belmonte-Ureña, L.J., Cortés-García, F.J. et al. (2020). Agricultural waste: review of the evolution, approaches and perspectives on alternative uses. *Global Ecology and Conservation* 22: e00902.

El-Naggar, A., El-Naggar, A.H., Shaheen, S.M. et al. (2019). Biochar composition-dependent impacts on soil nutrient release, carbon mineralization, and potential environmental risk: a review. *Journal of Environmental Management* 241: 458–467.

Gao, Y., Jiang, Z., Li, J. et al. (2019). A comparison of the characteristics and atrazine adsorption capacity of co-pyrolysed and mixed biochars generated from corn straw and sawdust. *Environmental Research* 172: 561–568.

Garrido-Herrera, F.J., González-Pradas, E., and Fernández-Pérez, M. (2006). Controlled release of isoproturon, imidacloprid, and cyromazine from alginate-bentonite-activated carbon formulations. *Journal of Agricultural and Food Chemistry* 54 (26): 10053–10060.

Ghormade, V., Deshpande, M.V., and Paknikar, K.M. (2011). Perspectives for nano-biotechnology enabled protection and nutrition of plants. *Biotechnology Advances* 29 (6): 792–803.

Guo, Z., Fan, J., Zhang, J. et al. (2016). Sorption heavy metal ions by activated carbons with well-developed microporosity and amino groups derived from Phragmites australis by ammonium phosphates activation. *Journal of the Taiwan Institute of Chemical Engineers* 58: 290–296.

Heitkötter, J. and Marschner, B. (2015). Interactive effects of biochar ageing in soils related to feedstock, pyrolysis temperature, and historic charcoal production. *Geoderma* 245: 56–64.

Herath, H.M.M.T.B., Pahalagamage, S.P., Yogendranathan, N. et al. (2017). Amitraz poisoning: a case report of an unusual pesticide poisoning in Sri Lanka and literature review. *BMC Pharmacology & Toxicology* 18 (1): 6.

Huang, P., Yue, L., Yu, H. et al. (2018). Adsorption and desorption behavior of avermectin on biochar prepared by cassava residue. *Chinese Journal of Tropical Crops* 39 (2): 361–371.

Jacob, M.M., Ponnuchamy, M., Kapoor, A. et al. (2020). Bagasse based biochar for the adsorptive removal of chlorpyrifos from contaminated water. *Journal of Environmental Chemical Engineering* 8 (4): 103904.

Jevrosimov, I., Kragulj Isakovski, M., Apostolović, T. et al. (2021). Mechanisms of alachlor and pentachlorobenzene adsorption on biochar and hydrochar originating from *Miscanthus giganteus* and sugar beet shreds. *Chemical Papers* 75 (5): 2105–2120.

Jin, J., Li, S., Peng, X. et al. (2018). HNO_3 modified biochars for uranium (VI) removal from aqueous solution. *Bioresource Technology* 256: 247–253.

Kaczorowska, M.A. and Bożejewicz, D. (2024). The application of chitosan-based adsorbents for the removal of hazardous pollutants from aqueous solutions-a review. *Sustainability* 16 (7): 2615.

Keiluweit, M., Nico, P.S., Johnson, M. et al. (2010). Dynamic molecular structure of plant biomass-derived black carbon (biochar). *Environmental Science & Technology* 44 (4): 1247–1253.

Khalid, S., Shahid, M., Murtaza, B. et al. (2020). A critical review of different factors governing the fate of pesticides in soil under biochar application. *Science of the Total Environment* 711: 134645.

Kim, K.H., Kim, J.Y., Cho, T.S. et al. (2012). Influence of pyrolysis temperature on physicochemical properties of biochar obtained from the fast pyrolysis of pitch pine. *Bioresource Technology* 118: 158–162.

Lee, S., Han, J., and Ro, H.M. (2018). Interpreting the pH-dependent mechanism of simazine sorption to Miscanthus biochar produced at different pyrolysis temperatures for its application to soil. *Korean Journal of Chemical Engineering* 35 (7): 1468–1476.

Lee, X.J., Ong, H.C., Gan, Y.Y. et al. (2020). State of art review on conventional and advanced pyrolysis of macroalgae and microalgae for biochar, bio-oil and bio-syngas production. *Energy Conversion and Management* 210: 112707.

Leng, L., Xiong, Q., Yang, L. et al. (2021). An overview on engineering the surface area and porosity of biochar. *Science of the Total Environment* 763: 144204.

Li, H., Dong, X., da Silva, E.B. et al. (2017). Mechanisms of metal sorption by biochars: biochar characteristics and modifications. *Chemosphere* 178: 466–478.

Liu, W., Jiang, H., and Yu, H. (2015). Development of biochar-based functional materials: toward a sustainable platform carbon material. *Chemical Reviews* 115 (22): 12251–12285.

Liu, Y., Lonappan, L., Brar, S.K. et al. (2018). Impact of biochar amendment in agricultural soils on the sorption, desorption, and degradation of pesticides: a review. *Science of the Total Environment* 645: 60–70.

Liu, Y., Yao, S., Wang, Y. et al. (2017). Bio- and hydrochars from rice straw and pig manure: inter-comparison. *Bioresource Technology* 235: 332–337.

Lü, J., Li, J., Li, Y. et al. (2012). Use of rice straw biochar simultaneously as the sustained release carrier of herbicides and soil amendment for their reduced leaching. *Journal of Agricultural and Food Chemistry* 60 (26): 6463–6470.

Lu, P., Qi, Z., Lu, Y. et al. (2024). Study on the treatment efficiency of imidacloprid wastewater by biochar soil filtration system. *Technology of Water Treatment* 50 (1): 61–65.

Mandal, A., Singh, N., and Purakayastha, T.J. (2017). Characterization of pesticide sorption behaviour of slow pyrolysis biochars as low cost adsorbent for atrazine and imidacloprid removal. *Science of the Total Environment* 577: 376–385.

Mandal, S., Kunhikrishnan, A., Bolan, N.S. et al. (2016). Application of biochar produced from biowaste materials for environmental protection and sustainable agriculture production. In: *Environmental Materials and Waste: Resource Recovery and Pollution Prevention*, 73–89. Academic Press.

Mayakaduwa, S.S., Herath, I., and Ok, Y.S. (2017). Insights into aqueous carbofuran removal by modified and non-modified rice husk biochars. *Environmental Science and Pollution Research* 24 (29): 22755–22763.

Mohan, D., Kumar, H., Sarswat, A. et al. (2014). Cadmium and lead remediation using magnetic oak wood and oak bark fast pyrolysis bio-chars. *Chemical Engineering Journal* 236: 513–528.

Naghdi, M., Taheran, M., Brar, S.K. et al. (2017). A green method for production of nanobiochar by ball milling-optimization and characterization. *Journal of Cleaner Production* 164: 1394–1405.

O'Connor, J., Hoang, S.A., Bradney, L. et al. (2021). A review on the valorisation of food waste as a nutrient source and soil amendment. *Environmental Pollution* 272: 115985.

Rallet, D., Paltahe, A., Tsamo, C. et al. (2022). Synthesis of clay-biochar composite for glyphosate removal from aqueous solution. *Heliyon* 8 (3): e09112.

Rani, L., Thapa, K., Kanojia, N. et al. (2021). An extensive review on the consequences of chemical pesticides on human health and environment. *Journal of Cleaner Production* 283: 124657.

Ren, D., Yu, H., Wu, J. et al. (2020). The study on adsorption behavior of 2,4-DCP in solution by biomass carbon modified with CTAB-KOH. *Water Science and Technology* 82 (8): 1535–1546.

Sbizzaro, M., Sampaio, S.C., dos Reis, R.R. et al. (2021). Effect of production temperature in biochar properties from bamboo culm and its influences on atrazine adsorption from aqueous systems. *Journal of Molecular Liquids* 343: 117667

Schommer, V.A., Vanin, A.P., Nazari, M.T. et al. (2023). Biochar-immobilized *Bacillus* spp. for heavy metals bioremediation: a review on immobilization techniques, bioremediation mechanisms and effects on soil. *Science of the Total Environment* 881: 163385.

Singh, A., Dhiman, N., Kar, A.K. et al. (2020). Advances in controlled release pesticide formulations: Prospects to safer integrated pest management and sustainable agriculture. *Journal of Hazardous Materials* 385: 121525.

Song, F., You, X., Ma, Y. et al. (2019). Rapid removal of triazine pesticides by P doped biochar and the adsorption mechanism. *Chemosphere* 235: 918–925.

Tan, X., Liu, S., Liu, Y. et al. (2017). Biochar as potential sustainable precursors for activated carbon production: multiple applications in environmental protection and energy storage. *Bioresource Technology* 227: 359–372.

Tang, L., Chen, S., Wand, N. et al. (2025). Comparative study of adsorption and slow-release performance and mechanism of phosphoric acid and ammonia-modified biochars on 2,4-dichlorophenoxyacetic acid. *Separation and Purification Technology* 358: 130470.

Tomczyk, A., Sokołowska, Z., and Boguta, P. (2020). Biochar physicochemical properties: pyrolysis temperature and feedstock kind effects. *Reviews in Environmental Science and Biotechnology* 19 (1): 191–215.

Tulun, Ş., Akgül, G., Alver, A. et al. (2021). Adaptive neuro-fuzzy interference system modelling for chlorpyrifos removal with walnut shell biochar. *Arabian Journal of Chemistry* 14 (12): 103443.

Twagirayezu, G., Chen, H., Wu, Y. et al. (2024). Insights into the influences of biochar on the fate and transport of pesticides in the soil environment: a critical review. *Biochar* 6 (1): 9.

Usman, M., Hao, S., Chen, H. et al. (2019). Molecular and microbial insights towards understanding the anaerobic digestion of the wastewater from hydrothermal liquefaction of sewage sludge facilitated by granular activated carbon (GAC). *Environment International* 133: 105257.

Wang, N., Pan, S., Li, S. et al. (2024). Combination of magnesium modified biochar and iron oxides down-regulates phosphates transport in porous media. *Chemical Engineering Journal* 498: 155151.

Wang, W., Tan, J., Li, S. et al. (2022). Transport, retention and release of phytate in soil with addition of Mg–Al layered double hydroxides. *Journal of Cleaner Production* 379: 134774.

Wang, Z., Burra, K.G., Zhang, M. et al. (2020). Co-pyrolysis of waste tire and pine bark for syngas and char production. *Fuel* 274: 117878.

Waqas, M., Aburiazaiza, A.S., Miandad, R. et al. (2018). Development of biochar as fuel and catalyst in energy recovery technologies. *Journal of Cleaner Production* 188: 477–488.

Wu, C., Liu, X., Wu, X. et al. (2019). Sorption, degradation and bioavailability of oxyfluorfen in biochar-amended soils. *Science of the Total Environment* 658: 87–94.

Xi, J., Li, H., Xi, J. et al. (2020). Preparation of high porosity biochar materials by template method: a review. *Environmental Science and Pollution Research* 27 (17): 20675–20684.

Xie, J., Zheng, Y., Liu, X. et al. (2019). Human health safety studies of a new insecticide: dissipation kinetics and dietary risk assessment of afidopyropen and one of its metabolites in cucumber and nectarine. *Regulatory Toxicology and Pharmacology* 103: 150–157.

Yadav, S., Singh, S.R., Bahadur, L. et al. (2023). Sugarcane trash ash affects degradation and bioavailability of pesticides in soils. *Sugar Tech* 25 (1): 77–85.

Yang, G. and Jiang, H. (2014). Amino modification of biochar for enhanced adsorption of copper ions from synthetic wastewater. *Water Research* 48: 396–405.

Yang, K., Feng, G., and Wang, X. (2023). Research progress and influencing factors of biochar on pesticide residue adsorption. *Journal of Zhejiang Agricultural Sciences* 64 (11): 2801–2806.

Yavari, S., Malakahmad, A., and Sapari, N.B. (2015). Biochar efficiency in pesticides sorption as a function of production variables-a review. *Environmental Science and Pollution Research* 22 (18): 13824–13841.

Yu, H., Zou, W., Chen, J. et al. (2019). Biochar amendment improves crop production in problem soils: a review. *Journal of Environmental Management* 232: 8–21.

Yu, K., Lao, B.F., Show, P.L. et al. (2017). Recent developments on algal biochar production and characterization. *Bioresource Technology* 246: 2–11.

Yu, X., Mu, C., Gu, C. et al. (2011). Impact of woodchip biochar amendment on the sorption and dissipation of pesticide acetamiprid in agricultural soils. *Chemosphere* 85 (8): 1284–1289.

Zhang, P., Sun, H., Ren, C. et al. (2018). Sorption mechanisms of neonicotinoids on biochars and the impact of deashing treatments on biochar structure and neonicotinoids sorption. *Environmental Pollution* 234: 812–820.

Zhang, P., Sun, H., Yu, L. et al. (2013). Adsorption and catalytic hydrolysis of carbaryl and atrazine on pig manure-derived biochars: impact of structural properties of biochars. *Journal of Hazardous Materials* 244: 217–224.

Zhang, Z., Zhou, C., Yang, J. et al. (2022). Preparation and characterization of apricot kernel shell biochar and its adsorption mechanism for atrazine. *Sustainability* 14 (7): 4082.

Zhao, X., Cui, H., Wang, Y. et al. (2018a). Development strategies and prospects of nano-based smart pesticide formulation. *Journal of Agricultural and Food Chemistry* 66 (26): 6504–6512.

Zhao, L., Yang, F., Jiang, Q. et al. (2018b). Characterization of modified biochars prepared at low pyrolysis temperature as an efficient adsorbent for atrazine removal. *Environmental Science and Pollution Research* 25 (2): 1405–1417.

Zhong, X., Wang, T., Yuan, W. et al. (2017). Progresses of preparation, modification and environmental behavior of biochar. *Journal of Natural Science of Hunan Normal* 40 (05): 44–50.

Zhou, J., Chen, H., Thring, R.W. et al. (2019). Chemical pretreatment of rice straw biochar: effect on biochar properties and hexavalent chromium adsorption. *International Journal of Environmental Research* 13 (1): 91–105.

7

Waste-derived Biochar as Targeting Plant Disease Suppressors

Mi Wei, Zhongwang Liu, Yuqing Sun, and Jinfang Tan*

School of Agriculture and Biotechnology, Sun Yat-sen University, Guangzhou, Guangdong, China

*Corresponding author: weim29@mail.sysu.edu.cn

7.1 Introduction

Plant disease outbreaks are threatening food security in many areas of the world (Ristaino et al. 2021). Plant diseases can be classified into soil-borne and air-borne diseases, and the reasons for inducing plant disease include fungal pathogens, bacterial pathogens, oomycete pathogens, and plant-parasitic nematodes (Poveda et al. 2021). Applying biochar for plant disease management is environmentally benign and is being widely investigated by researchers worldwide. Biochar is a black solid substance prepared by the pyrolysis of biomass (such as wood, manure, and straw) under anaerobic conditions (Lehmann 2009; Borthakur et al. 2019; Chen et al. 2019).

Some researchers have reviewed the research progress and potential mechanism of biochar for plant disease prevention (Alaylar et al. 2021; de Medeiros et al. 2021; Poveda et al. 2021; Hou et al. 2022). In summary, the disease resulting from plant pathogens has been considered a function of the interaction between the factors at the three vertices of the "disease triangle": plant, pathogens, and soil (environmental conditions). Biochar controls plant disease by improving soil quality, increasing the activities and abundance of beneficial microorganisms, absorbing phytotoxic and allelopathic compounds, inducing plant resistance, and exerting fungitoxic effects. The characteristics of biochar, such as pore structure, surface area, pH, available nutrients, labile C, and volatile organic compounds, can directly influence all factors of the disease triangle or indirectly influence

Waste-derived Biochar for Sustainable Rural Revitalization, First Edition. Edited by Yuqing Sun and Daniel C.W. Tsang.
Published 2026 by John Wiley & Sons, Inc.

through impacts on the rhizosphere microbiome to control plant diseases (Graber et al. 2014; Jaiswal et al. 2014). However, the influence of the characteristics of biochar on plant diseases is still unclear. A specific type of biochar should be optimum to control a specific type of plant disease.

In this section, we used meta-analysis to clarify the influence of different types of biochar on plant diseases, and the optimum control strategies of different types of biochar corresponding to plant diseases were summarized.

7.2 Methods

A meta-analysis containing 48 papers with 131 experimental case studies was conducted to analyze the influence of biochar with different characteristics on plant diseases. The effect size of biochar application on managing plant diseases was evaluated by Hedge's using Eqs. (7.1), (7.2), and (7.3) (Li et al. 2020):

$$\text{Hedge's}\, g = J\frac{\overline{X_T} - \overline{X_C}}{S_{\text{within}}} \tag{7.1}$$

$$S_{\text{within}} = \sqrt{\frac{(n_T - 1)S_T^2 + (n_C - 1)S_C^2}{n_T + n_C - 2}} \tag{7.2}$$

$$J = 1 - \frac{3}{4(n_T + n_C - 2) - 1} \tag{7.3}$$

where $\overline{X_T}$ and $\overline{X_C}$ represent the effectiveness of plant disease control in the biochar and control treatments, respectively; S_{within} represents SD between treatments; J represents standard sample size; n_T and n_C represent the sample size of biochar and control treatments, respectively; and S_T and S_C represent SD for biochar and control treatments, respectively.

7.3 Using Biochar for Managing Plant Diseases

The results of the meta-analysis show that the application of biochar significantly controlled plant diseases ($P < 0.001$) regardless of soil-borne ($P < 0.01$) or air-borne diseases ($P < 0.05$) compared to the control (Figure 7.1a). Significant heterogeneity existed in the soil-borne and air-borne pathogen meta-analyses because the type and concentration of biochar and the conditions that work in one plant-pathogen system may not necessarily work in the same manner in other systems (Copley et al. 2017). Thus, subgroup analysis was performed according to the reason for the induced plant disease. As shown in Figure 7.1b and c, biochar can significantly control the soil-borne diseases induced by fungi ($P < 0.01$)

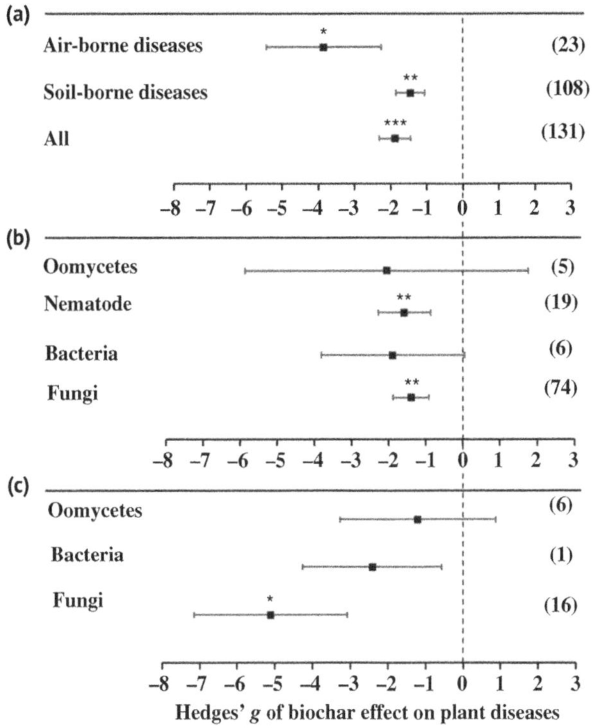

Figure 7.1 Mean effect size of random-effect models (Hedge's g ± 95% confidence interval) (a) for the effectiveness of plant soil-borne and air-borne disease control based on biochar amendment and (b) for the effectiveness of plant soil-borne and (c) foliar disease control as categorized by pathogen types based on biochar amendment.

and nematodes ($P < 0.01$) and the air-borne diseases induced by fungi ($P < 0.05$). Significant heterogeneity existed in the fungal pathogen and plant-parasitic nematode (induce plant soil-borne disease) and fungal pathogen (induce plant air-borne disease) meta-analyses. Drivers of the heterogeneity in responses are varied, including the feedstock, pyrolysis temperature, pH, and load ratio of biochar, because they are associated with the effectiveness of plant disease control. Hence, a submeta-analysis was conducted to reveal the function of different types of biochar in plant-pathogen systems.

7.3.1 Effect of Biochar Feedstock on the Plant Disease Control

Wood ($P < 0.01$) and residue ($P < 0.05$) biochars significantly control soil-borne diseases caused by pathogenic fungi (Figure 7.2a), straw ($P < 0.05$), and wood ($P < 0.05$) biochars significantly control soil-borne diseases caused by plant-parasitic

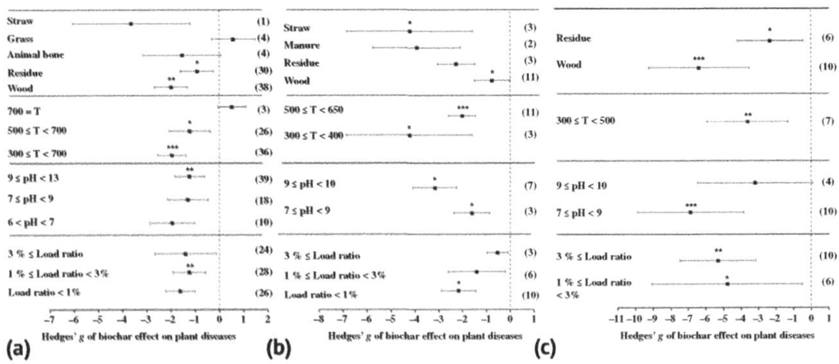

Figure 7.2 Mean effect size of random-effect models (Hedge's $g \pm 95\%$ confidence interval) for the effectiveness of (a) plant soil-borne fungi, (b) plant-parasitic nematode, and (c) air-borne fungi disease control based on biochar amendment in response to biochar as categorized by feedstock type, pyrolysis temperature, pH, and load ratio.

nematodes (Figure 7.2b), and wood ($P < 0.001$) and residue ($P < 0.05$) biochars significantly control air-borne diseases caused by pathogenic fungi (Figure 7.2c).

7.3.1.1 Soil-borne Diseases Induced by Pathogenic Fungi

The disease control of wood biochar is attributed to its high carbon (C) content, large specific surface area, and pore volume (Ippolito et al. 2020). A high C content can attract pathogenic fungi to obtain a C source on the surface of biochar. A large specific surface area and pore volume can adsorb root exudates, which induce chemotaxis of pathogens to biochar (Gu et al. 2017). In addition, the toxins produced by pathogens (such as organic acids and extracellular enzymes) can also be adsorbed by biochar (Poveda et al. 2021). Therefore, wood biochar can advantageously induce soil-borne fungal pathogens to stay away from plant roots and bind induced soil-borne fungal pathogens on its surface and to its pore, thus avoiding plant infection. In addition, wood biochar can reduce the deleterious effects of allelopathic agents, such as ferulic, coumaric, and cinnamic, on arbuscular mycorrhizal colonization by adsorption (Elmer and Pignatello 2011). Residue biochar contains high P (8 g/kg), K (40.9 g/kg), and Mg (6.06 g/kg) contents (Ippolito et al. 2020), which improves the soil nutrient supply. This makes plants maintain a high level of inhibitory compounds in tissues or enables quick plant responses to pathogen attack (Datnoff et al. 2007; Graber et al. 2014). In addition, it contains high cellulose, which helps in the formation of tar (a mixture of char, organic compounds, aldehydes, and discrete ketones) (Yu et al. 2014). The low levels of phytotoxic compounds in tar can induce plants to produce systemic resistance (Elad et al. 2010).

7.3.1.2 Soil-borne Diseases Induced by Plant-Parasitic Nematodes

The reduction in nematode infection rates and abundances depends on the type of feedstock, and not all feedstocks are suitable for utilization in nematode pest management with biochar (George et al. 2016). The high Ca (26.3 g/kg) content in wood biochar may be a crucial factor for controlling plant disease. In Huang's study, a direct toxic effect of wood biochar exudates on nematode development, viability, or infectivity was not observed, while increased plant resistance was observed (Huang et al. 2015). In response to pathogen effector manipulation, plants can evolve HeLo-domain and CC-domain nucleotide-binding/leucine-rich repeats to form effector-triggered oligomeric pores with Ca^{2+}-permeable ion channel activity to boost Ca^{2+} cascades and amplify pattern recognition receptor immunity, resulting in a more robust response called effector-triggered immunity (Parker et al. 2021). In addition, wood and straw biochar significantly increased the soil organic carbon content ($P < 0.05$) (Xu et al. 2021), which is another reason for the control of plant-parasitic nematodes. The increased soil organic carbon not only increases the population of beneficial soil microbes but also increases the population of nonplant-parasitic nematodes. They reduced the population of plant-parasitic nematodes through competition and antagonism (Zhang et al. 2013).

7.3.1.3 Air-borne Diseases Induced by Pathogenic Fungi

Biochar induces systemic plant defenses against air-borne diseases by improved nutrient supply, stimulation of a beneficial soil microbial consortium, input of chemical elicitors such as salts and organic chemicals, or improvement in soil physical qualities (Elad et al. 2010). Wood biochar has a good disease control effect, which is ascribed to its improvement in soil physical qualities because it can significantly ($P < 0.05$) decrease bulk density and increase soil porosity (Singh et al. 2022). For residual biochar, its mechanism for controlling plant air-borne diseases is similar to that for controlling plant soil-borne diseases induced by pathogenic fungi (Section 7.3.1.1).

7.3.2 Effect of Pyrolysis Temperature of Biochar on the Plant Disease Control

The effect of the pyrolysis temperature of biochar on the effectiveness of plant disease control is presented in Figure 7.2. Biochar pyrolysis temperatures between 500 and 700 °C ($P < 0.05$) and between 300 and 500 °C ($P < 0.001$) had significant inhibitory effects on soil-borne diseases caused by pathogenic fungi (Figure 7.2a). Biochar pyrolysis temperatures between 500 and 650 °C ($P < 0.001$) and between 300 and 400 °C ($P < 0.05$) also had significant inhibitory effects on soil-borne diseases caused by nematodes (Figure 7.2b). Biochar pyrolysis temperatures between

300 and 500 °C ($P < 0.05$) had a significant inhibitory effect on air-borne diseases caused by pathogenic fungi (Figure 7.2c).

7.3.2.1 Soil-borne Diseases Induced by Pathogenic Fungi

Biochar produced at low pyrolysis temperatures ($300 \leq T < 500$ °C) has a good control effect on soil-borne diseases, which is attributed to its positive effect on soil microbes. Suppressiveness against fungal pathogens and plant-parasite nematodes has been associated with diverse soil microbial communities that have more competitive, antagonistic fungi and bacteria or that are conducive to the activation of induced systemic resistance (Rahman et al. 2014). Biochar prepared at low temperatures contains a large amount of organic nutrients and nonpyrolyzed organic compounds (tar and volatile matter) (Zhang et al. 2018; Li et al. 2020), such as butyric acids and hydroxypropionic, propylene glycol and ethylene glycol, o-cresol and benzoic acid, quinones (resorcinol and hydroquinone), and 2-phenoxyethanol, which are unstable and bioavailable (Graber et al. 2010; Sun et al. 2015). Thus, biochar produced at lower pyrolytic temperatures (<500 °C) had a greater effect on microbial diversity (Li et al. 2020; Singh et al. 2022) and the shifts in plant rhizosphere and soil microbes toward beneficial plant growth-promoting rhizobacteria and fungi, many species of which are known to have the function of biocontrol (Graber et al. 2010). In addition, biochar can alter the chemistry of plant root exudates, thereby inhibiting pathogens (Akhter et al. 2015; Jin et al. 2022). For example, a low concentration of flavonoids suppressed mycelial growth of *Fusarium oxysporum f.* sp. *Lycopersici* (Steinkellner and Mammerler 2007). It has been proven that the alteration of the chemistry of plant exudates is attributed to the increase in the abundance of plant growth promoting rhizobacteria in the plant rhizosphere (Jin et al. 2022).

Biochar produced at high temperatures (<500 °C) decreases microbial diversity (Li et al. 2020), while it is more helpful for improving soil bulk densities and increasing porosity (Singh et al. 2022). In addition, biochar inhibition of microbial growth is mainly controlled by the availability of easily degradable carbon sources, with a minor role played by newly formed aromatic and phenolic compounds that are potentially fungitoxic (Bonanomi et al. 2015). The degradable carbon sources of biochar decreased with increasing pyrolysis temperature, which resulted in the inhibition of pathogenic fungi (Marra et al. 2018). Therefore, biochar produced between 500 and 700 °C has a significant effect on soil-borne diseases caused by pathogenic fungi.

7.3.2.2 Soil-borne Diseases Induced by Plant-parasitic Nematodes

In addition to increasing the diversity of soil microbial communities, the nematicidal effect of organic compounds, such as fatty acids and phenols, contained in low-temperature biochar is also another reason for the control of plant diseases induced by plant-parasitic nematodes. For example, phenolic compounds contained at a low pyrolysis temperature (300 °C) have an inhibitory effect on the eggs of plant-parasitic nematodes (Bonanomi et al. 2021). The toluene contained at low pyrolysis temperatures (Spokas et al. 2011) has nematicidal activity against *Meloidogyne incognita* (Poveda 2021). Conversely, the direct toxic effect of biochar produced at 500–650 °C on controlling plant-parasitic nematodes is not observed, but the induced resistance is a possible explanation for the significantly reduced infection rates (Huang et al. 2015; George et al. 2016). The ethylene contained in biochar produced at 500–650 °C increased the expression of ethylene genes that could induce resistance in plants to pathogens (Huang et al. 2015). In addition, biochar produced at 500–650 °C has a high nutrient concentration, which is another reason for the effective control of plant diseases induced by parasitic nematodes. A meta-analysis shows that increasing pyrolysis temperature increased P and Ca content and had mixed effects with respect to Mg content (Ippolito et al. 2020).

7.3.2.3 Air-borne Diseases Induced by Pathogenic Fungi

Systemic resistance induced by improving the structure of the soil microbial community (the mechanism is described in Section 7.3.2.1) is the main reason for the control of low pyrolysis temperature biochar for air-borne diseases because biochar is spatially distant from the air-borne pathogen attack site. Soil microbes can directly induce plant defense responses in the absence of an attacker (Van Wees et al. 2008; Pineda et al. 2010; Pineda et al. 2017). The study of Kolton et al. (2017) showed that the change in the soil microbial community was induced by the increase in bacterial diversity and communities associated with phenolic compound and carbohydrate metabolism (Kolton et al. 2017).

7.3.3 Effect of Biochar pH on the Plant Disease Control

The effect of the pH of biochar on the effectiveness of plant disease control is presented in Figure 7.2. The biochar with a pH over 9 had a significant ($P < 0.01$) inhibitory effect on soil-borne diseases caused by pathogenic fungi (Figure 7.2a). Both biochars with pH values over 9 ($P < 0.05$) and between 7 and 9 ($P < 0.05$) had significant inhibitory effects on soil-borne diseases caused by nematodes (Figure 7.2b). A biochar pH between 7 and 9 ($P < 0.001$) had a significant inhibitory effect on air-borne diseases caused by pathogenic fungi (Figure 7.2c).

7.3.3.1 Soil-borne Diseases Induced by Pathogenic Fungi

A high pH condition is not conducive to fungal growth (Pietri and Brookes 2009). Therefore, biochar with a pH over 9 has a significant inhibitory effect on soil-borne diseases caused by pathogenic fungi. In addition, the positive influence of biochar on increasing soil pH is more profound in acidic soils and soil with low soil organic matter (SOM) content, probably because SOM content is linked to the pH buffering capacity of soil (Gul et al. 2015). Among the 39 cases with a biochar pH greater than 9, 24 were about the acidic culture substrate of peat (Jaiswal et al. 2014, 2015, 2017), 5 were respectively 4 acidic sandy soil (Asadabadi et al. 2021) and 1 acidic eutric cambisol (Tian et al. 2021), and 10 were sandy soil with low organic matter content (Akhter et al. 2015; Copley et al. 2017; Rogovska et al. 2016). The application of biochar acid soil had a positive effect on P availability (Glaser and Lehr 2019; Yang et al. 2021).

7.3.3.2 Soil-borne Diseases Induced by Plant-parasitic Nematodes

The pH of biochar between 9 and 10 has a larger Hedges' g than that between 7 and 9 on the effect of pH on soil-borne diseases induced by plant-parasitic nematodes. Biochar with high pH can offer greater cation exchange capacity (CEC) provision to soil (Gul et al. 2015). The high CEC of biochar can chemically promote soil fertility (Wang et al. 2020).

7.3.3.3 Air-borne Diseases Induced by Pathogenic Fungi

pH is an important determinant of microbial community development, diversity, and structure (Lehmann et al. 2011; Husson 2013). Most soil bacteria develop between pH 4.5 and 10. The diversity was highest in neutral soils and lowest in acidic soils. In addition, the low pH of biochar corresponds to the low pyrolysis temperature. Therefore, we concluded that biochar with a pH between 7 and 9 effectively controls air-borne diseases induced by pathogenic fungi mainly by improving the soil community structure. In contrast, biochar with high pH (high content of K^+) resulted in the downregulation of genes related to plant immunity and defense, such as salicylic acid and jasmonic acid (Viger et al. 2015).

7.3.4 Effect of the Load Ratio of Biochar on the Plant Disease Control

Plant susceptibility/resistance to disease is dependent on the crucial factor of the biochar dose (Frenkel et al. 2017). The effect of the load ratio of biochar on the effectiveness of plant disease control is presented in Figure 7.2. A load ratio of biochar between 1 and 3% ($P < 0.01$) can significantly control soil-borne diseases caused by pathogenic fungi (Figure 7.2a). The load ratio of biochar under 1% ($P < 0.05$) can significantly control the soil-borne diseases caused by nematodes (Figure 7.2b). Both load ratios of biochar over 3% ($P < 0.01$) and between 1 and 3% ($P < 0.05$) can significantly control air-borne diseases caused by pathogenic fungi (Figure 7.2c).

7.3.4.1 Soil-borne Diseases Induced by Pathogenic Fungi

Jaiswal et al. (2014) show that the control effect of *Rhizoctonia solani* forms a U-shaped response curve biochar concentration. The low concentrations suppressed damping off by *R. solani*; however, the high concentrations were ineffective or even increased the disease incidence and severity. This viewpoint was confirmed again by meta-analysis in this section. This phenomenon is attributed to the toxic compounds in biochar that can damage plant roots and predispose them to pathogen attack (Graber et al. 2014). In addition, a high concentration of biochar (5%) induced the downregulation of genes associated with plant defense and immunity, such as the jasmonic acid and salicylic acid pathways, which resulted in a significant increase in disease severity for plants (Copley et al. 2017).

7.3.4.2 Soil-borne Diseases Induced by Plant-parasitic Nematodes

At present, there is no study about the dose-effect of biochar on the control of plant disease induced by plant-parasitic nematodes. The results of the meta-analysis show that a low load ratio (< 1%) can effectively control soil-borne disease induced by plant-parasitic nematodes. The middle load ratio and the high load ratio of biochar also show a certain prevention and control effect. The dose-effect should be further studied.

7.3.4.3 Air-borne Diseases Induced by Pathogenic Fungi

Our results show that a high load ratio (≥ 3%) is more effective in controlling plant air-borne diseases. Frenkel et al. (2017) concluded that the U-shaped response curve is not suitable for biochar concentration and foliar pathogens. Existing studies have shown that a high load ratio of biochar is conducive to inducing systemic resistance (Elad et al. 2010; Meller Harel et al. 2012). The improved nutrient supply, input of chemical elicitors (organic chemicals or salts), stimulation of beneficial soil microbial consortia, or improvement in soil physical qualities are responsible for the induced systemic resistance (Elad et al. 2010). However, the relationship between these effects and the load ratio of biochar is still unclear.

7.3.5 Optimal Application Methods of Biochar for Different Plant Diseases

Biochar shows a broad-spectrum capacity for plant disease suppression, which depends on its influence on soil, plant, and pathogens. According to the results of the meta-analysis, we found that biochars produced by wood feedstock, pyrolysis temperatures between 300 and 500 °C, pH values between 9 and 13, and load ratios between 1 and 3% are suitable for suppressing plant soil-borne diseases induced by pathogenic fungi; biochars produced by straw feedstock, pyrolysis temperatures

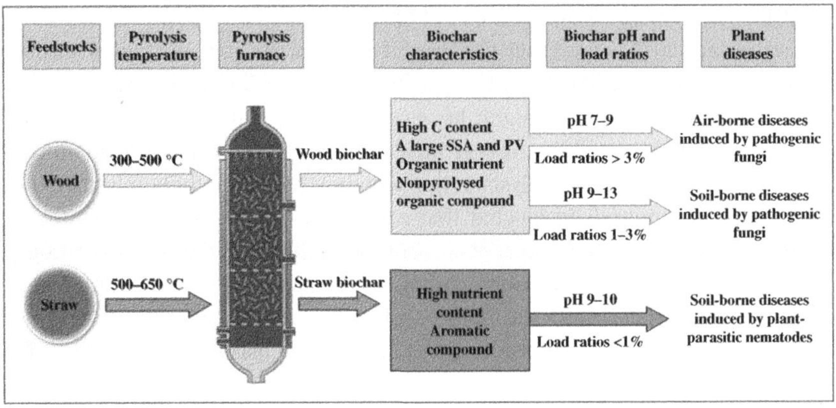

Figure 7.3 The optimal control results: different types of biochar correspond to different types of plant diseases.

between 500 and 650 °C, pH values between 9 and 10, and load ratios less than 1% are suitable for suppressing soil-borne diseases induced by plant-parasitic nematodes; and biochars produced by wood feedstock, pyrolysis temperatures between 300 and 500 °C, pH values between 7 and 9, and load ratios over 3% are suitable for suppressing plant air-borne diseases induced by pathogenic fungi (Figure 7.3).

7.4 Conclusion

The section highlights the efficacy of biochar in managing plant diseases by inhibiting pathogens and enhancing soil quality. It reveals that biochar, derived from various feedstocks and processed at specific temperatures, can significantly control a range of fungal, bacterial, nematode, and oomycete diseases. The study provides insights into the optimal biochar properties and application rates for different plant diseases, emphasizing its potential as an environmentally friendly disease management strategy.

Author Contribution

Study was selected by Zhongwang Liu and Mi Wei. The data was analyzed by Zhongwang Liu. This report was drafted by Zhongwang Liu and Mi Wei, and reviewed by Yuqing Sun, Jinfang Tan, and Mi Wei.

References

Akhter, A., Hage-Ahmed, K., Soja, G.et al. (2015). Compost and biochar alter mycorrhization, tomato root exudation, and development of Fusarium oxysporum f. sp lycopersici. *Frontiers in Plant Science* 6: 1–13. https://doi.org/10.3389/fpls.2015.00529.

Alaylar, B., Güllüce, M., Egamberdieva, D. et al. (2021). Biochar mediated control of soil-borne phytopathogens. *Environmental Sustainability* 4 (2): 329–334. https://doi.org/10.1007/s42398-021-00187-5.

Asadabadi, R.S., Hage-Ahmed, K., and Steinkellner, S. (2021). Biochar, compost and arbuscular mycorrhizal fungi: a tripartite approach to combat *Sclerotinia sclerotiorum* in soybean. *Journal of Plant Diseases and Protection* 128 (6): 1433–1445. https://doi.org/10.1007/s41348-021-00495-2.

Bonanomi, G., Ippolito, F., and Scala, F. (2015). A "Black" future for plant pathology? Biochar as a new soil amendment for controlling plant diseases. *Journal of Plant Pathology* 97 (2): 223–234.

Bonanomi, G., Jesu, G., Zotti, M. et al. (2021). Biochar-derived smoke-water exerts biological effects on nematodes, insects, and higher plants but not fungi. *Science of the Total Environment* 750: 142307. https://doi.org/10.1016/j.scitotenv.2020.142307.

Borthakur P.K., Bhattacharyya R.K., and Utpal, D. (2019). Biochar in organic farming. *Organic Farming* 109–134. https://doi.org/10.1007/978-3-030-04657-6_7.

Chen, W., Meng, J., Han, X. et al. (2019). Past, present, and future of biochar. *Biochar* 1 (1): 75–87. https://doi.org/10.1007/s42773-019-00008-3.

Copley, T., Bayen, S. and Jabaji, S. (2017). Biochar amendment modifies expression of soybean and *Rhizoctonia solani* genes leading to increased severity of Rhizoctonia foliar blight. *Frontiers in Plant Science* 8: 1–15. https://doi.org/10.3389/fpls.2017.00221.

Datnoff, L.E., Elmer, W.H. and Huber, D.M. (2007). *Mineral Nutrition and Plant Disease*. American Phytopathological Society (APS Press).

de Medeiros, E.V., Lima, N.T., de Sousa Lima, J.R. et al. (2021). Biochar as a strategy to manage plant diseases caused by pathogens inhabiting the soil: a critical review. *Phytoparasitica* 49 (4): 713–726. https://doi.org/10.1007/s12600-021-00887-y.

Elad, Y., David, D.R., Harel, Y.M. et al. (2010). Induction of systemic resistance in plants by biochar, a soil-applied carbon sequestering agent. *Phytopathology* 100 (9): 913–921. https://doi.org/10.1094/PHYTO-100-9-0913.

Elmer, W.H. and Pignatello, J.J. (2011). Effect of biochar amendments on mycorrhizal associations and Fusarium crown and root rot of asparagus in replant soils. *Plant Diseases* 95 (8): 960–966. https://doi.org/10.1094/PDIS-10-10-0741.

Frenkel, O., Jaiswal, A.K., Elad, Y. et al. (2017). The effect of biochar on plant diseases: What should we learn while designing biochar substrates? *Journal of*

Environmental Engineering and Landscape Management 25 (2): 105–113. https://doi.org/10.3846/16486897.2017.1307202.

George, C., Kohler, J. and Rillig, M.C. (2016). Biochars reduce infection rates of the root-lesion nematode *Pratylenchus penetrans* and associated biomass loss in carrot. *Soil Biology and Biochemistry* 95: 11–18. https://doi.org/10.1016/j.soilbio.2015.12.003.

Glaser, B. and Lehr, V.I. (2019). Biochar effects on phosphorus availability in agricultural soils: a meta-analysis. *Scientific Reports* 9 (1): 9338. https://doi.org/10.1038/s41598-019-45693-z.

Graber, E.R., Frenkel, O., Jaiswal, A.K. et al. (2014). How may biochar influence severity of diseases caused by soilborne pathogens? *Carbon Management* 5 (2): 169–183. https://doi.org/10.1080/17583004.2014.913360.

Graber, E.R., Meller Harel, Y., Kolton, M. et al. (2010). Biochar impact on development and productivity of pepper and tomato grown in fertigated soilless media. *Plant Soil* 337 (1–2): 481–496. https://doi.org/10.1007/s11104-010-0544-6.

Gu, Y., Hou, Y., Huang, D. et al. (2017). Application of biochar reduces *Ralstonia solanacearum* infection via effects on pathogen chemotaxis, swarming motility, and root exudate adsorption. *Plant Soil* 415 (1–2): 269–281. https://doi.org/10.1007/s11104-016-3159-8.

Gul, S., Whalen, J.K., Thomas, B.W. et al. (2015). Physico-chemical properties and microbial responses in biochar-amended soils: mechanisms and future directions. *Agriculture Ecosystems & Environment* 206: 46–59. https://doi.org/10.1016/j.agee.2015.03.015.

Hou, J., Pugazhendhi, A., Phuong, T.N. et al. (2022). Plant resistance to disease: using biochar to inhibit harmful microbes and absorb nutrients. *Environmental Research* 214 (Pt 2): 113883. https://doi.org/10.1016/j.envres.2022.113883.

Huang, W.K., Ji, H.L., Gheysen, G. et al. (2015). Biochar-amended potting medium reduces the susceptibility of rice to root-knot nematode infections. *BMC Plant Biology* 15. https://doi.org/10.1186/s12870-015-0654-7.

Husson, O. (2013). Redox potential (Eh) and pH as drivers of soil/plant/microorganism systems: a transdisciplinary overview pointing to integrative opportunities for agronomy. *Plant Soil* 362 (1): 389–417. https://doi.org/10.1007/s11104-012-1429-7.

Ippolito, J.A., Cui, L.Q., Kammann, C. et al. (2020). Feedstock choice, pyrolysis temperature and type influence biochar characteristics: a comprehensive meta-data analysis review. *Biochar* 2 (4): 421–438. https://doi.org/10.1007/s42773-020-00067-x.

Jaiswal, A.K., Elad, Y., Graber, E.R. et al. (2014). *Rhizoctonia solani* suppression and plant growth promotion in cucumber as affected by biochar pyrolysis temperature, feedstock and concentration. *Soil Biology and Biochemistry* 69: 110–118. https://doi.org/10.1016/j.soilbio.2013.10.051.

Jaiswal, A.K., Elad, Y., Paudel, I. et al. (2017). Linking the belowground microbial composition, diversity and activity to soilborne disease suppression and growth promotion of tomato amended with biochar. *Scientific Reports* 7: 44382. https://doi.org/10.1038/srep44382.

Jaiswal, A.K., Frenkel, O., Elad, Y. et al. (2015). Non-monotonic influence of biochar dose on bean seedling growth and susceptibility to *Rhizoctonia solani*: the "Shifted R-max-Effect". *Plant Soil* 395 (1–2): 125–140. https://doi.org/10.1007/s11104-014-2331-2.

Jin, X., Bai, Y., Khashi u Rahman, M. et al. (2022). Biochar stimulates tomato roots to recruit a bacterial assemblage contributing to disease resistance against *Fusarium* wilt. *iMeta* 1 (3): 1–13. https://doi.org/10.1002/imt2.37.

Kolton, M., Graber, E.R., Tsehansky, L. et al. (2017). Biochar-stimulated plant performance is strongly linked to microbial diversity and metabolic potential in the rhizosphere. *New Phytologist* 213 (3): 1393–1404. https://doi.org/10.1111/nph.14253.

Lehmann, J. (2009). Biological carbon sequestration must and can be a win-win approach. *Climate Change* 97 (3–4): 459–463. https://doi.org/10.1007/s10584-009-9695-y.

Lehmann, J., Rillig, M.C., Thies, J. et al. (2011). Biochar effects on soil biota – a review. *Soil Biology and Biochemistry* 43 (9): 1812–1836. https://doi.org/10.1016/j.soilbio.2011.04.022.

Li, X., Wang, T., Chang, S.X. et al. (2020). Biochar increases soil microbial biomass but has variable effects on microbial diversity: a meta-analysis. *Science of the Total Environment* 749: 141593. https://doi.org/10.1016/j.scitotenv.2020.141593.

Marra, R., Vinale, F., Cesarano, G. et al. (2018). Biochars from olive mill waste have contrasting effects on plants, fungi and phytoparasitic nematodes. *Plos One* 13 (6): 1–24. https://doi.org/10.1371/journal.pone.0198728.

Meller Harel, Y., Elad, Y., Rav-David, D. et al. (2012). Biochar mediates systemic response of strawberry to foliar fungal pathogens. *Plant and Soil* 357 (1–2): 245–257. https://doi.org/10.1007/s11104-012-1129-3.

Parker, J.E., Hessler, G. and Cui, H. (2021). A new biochemistry connecting pathogen detection to induced defense in plants. *New Phytologist* 234 (2022): 819–826. https://doi.org/10.1111/nph.17924.

Pietri, J.A. and Brookes, P. 2009. Substrate inputs and pH as factors controlling microbial biomass, activity and community structure in an arable soil. *Soil Biology and Biochemistry* 41 (7): 1396–1405. https://doi.org/10.1016/j.soilbio.2009.03.017.

Pineda, A., Kaplan, I., and Bezemer, T.M. (2017). Steering soil microbiomes to suppress aboveground insect pests. *Trends in Plant Science* 22 (9): 770–778. https://doi.org/10.1016/j.tplants.2017.07.002.

Pineda, A., Zheng, S.J., van Loon, J.J. et al. (2010). Helping plants to deal with insects: the role of beneficial soil-borne microbes. *Trends in Plant Science* 15 (9): 507–514. https://doi.org/10.1016/j.tplants.2010.05.007.

Poveda, J. (2021). Beneficial effects of microbial volatile organic compounds (MVOCs) in plants. *Applied Soil Ecology* 168: 104118. https://doi.org/10.1016/j.apsoil.2021.104118.

Poveda, J., Martinez-Gomez, A., Fenoll, C. et al. (2021). The use of biochar for plant pathogen control. *Phytopathology* 111 (9): 1490–1499. https://doi.org/10.1094/phyto-06-20-0248-rvw.

Rahman, L., Whitelaw-Weckert, M.A., and Orchard, B. (2014). Impact of organic soil amendments, including poultry-litter biochar, on nematodes in a Riverina, New South Wales, vineyard. *Soil Research* 52 (6): 604–619. https://doi.org/10.1071/sr14041.

Ristaino, J.B., Anderson, P.K., Bebber, D.P. et al. (2021). The persistent threat of emerging plant disease pandemics to global food security. *Proceedings of the National Academy of Sciences of the United State of America* 118 (23): e2022239118. https://doi.org/10.1073/pnas.2022239118.

Rogovska, N., Laird, D., Leandro, L., and Aller, D. (2016). Biochar effect on severity of soybean root disease caused by *Fusarium virguliforme. Plant and Soil* 413 (1–2): 111–126. https://doi.org/10.1007/s11104-016-3086-8.

Singh, H., Northup, B.K., Rice, C.W. et al. (2022). Biochar applications influence soil physical and chemical properties, microbial diversity, and crop productivity: a meta-analysis. *Biochar* 4 (1): 1–17. https://doi.org/10.1007/s42773-022-00138-1.

Spokas, K.A., Novak, J.M., Stewart, C.E. et al. (2011). Qualitative analysis of volatile organic compounds on biochar. *Chemosphere* 85 (5): 869–82. https://doi.org/10.1016/j.chemosphere.2011.06.108.

Steinkellner, S. and Mammerler, R. (2007). Effect of flavonoids on the development of Fusarium oxysporum f. sp. Lycopersici. *Journal of Plant Interactions* 2 (1): 17–23. https://doi.org/10.1080/17429140701409352.

Sun, D., Meng, J., Liang, H. et al. (2015). Effect of volatile organic compounds absorbed to fresh biochar on survival of *Bacillus mucilaginosus* and structure of soil microbial communities. *Journal of Soils and Sediments* 15 (2): 271–281. https://doi.org/10.1007/s11368-014-0996-z.

Tian, J.H., Rao, S., Yang, L., and Cai, Z.K. (2021). Wheat straw biochar amendment suppresses tomato bacterial wilt caused by *Ralstonia solanacearum*: Potential effects of rhizosphere organic acids and amino acids. *Journal of Integrative Agriculture* 20 (9): 2450–2462. https://doi.org/10.1016/s2095-3119(20)63455-4.

Van Wees, S.C., Van der Ent, S., and Pieterse, C.M. (2008). Plant immune responses triggered by beneficial microbes. *Current Opinion in Plant Biology* 11 (4): 443–448. https://doi.org/10.1016/j.pbi.2008.05.005.

Viger, M., Hancock, R.D., Miglietta, F. et al. (2015). More plant growth but less plant defence? First global gene expression data for plants grown in soil amended with biochar. *GCB Bioenergy* 7 (4): 658–672. https://doi.org/10.1111/gcbb.12182.

Wang, L., O'Connor, D., Rinklebe, J. et al. (2020). Biochar aging: mechanisms, physicochemical changes, assessment, and implications for field applications. *Environmental Science and Technology* 54 (23): 14797–14814. https://doi.org/10.1021/acs.est.0c04033.

Xu, H., Cai, A., Wu, D. et al. (2021). Effects of biochar application on crop productivity, soil carbon sequestration, and global warming potential controlled by biochar C:N ratio and soil pH: a global meta-analysis. *Soil and Tillage Research* 213: 105125. https://doi.org/10.1016/j.still.2021.105125.

Yang, L., Wu, Y., Wang, Y. et al. (2021). Effects of biochar addition on the abundance, speciation, availability, and leaching loss of soil phosphorus. *Science of the Total Environment* 758: 143657. https://doi.org/10.1016/j.scitotenv.2020.143657.

Yu, H., Zhang, Z., Li, Z. et al. (2014). Characteristics of tar formation during cellulose, hemicellulose and lignin gasification. *Fuel* 118: 250–256. https://doi.org/10.1016/j.fuel.2013.10.080.

Zhang, L., Jing, Y., Xiang, Y. et al. (2018). Responses of soil microbial community structure changes and activities to biochar addition: a meta-analysis. *Science of the Total Environment* 643: 926–935. https://doi.org/10.1016/j.scitotenv.2018.06.231.

Zhang, X.K., Li, Q., Liang, W.J. et al. (2013). Soil nematode response to biochar addition in a Chinese wheat field. *Pedosphere* 23 (1): 98–103. https://doi.org/10.1016/s1002-0160(12)60084-8.

8

Improvement of Soil Physical, Chemical, and Biological Properties by Waste-derived Biochar

Ying Zhao[1,2#], Zhuqing Liu[1,2#], Jiang Song[1,2#], Kui Cheng[2,3], and Fan Yang[1,2*]*

[1] *School of Water Conservancy and Civil Engineering, Northeast Agricultural University, Harbin, China*
[2] *International Cooperation Joint Laboratory of Health in Cold Region Black Soil Habitat of the Ministry of Education, Harbin, China*
[3] *College of Engineering, Northeast Agricultural University, Harbin, China*

*Corresponding authors: yangfan_neau@163.com; chengkui@neau.edu.cn

Soil is an important part of the biosphere and the basis of agriculture and natural ecosystems. Because of its important role in the sustainable development of agriculture, soil quality is receiving extensive attention from all quarters worldwide. In recent years, with the interference of human factors, problems of soil (acidification, salinization, and desertification) have become more and more prominent. Therefore, improving the physical, chemical and biological properties of degraded soils and enhancing soil quality is of great significance (Figure 8.1). Relevant research has proved that biochar added to soil can not only increase the number of soil aggregates and microorganisms but also improve the stability of soil aggregates and the interaction of mineral particles to avoid soil erosion. In addition, biochar can regulate soil acidity and alkalinity and increase soil cation exchange capacity. These properties can improve soil organic matter content, increase production and income, and promote sustainable agricultural development. This chapter will analyze the mechanism of adding biochar to soil physical, chemical, and biological properties.

#These authors contributed equally to this chapter.

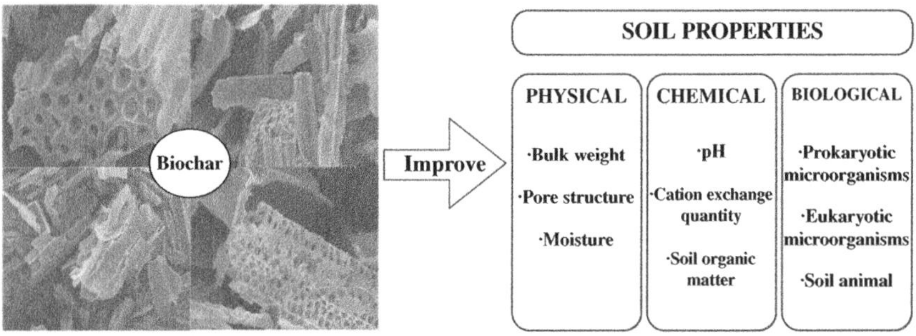

Figure 8.1 Biochar for soil improvement.

8.1 Biochar Improves Soil Physical Properties

Biochar, as a stable soil amendment, plays an important role in improving soil physical properties. Soil bulk weight, pore structure, and soil moisture are important indicators of soil physical properties (Gao et al. 2012; Li et al. 2019). Soil bulk weight is closely related to soil compactness, which reflects soil quality. Soil pore structure plays an important role in soil aeration, water retention, and nutrient release. Soil moisture is critical to the metabolism of plants and other soil organisms and is an important factor in soil formation and development. For this reason, this chapter discusses the effects of biochar on soil bulk density, pore structure, and soil moisture, respectively, to provide a scientific basis for improving soil physical properties.

8.1.1 Effect of Biochar on Soil Bulk Weight

Soil bulk weight is one of the important indicators of the physical and chemical properties of the soil, which is closely related to the degree of soil compactness and can reflect the soil quality (Håkansson and Lipiec 2000). As a stable soil conditioner, biochar added to the soil is important in reducing soil capacity (Table 8.1). There are two main reasons for biochar to reduce soil bulkiness. On the one hand, biochar can reduce the soil bulkiness by its great specific surface area and surface charge, on the other hand, biochar can improve the soil structure (Abel et al. 2013). Generally, low bulk weight soil with higher organic matter content is more favorable for soil nutrient retention. Biochar reduces soil compaction, which is conducive to seed germination and saves planting costs (Gonzaga et al. 2020; Joseph et al. 2018). Therefore, the application of biochar can reduce soil bulk weight and contribute to the improvement of soil productivity. However, the effect of biochar on soil bulk weight is also related to dilution and friction (Jien and Wang 2013).

Table 8.1 Effects of different biochar applications on soil bulk weight.

Materials	Temperature (°C)	Applications	Effects	References
Yellow pine	400	Acidic sandy soil	Soil bulk density decreased 40%	Chang et al. (2021)
Paper fiber and grain husk	600	Silt loam	Significant decrease in soil bulk density values	Makó et al. (2020)
Empty fruit bunch	600	Silty clay loam	Soil bulk density decreased	Wang et al. (2022)
Maple wood	500	Silt loam soil	A decline in bulk density	Ahmed et al. (2017)
Corncobs	360	Sandy loam	Decrease in soil bulk density	Zhao et al. (2016)

The application of biochar to compacted soil does not result in effective recovery with the addition of biochar due to the low elasticity of biochar, which results in the soil's ability to recover.

The high porosity of biochar makes it a good treatment for reducing soil bulk density. Moreover, the change in weight capacity is due to the mixing of biochar with good pore structure into the soil surface layer, which changes the original clay-heavy texture of saline soil and effectively improves the soil structure (Duan et al. 2021; Liang et al. 2021). Ge et al. (2014) study the effect of biochar on soil bulk weight and cation exchange in apple orchards, found that the addition of biochar could significantly reduce the soil bulk weight and the trend of the effect on the soil bulk weight in the two soil layers of 0–5 cm and 5–10 cm was the same. However, the densities of biochar prepared from different raw materials varied greatly and had different mechanical strengths (Zhang et al. 2020b; Akhtar and Sarmah 2018). Biochar may undergo mechanical fragmentation or become smaller particles due to other physicochemical effects, which can enter the soil pores and increase the soil capacity (Omondi et al. 2016). Although studies on biochar as a soil amendment to increase soil bulk weight have been reported, they deserve further attention, especially for their long-term effects on the soil.

8.1.2 Effect of Biochar on Soil Pore Structure

Soil pore space is the space between solid soil particles. The size and shape of soil pore and its stabilizing strength are closely related to soil structure. It is the space that holds water and air and the place for plant root growth, soil fauna, and microbial activity (Lucas et al. 2019). Good soil pore structure is important in soil aeration, water retention, and nutrient release. The pore distribution, connectivity, particle size of biochar, and its movement through the soil affect the soil pore structure, which can improve the overall adsorption capacity of the soil.

After adding biochar to soil, soil pore structure is affected by multiple aspects of biochar, such as intrinsic porosity, connectivity, and mechanical strength (Baiamonte et al. 2019). In addition, the transfer and interaction of biochar particles in the soil can lead to changes in total soil porosity. In addition, soil volume and porosity shows a linear decrease and increase, respectively, with the increase of biochar application (Glab et al. 2016). The ability of biochar to increase total soil porosity is mainly because biochar is loose and porous, which promotes the formation of soil aggregates. Although, in most cases, biochar can reduce soil bulk weight and increase total soil porosity. The degree of reduction or increase is related to soil type, source of biochar, cracking temperature, application rate, application method, etc. (Abel et al. 2013; Lusiba et al. 2017). The results of Zhou et al. (2019) showed that after applying biochar made from corn cobs pyrolyzed for 24 hours at 360 °C to sandy loam soils, sandy loam soils had significantly increased and their large porosity was significantly increased. Some studies concluded that applying biochar decreased soil macroporosity, with a gradual shift from large to small pores. Petersen et al. (2016) showed that the application of biochar produced by rapid cracking of wheat straw at 525 °C transformed the drained pores of sandy soil, which ranged from 60 to 300 μm pore sizes, into pores ranging from 0.2 to 60 μm. Fan et al. (2020a) found that the annual application of biochar made from straw cracking significantly reduced the porosity and pore number through successive years of rice–wheat rotation positioning experiments. The phenomenon may be attributed to small particle size biochar entering the soil macropore, which gradually reduces the pore size of the soil macropores and transforms them into relatively small pores. The application of biochar can change the distribution of soil macropores, and the actual extent to which biochar changes the structure of soil macropores needs to be further considered in the context of biochar source, the cracking temperature, the type of soil, and the number of years of application.

8.1.3 Effect of Biochar on Soil Moisture

Soil moisture is vital to the metabolism of plants and other soil organisms, and the various elements dissolved in the soil are an important source of effective nutrients required for plant growth (Table 8.2). Soil moisture indicator usually includes saturated water content, capillary water holding capacity, capillary fracture water content, field water holding capacity, and wilting coefficient (Evett et al. 2019). These indicators reflect the water-holding and water-retaining capacity of the soil

Table 8.2 Effects of different biochar applications on soil moisture.

Materials	Temperature (°C)	Applications	Effects	References
Willow stem wood	320	Heavy clay soil	Increased the amount of readily plant available water	Rasa et al. (2018)
Apple branch	550	Black clay soil	Water-holding capacity were significantly increased	An et al. (2016)
Oak feedstock	605	Low clay aggregate	Increase moisture retention (7–8.5%)	Licht et al. (2017)
Tobacco stem and rice husk	550	Paddy soil and red soil	The hygroscopicity of biochar-soil mixtures could be significantly improved	Lin et al. (2016)
Corsican Pine	1000	Low in silt soils	Available water were more pronounced	Peake et al. (2014)

and are important guides for improving soil moisture characteristics. However, excessive reclamation and agricultural activities have caused serious soil erosion phenomena. Therefore, reducing water loss in soil will provide an important guarantee for soil nutrient retention. The water-holding capacity of soil depends on the distribution and connectivity of soil pores, which is largely limited by soil particle size, structural characteristics, and soil organic matter content. When biochar is added to the soil, the water-holding capacity and water retention properties of the soil can be significantly improved due to the large specific surface area of biochar. Glaser et al. (2002) pointed out that the water-holding capacity of cultivated soils in biochar-rich Amazon River Basin was about 18% higher than that of the ones that did not contain biochar. Liang et al. (2014) proved through three years of experiments that adding a certain amount of biochar could significantly improve the soil's water-holding capacity, and the increase can reach more than 9%. In addition, Zhang et al. (2020a) found that the application of biochar increased the water-holding capacity of soil under low suction conditions. The water-holding capacity of sandy soils increases after biochar addition, but the enhancement effect diminishes with increasing suction force values. The water-holding capacity of loamy and clayey soils decreases significantly under medium and high suction conditions, and the reduction effect became more pronounced with increasing suction value and biochar addition. This is mainly due to the increase in the number and volume ratio of small pores in sandy soils with the addition of biochar.

The addition of biochar alters soil moisture constants, including field water-holding capacity, wilt coefficient, and maximum effective water content. The water-holding capacity of biochar is stronger than that of sandy soils, and the fine pore structure of biochar can lead to a significant increase in the field water-holding capacity of soils with the addition of biochar to the soil. In addition, the water-holding capacity of biochar is weaker than that of loamy and clayey soils, resulting in loamy and clayey soils draining more readily with the addition of biochar. The effect of biochar on soil water-holding capacity may be related to the nature of biochar itself and indirect factors such as influencing soil structure and properties. Thermal expansion leads to the formation of many pores of different pore sizes in biochar (Edeh et al. 2020). Due to stress, the pore surface is partially destroyed and the biochar skeleton looks incomplete. In addition, due to the presence of ash, the biochar surface tends to form abundant oxygen-containing functional groups (Amin and Chetpattananondh 2019). The presence of oxygen-containing functional groups gives biochar certain hydrophilicity, which enhances its ability to retain soil moisture. The number of oxygen-containing functional groups gradually increases as the surface of biochar in the soil oxidizes. At the same time, biochar usually contains hydrophobic functional groups with a certain hydrophobic effect. However, the number of hydrophobic functional groups and their presence or absence in biochar are also closely related to the feedstock and the temperature at which the biochar is prepared. Generally, biochar prepared at 400–600 °C has a high water-holding capacity and low hydrophobicity (Chen et al. 2008). The hydrophobicity of biochar tends to promote the occurrence of soil surface runoff and accelerate the loss of applied fertilizers in agricultural fields, thus increasing the risk of farmland surface pollution. Therefore, when specifically applying biochar in agricultural soils, special attention needs to be paid to the effect of hydrophobicity on the water-holding capacity of field soils.

8.2 Biochar Improves Soil Chemical Properties

Soil pH, soil cation exchange capacity, and soil organic matter content are all chemical properties of soil that are closely related to soil quality. Soil pH is one of the important indicators of soil, which directly or indirectly affect the crops growth and the migration of pollutants (Xie et al. 2022). Soil cation exchange quantity refers to the total amount of various cations that can be adsorbed by soil colloids, which is an important index that reflects the buffering capacity of the soil and is closely related to other indexes (Ouyang et al. 2023). Soil minerals and organic matter are known as the soil's "bone" and "muscle." Their content is closely related to the level of soil fertility, is the main source of crop nutrients, has the effect of improving soil quality and promoting crop growth and development, for microbial activity also has a significant impact (Melero et al. 2007; Xia et al. 2023). Therefore, improving the chemical properties of degraded soils is significant in upgrading soil quality.

8.2.1 Effect of Biochar on Soil pH

Soil pH is an important indicator of soil quality, and good soil acidity is a prerequisite for high crop yields. The chemical balance between acidic and alkaline substances determines the acidity of soil. Soil acidity is expressed in terms of both strength and capacity indicators. Depending on the form of H^+ in the soil, it can be categorized into active and latent acidity (Table 8.3). Soil acidity and alkalinity have an important influence on the respiration and growth of plant roots, as well as the absorption function of the root system. China has a huge area of acidic soil, and soil acidification has become an important cause of soil degradation and a major global environmental problem (Raza et al. 2020). The addition of biochar can significantly increase soil pH and improve the soil acidification problem. The effect of biochar on soil pH is also related to the nature of the soil itself (Xie et al. 2022). Liu et al. (2020) investigated the effect of biochar addition on saline soil improvement and found that the soil pH value did not increase

Table 8.3 Effects of different biochar applications on soil pH.

Materials	Temperature (°C)	Applications	Effects	References
Wheat straw	600	Acidified tea plantation	Significantly ($P < 0.05$) increased the average soil pH	Jiang et al. (2018)
Sewage sludge	550	Acidic sandy soil (pH = 4.9)	pH increased to 5.4	Buss et al. (2018)
Miscanthus straw	700		pH increased to 6.75	
Soybean stover	700	Acidic soil (pH = 5.02)	pH increased to 7.42	Moon et al. (2017)
Oak	700		pH increased to 5.37	
Peanut straw	400	Acidic soil (pH = 4.6)	pH 0.34 higher than $Ca(OH)_2$ amended soil	Shi et al. (2019)
Rice straw	400	Acidic soil (pH = 4.6)	pH 0.27 higher than $Ca(OH)_2$ amended soil	
Maize and sorghum straw	350	Acidic yellow-brown soil (pH = 4.6)	Biochar application significantly increased soil pH ($P < 0.05$)	Wu et al. (2020)

but slightly decreased with the increase of biochar application. The effect of biochar on soil pH is affected by a combination of factors such as its own physical and chemical properties, the amount of addition, and the physical and chemical properties of the soil itself (Omondi et al. 2016; Zhao et al. 2022).

8.2.2 Effect of Biochar on Soil Cation Exchange Quantity

The cation exchange capacity of a soil is the total amount of various exchangeable cations adsorbed by the soil colloid (e.g. calcium, magnesium, potassium, and sodium), which is the main source of the buffering capacity of the soil (Li et al. 2022; Jia et al. 2021). In general, the amount of cation exchange in different soils is determined by the amount of different colloidal structures. The soil cation exchange quantity is an important indicator for soil improvement, rational fertilization, and soil classification. Soil with high cation exchange capacity has a greater ability to bind plant nutrient cations, thus retaining these cations in the root zone for plant uptake and preventing them from leaching below the root zone (Cui et al. 2021). In addition, soil cation exchange quantity mainly comes from various minerals in the soil, in which the finer texture of the soil has a higher cation exchange quantity (Droge and Goss 2013).

The current study found that the effect of biochar on soil cation exchange is largely dependent on the nature of biochar itself (Hardy et al. 2017). Therefore, it is extremely important to master the carbonization time and temperature of biochar. Relevant studies have pointed out that the improvement of soil cation exchange by biochar is also closely related to the time of biochar in the soil (Borchard et al. 2014). The longer the biochar exists in the soil, the more oxygen-containing functional groups will be formed on its surface, thus increasing its surface charge and leading to higher soil cation exchange (Fan et al. 2018). In addition, the cation exchange capacity of the applied biochar itself is also a key influencing factor; in general, biochar that has a higher cation exchange capacity of its own will play a more pronounced role if it is added to the soil, and vice versa the role will be reduced (Hailegnaw et al. 2019). In addition, biochar is more sensitive to the effects of acidic soils with low cation exchange, while it has less effect on soils with high cation exchange (Zhao et al. 2022). However, there are fewer long-term field trial studies on the effect of biochar application on soil cation exchange, and further research and advancement are needed.

8.2.3 Effect of Biochar on Soil Organic Matter

Soil organic matter refers to a variety of carbon-containing organic compounds that exist in the soil in various forms and states and is an important component of the soil (Jensen et al. 2020). Specifically, it includes various products from different stages of decomposition and synthesis of animal, plant, and microbial residues in the soil (Cayuela et al. 2009). It plays an important role in soil fertility, environmental protection, and sustainable agricultural development (Table 8.4). It can be combined with soil nutrients and used as a carrier to transport soil nutrients, and it can also be used as a raw material for plant growth and microbial reproduction and development. The high pH, porous nature, high specific

Table 8.4 Effect of different biochar application rates on soil organic matter.

Materials	Temperature (°C)	Experimental site	Effects	References
Rice straw	500	Urumqi (86.55 °E, 44.41 °N)	With the 12, 60, 120, and 170 g/kg addition, the soil organic matter increased by 1.52, 19.47, 45.58, and 84.12%, respectively	Yang et al. (2024)
Wheat straw	350	Hua Jia Ji, Hangzhou (120.26 °E, 30.17 °N)	Soil organic carbon was higher in soil with 1% biochar than in soil with 2% straw	Zhang et al. (2012)
Rice husk	550	Scientific Research Base of the Zhejiang Academy of Agricultural Sciences (120.20 °E, 30.31 °N)	5% biochar inhibited SOC degradation by 7.95–10.7% after 120 days of incubation	Liu et al. (2018)
Pecan shell	550			
Bamboo	550			
Aubergine clay—rice husk	500	Experimental base of Zhejiang Academy of Agricultural Sciences (120 °11′34″E, 39 °18′27″N)	Soil organic matter: rice husk > Aubergine clay–rice husk > sludge > blank control	Li et al. (2023b)
Rice husk	500			
Sludge	500			
Straw	500	Hetao Irrigation Area (107 °6′–107 °4′E, 40 °31′–44 °17′N)	Organic matter content increased by 35.4% when biochar was applied at 50 t/ha	Cen et al. (2021)
Maize straw	550	Kunming City, Yunnan Province (102 °52′1.99″E, 24 °51′38.53″N)	The addition of biochar retained easily degradable fractions	Li et al. (2023a)

surface area, and abundant negative surface charge of biochar can influence the mineralization process of soil organic matter (Qian et al. 2015). The inhibitory effect of biochar input on soil organic matter decomposition and transformation has been widely reported. Yu et al. (2014) found that woody biochar addition prepared under different temperature conditions could effectively inhibit the mineralization of exogenous organic matter. Therefore, due to its huge specific surface area and pore structure, biochar can preserve organic matter in its pores and inhibit the mineralization of soil organic matter.

In addition to being present in the soil as solid particles, biochar can also migrate through the soil in a soluble state. Although the proportion of biochar is small, it has an impact on soil organic carbon pool turnover. Analysis of soil organic matter in tropical rain forests, which cover 8% of the global land area, showed that many years after the disappearance of forest fires, there is still a large amount of soluble biochar that can be effectively transported in the soil, which is characterized by strong physical and chemical transport effects. Some of these biochar will be maintained in the soil by migratory transformation, which plays an important role in sequestering the grassland soil organic carbon pool. Based on the above study, it can be hypothesized that biochar imported into the soil can not only be stored in the soil in the solid form, but also migrate in the soil solution in the dissolved form, which may play an important role in enriching the organic matter content of soils in other zones.

The inhibitory effect of biochar inputs on the mineralization of soil organic matter and the promotion of soil humification process will help to increase the soil organic matter content. Sui et al. (2022) studied the trend of soil organic matter content after biochar application. The results of the study found that soil organic matter can combine with soil nutrients and is a carrier for transporting soil nutrients, and the application of biochar significantly increased the soil organic matter content. In addition, after 90 days of incubation at ambient temperature and 90 freeze–thaw cycles, the soil organic matter content decreased compared with that of the control soil. However, the soil organic matter content of the treatment groups incubated for 90 days showed different increases after application of 1, 2, 3 and 4% biochar.

Through an indoor incubation experiment, Kuang et al. (2012) investigated the effects of biochar addition on organic carbon mineralization and microbial mass carbon and nitrogen in red soil rice soil. The results showed that biochar addition could effectively reduce the soil organic carbon mineralization rate and mineralization accumulation rate, and significantly improve the soil microbial quality (Melero et al. 2007). In addition, the addition of rice straw biochar could increase the content of soluble organic carbon and soil organic matter in rice soil, and the content of soil soluble organic carbon and soil organic matter would increase with the increase of the addition amount.

8.3 Biochar Improves Soil Biological Properties

Soil organisms are the core of the whole soil system, playing the roles of consumers and decomposers in the soil ecosystem, and playing an irreplaceable role in soil formation, soil organic matter and nutrient turnover and plant growth (Choudhary et al. 2016). Biochar has been increasingly emphasized as a means of regulating soil biological properties because of its special physical and chemical properties (Figure 8.2). In addition, community structure of soil organisms provides rich organic matter to the soil, thereby promoting the growth of soil organisms and the secretion of enzymes (Ali et al. 2020). Accordingly, this section describes the mechanisms related to the regulation of soil biological functions by biochar in terms of the succession of soil biological community structure and changes in microbial ecological functions.

8.3.1 Effect of Biochar on Soil Prokaryotic Microorganisms

There are many prokaryotic microorganisms in the soil that are beneficial to plant growth or related to material metabolism, including biocontrol bacteria, nitrogen-fixing bacteria, methanogenic bacteria, methane-oxidizing bacteria, and ammonia-oxidizing bacteria (Wang and Bao 2022). These microorganisms can

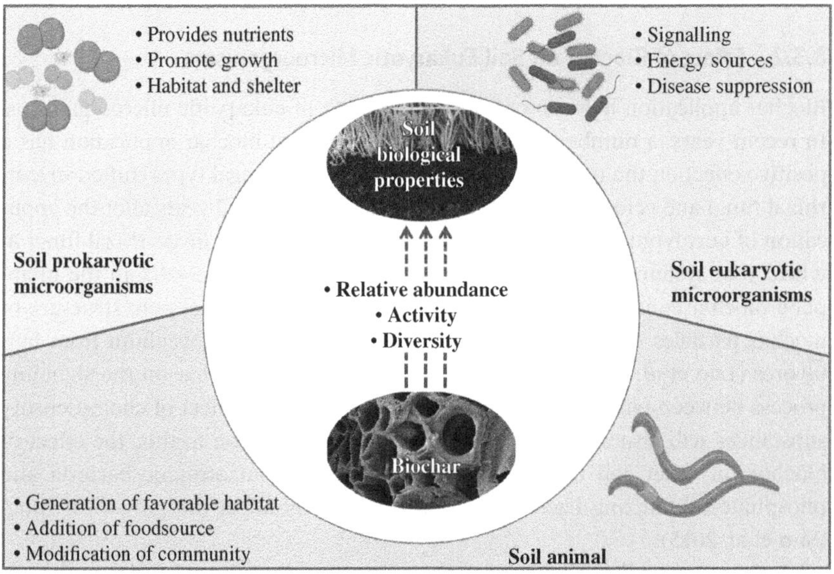

Figure 8.2 Effect of biochar application on soil biological properties.

promote soil nutrient and material cycling, inhibit soil pests and diseases, and stimulate plant growth. In recent years, it has been shown that the imposition of biochar can affect the growth and metabolism of prokaryotic microorganisms. On the one hand, biochar stimulates the growth of soil microorganisms mainly by changing the soil physicochemical properties (Fan et al. 2020b). In general, the application of biochar causes an increase in soil redox potential, and high redox potential helps the growth of methane-oxidizing and nitrogen-fixing bacteria (Spokas 2010). On the other hand, the application of biochar can also increase the carbon to nitrogen ratio of the soil, reduce the effective nitrogen content of the soil and stimulate the growth of nitrogen-fixing bacteria (Rondon et al. 2007). Since biochar is rich in nutrients such as phosphorus, potassium, calcium, magnesium and zinc, and soluble organic carbon, it can provide nutrients for soil microorganisms. In addition, biochar pores can serve as a habitat and refuge for soil microorganisms, protecting bacteria, fungi, and protozoa from attack. Macroporous (>200 nm) biochar is a habitat for most microorganisms because they are sized appropriately to hold bacteria and can store water and dissolved substances needed for microbial metabolism. Biochar application can cause changes in the structural diversity, abundance and activity of soil microbial populations, which can affect the biological properties of the soil (Jianjing et al. 2019). Pietikainen et al. (2000) found that bacteria attached to biochar were difficult to leach, which increased the abundance of bacteria. In addition, biochar had a significant effect on actinomycetes, and the response of actinomycetes to biochar varied in different soil types.

8.3.2 Effect of Biochar on Soil Eukaryotic Microorganisms

Biochar application helps to promote the growth of eukaryotic microorganisms. In recent years, a number of studies have shown that biochar application has a positive effect on the two most common mycorrhizal fungal types (tufted mycorrhizal fungi and ectomycorrhizal fungi). It was found that 2 years after the application of eucalyptus biochar, the colonization rate of tufted mycorrhizal fungi in wheat root systems increased to 20–40%, compared to only 5–20% in the unapplied biochar control group (Makoto et al. 2010). The internal pore structure of biochar particles can provide a "refuge" for the extra-root mycelium from herbivores (Luo et al. 2013). Secondly, the interference of biochar on the signaling process between plants and fungi, and the detoxification effect of chemosensory substances will also affect the growth of fungi. In addition to this, the effect of biochar on other soil microorganisms (mycorrhizal auxotrophic bacteria and phosphate-solubilizing bacteria) will also indirectly affect the growth of fungi (Sun et al. 2015).

Biochar is equally capable of influencing the diversity of fungi and their community structure. Analyzed from a functional point of view, fungi include three functional groups, namely saprophytic fungi, pathogenic fungi, and symbiotic fungi (Marlène et al. 2010). Saprophytic fungi rely on carbon substrates as an energy source, so the unstable carbon component in biochar can be used as a carbon source to promote the growth of saprophytic fungi, thus enhancing the competition between saprophytic fungi and pathogenic and symbiotic fungi, and ultimately decreasing fungal diversity (Ascough et al. 2010). Several studies have found that saprophytic fungi have the ability to utilize the stabilized carbon portion of biochar when unstable carbon is used. In addition, pathogenic bacteria in the soil can cause crop diseases, while biochar can reduce plant diseases by inhibiting the spread of pathogens in the soil (Poveda et al. 2021). Studies have found that biochar application can reduce the relative abundance of potential plant pathogens (e.g. Fusarium) in the soil, thereby suppressing crop diseases in black soils. For example, coconut biochar suppresses Fusarium infection in asparagus roots due to the application of biochar promoting soil microbes to compete with pathogens for carbon resources and produce toxic compounds against parasitic pathogens.

8.3.3 Effect of Biochar on Soil Animal

The effect of biochar on soil fauna is related to its preparation temperature. For example, the reproduction rate of elasmobranch soil fauna was lower after the application of biochar prepared at 550 °C compared to biochar prepared at 350 °C (Domene et al. 2015). It was shown that corn stover biochar prepared at 700 °C was more toxic to earthworms than biochar prepared at 350 °C. This is due to the fact that the higher the preparation temperature of biochar, which contains a higher percentage of ash, leads to high salt content of the soil, which has a certain toxic effect on earthworms.

Biochar is prepared from different biomass raw materials and contains a wide range of different elements, which can also have different effects on soil fauna. Biochar made from poultry manure can be toxic to earthworms due to high pH, salt content, and ammonium content, while biochar made from pine wood is not toxic to earthworms (Marchand et al. 2017). In addition, the size of biochar particles also varies, which has different effects on soil fauna, and either too large or too small particles are difficult to be utilized by the soil fauna. Studies have shown that biochar particles of pine wood chips smaller than 0.5 mm have an inhibitory effect on soil fauna (Prodana et al. 2019); this is because large particle sizes are usually not easy for soil fauna to feed on or transport, whereas smaller biochars are more likely to be ingested by soil fauna, and excessive feeding on contaminant-laden biochars can have a negative effect on soil organisms.

The properties of biochar and its stability in soil tend to diminish over time. The pH of biochar changes differently over time depending on the type of feedstock and its particle size decreases with time. Therefore, long-term monitoring of the effects of biochar on soil fauna (earthworms, nematodes, etc.) and soil ecosystems will be needed in the future.

References

Abel, S., Peters, A., Trinks, S. et al. (2013). Impact of biochar and hydrochar addition on water retention and water repellency of sandy soil. *Geoderma* 202: 183–191.

Ali, A., Shaheen, S.M., Guo, D. et al. (2020). Apricot shell- and apple tree-derived biochar affect the fractionation and bioavailability of Zn and Cd as well as the microbial activity in smelter contaminated soil. *Environmental Pollution* 264: 114773.

Ahmed, A., Gariepy, Y., and Raghavan, V. (2017). Influence of wood-derived biochar on the compactibility and strength of silt loam soil. *International Agrophysics* 31 (2): 149–155.

Akhtar, A. and Sarmah, A.K. (2018). Novel biochar-concrete composites: manufacturing, characterization and evaluation of the mechanical properties. *Science of the Total Environment* 616: 408–416.

Amin, M. and Chetpattananondh, P. (2019). Biochar from extracted marine *Chlorella* sp. residue for high efficiency adsorption with ultrasonication to remove Cr(VI), Zn(II) and Ni(II). *Bioresource Technology* 289: 121578.

An, Y., Ji, Q., Zhao, S.-X. et al. (2016). Effect of biochar application on soil aggregates distribution and moisture retention in orchard soil. *Huan jing ke xue=Huanjing kexue* 37 (1): 293–300.

Ascough, P.L., Sturrock, C.J., and Bird, M.I. (2010). Investigation of growth responses in saprophytic fungi to charred biomass. *Isotopes in Environmental and Health Studies* 46: 64–77.

Baiamonte, G., Crescimanno, G., Parrino, F. et al. (2019). Effect of biochar on the physical and structural properties of a sandy soil. *Catena* 175: 294–303.

Borchard, N., Ladd, B., Eschemann, S. et al. (2014). Black carbon and soil properties at historical charcoal production sites in Germany. *Geoderma* 232: 236–242.

Buss, W., Shepherd, J.G., Heal, K.V. et al. (2018). Spatial and temporal microscale pH change at the soil-biochar interface. *Geoderma* 331: 50–52.

Cayuela, M.L., Sinicco, T., and Mondini, C. (2009). Mineralization dynamics and biochemical properties during initial decomposition of plant and animal residues in soil. *Applied Soil Ecology* 41: 118–127.

Cen, R., Feng, W.Y., Yang, F. et al. (2021). Effect mechanism of biochar application on soil structure and organic matter in semi-arid areas. *Journal of Environmental Management* 286: 112198.

Chang, Y.R., Rossi, L., Zotarelli, L. et al. (2021). Biochar improves soil physical characteristics and strengthens root architecture in Muscadine grape *Vitis rotundifoli* L. *Chemical and Biological Technologies in Agriculture* 8 (1): 1–11.

Chen, B., Zhou, D., and Zhu, L. (2008). Transitional adsorption and partition of nonpolar and polar aromatic contaminants by biochars of pine needles with different pyrolytic temperatures. *Environmental Science & Technology* 42: 5137–5143.

Choudhary, S.K., Gupta, S.K., Singh, M.K. et al. (2016). Role and its utilization of beneficial micro-organisms for sustainable crop production. *International Journal of Agriculutural Science* 12 (2): 370–378.

Cui, X., Mao, P., Sun, S. et al. (2021). Phytoremediation of cadmium contaminated soils by *Amaranthus Hypochondriacus* L.: the effects of soil properties highlighting cation exchange capacity. *Chemosphere* 283: 131067.

Domene, X., Enders, A., Hanley, K. et al. (2015). Ecotoxicological characterization of biochars: role of feedstock and pyrolysis temperature. *Science of the Total Environment* 512: 552–561.

Droge, S.T.J. and Goss, K.-U. (2013). Development and evaluation of a new sorption model for organic cations in soil: contributions from organic matter and clay minerals. *Environmental Science & Technology* 47: 14233–14241.

Duan, M., Liu, G., Zhou, B. et al. (2021). Effects of modified biochar on water and salt distribution and water-stable macro-aggregates in saline-alkaline soil. *Journal of Soils and Sediments* 21: 2192–2202.

Edeh, I.G., Masek, O., and Buss, W. (2020). A meta-analysis on biochar's effects on soil water properties–new insights and future research challenges. *Science of the Total Environment* 714: 136857.

Evett, S.R., Stone, K.C., Schwartz, R.C. et al. (2019). Resolving discrepancies between laboratory-determined field capacity values and field water content observations: implications for irrigation management. *Irrigation Science* 37: 751–759.

Fan, Q., Sun, J., Chu, L. et al. (2018). Effects of chemical oxidation on surface oxygen-containing functional groups and adsorption behavior of biochar. *Chemosphere* 207: 33–40.

Fan, R., Zhang, B., Li, J. et al. (2020a) Straw-derived biochar mitigates CO_2 emission through changes in soil pore structure in a wheat-rice rotation system. *Chemosphere* 243: 125329.

Fan, S., Zuo, J., and Dong, H. (2020b). Changes in soil properties and bacterial community composition with biochar amendment after six years. *Agronomy* 10: 746.

Gao, H., He, X., Chen, X. et al. (2012). Effect of biochar and biochar-based ammonium nitrate fertilizers on soil chemical properties and crop yield. *Journal of Agro-Environment Science* 31: 1948–1955.

Ge, S., Peng, L., Ren, Y. et al. (2014). Effect of straw and biochar on soil bulk density, cation exchange capacity and nitrogen absorption in apple orchard soil. *Scientia Agricultura Sinica* 47: 366–373.

Glab, T., Palmowska, J., Zaleski, T. et al. (2016). Effect of biochar application on soil hydrological properties and physical quality of sandy soil. *Geoderma* 281: 11–20.

Glaser, B., Lehmann, J., and Zech, W. (2002). Ameliorating physical and chemical properties of highly weathered soils in the tropics with charcoal–a review. *Biology and Fertility of Soils* 35: 219–230.

Gonzaga, M.I.S, Matias, M.I.A.S., Andrade, K.R. et al. (2020). Aged biochar changed copper availability and distribution among soil fractions and influenced corn seed germination in a copper-contaminated soil. *Chemosphere* 240: 124828.

Hailegnaw, N.S., Mercl, F., Pracke, K. et al. (2019). Mutual relationships of biochar and soil pH, CEC, and exchangeable base cations in a model laboratory experiment. *Journal of Soils and Sediments* 19: 2405–2416.

Håkansson, I. and Lipiec, J. (2000). A review of the usefulness of relative bulk density values in studies of soil structure and compaction. *Soil & Tillage Research* 53: 71–85.

Hardy, B., Cornelis, J.T., Houben, D. et al. (2017). Evaluation of the long-term effect of biochar on properties of temperate agricultural soil at pre-industrial charcoal kiln sites in Wallonia, Belgium. *European Journal of Soil Science* 68: 80–89.

Jensen, J.L., Schjonning, P., Watts, C.W. et al. (2020). Soil degradation and recovery–changes in organic matter fractions and structural stability. *Geoderma* 364: 114181.

Jia, M., Yu, J., Li, Z. et al. (2021). Effects of biochar on the migration and transformation of metal species in a highly acid soil contaminated with multiple metals and leached with solutions of different pH. *Chemosphere* 278: 130344.

Jiang, M., He, Z., Sun, Y. et al. (2018). The effect of wheat-straw derived biochar on the soil pH and emissions of CO_2 and CH_4 from tea garden soil. *Journal of Agro-Environment Science* 37: 196–204.

Jianjing, X., Zhuzhu, L., and Ying, C. (2019). Response of soil microbial functional diversity to an intercropping system on slope land in the Loess Plateau. *Pratacultural Science* 195: 104335.

Jien, S.-H. and Wang, C.-S. (2013). Effects of biochar on soil properties and erosion potential in a highly weathered soil. *Catena* 110: 225–233.

Joseph, S., Kammann, C.I., Shepherd, J.G. et al. (2018). Microstructural and associated chemical changes during the composting of a high temperature biochar: mechanisms for nitrate, phosphate and other nutrient retention and release. *Science of the Total Environment* 618: 1210–1223.

Kuang, C., Jiang, C., Li, Z. et al. (2012). Effects of biochar amendments on soil organic carbon mineralization and microbial biomass in red paddy soils. *Soils* 44: 570–575.

Li, F.F., Zhou, J.H., Guo, X.R. et al. (2023a). Effect of biochar and earthworms on mineralization of organic matter in top soil and deep soil. *Environmental Engineering Science* 40 (8): 340–346.

Li, X., Wang, Y., Lu, H. et al. (2023b). Effects of biochar-based organic fertilizer on the growth and soil properties of greenhouse tomatoes. *Journal of Agro-Environment Science* 42: 568–577.

Li, Y., Zhang, F., Yang, M. et al. (2019). Impacts of biochar application rates and particle sizes on runoff and soil loss in small cultivated loess plots under simulated rainfall. *Science of the Total Environment*: 649: 1403–1413.

Li, Y., Zhao, Y., Cheng, K. et al. (2022). Effects of biochar on transport and retention of phosphorus in porous media: laboratory test and modeling. *Environmental Pollution* 297: 118788.

Licht, J., Smith, N., Mitchell, P. et al. (2017). Impact of lignocellulosic and hemicellulosic biochar on soil moisture in low clay soils. *Journal of Plant Nutrition and Soil Science* 180 (5): 576–584.

Lin, G.Y., Yang, H.P., Hu, J.H. et al. (2016). Effects of the physicochemical properties of biochar and soil on moisture sorption. *Journal of Renewable and Sustainable Energy* 8 (6).

Liang, F., Li, G.-T., Lin, Q.-M. et al. (2014). Crop yield and soil properties in the first 3 years after biochar application to a calcareous soil. *Journal of Integrative Agriculture* 13: 525–532.

Liang, J., Li, Y., Si, B. et al. (2021). Optimizing biochar application to improve soil physical and hydraulic properties in saline-alkali soils. *Science of the Total Environment* 771: 144802.

Liu, Y., Chen, Y., Wang, Y. et al. (2018). Negative priming effect of three kinds of biochar on the mineralization of native soil organic carbon. *Land Degradation & Development* 29: 3985–3994.

Liu, Z.J., Li, P.F., Huang, S.W. et al. (2020). Effects of wheat straw biochar application on soil carbon content changes in different tillage practices. *Environmental Science* 42 (6): 3000–3009.

Lucas, M., Schlueter, S., Vogel, H.-J. et al. (2019). Soil structure formation along an agricultural chronosequence. *Geoderma* 350: 61–72.

Lusiba, S., Odhiambo, J., and Ogola, J. (2017). Effect of biochar and phosphorus fertilizer application on soil fertility: soil physical and chemical properties. *Archives of Agronomy and Soil Science* 63: 477–490.

Luo, Y., Durenkamp, M., De Nobili, M. et al. (2013). Microbial biomass growth, following incorporation of biochars produced at 350 °C or 700 °C, in a silty-clay loam soil of high and low pH. *Soil Biology & Biochemistry* 57: 513–523.

Makó, A., Barna, G., and Horel, A. (2020). Soil physical properties affected by biochar addition at different plant phaenological phases. Part II. *International Agrophysics* 34 (1): 1–7.

Makoto, K., Tamai, Y., Kim, Y.S. et al. (2010). Buried charcoal layer and ectomycorrhizae cooperatively promote the growth of *Larix gmelinii* seedlings. *Plant & Soil* 327: 143–152.

Marchand, L., Brunel-Muguet, S., Lamy, I. et al. (2017). Modulation of trace element bioavailability for two earthworm species after biochar amendment into a contaminated technosol. *Ecotoxicology* 26: 1378–1391. https://doi.org/10.1007/s10646-017-1862-8.

Marlène, J., Rasconi, S., and Sime-Ngando, T. (2010). Diversity and functions of microscopic fungi: a missing component in pelagic food webs. *Aquatic Sciences-Research Across Boundaries* 72: 255–268.

Melero, S., Madejon, E., Carlos Ruiz, J. et al. (2007). Chemical and biochemical properties of a clay soil under dryland agriculture system as affected by organic fertilization. *European Journal of Agronomy* 26: 327–334.

Moon, D.H., Hwang, I., Chang, Y.Y. et al. (2017). Quality improvement of acidic soils by biochar derived from renewable materials. *Environmental Science and Pollution Research* 24: 4194–4199.

Omondi, M.O., Xia, X., Nahayo, A. et al. (2016). Quantification of biochar effects on soil hydrological properties using meta-analysis of literature data. *Geoderma* 274: 28–34.

Ouyang, N., Zhang, P., Zhang, Y. et al. (2023). Cation exchange properties of subsurface soil in mid-subtropical China: variations, correlation with soil-forming factors, and prediction. *Agronomy-Basel* 13: 741.

Peake, L.R., Reid, B.J., and Tang, X.Y. (2014). Quantifying the influence of biochar on the physical and hydrological properties of dissimilar soils. *Geoderma* 235: 182–190.

Petersen, C.T., Hansen, E., Larsen, H.H. et al. (2016). Pore-size distribution and compressibility of coarse sandy subsoil with added biochar. *European Journal of Soil Science* 67: 726–736.

Pietikainen, J., Kiikkila, O., and Fritze, H. (2000). Charcoal as a habitat for microbes and its effect on the microbial community of the underlying humus. *Oikos* 89 (2): 231–242.

Poveda, J., Martínez-Gómez, A., Fenoll, C. et al. (2021). The use of biochar for plant pathogen control. *Phytopathology* 111: 1490–1499.

Prodana, M., Silva, C., Gravato, C. et al. (2019). Influence of biochar particle size on biota responses. *Ecotoxicology and Environmental Safety* 174: 120–128.

Qian, K., Kumar, A., Zhang, H. et al. (2015). Recent advances in utilization of biochar. *Renewable & Sustainable Energy Reviews* 42: 1055–1064.

Rasa, K., Heikkinen, J., Hannula, M. et al. (2018). How and why does willow biochar increase a clay soil water retention capacity? *Biomass & Bioenergy* 119: 346–353.

Raza, S., Miao, N., Wang, P. et al. (2020). Dramatic loss of inorganic carbon by nitrogen-induced soil acidification in Chinese croplands. *Global Change Biology* 26: 3738–3751.

Rondon, M.A., Lehmann, J., Ramírez, J. et al. (2007). Biological nitrogen fixation by common beans (*Phaseolus vulgaris* L.) increases with bio-char additions. *Biology & Fertility of Soils* 43: 699–708.

Shi, R.Y., Ni, N., Nkoh, J.N. et al. (2019). Beneficial dual role of biochars in inhibiting soil acidification resulting from nitrification. *Chemosphere* 234: 43–51.

Spokas, K.A. (2010). Review of the stability of biochar in soils: predictability of O:C molar ratios. *Carbon Management* 1: 289–303.

Sui, L., Tang, C., Cheng, K. et al.(2022). Biochar addition regulates soil phosphorus fractions and improves release of available phosphorus under freezing-thawing cycles. *Science of the Total Environment* 848: 157748.

Sun, D., Meng, J., Liang, H. et al. (2015). Effect of volatile organic compounds absorbed to fresh biochar on survival of *Bacillus mucilaginosus* and structure of soil microbial communities. *Journal of Soils & Sediments* 15: 271–281.

Wang, Y. and Bao, G. (2022). Diversity of prokaryotic microorganisms in alkaline saline soil of the Qarhan Salt Lake area in the Qinghai–Tibet Plateau. *Scientific Reports* 12: 3365.

Wu, S.W., Zhang, Y., Tan, Q.L. et al. (2020). Biochar is superior to lime in improving acidic soil properties and fruit quality of Satsuma mandarin. *Science of the Total Environment* 714: 136722.

Xia, L., Cao, L., Yang, Y. et al. (2023). Integrated biochar solutions can achieve carbon-neutral staple crop production. *Nature Food* 4: 236–246.

Xie, Y., Wang, L., Li, H. et al. (2022). A critical review on production, modification and utilization of biochar. *Journal of Analytical and Applied Pyrolysis* 161: 105405.

Yang, B.B., Chen, Y., and Yuan, S.C. (2024). Investigation on structure, evaporation, and desiccation cracking of soil with straw biochar. *Land Degradation & Development* 35: 3126–3135.

Yu, H., Chen, M., Deng, H. et al. (2014). Preparation, characterization and adsorption performance of bagasse-based biochar. *Chinese Journal of Tropical Crops* 35: 595–602.

Zhang, H., Li, Q., Zhang, X. et al. (2020a). Insight into the mechanism of low molecular weight organic acids-mediated release of phosphorus and potassium from biochars. *Science of the Total Environment* 742: 140416.

Zhang, M., Walelign, D.B., and Tang, H. (2012). Effects of biochars application on active organic carbon fractions in soil. *Journal of Soil and Water Conservation* 26: 127–131.

Zhang, Q., Xu, H., Lu, W. et al. (2020b). Properties evaluation of biochar/high-density polyethylene composites: emphasizing the porous structure of biochar by activation. *Science of the Total Environment* 737: 139770.

Zhao, Y., Fan, D., Li, Y. et al. (2022). Application of machine learning in predicting the adsorption capacity of organic compounds onto biochar and resin. *Environmental Research* 208: 112694.

Zhao, J.K., Ren, T.S., Zhang, Q.Z. et al. (2016). Effects of biochar amendment on soil thermal properties in the North China plain. *Soil Science Society of America Journal* 80 (5): 1157–1166.

Zhou H., Fang, H, Zhang, Q. et al. (2019). Biochar enhances soil hydraulic function but not soil aggregation in a sandy loam. *European Journal·of Soil Science* 70: 291–300.

9

Impact of Biochar on Pesticides Transportation, Bioavailability, Performance, and Degradation in Soil Environment

Xin Liu[1], *Lingfeng Zeng*[1], *Yuqing Sun*[2], *Yaoyu Zhou*[1], *and Daniel C.W. Tsang*[3]

[1] *College of Environment and Ecology, Hunan Agricultural University, Changsha, China*
[2] *School of Agriculture and Biotechnology, Sun Yat-sen University, Guangzhou, Guangdong, China*
[3] *Department of Civil and Environmental Engineering, The Hong Kong University of Science and Technology, Clear Water Bay, Hong Kong, China*

9.1 Introduction to Biochar and Pesticides

9.1.1 Biochar Properties and Application in Agriculture

Biochar is made by heating biomass in low-oxygen environments. The material's characteristics depend on the raw material and the temperatures used. A study with different temperatures (200–650 °C) and 12 feedstocks showed that raw material type mainly changes total organic carbon, fixed carbon, and mineral content (Zhao et al. 2013). Treatment temperature mostly affects surface area and pH levels. Various production methods create biochars with different properties (Wang et al. 2024c). Hydrothermal carbonization produces more aromatic structures with functional groups, while anaerobic pyrolysis creates materials with better pore structure and larger surface areas.

Biochar use in farming has grown as people recognize its benefits for soils. Mixing biochar with traditional fertilizers shows promise in agriculture (Osman et al. 2022). It also works well as animal feed, improving growth, gut health, methane reduction, egg production, and lowering toxins. For climate management, adding biochar to soil is seen as a possible solution. A machine learning study analyzed how biochar types (from wood, straw, or manure) affect nitrogen loss and crop yields. Results showed strategic placement could increase global crop yields by 4–16%, cut farm N_2O emissions by 6–30%, and reduce nitrogen leaching by

Waste-derived Biochar for Sustainable Rural Revitalization, First Edition. Edited by Yuqing Sun and Daniel C.W. Tsang.
Published 2026 by John Wiley & Sons, Inc.

12–29% (Liu et al. 2019). This suggests biochar helps sustainable farming through better soil health and productivity, not just pesticide management.

Biochar production can create environmental risks. Some materials release harmful substances due to their surface properties, pH changes, or reactions with environment (Xiang et al. 2021). These production factors must be controlled when developing biochar for specific uses like pesticide management, where material properties strongly influence pesticide interactions.

9.1.2 Types of Pesticides and Their Environmental Impact

Pesticides are widely used in agriculture to protect crops from pests, but they can have significant environmental impacts. There are various types of pesticides, including insecticides, herbicides, and fungicides. A study involving a chemical library of 1024 molecules from different pesticide classes revealed that 57% of the substances altered larval behavior in *Drosophila melanogaster*, and an even greater proportion affected their long-term viability (Gandara et al. 2024). Pesticides can also have negative effects on non-target organisms across trophic levels. A comprehensive review of 1705 studies, which included 20,212 effect sizes, indicated negative impacts on the growth, reproduction, behavior, and other physiological indicators of non-target plants, animals, and microorganisms in terrestrial and aquatic settings (Wan et al. 2025).

The distribution of pesticide residues in the environment can pose risks to endangered species. In a farming region in western Taiwan, 67 different pesticides were found in 79 soil samples, with the hazard index linked to land use patterns, posing a possible risk to the leopard cat (Liao et al. 2020). The use of pesticides also affects soil microbial communities. For example, the application of chemical pesticides (chlorpyrifos and cypermethrin) and a biological pesticide (azadirachtin) in *Vigna radiata* fields led to a shift in the rhizospheric bacterial community structure, despite no significant effect on plant parameters (Singh et al. 2015). These environmental impacts highlight the need for more sustainable approaches to pesticide use, such as the use of biochar to mitigate some of these effects.

9.1.3 Interaction Mechanisms Between Biochar and Pesticides

Biochar and pesticides interact through several processes. Adsorption is a key process, where pesticides stick to biochar surfaces. The study explored the adsorption of carbaryl and atrazine on biochars derived from pig manure. The study revealed that both pesticides were significantly adsorbed, with carbaryl having organic carbon normalized sorption coefficient (K_{oc}) values increased at atrazine concentration of 0.5 mg/L (Zhang et al. 2013). The sorption process cannot be solely attributed to the hydrophobic effect, as it also involves mechanisms such as pore-filling and π–π electron donor–acceptor interactions.

9.2 Biochar Application for Pesticide Control

9.2.1 Early Studies on Biochar and Pesticide Interaction

The use of biochar in soil will alter the bioavailability and persistence of pesticides, reduce environmental pollution, and improve soil health (Khalid et al. 2020). Biochar also improves the efficacy of biopesticides such as coumarins while minimizing their leaching to water. As a general-purpose adsorbent, biochar can reduce the bioaccessibility and bioaccumulation of hazardous pesticides, such as organochlorine compounds, in soil through adsorption (Ali et al. 2019b). The addition of different types of biochar (from sewage sludge biochar, soybean straw biochar, rice straw biochar, and peanut hull biochar) resulted in a significant reduction (around 85%) in the accumulation of organochlorine compounds in growing vegetables planted in contaminated soil. Risk assessment showed that biochar reduced the hazard quotient index and increased lifetime cancer risk values for organochlorine pesticides (OCPs) associated with consumption of vegetable crops. This reduces health risks from pesticide residues in crops and promotes a healthier soil ecosystem by altering microbial dynamics.

Soil properties strongly affect how biochar interacts with pesticides, changing how they stick to soil or break down. A large global study found that biochar doesn't increase pesticide levels in soil overall but does lower pesticide amounts in plants and earthworms, which reduces environmental harm (Wang et al. 2024a). Biochar also improves water filtration systems like biofilters by helping them capture and break down more pesticides, leading to cleaner water and less pollution (Mukherjee et al. 2016). Research on specific pesticides, such as acetamiprid, demonstrates that biochar enhances pesticide sorption, thereby decreasing their mobility and leaching potential, particularly in soils with low organic matter content (Yu et al. 2011). This underscores the importance of soil properties in determining the effectiveness of biochar. In summary, the application of biochar in agricultural soils represents a sustainable strategy for pesticide management, offering the dual advantages of improving soil fertility and diminishing the environmental impact of pesticide usage (Lykogianni et al. 2021). By altering the fate and transport of pesticides within the soil, biochar contributes to the maintenance of soil and water quality, thereby supporting sustainable agricultural practices.

9.2.2 Adsorption Capabilities of Biochar for Pesticides

Biochar has large surface area and strong adsorption ability. This adsorption action affects how pesticides move through soil. The main factor controlling pesticide movement is this surface-based adsorption process. The adsorption behavior is a physicochemical process where pesticide molecules are held on a solid surface through various types of bonding, including hydrogen, ionic, and covalent bonds, between charged sites on soil colloids. The accumulation and movement of pesticides in the environment are determined, impacting environmental biology and the safety of food. Relevant studies have confirmed that the addition of biochar to soil can significantly improve the soil's adsorbability to pesticides. Biochar contains hydrophobic areas, pores, and a range of minerals that aid in maintaining soil properties and microorganisms. Soil pesticides exhibit solubility, migration volatility, degradation, bioaccumulation, and environmental fate. The relationship between soils, their elements, and pesticides can lead to both beneficial and harmful effects on the soil ecosystem. Various biochar types (unmodified, modified, and aged) differ based on the pyrolysis temperature and raw material. Ultimately, the characteristics of biochar can boost the ability to adsorb pesticides by increasing microporosity. Variations in adsorption might be due to differing chemical and physicochemical properties and the nature of the contaminants.

For instance, studies have shown that biochar amendments can significantly reduce the bioaccessibility of OCPs in contaminated soils, leading to decreased bioaccumulation in plants and reduced risk to human health (Ali et al. 2019a). A study on the sorption properties of green waste biochar for atrazine and simazine found that the biochar could adsorb these two triazine pesticides, with the sorption affinity increasing with decreasing solid/solution ratio (Zheng et al. 2010). Reducing the solid-to-solution ratio from 1 : 50 to 1 : 1000 (g/mL) increased the adsorbed amounts of atrazine and simazine from 451 to 1158 mg/kg and from 243 to 1066 mg/kg, respectively. The adsorption isotherms followed a Freundlich model, and when the two pesticides co-existed, competitive sorption occurred. The discovery of the sorption capabilities of biochar for pesticides, as demonstrated in early studies like the one on green waste biochar and triazine pesticides (Zheng et al. 2010), was an important starting point. This led to further research on how biochar can be used to reduce the mobility and bioavailability of pesticides in soil.

Pesticide molecules are retained on the surface of solid particles through a physicochemical process involving bonding types such as hydrogen, ionic, covalent, and van der Waals forces at charged sites on soil colloids. In a microcosmic experiment with lettuce and the chiral pesticide metalaxyl. Adding biochar made from wood waste significantly reduced the pesticide accumulation in shoot and their metabolites by 57.7–86.3% and 13.3–32.5%, respectively (You et al. 2021). These early studies demonstrated the potential of biochar in influencing pesticide behavior in soil-plant ecosystems but also highlighted the need for further research to fully understand the underlying mechanisms.

9.2.3 Biochar Influence the Degradation of Pesticides

Biochar can influence the degradation of pesticides. For example, biochar accelerated the polycyclic aromatic hydrocarbons (PAHs) biodegradation in petroleum-polluted soil, and the biochars prepared at 500 °C exhibited the best performance (Kong et al. 2018). The identification of the role of biochar in shifting bacterial communities, as seen in the study with the chiral pesticide metalaxyl (You et al. 2021), also provided new insights. The enhanced abundance of pesticide-degrading bacteria in biochar-amended soil suggests that biochar can create a more favorable environment for the degradation of pesticides.

Biochar also affects the hydrolysis of pesticides. The hydrolysis of carbaryl and atrazine was accelerated in the presence of biochar. Carbaryl and atrazine were decomposed by 71.8 and 27.9%, respectively, in the presence of high-temperature preparation (700 °C) of swine manure-derived biochar within 12 hours (Zhang et al. 2013). The elevated solution pH and minerals released from the surface of biochar are the main reason for the enhanced hydrolysis; the dissolved metal ions are also able to act as catalysts for the pesticide degradation process. Therefore, the effect of biochar on the pesticide degradation process is the result of the combination of several mechanisms of action.

Moreover, biochar can alter the microbial community structure in soils, and then influence pesticide degradation. The presence of biochar has been shown to enhance the abundance of abundance-related pesticide degrading, such as *Luteimonas, Actinobacteria, Proteobacteria, Methylophilus*, and *Hydrogenophaga*, thereby promoting the breakdown of certain pesticides (You et al. 2021; Ali et al. 2019b). The microbial influence was strongly dependent on the biochar type applicated in soil. The degradation of tebuconazole was accelerated by combining a strain that degrades it (*Alcaligenes faecalis* WZ-2) with biochar derived from wheat straw, cutting its half-life in soil from 40.8 to 13.3 days (Sun et al. 2020). By immobilizing the strain on biochar, the degradation of tebuconazole was accelerated, and the native soil microbial enzyme activities and microbiome community composition were restored, promoting soil health and enhancing pesticide performance.

Biochar in pesticide degradation is also affected by its chemical properties, which can be tailored through modifications. Biochar engineered with specific metals or activated through chemical treatments can enhance its ability to mediate Fenton-like reactions, thereby promoting pesticide degradation (Zeng et al. 2021). Tailored biochars can effectively degrade 2,4-dichlorophenol through enhancing electron transfer and reactive oxygen species generation.

9.2.4 Influencing Factor for Biochar Application in Pesticides Control

The interaction between biochar and pesticides is further influenced by environmental conditions such as soil moisture and temperature (Figure 9.1). Biochar amendments have been found to enhance the degradation of pesticides like phorate and terbufos, particularly under conditions of optimal soil moisture through meta-analysis (Wang et al. 2024a). Wang and colleagues found that the content of soil organic matter plays a crucial role in biochar–pesticide interactions, indicating that adding biochar to soils with high organic matter content is less likely to enhance pesticide longevity.

Additionally, the aging process of biochar in soil can affect its sorption properties and, consequently, its impact on pesticide degradation. Aged biochar may exhibit different sorption characteristics compared to fresh biochar, influencing the persistence and transformation of pesticides in the soil (Fan et al. 2025). However, the effectiveness of biochar in pesticide degradation is highly dependent on its properties and the specific environmental context, necessitating careful consideration in its application (Holanda et al. 2023). These milestones have set the stage for further research and the development of biochar-based strategies for

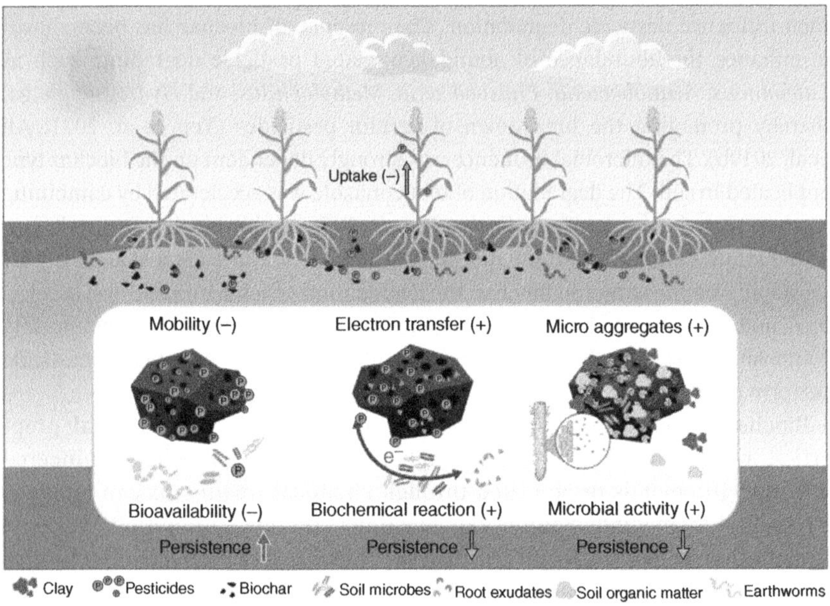

Figure 9.1 Mechanism of the interaction between biochar and pesticides in the soil-plant system. *Source:* Wang et al. 2024a/with permission of Elsevier.

more effective pesticide control. Biochar addition exhibited a positive correlation with the adsorption performance. The specific surface area and microporosity of biochar improved soil properties with the increase of the amount of biochar added, resulting in an increase in adsorption sites and better adsorption effects.

9.3 Biochar's Impact on Pesticide Transportation

9.3.1 Biochar's Role in Pesticide Adsorption and Mobility

Biochar plays a significant role in pesticide adsorption, which in turn affects pesticide mobility in the environment. Adsorption of pesticides onto biochar is influenced by various factors related to biochar properties. For instance, biochars with high porosity and large surface areas, such as those produced by anaerobic pyrolysis, have been shown to have a greater capacity to adsorb pesticides (Wang et al. 2024c). The adsorption of pesticides by biochar can reduce their mobility in soil. When pesticides are adsorbed onto biochar, they are less likely to leach through the soil profile or be transported by runoff. This has been demonstrated in studies where biochar addition to soil decreased the concentration of pesticides in soil pore water, thus reducing their availability for leaching (Malheiro et al. 2020). By immobilizing pesticides, biochar can help to contain them in the soil, minimizing their potential to contaminate groundwater and surface water bodies.

The effect of biochar on soil physicochemical properties also indirectly affects the pesticide transport process in soil (Dong et al. 2019). It has been reported to prove that the organic and inorganic carbon content of soil is an important parameter affecting the pesticide migration properties (Cabrerizo et al. 2018). Thus, the use of biochar materials can influence soil carbon content and thus further affect biochar fate process. Biochar application increased the soil inorganic carbon content (Sun et al. 2021). Soil inorganic carbon or carbonate minerals, such as calcite and dolomite, may provide additional adsorption sites for pesticide adsorption processes. Specific interactions between the surfaces of these minerals and the pesticides affect the adsorption behavior.

9.3.2 Factors Influencing Pesticide Transportation with Biochar

Biochar type and soil properties influence the transportation of pesticides. Biochar type is determined by its feedstock and production temperature. Biochars derived from different feedstocks, such as wood, straw, or manure, have different chemical and physical properties that can affect pesticide adsorption and transportation. Biochars with high ash contents may have different adsorption mechanisms for pesticides compared to biochars with lower ash contents (Zhang et al. 2013). Through a global meta-analysis and machine learning approach, soil organic matter content was identified as a key determinant in biochar–pesticide interactions (Wang et al. 2024a). Soils rich in organic matter may have different sorption capacities for pesticides even in the presence of biochar. The pH of the soil can affect the ionization state of pesticides, then influence their adsorption onto biochar.

9.3.3 Case Studies on Biochar Application in Different Soil Types

Biochar application in different soil types has provided valuable insights into its effectiveness in pesticide management. The addition of biochars from different sources (wood pellets and macadamia nut shells) had different effects on the sorption of herbicides (Cabrera et al. 2014). Soils amended with biochars from wood pellets nearly fully absorbed aminocyclopyrachlor and bentazone, whereas macadamia nut shells showed less sorption. The most significant cause is the competition for sorption sites between biochar-derived dissolved organic carbon (DOC) and herbicides.

The adsorption of herbicides (azimsulfuron and penoxsulam) in paddy soils of different ages (100 and 700 years) was impacted by biochar addition (García-Jaramillo et al. 2020). The adsorption of azimsulfuron was much stronger to biochar than to the soils, and the presence of biochar water-extractable substances (BWES) influenced the adsorption behavior. These case studies highlight the importance of considering soil type and biochar properties when applying biochar for pesticide management, as the interactions can vary significantly depending on these factors.

9.4 Bioavailability of Pesticides in the Presence of Biochar

9.4.1 Mechanisms of Biochar Affecting Pesticide Bioavailability

Biochar can affect the bioavailability of pesticides through several mechanisms. One of the main mechanisms is adsorption. When pesticides are adsorbed onto biochar, their availability for uptake by organisms is reduced. Wheat straw biochar incorporation into soil increased the sorption of Hexachlorobenzene (HCB), and the results about chemical extraction and earthworm bioassay both indicated a rapid reduction in bioavailability (Song et al. 2012). The absorption of HCB by biochar was 42 times more effective than by soil, with a noticeable decrease in bioavailability at just a 0.1% biochar application.

Biochar can also influence the biodegradation of pesticides, which in turn affects their bioavailability. A study on the chiral pesticide metalaxyl found that amending soil with biochar derived from wood waste enhanced bacterial diversity, changed the bacterial community structure, and increased the presence of pesticide-degrading bacteria (You et al. 2021). This led to the enantioselective degradation of metalaxyl in the soil, reducing its bioavailability. Additionally, biochar can modify the soil environment, such as changing the pH and soil organic matter content, which can affect the solubility and mobility of pesticides, thereby influencing their bioavailability.

9.4.2 Experimental Studies on Bioavailability Changes

Experimental studies have been conducted to investigate the influence of biochar on the pesticide bioavailability in soil animals and plants (You et al. 2021; Bošković et al. 2021). In a study on the bioavailability of dimethoate, biochar application to soil decreased the concentration of dimethoate in soil pore water, influencing its bioavailability and consequent toxicity to soil organisms (Malheiro et al. 2020). Applying biochar prevented dimethoate from affecting the survival rate and offspring production of the collembolan *Folsomia candida*. The shoot length and fresh and dry weights of *Brassica rapa* were less influenced by dimethoate when biochar was added, regardless of the application rate of 2.5 or 5% w/w. Another study focused on the uptake of pharmaceuticals by radish in soil amended with biochar (Li et al. 2020). The addition of 1.0% biochar significantly decreased the accumulation of several pharmaceuticals (such as acetaminophen, carbamazepine, sulfadiazine, etc.) in radish, mainly due to the lowered concentrations of these pharmaceuticals in pore water. These experimental studies provide evidence of the impact of biochar on pesticide (and pharmaceutical, which can be considered a type of pesticide in the context of environmental contaminants) bioavailability and highlight the potential of biochar in reducing the exposure of organisms to these contaminants. Biochar amendments can lower the enantioselective uptake of the chiral pesticide metalaxyl by lettuce, while simultaneously enhancing the abundance of pesticide-degrading bacteria in the soil (You et al. 2021).

9.4.3 Implications for Pesticide Efficacy and Environmental Safety

The changes in pesticide bioavailability due to biochar have important implications for pesticide efficacy and environmental safety. Reduced bioavailability may lead to a decrease in the immediate efficacy of pesticides if they are not able to reach their target pests in sufficient concentrations. It may be necessary to adjust the dosage of pesticides to ensure effective pest control, and careful consideration needs to be given to the application rate of biochar. The decreased bioavailability due to biochar addition mitigated the hazard of dimethoate to soil pore water-exposed biota (Malheiro et al. 2020). The reduced bioavailability can contribute to a more sustainable use of pesticides, minimizing their negative impacts on the environment and human health.

9.5 Performance Enhancement of Pesticides with Biochar

9.5.1 Synergistic Effects of Biochar on Pesticide Performance

Biochar application has synergistic effects on pesticide performance, enhancing their effectiveness in pest control. Biochar helps pesticides stick better to plant surfaces as a carrier for pesticides. The combination of straw ash-based biochar and biosilica in a loss-control pesticide (LCP) formulation can hold more pesticide molecules, keeping them on the leaf surface (Cai et al. 2013). This better adhesion stops them from being washed off by rain or wind, making the pesticide work longer. Biochar also helps plants grow better and resist pests, with stronger roots and better nutrient absorption, because better soil conditions make plants tougher against bugs (Wen et al. 2024). When combined with fertilizers, biochar helps crops use nutrients more efficiently, leading to healthier plants that resist pests better (Xia et al. 2022).

9.5.2 Biochar as a Carrier for Controlled Pesticide Release

Biochar can act as a slow-release pesticide carrier, making pesticides work longer and reducing environmental harm. This happens because biochar holds pesticides and controls how they're released over time. Studies show rice husk biochar mixed with fertilizer (pristine biochar [PBC]) adsorbs twice as much pesticide (55.90 vs. 42.71 mg/g) compared to plain biochar (An et al. 2021). However, higher pesticide binding can slow down controlled release. Similar slow-release methods using coated silica particles work through electric charges, a principle that could apply to biochar systems (Shan et al. 2019). Biochar also improves soil structure, boosts nutrient availability, and supports helpful soil microbes, which strengthens plants and reduces chemical needs (Shi et al. 2024). Controlled-release biochar pesticide formulas provide steady pest protection, needing fewer applications while avoiding sudden pesticide surges that pollute the environment.

9.5.3 Field Trials Demonstrating Enhanced Pesticide Efficiency

Field tests show biochar helps pesticides work better. In one trial combining biochar with organic fertilizer to clean neonicotinoid-contaminated soil, peanut hull and chicken manure biochar boosted soil's ability to trap these pesticides (Cheng et al. 2022). The mix increased pesticide-binding strength (Kf values) by 7.2–12 times compared to untreated soil. The pesticide breakdown was mainly due to natural microbes, with the biochar–fertilizer combo increasing populations of pesticide-degrading organisms. Another field test used a self-mixing capsule system for targeted fungus control. This system adjusted nanocapsule ratios between 18.81 and 85.21% to optimize delivery. The best results came from a 65 : 35 nanocapsule-to-microcapsule mix that effectively blocked *Phytophthora capsici* on leaves and stopped new invasions quickly (Wang et al. 2024b). These real-world trials prove biochar can make pesticide applications more effective.

9.6 Degradation of Pesticides Influenced by Biochar

9.6.1 Biochar's Role in Pesticide Degradation Pathways

Biochar helps break down pesticides. For example, it changes the soil's chemical makeup. Biochar makes soil more alkaline, which speeds up the breakdown of certain pesticides. Biochar made at 700 °C broke down 71.8% of carbaryl and 27.9% of atrazine within 12 hours. Tests showed the higher pH from the biochar was the main cause of faster pesticide breakdown (Zhang et al. 2013). Biochar can act as a catalyst or provide a surface for chemical reactions, along with minerals or functional groups that can participate in redox reactions, leading to the decomposition of pesticides. The degradation level of *p*-nitrophenol by biochar was linked to the strength of electron paramagnetic resonance (EPR) signals, suggesting a potential role for environmentally persistent free radicals (EPFRs) (Yang et al. 2016).

9.6.2 Microbial Interactions Facilitated by Biochar

Biochar can facilitate microbial interactions that contribute to pesticide degradation. Biochar provides a habitat for microorganisms, increasing their abundance and diversity in the soil. In a study on petroleum-polluted soil, the addition of sawdust and wheat straw biochars prepared at 300 and 500 °C promoted the growth of some specific taxa, including PAHs-degrading bacteria (Kong et al. 2018). Although the predominant bacterial diversity slightly decreased, the abundance of these degraders was enhanced, leading to an acceleration of PAHs biodegradation.

Biochar promotes microbial interactions that enhance pesticide degradation. In a study on nitrobenzene (NB) reduction. Biochar made from Fe-bearing sludge improved both biological and non-biological reduction to aniline via microbial dissimilatory iron reduction (Lu et al. 2021). Graphitized regions and quinone-like formations in biochar act as redox-active sites, which may be vital for boosting electron transfer during microbial dissimilatory iron reduction and NB breakdown. The redox-active sites in biochar, consisting of graphitized domains and quinone-like structures, facilitated electron transfer for reducing iron and decomposing NB. These processes are vital for the degradation of pesticides in soil.

9.6.3 Long-term Effects of Biochar on Pesticide Residues

Some studies indicate that biochar can reduce pesticide bioavailability and their long-term soil residues. In a study with aged biochar and lindane, biochar reduced earthworm exposure to lindane despite its altered toxicity mechanism (Xu et al. 2020). Biochar aging may change its properties and interactions with pesticides. A study found that aged biochar could adversely affect earthworms and lose its ability to adsorb or degrade pesticides (Shi et al. 2023). Understanding these long-term effects is crucial for assessing the sustainability and environmental impact of biochar-based pesticide management.

9.7 Future Prospects and Challenges in Biochar–Pesticide Research

9.7.1 Emerging Technologies in Biochar Production for Pesticide Management

Innovations in biochar preparation methods offer better ways to manage pesticides. Microwave-assisted heating works better than traditional biomass-heating methods (Hadiya et al. 2022). New methods in making biochar offer better ways to manage pesticides. Microwave-assisted heating works better than traditional biomass-heating methods. Mixing different biomasses during heating creates improved biochar. Combining plant waste and seaweed biomass makes biochar that performs better in heat treatment (Fakayode et al. 2020). Mixing different biomasses during heating creates improved biochar. Combining plant waste and seaweed biomass makes biochar that performs better in heat treatment.

9.7.2 Potential Environmental and Regulatory Challenges

Using biochar for pesticide cleanup might face environmental regulation hurdles. Studies show biochar application may generate EPFRs, which are stable and emerging pollutants (Odinga et al. 2020). EPFRs can trigger toxic oxygen compounds that harm farmland and human health. The half-life and persistence in biochar residues and the atmosphere could cause years of environmental damage. Current laws also struggle to measure biochar's long-term effects, lacking clear guidelines for safety assessment. These challenges need to be addressed through the development of more comprehensive regulatory policies and environmental impact assessment methods.

9.7.3 Future Research Directions and Innovations in Biochar Applications

Future biochar research should focus on three key areas. First, we need better long-term studies on how biochar affects soil health. This includes impacts on soil animals, microbes, and nutrient cycles. One review found biochar can both help and harm ecosystems by changing soil chemistry, greenhouse gas emissions, crop growth, and pollutant behavior (Kuppusamy et al. 2016). Most studies reckon without biochar's harmful effects on soil life, so future work must track these unintended consequences. Second, research should develop better ways to customize biochar for specific pesticide needs. This means finding the best production methods and materials to create biochar that trap pesticides better, break them down faster, or release them slowly. For example, mixing biochar with nano-sized particles could boost pesticide delivery or speed up breakdown through combined actions. Third, combining biochar with other new technologies needs more study. Working with smart materials or sensors might create smarter pesticide systems that only activate when needed. These approaches could make biochar treatments both more effective and better for the environment.

References

Ali, N., Khan, S., Li, Y. et al. (2019a). Influence of biochars on the accessibility of organochlorine pesticides and microbial community in contaminated soils. *Science of the Total Environment* 647: 551–560.

Ali, N., Khan, S., Yao, H. et al. (2019b). Biochars reduced the bioaccessibility and (bio)uptake of organochlorine pesticides and changed the microbial community dynamics in agricultural soils. *Chemosphere* 224: 805–815.

An, X., Wu, Z., Shi, W. et al. (2021). Biochar for simultaneously enhancing the slow-release performance of fertilizers and minimizing the pollution of pesticides. *Journal of Hazardous Materials* 407: 124865.

Bošković, N., Bílková, Z., Šudoma, M. et al. (2021). Conazole fungicides epoxiconazole and tebuconazole in biochar amended soils: degradation and bioaccumulation in earthworms. *Chemosphere* 274: 129700.

Cabrera, A., Cox, L., Spokas, K. et al. (2014). Influence of biochar amendments on the sorption-desorption of aminocyclopyrachlor, bentazone and pyraclostrobin pesticides to an agricultural soil. *The Science of the Total Environment* 470–471: 438–443.

Cabrerizo, A., Muir, D.C.G., De Silva, A.O. et al. (2018). Legacy and emerging persistent organic pollutants (POPs) in terrestrial compartments in the high arctic: sorption and secondary sources. *Environmental Science & Technology* 52: 14187–14197.

Cai, D., Wang, L., Zhang, G. et al. (2013). Controlling pesticide loss by natural porous micro/nano composites: straw ash-based biochar and biosilica. *ACS Applied Materials & Interfaces* 5: 9212–9216.

Cheng, H., Tang, G., Wang, S. et al. (2022). Combined remediation effects of biochar and organic fertilizer on immobilization and dissipation of neonicotinoids in soils. *Environment International* 169: 107500.

Dong, X., Singh, B.P., Li, G. et al. (2019). Biochar increased field soil inorganic carbon content five years after application. *Soil and Tillage Research* 186: 36–41.

Fakayode, O.A., Aboagarib, E.A.A., Zhou, C. et al. (2020). Co-pyrolysis of lignocellulosic and macroalgae biomasses for the production of biochar–a review. *Bioresource Technology* 297: 122408.

Fan, J., Duan, T., Wu, X. et al. (2025). Can the aging process necessarily weaken the effect of biochar on cadmium-contaminated soil remediation: considering biochar at different pyrolysis temperatures and aging treatment. *Environmental Geochemistry and Health* 47: 66.

Gandara, L., Jacoby, R., Laurent, F. et al. (2024). Pervasive sublethal effects of agrochemicals on insects at environmentally relevant concentrations. *Science (New York, N.Y.)* 386: 446–453.

García-Jaramillo, M., Trippe, K.M., Helmus, R. et al. (2020). An examination of the role of biochar and biochar water-extractable substances on the sorption of ionizable herbicides in rice paddy soils. *The Science of the Total Environment* 706: 135682.

Hadiya, V., Popat, K., Vyas, S. et al. (2022). Biochar production with amelioration of microwave-assisted pyrolysis: current scenario, drawbacks and perspectives. *Bioresource Technology* 355: 127303.

Holanda, M.A.S., Coelho Menezes, J.M., Coutinho, H.D.M., et al. (2023). Effectiveness of biochar as an adsorbent for pesticides: systematic review and meta-analysis. *Journal of Environmental Management* 345: 118719.

Khalid, S., Shahid, M., Murtaza, B. et al. (2020). A critical review of different factors governing the fate of pesticides in soil under biochar application. *Science of the Total Environment* 711: 134645.

Kong, L., Gao, Y., Zhou, Q. et al. (2018). Biochar accelerates PAHs biodegradation in petroleum-polluted soil by biostimulation strategy. *Journal of Hazardous Materials* 343: 276–284.

Kuppusamy, S., Thavamani, P., Megharaj, M. et al. (2016). Agronomic and remedial benefits and risks of applying biochar to soil: current knowledge and future research directions. *Environment International* 87: 1–12.

Li, Y., He, J., Qi, H. et al. (2020). Impact of biochar amendment on the uptake, fate and bioavailability of pharmaceuticals in soil-radish systems. *Journal of Hazardous Materials* 398: 122852.

Liao, J.Y., Fan, C., Huang, Y.Z. et al. (2020). Distribution of residual agricultural pesticides and their impact assessment on the survival of an endangered species. *Journal of Hazardous Materials* 389: 121871.

Liu, Q., Liu, B., Zhang, Y., et al. (2019). Biochar application as a tool to decrease soil nitrogen losses (NH_3 volatilization, N_2O emissions, and N leaching) from croplands: options and mitigation strength in a global perspective *Global Change Biology* 25: 2077–2093.

Lu, Y., Xie, Q., Tang, L. et al. (2021). The reduction of nitrobenzene by extracellular electron transfer facilitated by Fe-bearing biochar derived from sewage sludge. *Journal of Hazardous Materials* 403: 123682.

Lykogianni, M., Bempelou, E., Karamaouna, F. et al. (2021). Do pesticides promote or hinder sustainability in agriculture? The challenge of sustainable use of pesticides in modern agriculture. *The Science of the Total Environment* 795: 148625.

Malheiro, C., Cardoso, D.N., Neves, J. et al. (2020). Biochar in soil mitigates dimethoate hazard to soil pore water exposed biota. *Journal of Hazardous Materials* 400: 123304.

Mukherjee, S., Tappe, W., Weihermueller, L. et al. (2016). Dissipation of bentazone, pyrimethanil and boscalid in biochar and digestate based soil mixtures for biopurification systems. *The Science of the Total Environment* 544: 192–202.

Odinga, E.S., Waigi, M.G., Gudda, F.O. et al. (2020). Occurrence, formation, environmental fate and risks of environmentally persistent free radicals in biochars. *Environment International* 134: 105172.

Osman, A.I., Fawzy, S., Farghali, M. et al. (2022). Biochar for agronomy, animal farming, anaerobic digestion, composting, water treatment, soil remediation, construction, energy storage, and carbon sequestration: a review. *Environmental Chemistry Letters* 20: 2385–2485.

Shan, Y., Cao, L., Xu, C. et al. (2019). Sulfonate-functionalized mesoporous silica nanoparticles as carriers for controlled herbicide diquat dibromide release through electrostatic interaction. *International Journal of Molecular Sciences* 20: 1330.

Shi, G., Li, H., Fu, Q. et al. (2024). Effects of biochar and compost on the abundant and rare microbial communities assembly and multifunctionality in pesticide-contaminated soil under freeze–thaw cycles. *Environmental Pollution (Barking, Essex: 1987)* 362: 125003.

Shi, Z., Wen, M., Zhao, Y. et al. (2023). Vermitoxicity of aged biochar and exploring potential damage factors. *Environment International* 172: 107787.

Singh, S., Gupta, R., and Sharma, S. (2015). Effects of chemical and biological pesticides on plant growth parameters and *rhizospheric* bacterial community structure in *Vigna radiata*. *Journal of Hazardous Materials* 291: 102–110.

Song, Y., Wang, F., Bian, Y. et al. (2012). Bioavailability assessment of hexachlorobenzene in soil as affected by wheat straw biochar. *Journal of Hazardous Materials* 217–218: 391–397.

Sun, T., Miao, J., Saleem, M. et al. (2020). Bacterial compatibility and immobilization with biochar improved tebuconazole degradation, soil microbiome composition and functioning. *Journal of Hazardous Materials* 398: 122941.

Sun, Y., Xiong, X., He, M. et al. (2021). Roles of biochar-derived dissolved organic matter in soil amendment and environmental remediation: a critical review. *Chemical Engineering Journal* 424: 130387.

Wan, N.F., Fu, L., Dainese, M. et al. (2025). Pesticides have negative effects on non-target organisms *Nature Communications* 16: 1360.

Wang, J., Norgaard, T., Pugliese, L. et al. (2024a). Global meta-analysis and machine learning reveal the critical role of soil properties in influencing biochar-pesticide interactions. *Environment International* 193: 109131.

Wang, R., Luan, F.X., Wang, Z.S. et al. (2024b). Spontaneously formed ratio-tunable micro- and nano-capsule coexist system for precision and on-demand fungicide delivery in crop leave. *Advanced Materials (Deerfield Beach, Fla.)* 36: e2409839.

Wang, Y., Wang, K., Wang, X. et al. (2024c). Effect of different production methods on physicochemical properties and adsorption capacities of biochar from sewage sludge and kitchen waste: mechanism and correlation analysis. *Journal of Hazardous Materials* 461: 132690.

Wen, Y., Ma, Y., Wu, Z. et al. (2024). Enhancing rice ecological production: synergistic effects of wheat-straw decomposition and microbial agents on soil health and yield. *Frontiers in Plant Science* 15: 1368184.

Xia, H., Riaz, M., Zhang, M. et al. (2022). Biochar-N fertilizer interaction increases N utilization efficiency by modifying soil C/N component under N fertilizer deep placement modes. *Chemosphere* 286: 131594.

Xiang, L., Liu, S., Ye, S. et al. (2021). Potential hazards of biochar: the negative environmental impacts of biochar applications. *Journal of Hazardous Materials* 420: 126611.

Xu, T., Miao, J., Chen, Y. et al. (2020). The long-term environmental risks from the aging of organochlorine pesticide lindane. *Environment International* 141: 105778.

Yang, J., Pan, B., Li, H. et al. (2016). Degradation of p-nitrophenol on biochars: role of persistent free radicals. *Environmental Science & Technology* 50: 694–700.

You, X., Suo, F., Yin, S. et al. (2021). Biochar decreased enantioselective uptake of chiral pesticide metalaxyl by lettuce and shifted bacterial community in agricultural soil. *Journal of Hazardous Materials* 417: 126047.

Yu, X.Y., Mu, C.L., Gu, C. et al. (2011). Impact of woodchip biochar amendment on the sorption and dissipation of pesticide acetamiprid in agricultural soils. *Chemosphere* 85: 1284–1289.

Zeng, L., Chen, Q., Tan, Y. et al. (2021). Dual roles of biochar redox property in mediating 2,4-dichlorophenol degradation in the presence of Fe^{3+} and persulfate. *Chemosphere* 279: 130456.

Zhang, P., Sun, H.L., Yu, L. et al. (2013). Adsorption and catalytic hydrolysis of carbaryl and atrazine on pig manure-derived biochars: impact of structural properties of biochars. *Journal of Hazardous Materials* 244–245: 217–224.

Zhao, L., Cao, X., Mašek, O. et al. (2013). Heterogeneity of biochar properties as a function of feedstock sources and production temperatures. *Journal of Hazardous Materials* 256–257: 1–9.

Zheng, W., Guo, M., Chow, T. et al. (2010). Sorption properties of greenwaste biochar for two triazine pesticides. *Journal of Hazardous Materials* 181: 121–126.

10

Waste-derived Biochar as Adsorbent for Agriculture Wastewater Treatment

Jianhua Qu[1] *and Ying Zhang*[1,2]

[1] *School of Resources and Environment, Northeast Agricultural University, Harbin, China*
[2] *Northeast Institute of Geography and Agroecology, Chinese Academy of Sciences, Changchun, China*

10.1 Introduction

The agricultural wastewater, generated during the production of crops and livestock, contains high concentrations of inorganic salts such as nitrogen and phosphorus, as well as pollutants including pesticides, veterinary drugs, and heavy metals (Preetham and Vengala 2021). It primarily originates from excessive use of fertilizers and pesticides, runoff and leaching from crop fields, discharge of livestock manure and farm wastewater, as well as the effluent from agricultural product processing. If not adequately treated prior to discharge, it can cause severe contamination to water bodies, soil quality degradation, and disruption to ecosystems.

Pesticide residues and heavy metals present in agricultural wastewater can enter the human body through the food chain, potentially leading to acute or chronic poisoning that adversely affects the nervous system, endocrine system, and immune system of individuals (Gallo et al. 2020). Heavy metals such as cadmium, lead, and mercury exhibit significant bioaccumulation and toxicity levels, which not only impede human growth and development, but also pose serious health risks including cancer and neurological disorders. Furthermore, agricultural wastewater harbors a multitude of pathogens encompassing bacteria, viruses, and parasites that can infiltrate the human body via consumption of contaminated water or food sources thereby causing various infectious diseases as well as parasitic ailments (Abdoli 2019). Water pollution is closely associated with diseases like typhoid fever, cholera, gastroenteritis among others, thus posing a

Waste-derived Biochar for Sustainable Rural Revitalization, First Edition. Edited by Yuqing Sun and Daniel C.W. Tsang.
Published 2026 by John Wiley & Sons, Inc.

substantial threat to human health. At the same time, with the rapid development of modern agriculture, the treatment and resource utilization of agricultural waste has become a key problem that needs urgent attention and solution in the world. Many agricultural wastes, such as straw, rice husks, bark, peanut shells, etc., are not only diverse and huge, but also contain extremely rich biomass resources. After treatment, these agricultural wastes can become highly efficient adsorption materials, which are widely used in water purification, soil remediation, and other fields (Gao et al. 2022). These adsorption materials have the advantages of low cost, good environmental performance, and huge market demand. Among which, biochar is extensively utilized in the environmental sector and can effectively eliminate water pollutants through various mechanisms. To enhance its adsorption capacity, modification techniques are frequently employed. Physical modifications such as gas activation and ball milling can alter the pore structure of biochar and increase its specific surface area. Chemical modifications, including acid–base treatment, oxidation, and metal loading, can improve the adsorption rate and oxidation capability of biochar. However, attention should be paid to potential issues related to secondary pollution and residual metal ions. Metal-supported modifications can also confer magnetic properties on biochar, facilitating easy separation and regeneration (Yao et al. 2024). Biological modification aims to enhance adsorption capacity through microbial attachment while promoting energy conservation and environmental protection; nevertheless, it requires further investigation due to its lengthy cycle time and material toxicity concerns. Additionally, biochar has the ability to ameliorate soil properties by enhancing soil fertility and crop yields.

10.2 Preparation of Biochar-based Agriculture Wastewater Adsorbent

10.2.1 Preparation Methods of Biochar

Biomass can be subjected to pyrolysis at elevated temperatures to produce biochar, and different pyrolysis ways can have a pronounced influence on the physical and chemical properties of the resulting biochar. Generally, the main techniques for the production of biochar include conventional pyrolysis, hydrothermal carbonization, and microwave-assisted pyrolysis (Figure 10.1).

10.2.1.1 High-temperature Pyrolysis

Pyrolysis is a frequently employed technique for the synthesis of biochar through the treatment of waste materials under anoxic conditions, typically within the temperature range of 300–900 °C. The structure and properties of the resulting

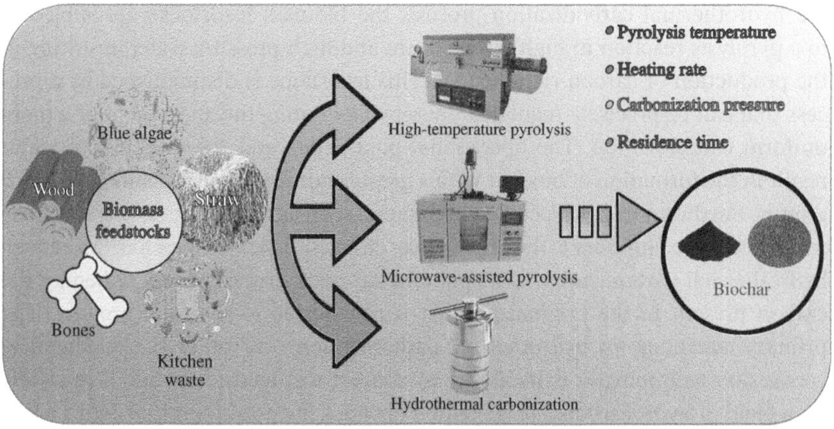

Figure 10.1 Preparation process and influencing factors of biochar.

biochar products can be tailored by varying the pyrolysis conditions. Pyrolysis methods are typically classified as either fast pyrolysis or slow pyrolysis, depending on the heating rate, pyrolysis temperature, residence time, and heating method employed. Among these pyrolysis techniques, fast pyrolysis has gained considerable attention as a promising technology for the production of renewable biofuels, biogas, and select commercially important products. The fast pyrolysis process entails the rapid decomposition of the biomass feedstocks under conditions of high temperature and rapid warming, resulting in the generation of a substantial quantity of bio-oil and a relatively modest amount of biochar. This process typically occurs at temperatures below 600 °C and is characterized by a high heating rate (≥200 °C/min), resulting in a relatively short processing time. This method is distinguished by the production of primarily bio-oil and a modest quantity of biochar, with biochar yields typically in the range of 12% or less. Furthermore, slow pyrolysis entails the heating and complete decomposition of biomass at medium temperatures (typically 350–500 °C), low warming rates, and extended residence times, thereby enhancing the degree of secondary pyrolysis of biomass. In the slow pyrolysis process, the biomass feedstocks can produce stable and carbon-rich biochar after a prolonged pyrolysis reaction (Liu et al. 2023).

10.2.1.2 Hydrothermal Carbonization

Hydrothermal carbonization is a process whereby biomass is subjected to temperatures of 200–300 °C in a subcritical or supercritical state of water within a sealed system, allowing for a controlled chemical reaction to occur. This method does not necessitate the drying or pretreatment of the feedstocks and is therefore well-suited to the processing of biomass with a high moisture content. In the hydrothermal carbonization process, the biomass feedstocks are subjected to a pyrolysis reaction in high-temperature and high-pressure water, resulting in the production of carbon-rich biochar. This technique is distinguished by a process that can be precisely regulated, thereby facilitating the attainment of a more uniform carbonization. The operational parameters and the presence of water result in the formation of biochar with a greater number of chemically functional groups. Furthermore, the specific parameters, such as temperature, pressure, and residence time, influence the distinctive characteristics of biochar. Moreover, hydrothermal carbonization is a spontaneous exothermic process, whereby the carbon present in the original product is transferred to the final product. The primary advantage of hydrothermal carbonization over other thermochemical conversion technologies is its ability to convert wet feedstocks into a relatively high yield of hydrocarbons. This method obviates the need for initial dehydration and drying, thereby reducing the energy requirements (Ji et al. 2022).

10.2.1.3 Microwave-assisted Pyrolysis

Microwave-assisted pyrolysis is a process that employs microwave fields to provide energy for the pyrolysis of materials. The process offers faster overall efficiency and a more controllable heating process and is considered a promising technology for the safe production of biochar. Microwave pyrolysis employs microwave energy to directly heat the biomass, resulting in rapid warming and ease of operation. During the process, the biomass feedstocks is rapidly heated by microwaves and undergoes a pyrolysis reaction. Despite the lower biochar yield of microwave pyrolysis, it exhibits enhanced energy utilization and a rapid reaction speed. This method is well-suited for treating biomass with high moisture content, eliminating the need for feedstocks drying pretreatment.

10.2.2 Influencing Factors in the Preparation of Biochar

The preparation process of biochar is influenced by a number of key factors, including the pyrolysis temperature, heating rate, and pressure at which the carbonization occurs, reaction time is allowed to proceed, and the type of biomass utilized. The collective influence of these factors determines the nature and yield of the final produced biochar.

10.2.2.1 Feedstocks Type

The type of biomass feedstocks is a critical factor influencing the performance and structure of biochar. The chemical composition and physical properties of different types of biomass vary, which consequently affects the properties of biochar produced during pyrolysis. For instance, wood-based biomass typically exhibits elevated carbon content and diminished ash content, thereby producing superior-quality biochar. Conversely, agricultural waste-based biomass may display augmented ash and impurity levels, resulting in the production of inferior-quality biochar. Accordingly, the type and nature of the biomass feedstocks must be taken into account when selecting the feedstocks in order to obtain the optimal biochar yield and quality (Qu et al. 2024).

10.2.2.2 Pyrolysis Temperature

Carbonization temperature is a significant factor influencing the yield and quality of biochar. Within a certain range, an increase in carbonization temperature can result in an increase in the carbon content of the resulting biochar, while simultaneously decreasing the hydrogen and oxygen content. However, an excessive carbonization temperature may result in a reduction in biochar yield, accompanied by an increase in energy consumption and cost. Therefore, it is necessary to exercise reasonable control over the carbonization temperature in order to obtain an optimal yield and quality of biochar (Qu et al. 2023a).

The hydrophobicity, aromaticity, and specific surface area of carbon are markedly influenced by pyrolysis temperature. In general, biochar prepared at high temperatures exhibits a high specific surface area and carbon content. This phenomenon may be attributed to the removal of volatile organic compounds at elevated temperatures, which results in an increase in micropore volume. Surprisingly, the adsorption capacity of biochar prepared at high temperatures for organic pollutants is superior to that of biochar prepared at low temperatures, which is associated with the enhanced hydrophobicity and aromaticity of biochar. Specifically, as the temperature increased, the aromatic content, specific surface area, ash content, and hydrophobicity of adsorbent increased, while the hydrophilicity and yield of the biochar decreased. Furthermore, the pyrolysis temperature is a crucial factor in determining the functional groups that are required. At the optimal temperature, aliphatic functional groups, benzene derivatives, and polycyclic structures begin to form. As the pyrolysis temperature was increased further, the O-containing functional groups decreased gradually. Consequently, biochar produced at low temperatures is predominantly hydrophilic due to the presence of rich O-containing functional groups of biochar. Conversely, biochar obtained at elevated temperatures is more prone to contain an aromatic structure and skeleton, resulting in the formation of hydrophobic functional groups. This renders the material more effective in the adsorption of organic pollutants.

10.2.2.3 Heating Rate

The heating rate is similarly a significant factor influencing the yield and quality of biochar. In the fast pyrolysis process, the rapid heating and short reaction time result in a lower biochar yield, but a higher bio-oil yield. In contrast, the slow pyrolysis process involves a gradual heating rate and a longer reaction time, resulting in a higher biochar yield. Accordingly, the optimal charring rate can be selected as needed to achieve the desired biochar yield and quality.

10.2.2.4 Carbonization Pressure

Carbonization pressure also exerts a certain influence on the yield and quality of biochar. In the hydrothermal carbonization method, the degree of pressure exerted affects the state of water present during the reaction, which in turn affects the production of biochar. In general, higher pressure is conducive to the generation and quality of biochar. However, excessive pressure may lead to increased complexity and cost associated with the equipment, necessitating the need for reasonable control of the carbonization pressure.

10.2.2.5 Residence Time

Reaction residence time is defined as the time that the biomass remains in the pyrolysis process. The duration of the residence time has an obvious impact on the yield and quality of the resulting biochar. In general, an increase in residence time results in a higher yield of biochar, while it also may lead to a reduction in quality. Accordingly, it is essential to select an appropriate residence time in accordance with the specific circumstances to ensure the optimal biochar yield and quality (Wang et al. 2019). Studies showed that a significant decrease in biochar yield and an evident increase in specific surface area were observed when the residence time was 30–90 minutes. This might be attributed to the decomposition of organic matter and the release of volatile substances, which promoted the formation and expansion of the biochar's pore structure. When the residence time was extended to 90–150 minutes, there was a slight decline in specific surface area, while the yield remained relatively stable. It indicated that an extension of the residence time beyond a certain point led to the collapse of the pore structure, resulting in a decline in specific surface area. On the other hand, as the residence time increased, the carbon content of biochar exhibited a slight decline, accompanied by a comparable reduction in the contents of H, N, and O. This decline was mainly due to the loss of carbon-bonded oxygen and hydrogen atoms resulting from the degradation of the structural core over longer residence time. At the same time, the H/C and N/C ratios of biochar decreased with the increasing residence time, indicating that the longer the residence time, the higher the aromaticity of biochar. Furthermore, both ash content and pH increased over time, as more organic matter was broken down and more alkali metals were released and converted to oxides.

10.3 Efficacy of Biochar-adsorption on Agriculture Wastewater Treatment

As a new multifunctional material, biochar has a broad application prospect in environmental pollution control because of its unique physical and chemical properties and decent adsorption properties. However, the adsorption capacity of biochar is affected by many factors, including the properties of the adsorbent itself, and external environmental conditions, as described in Figure 10.2.

10.3.1 Effect of Adsorbent Properties on Adsorption

10.3.1.1 Specific Surface Area and Pore Structure

Biochar is an efficient adsorptive material, and its adsorption capacity depends largely on the specific surface area and pore structure. The larger the specific surface area and the more complicated the pore structure, the more adsorption sites biochar can provide, which can adsorb pollutants more effectively. For example, Zhi et al. (2023) prepared luffa derived activated biochar through pyrolysis for adsorbing dyes in wastewater, and the result showed that the biochar possessed a high specific surface area (2262.67 m^2/g) and abundant micropores, which effectively enhanced the adsorption performance of biochar for pollutants, achieving 953.53 mg/g.

10.3.1.2 Surface Charges and Functional Groups

The adsorption performance of biochar is significantly affected by the surface charge properties and the type and number of functional groups of biochar. O-containing functional groups such as hydroxyl (—OH) and carboxyl (—COOH), which are rich in biochar surface, have strong complexation ability. And these functional groups can have specific chemical reactions with heavy metal ions such as lead (Pb), cadmium (Cd), and mercury (Hg) to form stable complexes or chemical bonds. This chemical reaction enables biochar to effectively adsorb heavy metal ions in wastewater, thereby reducing the environmental pollution.

Similarly, these O-containing functional groups are also able to interact with organic pollutants. For example, —COOH and —OH groups on the biochar surface can bind to organic pollutants such as pesticides, antibiotics, and phenolic compounds through hydrogen bonding, π–π stacking, or electron donor–acceptor interactions to achieve effective adsorption of organic pollutants. In the case of atrazine (a common pesticide), functional groups on the surface of biochar can form hydrogen bonds with the amino and chlorine atoms in the atrazine molecule, thereby removing it from agricultural wastewater (Qiu et al. 2022).

10.3.2 Effect of Environmental Conditions on Biochar Adsorption

10.3.2.1 Effect of Reaction Temperature on Biochar Adsorption

Temperature is a key factor affecting the adsorption performance of biochar, and its effect on the adsorption process is multifaceted. On the one hand, an increase in temperature increases the speed of movement and diffusion coefficient of

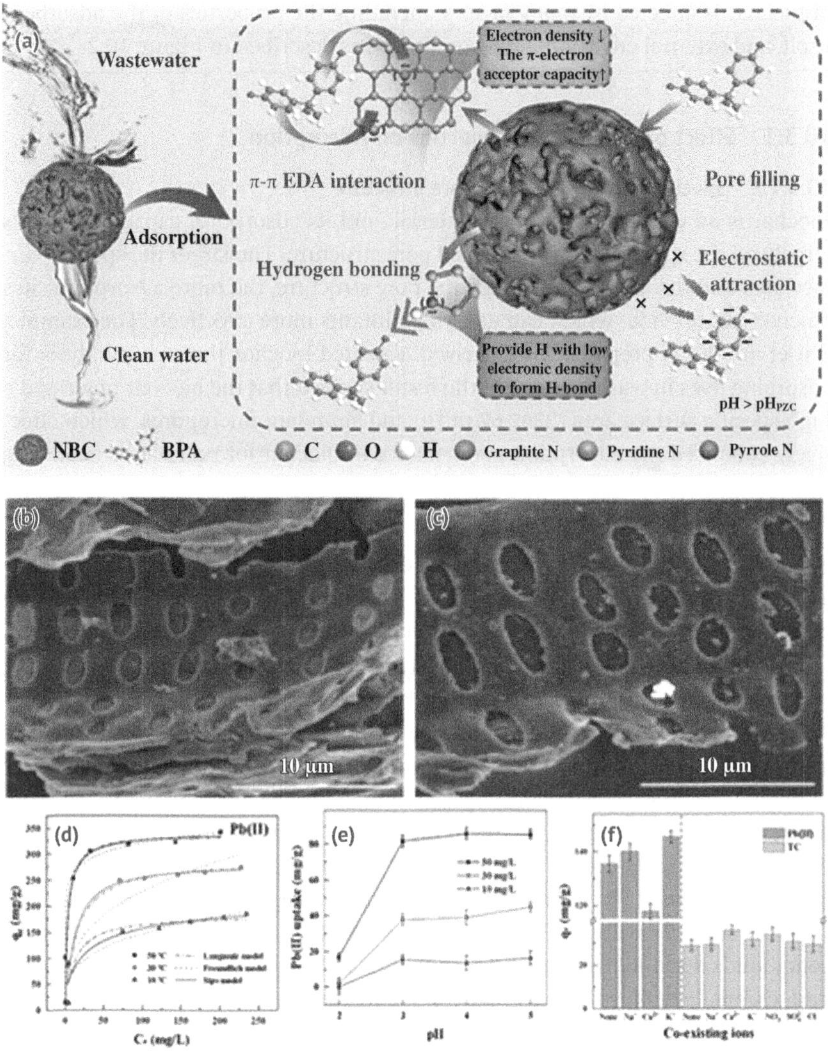

Figure 10.2 The effects of (a) surface charges and functional groups (Li et al. 2025/with permission of Elsevier), (b, c) specific surface area and pore structure (Wang et al. 2024/with permission of Elsevier), (d) reaction temperature, (e) pH, and (f) co-existing pollutants (Qu et al. 2023b/with permission of Elsevier) on adsorption performance of biochar for different pollutants.

adsorbent molecules, thus accelerating the rate of adsorption and enhancing the adsorption capacity of pollutants. On the other hand, the change of temperature also affects the functional group activity, so that the interaction force between pollutants and biochar is enhanced, which is conducive to the adsorption process. At the same time, it should be noted that excessive temperatures may also lead to a decrease in active sites on surface of the adsorbent or an increase in the desorption rate of adsorbent molecules, thus decreasing the adsorption efficiency (Li et al. 2017).

During adsorption, the endothermic or exothermic properties of the chemical reaction are also affected by temperature. For endothermic reactions, an increase in temperature is beneficial to the reaction, and according to Le Chatelier's principle, the system adsorbs heat to offset the increase in temperature, thus pushing the reaction in the direction of the product. This means that in the endothermic adsorption process, increasing the temperature can promote the chemical reaction between pollutants and biochar, thus improving the adsorption efficiency. However, for exothermic reactions, an increase in temperature causes the reaction to move in the direction of the reactants, reducing the adsorption efficiency. Therefore, in the exothermic adsorption process, too high temperature may not be conducive to the adsorption of pollutants.

10.3.2.2 Effect of pH Value on Biochar Adsorption

The pH value of the solution has a significant impact on the adsorption performance of biochar, because the change of pH value can directly change the charge properties of biochar surface and the activity of functional groups, and then affect its adsorption capacity of pollutants. The surface of biochar usually contains a variety of functional groups (e.g. —COOH, —OH, amine, etc.), which show different charge states at different pH values. In concrete terms, under acidic conditions, functional groups including —COOH and —OH groups on the surface of biochar may be protonated and positively charged, which makes it easier for biochar to adsorb negatively charged pollutants such as Cr(VI) and phosphate (Bilgic et al. 2023). Whereas, under alkaline conditions, the opposite may be true. Functional groups on the surface of biochar may lose protons and take on a negative charge, resulting in a higher affinity for positively charged contaminants. In addition, changes in pH also affect the presence of pollutants. For example, Pb(II) may exist in univalent form under acidic conditions, while it may form hydroxyl complexes or polyvalent ions under alkaline conditions. These morphological changes will directly affect the solubility and migration of pollutants and thus affect the adsorption effect of biochar. Besides, changes in pH also affect the degree of ionization of contaminants in the solution. For some organic pollutants, a change in pH value may lead to a change in the ionized state of their molecular structure, which affects their solubility in water and adsorption behavior on the surface of adsorbent. Similarly, under extreme pH conditions, the molecular structure of pollutants may change, and even lead to biodegradation or chemical decomposition, which has an important impact on the adsorption process.

Therefore, in practical applications, controlling the pH value of solution is one of the key factors to optimize the adsorption performance of biochar. By adjusting the pH value, the charge state and functional groups activity on adsorbent's surface can be effectively regulated, as well as the existence form and degree of ionization of pollutants, so as to improve the adsorption efficiency of biochar for specific pollutants.

10.3.2.3 Effect of Co-existing Pollutants on Biochar Adsorption

In the actual environment, biochar often faces a complex pollution situation, requiring simultaneous treatment of multiple types of pollutants. When these pollutants co-exist in the solution, they may compete with each other for the limited adsorption sites on the surface of biochar, which directly affects the adsorption efficiency of biochar for pollutants (Ding et al. 2022). For example, heavy metal ions such as Pb(II) and Cd(II) may co-exist in water and they both compete active sites on biochar, leading to a reduction in the amount of adsorption of a particular heavy metal. The change in ionic strength is also an important factor. When high concentrations of Na(I) or Ca(II) are present in solution, which compete with pollutants for adsorption sites on the biochar. This situation is particularly common in the treatment of saline wastewater, where high concentrations of salt may significantly affect the adsorption of biochar.

Specifically, contaminated water body usually contains both organic pollutants such as polycyclic aromatic hydrocarbons (PAHs) and heavy metal ions (such as Cd). In this case, PAHs may be strongly adsorbed in the hydrophobic region of the biochar by hydrophobic forces, while Cd(II) may bind to negatively charged functional groups on the biochar surface by electrostatic forces. Competition between the two contaminants may result in a decrease in the ability of biochar to adsorb one of the contaminants, therefore requiring optimization of the adsorption effect by adjusting the properties of the biochar or treatment conditions.

10.4 Effects of Modification Methods on Biochar-enhanced Adsorption Agriculture Wastewater Treatment

The modification method has a significant effect on the physicochemical properties of biochar, which not only affects the number of surface functional groups, but also changes the pore structures. Therefore, it is very important to select suitable modification technology to improve the adsorption performance of biochar to pollutants in agriculture wastewater. The commonly used modification methods mainly include the following three categories: physical activation, chemical treatment, and biological modification.

10.4.1 Physical Activation

Physical activation refers to the method of changing the surface morphology and pore structure of biochar by means of physical principles. Common activation methods involve mechanical ball milling, gas modification, as well as plasma treatment, as seen in Figure 10.3.

10.4.1.1 Mechanical Ball Milling

Mechanical ball milling can produce nanoscale ultrafine particle materials, and use mechanical energy to chemically react with the materials in milling process, thereby introducing O-containing functional groups (e.g., —COOH and —OH), which has the advantages of economic benefit, environmental protection, and low energy consumption. It is worth mentioning that during the milling process, various operational factors can obviously change the properties of biochar.

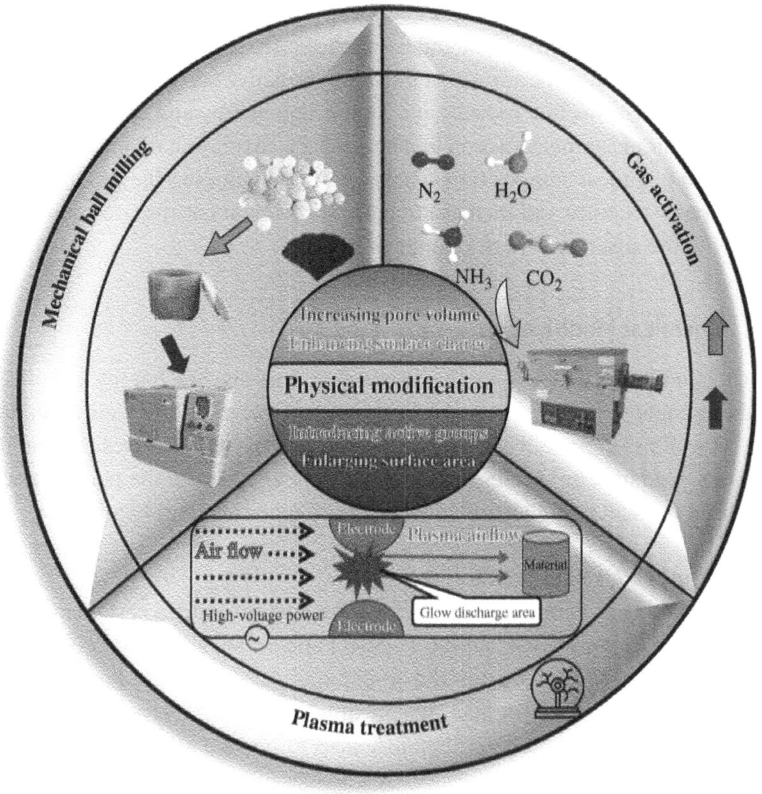

Figure 10.3 The process of physical modification and effect for biochar structure.

Therefore, the optimization of milling parameters (e.g., grinding time, ball particle size, grinding speed, and dry or wet grinding) is crucial to adjust the function and performance of ball milling biochar.

10.4.1.2 Gas Modification

Gas activation mainly adopts multifarious gasifying reagents (such as steam, CO_2, NH_3, and O_2) to introduce O-containing functional groups including —COOH, —OH, and —C=O groups onto biochar, which enhance its surface chemical activity. Meanwhile, during the activation process, the gas and biochar undergo redox reaction, which erodes the adsorbent surface and makes the pore structure more developed, therefore enlarging the specific surface area and pore size of biochar. This method can effectively improve adsorption property of biochar for pollutants with the advantages of cheap, convenient benefits, and no risk of secondary pollution.

In addition, plasma modification also is a prominent method. This technique can change the surface structure and chemical properties of biochar by putting biochar in the plasma environment and using the active particles in the plasma to etch, oxidize, and reduce the surface of biochar. The active particles in the plasma (such as ions, electrons, free radicals, etc.) have high energy and can collide and react with the atoms or molecules on biochar surface, thereby changing its surface structure and chemical composition. These changes, including the optimization of surface pore structure and increase of active functional groups, are conducive to improving the adsorption performance of biochar.

10.4.2 Chemical Modification

Chemical modification refers to the process of treating biochar by chemical methods to change its surface chemical properties, increase surface functional groups, as well as improve its porosity. This method usually involves the use of acids, bases, oxidants, or specific surface modifiers to optimize the performance of biochar in particular application, which is fully represented in Figure 10.4.

10.4.2.1 Acid and Alkaline Modification

Acid modification can achieve a porous structure and introduce a certain amount of acid functional groups (e.g., amino (—NH_2) and —COOH) on adsorbent surface. Broadly utilized acids during the functional process mainly include HCl, H_2SO_4, HNO_3, H_3PO_4, and citric acid. Specifically, Cao et al. (2024) prepared biogas residue biochar (CA-BC) modified by citric acid to remove atrazine (ATZ). The results showed that the specific surface area increased from 96.60 to 266.2 m^2/g, and the uptake amount of CA-BC increased from 3.67 ± 0.31 to 5.94 ± 0.07 mg/g

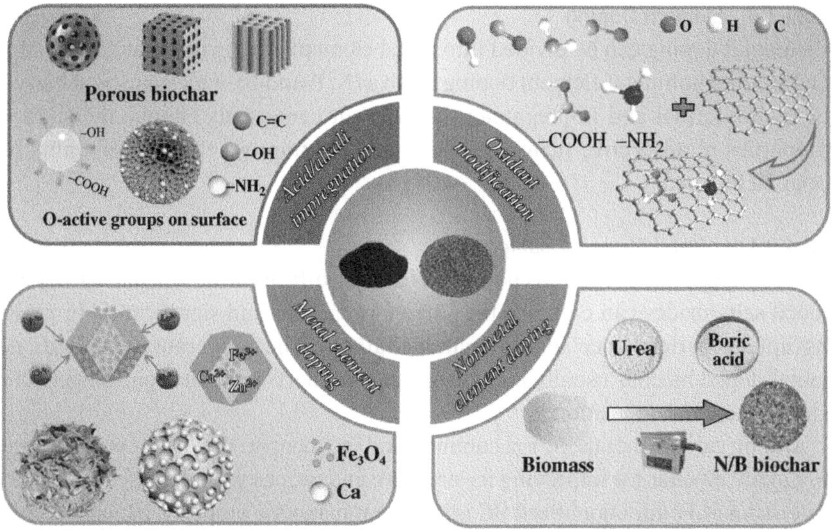

Figure 10.4 Different chemical modification methods of biochar for improving its adsorption performance.

after the decoration of citric acid. The main reasons could be attributed to the enhanced microporous structure, the increased O-containing functional groups, and the improved adaptability achieved by reducing the zero-charge point.

Similarly, alkali modification is also a valid technique to enhance biochar adsorption ability. The major purpose is to enlarge specific surface area and increase O-containing functional groups. Deng et al. (2024) adopted $KHCO_3$ to decorate biochar, and found that modified biochar produced abundant porous structures and an obvious layered structure with a noticeable specific surface area (1648.89 m^2/g), which facilitated the transport of organic pollutant, and thus increased the adsorption capacity (147.8 mg/g).

10.4.2.2 Oxidation

Surface oxidation is a common and the easy technique to generate O-containing functional groups (e.g., —OH, —COOH, —C=O, and —C—O) onto the surface of biochar. Oxidation is performed under a refluxing condition in the presence of a single or a mixture of inorganic acids (e.g., HNO_3 and H_2SO_4) and an oxidizing agent (e.g., H_2O_2, $KMnO_4$, and NaOCl). Therefore, it is important to regulate the amount of O onto biochar with the right dosage of oxidation agents and experimental conditions such as reaction time and temperature, which are essential for building desired functional structures.

10.4.2.3 Element Doping

Elemental doping can be divided into metal element doping (such as Ca, Fe, Mg, and La) and nonmetal element doping (such as N, P, and B), which can be achieved by hydrothermal and impregnation methods. It can greatly change the surface properties of adsorptive materials and effectively improve the uptake ability of original biochar.

10.4.2.3.1 Metal Element Doping

Studies have shown that the new structural modification of biochar by doping metal salts/oxides can enhance the surface properties and significantly improve its uptake performance for different pollutants. The preparation methods of metal-doped biochar usually include impregnation, co-precipitation, sol-gel, and electrochemical deposition.

Among various metals, Fe is a common and inexpensive metal that is often used to modify biochar for improving its structure and property. One of the main characteristics of Fe-functionalized BC is its decent magnetic property, which not only improves the removal capacity of heavy metals, nutrients, and organic pollutant in wastewater, but also effectively recovers adsorbent from water environment. Similarly, Mg as an alkaline earth element, is widely found in the earth crust and has the advantages of low price, easy availability, and environmental friendliness, which is used widely to decorate with biochar to remove heavy metals. Besides, La, a rare earth element, is often modified with biochar to remove phosphate from wastewater, which can achieve a good adsorption performance. For example, Chen et al. (2024) used cyanobacteria as biomass for resource utilization to remove excess phosphate in water and alleviate the eutrophication problem. This study proved that Mg/La modified biochar owned rich O-containing functional groups (—COOH and —OH) and high specific surface area (82.26–233.80 m^2/g). The maximum removal efficiency of Mg/La modified biochar for phosphate could reach 97.8%. The main adsorption mechanism was due to the electrostatic interaction between hydroxylated metal oxides and negatively charged phosphate ions, as well as the phosphate ions precipitated by chemical reaction with metal oxides in solution.

10.4.2.3.2 Nonmetal Element Doping

By introducing nonmetallic elements into the structure of biochar, the physical, chemical, and electrochemical properties of biochar can be changed. The preparation of biochar with nonmetallic elements usually involves mixing compounds containing nonmetallic elements with biochar precursors, and then achieving doping through pyrolysis, carbonization, and other processes. Among the doped elements, nitrogen (N), sulfur (S), boron (B), and phosphorus (P) are widely used.

Among them, the doping of N can introduce additional electrons, improve the electrical conductivity of biochar, and at the same time, N atoms can effectively promote the electron transfer of neighboring C atoms, thus affecting its charge density. Shaarani and Hameed (2011) used ammonia water to modify biochar derived from oil palm fruits and explored the effect of N-modified biochar on the adsorption capacity of 2,4-dichlorophenol (2,4-DCP). The results indicated that the incorporation of N produced basic groups (e.g., cyclic amides and amines groups), which strengthened alkalinity of biochar, and created a surface with positive charges, thus increasing the uptake amount for 2,4-DCP to 285.71 mg/g. Moreover, B modified biochar has a large specific surface area and good thermal and chemical stability, contributing to the high adsorption performance for heavy metals. In addition, P atom can activate biochar effectively due to its electronegativity. Suo et al. (2019) prepared P-doped biochar by one-step method and found that the specific surface area of biochar was enlarged, and the amounts of —COOH and $-PO_3$ were enriched after P doping, thus promoting the adsorption process. P-doped biochar still had a strong adsorption efficiency (> 96%) for triazine pesticides after five cycles through such mechanisms as van der Waals forces and electrostatic attractions, and it could quickly reach adsorption equilibrium.

Multi-element doping also exists, for instance, Guo et al. (2020) investigated that biochar doped with N and S elements could introduce pyridinic N and oxidized S groups on the surface, which was utilized to eliminate diethyl phthalate (DEP). By introducing N and S elements, the surface functional groups such as —NH and —C=O were dramatically boosted, and the modified biochar could generate nanoscale pore structures at low temperature, which significantly improved the adsorption rate of DEP.

Overall, acid/alkaline modification introduces a mass of O-containing acidic surface groups, contributing to the significant increase of —COOH, —OH, and phenolic hydroxyl groups, thereby strengthening the adsorption of heavy metals and organic pollutants. However, it is necessary to control the reaction conditions and dosages of acid and alkali accurately since the improper modification degree control is easy to cause the corrosion of the pore wall and the collapse of micropores. On the other hand, oxidation modification can also substantially enhance biochar pore structures and enrich the kinds and number of active groups on adsorbent surface, making the distribution of micropores concentrated and pore structure controllable. Moreover, metal element doping can enhance the dispersion of carbon material in water, reduce the agglomeration of material, and make the material attach with a certain magnetism, which is convenient for magnetic separation and recovery after treatment. Similarly, the doping of nonmetals can introduce corresponding element groups and increase the active site on the surface of the adsorbent, thus significantly improving the environmental pollutant removal performance of the modified biochar.

10.4.3 Biological Modification

Biological modification involves the treatment of biomass feedstocks or biochar using microorganisms, enzymes, or other bioactive substances, which are natural and environmentally friendly and do not pollute the environment. These treatments can change the surface properties of biochar, such as pore structure, kind and number of functional groups, so as to enhance its adsorption performance or give it new functional properties. The corresponding modification process is shown in Figure 10.5.

Zhu et al. (2024) carried out heavy metal repair by loading the heavy metal-resistant organic phosphorus-degrading bacteria OPB3-6-1 onto biochar. They found that biochar loaded with OPB3-6-1 (IOPB) possessed a synergistic effect on the adsorption of Pb(II), Cu(II), and Cd(II) with the highest adsorption capacities of 591, 300, and 20 mg/g, respectively. The main reason was that, on the one hand, the excellent porous structure and adsorbability of biochar provided growth space for supporting bacteria, and reduced the toxic effect of IOPB on heavy metals under pollution conditions, thus inducing the synergic adsorption of triple metals by biochar—bacteria coupling system. On the other hand, the addition of OPB3-6-1 to biochar increased amount of surface functional groups of the coupled system, therefore enhancing the complexation of heavy metals.

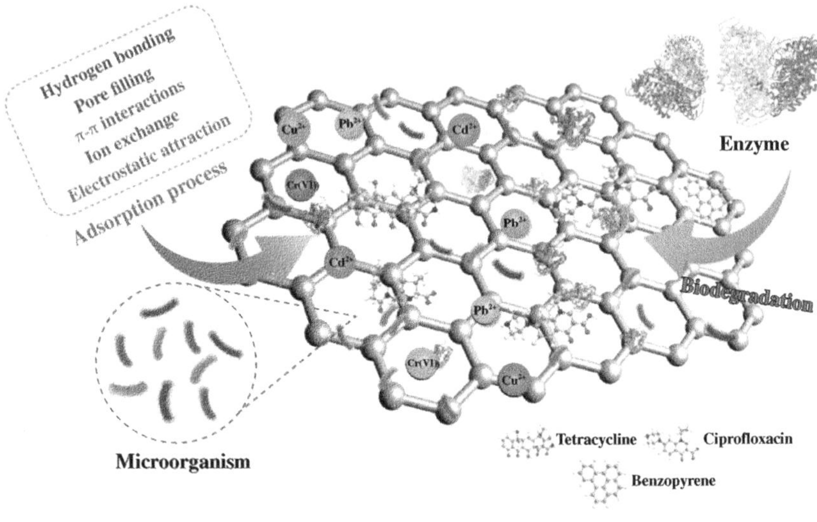

Figure 10.5 Biological modification techniques of biochar and removal process for different pollutants.

10.5 Mechanisms of Adsorption and Future Prospects

10.5.1 Mechanisms of Biochar for Adsorption of Inorganic Contaminants in Agriculture Wastewater

The removal of inorganic contaminants by biochar cannot be attributed to a single and straightforward mechanism. This complexity arises from the diverse chemical forms of surface functionalities on biochar. It is generally accepted that inorganic contaminants adsorption by biochar is not only influenced by their porous characteristics. Consequently, the adsorption process primarily relies on surface functionalities, which interact with inorganic contaminants through various mechanisms such as physical adsorption, surface complexation, precipitation, electrostatic interaction, ion exchange, and cation-π bonding (Figure 10.6).

10.5.1.1 Physical Adsorption

Biochar possesses a complex internal and surface structure by a combination of macropores, mesopores, and micropores. These features enable biochar to retain contaminants through physical adsorption, a process that is relatively weak and involves the migration and permeation of contaminants into the porous network of biochar, followed by deposition on the surface without the formation of chemical bonds. In general, the performance of physical adsorption is significantly influenced by the surface area, pore distribution, and pore size of the biochar. Increasing the number of micropores can substantially enhance the surface

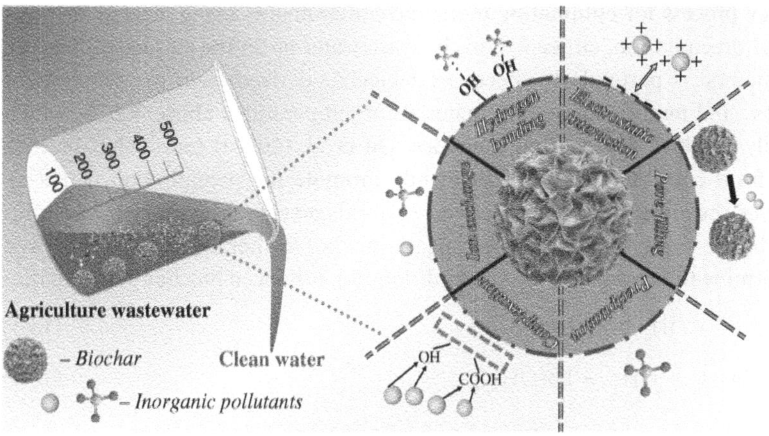

Figure 10.6 Adsorption mechanisms diagram of biochar for inorganic pollutants in agricultural wastewater.

area of biochar, thereby facilitating greater physical adsorption. Conversely, an increase in mesopores can improve the diffusion of contaminants, which in turn accelerates the adsorption process.

The selection of organic raw materials and the synthesis method including the temperature used during carbonization or pyrolysis, are fundamental factors that determine the pore structures of biochar. Consequently, the polarity and heterogeneity of the biochar surface influenced by various functional groups play a crucial role in physical adsorption. This is particularly evident in the movement of contaminants toward the biochar surface, which is driven by ionic dipole forces and electrostatic attractions.

10.5.1.2 Surface Complexation

Surface complexation is the most widely recognized mechanism for inorganic contaminants removal by biochar. This process involves forming multi-atom structures through specific metal–ligand interactions. Transition metals such as Cd, Cu, and Pb, with their partially filled d-orbitals, have a high affinity for ligands. Biochar can share its lone pair electrons with the partially filled d-orbitals of heavy metals, creating a complex–coordinate structure. Additionally, O-containing functional groups on biochar surface (e.g., —COOH, and —OH) also bind with inorganic contaminants through complexation.

$$R{-}OH + H_2PO_4^- \rightarrow R{-}H_2PO_4 + OH^- \tag{10.1}$$

$$R{-}COOH + Cd^{2+} \rightarrow R{-}COO^-{\text -}Cd^{2+} + H^+ \tag{10.2}$$

10.5.1.3 Precipitation

One key process for eliminating inorganic contaminants using biochar involves mineral precipitation, either within the wastewater or on the surface of biochar. This method is particularly effective for biochar derived from the pyrolysis of cellulose and hemicellulose-rich materials at temperatures above 300 °C, which typically exhibit alkaline characteristics. Qu et al. (2022a) found that biochar made from corn straw could significantly promote the precipitation of Pb(II). However, they noted that the effectiveness of this precipitation process was influenced by the pyrolysis temperature, suggesting that further studies were required to determine the optimal pyrolysis conditions for enhanced biochar performance.

$$Pb^{2+} + OH^- \rightarrow PbOH^+ \tag{10.3}$$

$$PbOH^+ + OH^- \rightarrow Pb(OH)_2\downarrow \tag{10.4}$$

10.5.1.4 Electrostatic Interaction

Electrostatic interactions occur between the negatively charged surfaces of biochar and the positively charged contaminants, especially when functional groups are present. Although this process is relatively weak, electrostatic interactions still contribute as a vital mechanism in the adsorption of contaminants by biochar. The variable surface charges of biochar mean that the point of zero charge (pH_{zpc}) and the solution pH play critical roles in determining the prevalence of these interactions. The ionization of surface groups largely influences the charged interface between the biochar and the solution (Qu et al. 2022b).

The surface charge characteristics of biochar are crucial in assessing the strength of electrostatic adsorption between biochar and contaminants. The solution pH is also vital for contaminants adsorption in agriculture wastewater. The zeta potential of biochar at different solution pH conditions can identify the point of zero charge (pH_{zpc}) of biochar. On the one hand, when the solution pH is lower than pH_{zpc}, a positively charged surface forms due to the protonation of the hydration surface, which might lead to the lower adsorption capacity. Additionally, H^+ in the solution can compete with contaminants for vacant adsorption sites on biochar. On the other hand, when the solution pH is higher than the pH_{zpc}, the biochar surface becomes negatively charged due to the deprotonation of the hydration surface, enhancing the electrostatic attraction between contaminants like Cd^{2+} and NH_4^+ and the biochar surface.

10.5.1.5 Ion Exchange

An alternative pathway for the removal of contaminants by biochar is the interaction between ionizable protons/cations and dissolved pollutants. The capacity for ion exchange is affected by both the size of the contaminant and the biochar surface properties. These surface properties are responsible for determining the cation exchange capacity (CEC) of biochars, with a higher CEC suggesting a greater ability to remove contaminants. Mia et al. (2017) documented the removal of phosphate and ammonium using aging of biochar, which was facilitated by the ion exchange process.

10.5.2 Mechanisms of Biochar for Adsorption of Organic Pollutants in Agriculture Wastewater

It is evident that multiple mechanisms are responsible for the adsorption of organic pollutants onto biochar. Notably, pore-filling, hydrogen bonding, and electrostatic Interaction are the primary factors influencing the adsorption process between biochar and organic contaminants. In addition to these, Lewis acid–base complexes, π–π interactions, and other mechanisms also contribute to the adsorption phenomenon, as displayed in Figure 10.7.

10.5.2.1 Pore Filling

The network of pores within biochar consists of macropores larger than 9 nm, mesopores ranging from 2 to 9 nm, and micropores less than 2 nm in size. The

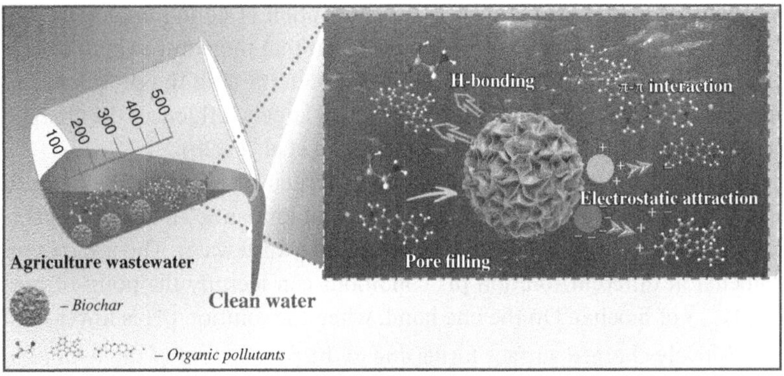

Figure 10.7 The corresponding mechanisms of biochar adsorption for organic pollutants in agricultural wastewater.

presence of micropores (< 2 nm) and mesopores (2~50 nm) on the biochar surface plays a crucial role in the pore-filling mechanism for the adsorption of organic pollutants due to their contribution to the specific surface area. According to Xiao and Pignatello (2015), the adsorption capacity of biochar for triazine herbicides was predominantly determined by its microporous and mesoporous structures.

10.5.2.2 Hydrogen Bonding

Hydrogen bonding represents a specific form of dipole–dipole interaction that takes place between a hydrogen atom and an electronegative atom. In this context, hydrogen donors are molecules like nitrogen, oxygen, and fluorine, which are covalently linked to hydrogen. Meanwhile, hydrogen acceptors are electronegative molecules possessing a lone pair of electrons. Functional groups on the surface of biochar such as —OH and —COOH can engage in hydrogen bonding with organic pollutants, thereby facilitating the adsorption performance of biochar (Souza and Ferreira 2024).

10.5.2.3 Electrostatic Interaction

The process of electrostatic interaction entails the mutual attraction between the oppositely charged ionizable functional groups present on both organic pollutants and the biochar surface. Like the case with inorganic contaminants, the mechanism of electrostatic removal of organic substances via biochar is also contingent upon the pH level of the solution.

10.5.2.4 Other Mechanisms

Beyond the previously discussed adsorption mechanisms, other potential processes, such as π–π EDA interactions, nucleophilic additions, cation bridging, and acid–base reactions, must not be overlooked. For instance, π–π EDA interactions are considered a primary mechanism for the adsorption of sulfonamides onto biochar. Functional groups like carboxylic acids and nitro groups from pollutants can serve as π-electron acceptors, while the aromatic structures on biochar can act as π-electron donors. Typically, the adsorption process involves the electrostatic attraction of organic functional groups and the entrapment within pores. Consequently, the presence of specific functional groups in organic contaminants, such as chlorine substitutions and protonated aniline rings, can influence both the adsorption capacity and the mechanisms involved. In summary, the adsorption of organic pollutants by biochar is a complex process that encompasses a variety of specific mechanisms.

10.5.3 Future Prospects

As a kind of adsorbent with extensive sources and excellent adsorption properties, biochar has a broad application prospect in the purification of agricultural wastewater. It is worth noting that compared with the simulated environment in the laboratory, due to the complex water environment of agricultural wastewater, there is a lack of systematic research on the practical application of biochar and the treatment of secondary pollution in the synthesis process. To facilitate the extensive use of biochar, additional research is imperative. This research should focus on several key areas: Firstly, there is a need for in-depth studies on biochar tailored for specific applications and functionalities, with the aim of optimizing its adsorption mechanisms to efficiently capture and release a variety of pollutants. Secondly, it is crucial to address the potential presence of toxic substances in biochar during pyrolysis by ensuring these are leached out before its application in contaminated water. Additionally, enhancing the recovery of pollutants from biochar is essential to prevent the long-term accumulation of hazardous substances that can turn into latent pollution sources. Thirdly, the technology for biochar adsorption of different pollutants must undergo continuous industrial-scale testing under real-world water conditions to manage unforeseen challenges. Lastly, for the sustainable application of biochar in environmental remediation, further studies are needed to evaluate production costs, economic viability, and technical constraints, and to develop strategies aimed at enhancing production efficiency.

References

Abdoli, A. (2019). Parasitic encephalitis in immunocompetent individuals. *The Lancet* 394 (10202): 914–915.

Bilgic, A., Cimen, A., and Kursunlu, A.N. (2023). A novel biosorbent functionalized pillar[5]arene: synthesis, characterization and effective biosorption of Cr(VI). *Science of the Total Environment* 857: 159312.

Cao, B.X., Li, M., Zhang, T. et al. (2024). Dynamics and mechanisms of atrazine adsorption on biogas-residue biochar with citric acid modification. *Separation and Purification Technology* 337: 126151.

Chen, H.B., Zhang, X.Q., Xiao, Y.X. et al. (2024). Metal-modified biochars prepared from blue algae and their ability of adsorbing phosphates from water and utilization as soil amendment. *Journal of Environmental Chemical Engineering* 12: 114561.

Deng, Y., Chen, J., Xiao, Z.D. et al. (2024). A novel N self-doped porous biochar synthesized by $KHCO_3$-activated chicken feather for the remediation of tetracycline-contaminated water and soil. *Water, Air, & Soil Pollution* 235: 133.

Ding, H., Zhang, Z.L., Li, Y. et al. (2022). Fabrication of novel Fe/Mn/N co-doped biochar and its enhanced adsorption for bisphenol a based on π–π electron donor–acceptor interaction. *Bioresource Technology* 364: 128018.

Gallo, M., Ferrara, L., Calogero, A. et al. (2020). Relationships between food and diseases: what to know to ensure food safety. *Food Research International* 137: 109414.

Gao, N., Du, W., Zhang, M. et al. (2022). Chitosan-modified biochar: preparation, modifications, mechanisms and applications. *International Journal of Biological Macromolecules* 209: 31–49.

Guo, R.S., Yan, L.L., Rao, P.H. et al. (2020). Nitrogen and sulfur co-doped biochar derived from peanut shell with enhanced adsorption capacity for diethyl phthalate. *Environmental Pollution* 258: 113674.

Ji, J.Q., Yuan, X.Z., Zhao, Y.L. et al. (2022). Mechanistic insights of removing pollutant in adsorption and advanced oxidation processes by sludge biochar. *Journal of Hazardous Materials* 430: 128375.

Li, H.B., Dong, X.L., Evandro B. et al. (2017). Mechanisms of metal sorption by biochars: biochar characteristics and modifications. *Chemosphere* 178: 466–478.

Li, Z.W., Tong, W.Y., Li, C. et al. (2025). Advances on nitrogen-doped biochar for adsorption and degradation of organic pollutants from aquatic environment: mechanisms and applications. *Separation and Purification Technology* 354: 129017.

Liu, Y., Weng, Z.L., Han, B. et al. (2023). Recent studies on the comprehensive application of biochar in multiple environmental fields. *Journal of Cleaner Production* 421: 138495.

Mia, S., Dijkstra, F.A., and Singh, B. (2017). Aging induced changes in biochar's functionality and adsorption behavior for phosphate and ammonium. *Environmental Science & Technology* 15 (51): 8359–8367.

Preetham, V. and Vengala, J. (2021). Removal of agricultural wastewater pollutants by integrating two waste materials, fish scales and neem leaves, as novel potential adsorbent. *Water Science and Technology* 84 (10–11): 2980–2996.

Qiu, B.B., Shao, Q.N., Shi, J.C. et al. (2022). Application of biochar for the adsorption of organic pollutants from wastewater: modification strategies, mechanisms and challenges. *Separation and Purification Technology* 300: 121925.

Qu, J.H., Bi, F.X., Li, S.Z. et al. (2022a). Microwave-assisted synthesis of polyethylenimine-grafted nanocellulose with ultra-high adsorption capacity for lead and phosphate scavenging from water. *Bioresource Technology* 362: 127819.

Qu, J.H., Meng, Q.J., Peng, W. et al. (2023a). Application of functionalized biochar for adsorption of organic pollutants from environmental media: synthesis strategies, removal mechanisms and outlook. *Journal of Cleaner Production* 423: 138690.

Qu, J.H., Peng, W., Wang, M.N. et al. (2024). Metal-doped biochar for selective recovery and reuse of phosphate from water: modification design, removal mechanism, and reutilization strategy. *Bioresource Technology* 407: 131075.

Qu, J.H., Wu, Z.H., Liu, Y. (2022b). Ball milling potassium ferrate activated biochar for efficient chromium and tetracycline decontamination: insights into activation and adsorption mechanisms. *Bioresource Technology* 360: 127407.

Qu, J.H., Zhang, B., Tong, H. et al. (2023b). High-efficiency decontamination of Pb(II) and tetracycline in contaminated water using ball-milled magnetic bone derived biochar. *Journal of Cleaner Production* 385: 135683.

Shaarani, F. and Hameed, B. (2011). Ammonia-modified activated carbon for the adsorption of 2,4-dichlorophenol. *Chemical Engineering Journal* 169: 180–185.

Souza, T.F.D. and Ferreira, G.M.D. (2024). Biochars as adsorbents of pesticides: laboratory-scale performances and real-world contexts, challenges, and prospects. *ACS ES&T Water* 4 (10): 4264–4282.

Suo, F.Y., You, X.W., Ma, Y.Q. et al. (2019). Rapid removal of triazine pesticides by P doped biochar and the adsorption mechanism. *Chemosphere* 235: 918–925.

Wang, Y., Li, J., Li, Q. et al. (2024). Effective amendment of cadmium in water and soil before and after aging of nitrogen-doped biochar: preparation optimization, removal efficiency and mechanism. *Journal of Hazardous Materials* 477: 135356.

Wang, Z.P., Liu, K., Xie, L.K. et al. (2019). Effects of residence time on characteristics of biochars prepared via co-pyrolysis of sewage sludge and cotton stalks. *Journal of Analytical and Applied Pyrolysis* 142: 104659.

Xiao, F. and Pignatello, J.J. (2015). Interactions of triazine herbicides with biochar: steric and electronic effects. *Water Research* 80 (1): 179–188.

Yao, T., Song, J., Zhou, C. et al. (2024). Recent progress of the applications of functionalized magnetic ionic liquids in sample pretreatment. *Separation and Purification Technology* 341: 126979.

Zhi, Z.F., Zhou, W.J., Chen, J.R. et al. (2023). Adsorption properties of active biochar: overlooked role of the structure of biomass. *Bioresource Technology* 387: 129695.

Zhu, X.L., Wang, K., Ma, X.J. et al. (2024). Loading organic phosphorus-degrading bacteria enhanced biochar performance for heavy metals adsorption. *Environmental Technology & Innovation* 34: 103585.

11

Waste-derived Biochar as Catalyst for Agriculture Wastewater Treatment

Xiaofei Tan, Hailan Yang, Qiang Chen, and Qianzhen Fang

College of Environmental Science and Engineering, Hunan University, Changsha, Hunan, China

Agricultural wastewater, originating from crop cultivation, livestock rearing, and agricultural processing, constitutes sewage or liquid waste containing a diverse range of pollutants generated during these production processes. Its implications for human health and ecological integrity cannot be disregarded. The primary sources of agricultural wastewater can be attributed to runoff from cultivated fields, effluent from daily farm operations, and the discharge of wastewater during agricultural product processing. These wastewaters encompass pathogens, suspended particulate matter, excessive fertilizer residues, antibiotics, pesticides, endocrine disruptors, insoluble solids, and salts alongside other detrimental substances.

The innovative practice of utilizing agricultural waste for wastewater treatment has garnered attention. This approach mitigates air pollution resulting from waste incineration and minimizes direct environmental contamination caused by agricultural waste. Furthermore, it effectively prevents the proliferation of bacteria and viruses that may arise from the natural decomposition of agricultural waste, safeguarding the ecosystem against potential health hazards. Crucially, it averts water source pollution caused by livestock manure, ensuring the availability of clean and safe water resources while promoting sustainable agriculture and environmental preservation (Kumar et al. 2023). The use of agricultural biomass/waste-derived biochar as catalysts for wastewater treatment is receiving attention due to their special chemical composition, abundant availability, cost effectiveness, and renewability (Figure 11.1) (Yang et al. 2020; Ye et al. 2020). Biochar as catalysts in advanced oxidation technology based on photocatalysis, hydrogen peroxide (H_2O_2), persulfate (PS), periodate (PI), O_3, and peracetic acid (PAA) have been widely applied in agricultural wastewater treatment.

Waste-derived Biochar for Sustainable Rural Revitalization, First Edition. Edited by Yuqing Sun and Daniel C.W. Tsang.
Published 2026 by John Wiley & Sons, Inc.

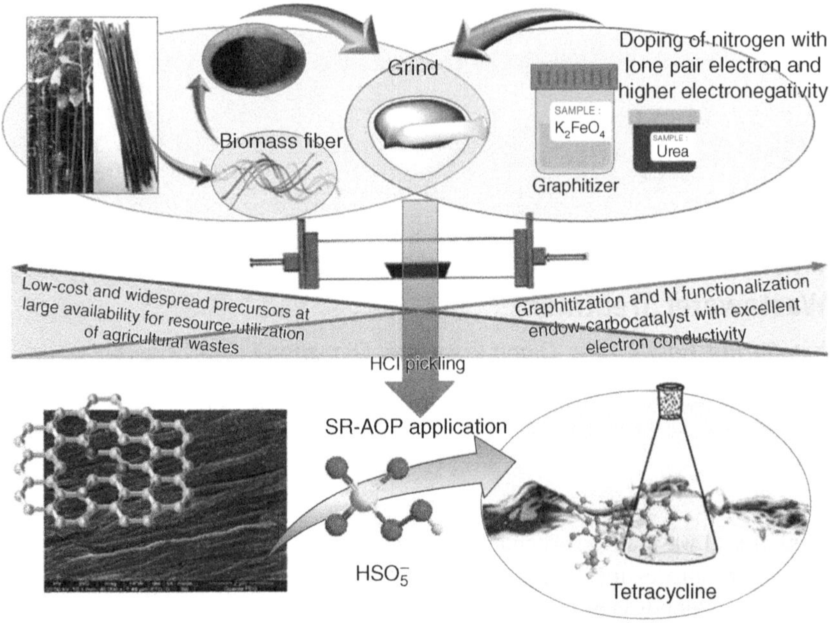

Figure 11.1 Schematic illustration of the synthesis of biochar-based catalyst using agricultural waste and application for organic pollutants degradation. *Source:* Ye et al. 2020/with permission of Elsevier.

11.1 Photocatalysis

11.1.1 Organic Pollutants Removal

Recently, biochar has been discovered to possess a semiconductor-like structure that can function as an active carbonaceous photocatalyst (Zhou et al. 2021). When employed as photocatalysts, oxygen-containing functional groups (e.g., C=O, C—OH) within biochar act as photosensitizers by facilitating the transfer of excited electrons to adsorbed electron acceptors (e.g., O_2), thereby promoting reactive oxygen species (ROS) generation (Gasim et al. 2022). Chen et al. (2017) have demonstrated that hydrocarbon particles presented in aquatic environments exhibited remarkable efficiency in producing ROS through oxygen molecular activation for antibiotic degradation under sunlight exposure. Furthermore, hydrochar with a higher density of photosensitive surface oxygenated functional groups (OFGs) generated more ROS than pyrochar when exposed to daylight; however, even trace amounts of ROS can be produced in darkness due to the presence of persistent free radicals (PFRs). In contrast to pyrochar derived from identical biomass sources, hydrochar exhibited significantly enhanced production rates of H_2O_2 and •OH under daylight irradiation conditions, leading to a sixfold increase in sulfadimidine degradation compared to dark conditions. The high capacity for active oxygen generation observed in hydrocarbon was attributed to its abundant oxygen-containing functional groups on the photoactive surface. Photocatalytic degradation of enrofloxacin (EFA) was observed in ball-milled biochar (Xiao et al. 2020b). Their research proposed that visible light could excite electrons in carbon defects (as the valence band [VB]), which then moved to oxygen-containing functional groups (as the conduction band [CB]). Simultaneously, h could be formed in the VB while electrons in the CB were further transferred to dissolved oxygen to produce $•O_2^-$. Among all samples, ball-milled biochar pyrolyzed at 300 °C (BM300) exhibited the highest EFA degradation rate (80.2% compared with 13.9% for unmilled biochar) and mineralization ability (66.4% compared with 0% for unmilled biochar).

In more research, biochar-based materials are increasingly becoming a key substrate role in the preparation of photocatalytic materials, not only as a supporting matrix for photocatalysts, but also playing an important role as an electron transport medium. What is more remarkable is that the synergistic effect between biochar and various types of metals or metal oxides significantly enhances its adsorption capacity of pollutants, broadens the absorption range of visible light spectrum, and effectively promotes the separation of photogenerated electrons and holes, as well as reduces the bandgap width of the material, which are a series of effects that together enhance the overall performance of the biochar-based catalysts in the photocatalytic reaction. Biochar and its composites have gained significant attention and extensive research in the field of environmental remediation, particularly for treating environmental pollutants. These materials not only exhibit remarkable enhancements in specific surface area but also effectively optimize charge separation and complexation processes, thereby substantially improving their dual efficiency in adsorption and photocatalysis. By innovatively combining biochar with semiconductor materials such as TiO_2 and ZnO, the composite catalysts not only inherit the advantages of each component material but also demonstrate exceptional performance characteristics through synergistic interactions. These composite catalysts can be prepared using various methods, offering customization possibilities for different application scenarios. Notably, they exhibit non-selective and efficient photodegradation of diverse pollutants present in agricultural wastewater, presenting a novel approach to address agricultural pollution issues while safeguarding water resources and the environment.

Traditionally, metal oxides such as TiO_2, ZnO, and $BiVO_4$ have been widely employed as photocatalysts for environmental applications. In the process of photocatalytic reaction, the absorption of electromagnetic radiation by the photocatalyst plays a crucial role in activating it and generating ROS, which can effectively degrade organic pollutants. To excite electrons from the CB to the VB, it is essential that the energy of absorbed light irradiation exceeds or equals the band gap energy (Eg) of the photocatalyst (Ye et al. 2019). Various synthetic methods including sol-gel, co-precipitation, solvothermal, hydrolysis, and self-assembly are utilized in developing biochar-based photocatalytic materials. These techniques involve using metal precursors to grow nanomaterials on prepared biochar surfaces (Subramaniam et al. 2023). Kumar et al. (2017) fabricated a hybrid hetero-assembly of g-C_3N_4/$FeVO_4$/Fe@NH_2-biochar and successfully applied it for visible-assisted photodegradation of methyl paraben and 2-chlorophenol. Furthermore, incorporation of magnetism through $FeVO_4$ and Fe-loaded biochar enables magnetic separation from aqueous systems. The withered magnolia blossom was selected for biochar preparation in the study conducted by Zhu et al. (2018). A mesoporous structure biochar-based g-C_3N_4 was synthesized using a one-step in situ together growth method. The incorporation of biochar effectively impedes the reunion and stacking of g-C_3N_4. A visible active assembly of Biochar@$Bi_2O_2CO_3$/g-C_3N_4/$CoFe_2O_4$, as reported by Kumar et al. (2018), demonstrated efficient degradation of the herbicide paraquat under both visible and solar light irradiation. Remarkably, a high degradation efficiency of 99.3% was achieved within 90 minutes under visible radiation and 92.1% within 120 minutes under solar light exposure. In another study conducted by Kumar et al. (2019), various substances including bare g-C_3N_4, acidified g-C_3N_4, polyaniline, and reduced graphene oxide (RGO) were supported on biochar for comparison purposes. Among these catalysts, the biochar-based photocatalyst exhibited excellent performance in the visible-assisted degradation of ibuprofen (IBN) and 2,4-Dichlorophenoxy acetic acid (2,4-D) when compared to single or binary catalyst systems. Notably, acidified g-C_3N_4/polyaniline/reduced graphene oxide@Biochar displayed an outstanding degradation rate of 20 mg/L for both 2,4-D (99.7%) and IBN (98.4%) after only 50 minutes of Xe lamp exposure while maintaining its high activity even under natural sunlight conditions. Luo and his team have initially synthesized bismuth oxycarbonate ($Bi_2O_2CO_3$) and biochar nanocomposites for synergistic adsorption-photodegradation of tetracycline (TC) (Luo et al. 2023). Among these composites, $Bi_2O_2CO_3$/2.5 wt% biochar exhibited the highest TC adsorption capacity (173.0 mg/L) and the most efficient combined removal rate (84.7%) within 60 minutes through adsorption and photodegradation processes, surpassing $Bi_2O_2CO_3$ alone by 1.9-fold in terms of adsorption capacity and 1.5-fold in terms of removal rate via photodegradation alone. The incorporation of biochar into the composite enhanced specific surface area, facilitated TC adsorption, promoted charge transfer capability, thereby augmenting the photocatalytic activity of the composites.

In general, the activation process of biochar and biochar-based photocatalysts involves the following steps (Eqs. 11.1–11.4): (1) irradiation to facilitate electron transfer from the VB to the CB, (2) generation of electron–hole pairs, and (3) production of •OH and $O_2^{\bullet -}$ species, subsequently leading to degradation of contaminants on the surface of the photocatalyst (Lyu 2020).

$$\text{Photocatalytic} + (\text{UV/visible light}) \rightarrow \text{photocatalytic}\ (e_{CB}^{-} + h_{VB}^{+}) \tag{11.1}$$

$$\text{Photocatalytic}\ (h_{VB}^{+}) + H_2O \rightarrow \text{photocatalytic} + H + OH \tag{11.2}$$

$$\text{Photocatalytic}\ (h_{VB}^{+}) + OH^{-} \rightarrow \text{photocatalytic}\ + \bullet OH \tag{11.3}$$

$$\text{Photocatalytic}\ (e_{CB}^{-}) + O_2 \rightarrow \text{photocatalytic} + O_2^{\bullet -} \tag{11.4}$$

11.1.2 Inorganic Pollutants Removal

Photocatalytic degradation of organic pollutants or in situ coagulation/immobilization of heavy metal ions in the solid phase represents the most suitable strategy for decontamination (Lu et al. 2022). In the study of Li et al. (2020), dissolved black carbon collected from biochar water extracts derived from pyrolytic crop residues (rice, soybeans, and peanuts) were studied for Hg(II) reduction and detoxification under sunlight irradiation. The diazotization-modified biochar adsorbing Ni^{2+} was employed as a precursor for the synthesis of heterogeneous NiS/NiSe/3D porous biochar (DAC-Ni/NiS/NiSe) via vulcanization, followed by in situ competitive hydrogen reduction, as reported by Yan et al. (2021). Under visible-light irradiation with 20 mg catalyst, the photocatalytic oxidation removal process achieved complete removal efficiency of 100% within 110 minutes for 10 mg/L As(III). The biochar (BC)/$ZnFe_2O_4$ magnetic nanocomposites were synthesized by the calcination-hydrothermal two-step method, as reported by Shen et al. (2020). When exposed to visible light irradiation, the BC/$ZnFe_2O_4$ photocatalyst exhibited a significantly higher Cr(VI) removal efficiency of 80%, surpassing that of BC(0) or $ZnFe_2O_4$ (44%). Furthermore, the BC/$ZnFe_2O_4$ composites demonstrated exceptional cycling stability and could be easily separated for recycling due to their inherent stability and distinctive magnetic properties. The TiO_2-modified biochar composites were prepared by the impregnation method, as reported by Wang et al. (2020). The introduction of TiO_2 significantly enhanced the specific surface area of the biochar composites. Among different initial concentrations of EFA, the TiO_2-alkaline biochar composites exhibited superior degradation efficiency. At a solution pH of 5.0 and a Ti-KBC dosage of 2.5 g/L, the degradation rate of EFA at an initial concentration of 100 mg/L reached 85.25%. In the study conducted by Geng et al. (2020), industrial wood flour waste was utilized for the preparation of biochar/BiOBr visible-light photocatalysts, which exhibited remarkable efficacy in degrading organic pollutants and reducing inorganic Cr(VI) under visible light irradiation. Notably, the highest graphitization degree was achieved at a pyrolysis temperature of 600 °C for the prepared biochar (600-WFB), subsequently leading to a significant enhancement in the visible-light photocatalytic performance of BiOBr.

In photocatalytic systems, biochar has traditionally occupied a subordinate position with undefined specific functionalities in various structural forms. Generally, biochar can serve as a carrier to prevent semiconductor aggregation and act as an economically viable adsorbent that enhances pollutant exposure to semiconductors. Moreover, it facilitates the narrowing of semiconductor band gaps and promotes the efficient separation of photo-induced electrons and holes (Zhou et al. 2021). In general, oxygen-containing groups, PFRs, and doped heteroatoms of biochar may exert positive influences photocatalytic processes.

11.2 H_2O_2-based Catalysis Processes

H_2O_2 is made up of two hydrogen atoms and two oxygen atoms, of which the molecular structure contains a O—O bond. H_2O_2 ($E^{\circ} = 1.78\,V$) was widely acknowledged as a crucial oxidant in the advanced oxidation processes (AOPs) for the removal of recalcitrant organic pollutants because of environment-friendly decomposition products of H_2O_2 (O_2 and H_2O), mild reaction conditions, and simple device requirements (Liu et al. 2021; Li et al. 2022; Lai et al. 2023). The direct reaction between H_2O_2 and recalcitrant organic pollutants is limited by the low oxidation capacity of H_2O_2, which needs to be activated via different methods, thus enhancing reaction rate (Grisales-Cifuentes et al. 2022). The activation of H_2O_2 can be realized by the cleavage of peroxide bond (O—O) in the symmetric structure. The classical Fenton reaction refers to the activation of H_2O_2 via ferrous ions (Fe^{2+}) to produce hydroxyl radicals (•OH) through a complex reaction sequence (Bokare and Choi 2014). •OH (1.8–2.7 V vs. normal hydrogen electrode [NHE]) possesses stronger oxidation capacity than that of H_2O_2 and can effectively decompose refractory organic pollutants (Li et al. 2021; Xiao et al. 2020b). Nevertheless, this traditional homogeneous Fenton reaction is limited by some factors, such as risk of secondary pollution, high dependence on pH, and difficult separation from solution (Fang et al. 2024; Ghanbari and Moradi 2017). In recent years, the heterogeneous reaction has become a research hotspot in AOPs due to the reusability of catalysts, simple separation process, and wide range of pH applicability. In heterogeneous H_2O_2-based AOPs, agriculture-based BC (ABC) has been widely used as a catalyst to activate H_2O_2 for the treatment of agriculture wastewater. At present, the types of ABC applied in H_2O_2 activation for the treatment of agricultural wastewater includes metal-free ABC and metal-ABC composites (Table 11.1). Some metals (e.g., Fe, Co, Cu, and Mn) have been used to modify ABC and improve its catalytic performance. The ABC/H_2O_2 systems can effectively eliminate many organic pollutants in agricultural wastewater, such as herbicides (e.g., 2,4-dichlorophenolic acid), antibiotics (e.g., sulfamethazine [SMZ] and ciprofloxacin [CIP]), and estrogens (e.g., 17 β-estradiol).

In the application of metal-free ABC for H_2O_2 activation, metal-free ABC can not only activate H_2O_2 alone (metal-free ABC/H_2O_2 systems) but also cooperate with co-existing substances to activate H_2O_2 (metal-free ABC/coexisting substances/H_2O_2 systems) for the elimination of organic pollutants (Table 11.1). The introduction of heteroatom doping and the regulation of pyrolysis temperature can improve the catalytic performance of metal-free ABC in metal-free ABC/H_2O_2 systems. Taking N-doped biochar geopolymer composite pyrolyzed at 600 °C (NBGC-600) as an example, the NBGC-600/H_2O_2 system exhibited higher catalytic efficiency in activating H_2O_2 to degrade TC in comparison with BC geopolymer composite pyrolyzed at 600 °C, which was attributed to greater content of PFRs, higher graphitization degree, graphitic N, and C=O groups of NBGC-600 (Huang et al. 2023a). The pyrolysis temperature has a great influence on the formation of environmentally PFRs and graphitization degree of BC, thus changing the catalytic performance of BC (Zhang et al. 2021b; Deng et al. 2020).

In the metal-free ABC/coexisting substances/H_2O_2 systems, there is a synergistic effect between metal-free ABC and coexisting substances, which is conducive to improving the activation efficiency of H_2O_2. For instance, the addition of pinewood sawdust-derived BC could improve the degradation rate of herbicide 2,4-dichlorophenoxyacetic acid in the natural pyrite/H_2O_2 system, and the relevant rate constant was enhanced 1.98–2.39 times under the materials loading of 0.1 g/L (pinewood sawdust-derived BC) and 1.5 g/L (Pyrite) (Zhu et al. 2020). And the addition of pinewood sawdust-derived BC could aggrandize the cumulative production of •OH. In summary, metal-free ABC could directly activate H_2O_2 to generate more •OH (e.g., the role of PFRs), and the surface functional groups (e.g., carboxyl group) of metal-free ABC could reduce Fe(III) and accelerate the cycling of Fe(III)/Fe(II), thereby enhancing the production of •OH (Lai et al. 2023; Zhu et al. 2020; Liu et al. 2020b).

In past decades, metal-BC composites have showed excellent performance as catalysts in activating H_2O_2. First, the combination of BC and metal materials could reduce the aggregated problem of metal materials (e.g., layered double hydroxide [LDH]) during the reaction (Li et al. 2022a; Fang et al. 2021). On the other hand, the synergistic effect between metal materials and BC also contributed to the formation of ROS, hence improving catalytic performance (Li et al. 2021). For example, the degradation rate of CIP in the FeS_2-ABC composite/H_2O_2 system could reach 96.8% within 20 minutes, which was higher than that of FeS_2/H_2O_2 system (83.2%) and ABC/H_2O_2 system (about 30%) (Sang et al. 2022). In summary, ABC could be regarded as an effective matrix for the dispersion of diverse metal materials (e.g., zero-valent iron, FeMn oxides, and CuCo-LDH) (Li et al. 2022a; Li et al. 2021; Mao et al. 2019). And •OH played a vital role in the degradation of many organic pollutants (Table 11.1). The density functional theory (DFT) calculation was usually utilized to analyze the degradation pathways of •OH-mediated oxidation of organic contaminants. Taking CIP as an example, single electron transfer, hydrogen atom abstraction, and radical adduct formation were considered as main degradation pathways of CIP in the zero-valent iron-ABC/H_2O_2 systems based on the results of DFT calculation (Mao et al. 2019). Although ABC was known as famous catalysts to promote H_2O_2 activation for the degradation of organic pollutants in agricultural wastewater, the existing research is still mainly at the theoretical level. And more attention should be paid to practical applications.

11.3 PS-based Catalysis Processes

In the past decades, PS has received widespread attention as a viable alternative to H_2O_2 for wastewater treatment in AOPs. Compared with H_2O_2-based AOPs, PS-based AOPs possess some following merits: (1) PS with solid form is more stable in the storage and transportation process, (2) corresponding reactions are less disturbed by the operating parameters, (3) the higher yield of radical formation can be achieved in the activation (Chen et al. 2018; Lee et al. 2020; Fang et al. 2023; Li et al. 2023a). Similar to H_2O_2 activation, the activation of PS is also mainly realized by the cleavage of the O—O bond. PS includes peroxymonosulfate (PMS, HSO_5^-) and peroxydisulfate (PDS, $S_2O_8^{2-}$), and PDS has a symmetric molecular structure, while PMS is asymmetrical (Xiao et al. 2020a). PMS and PDS have different redox potential of 1.82 and 2.01 V, respectively (Nidheesh et al. 2021). Besides, PMS (1.453 Å) has a shorter O—O band length than PDS (1.497 Å), which may be more difficult to be cleaved by a catalyst than PDS to generate ROS due to higher O—O bond energy of PMS (140–213.3 kJ/mol) than that of PDS (140 kJ/mol) (Ghanbari and Moradi 2017; Wang and Wang 2018; Ren et al. 2020).

At present, metal-free ABC and metal-ABC composites have been widely acknowledged as efficient catalysts for the activation of PDS and PMS in the agricultural wastewater treatment. As displayed in Table 11.1, common metal-free catalysts mainly include porous ABC, N-doped ABC, P-doped ABC, S-doped ABC, and B-doped ABC. And metal-ABC composites dominantly contain Fe-ABC composites, Mn-ABC composites, and Co-ABC composites. These metal-free ABC/PS systems and metal-ABC composites/PS systems can effectively degrade some refractory organic contaminants in the agricultural wastewater treatment, mainly including insecticide (e.g., thiamethoxam and imidacloprid), herbicide (e.g., quinclorac and atrazine), and antibiotics (e.g., sulfamethazine, tetracycline, and ciprofloxacin) (Table 11.1). For example, the nitrogen and sulfur co-doped porous ABC/PDS system could effectively decompose thiamethoxam via three degradation pathways, and relative degradation intermediates showed lower toxicity than thiamethoxam in the acute and chronic toxicity assessment (Wang et al. 2023f).

In ABC/PS systems, both radical pathways and nonradical pathways play a crucial role in the degradation of organic pollutants. Radical pathways (e.g., •OH and $SO_4^{\bullet-}$) possessed strong oxidation capacity and can realize high mineralization efficiency for refractory organic contaminants. Metal sites in metal-ABC composites often serve as main active sites for activating PS to generate •OH and $SO_4^{\bullet-}$. For instance, the transformation of Co(II) to Co(III) could promote the breakage of O—O bond in PMS, thus forming $SO_4^{\bullet-}$ (Liu et al. 2020a). And $SO_4^{\bullet-}$ could further react with H_2O to produce •OH (Zhong et al. 2022). In addition to metal sites, PFRs on the surface of ABC could also induce the breakage of O—O bond of PS to generate radicals via contributing electrons (Huang et al. 2022). Superoxide radical ($O_2^{\bullet-}$) could also contribute to the degradation of organic contaminants in the agricultural wastewater. Du et al. (2020b) applied a graphitized biochar to effectively activate PDS for the removal of sulfamethoxazole (SMX) and found that $O_2^{\bullet-}$ played a dominant role. It is worth noting that PFRs could also react with O_2 to produce $O_2^{\bullet-}$ (Liu et al. 2020a; Huang et al. 2022). The representative nonradical pathways mainly include electron transfer regime, singlet oxygen ($^1O_2$)-dominated oxidation, and high-valent metal-oxo species induced oxidation. Nitrogen-doped ABC derived from pomelo peel was successfully used to activate PMS for the elimination of SMX, which was predominantly degraded by the $^1O_2$ and electron transfer (Wang and Chen 2022). And it is noteworthy that the carbonyl group and defects served as important active sites in the transformation from PMS to $^1O_2$. In the SG650/PDS system, electron transfer mainly contributed to the degradation of sulfadiazine (SDZ), and the graphitic structure of SG650 could be the bridge to mediate electron transfer from organic pollutants to PDS (Feng et al. 2022). In the study of Ni-Fe LDH-biochar (BC) composite-induced PMS-based AOP, the results indicated that the electron-shuttle mechanism played the dominant role in the removal of tetracycline hydrochloride (TCH) (Figure 11.2) (Fang et al. 2024).

In summary, radical pathways have strong oxidation efficiency, but their catalytic effects are often affected by existing substances (e.g., inorganic anions) in complex water matrices (Lee et al. 2020; Wang and Chen 2022). These nonradical pathways show broad pH tolerance, possess high reaction selectivity, and exhibit strong resistance to existing substances, but the moderate oxidation capacity of these nonradical pathways may lead to insufficient mineralization (Fang et al. 2023; Ren et al. 2021). For the treatment of practical agricultural wastewater, the combination of radical pathways-dominated systems and nonradical pathways-dominated systems may result in unexpected synergistic effects.

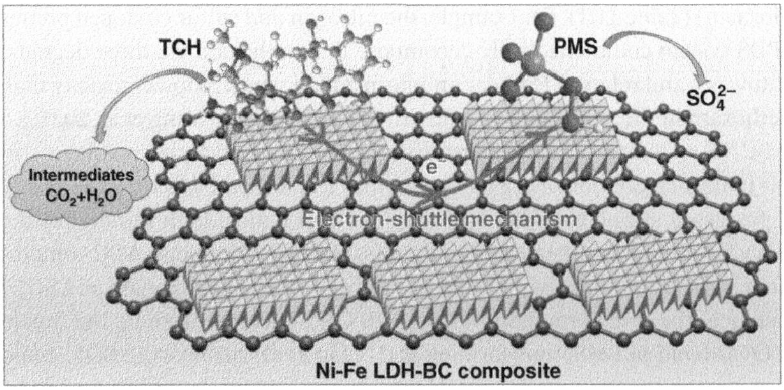

Figure 11.2 Degradation mechanism of TCH in the Ni-Fe LDH-BC/PMS system. *Source:* Adapted from Fang et al. 2024.

11.4 PI-based Catalysis Processes

In recent years, PI (IO_4^-) has received more and more attention and is regarded as an ideal candidate oxidant in AOPs. PI is an anion composed of oxygen and iodine with an oxidation state +7, and analogous to PS, PI exists in solid form and has advantages in transportation and storage (Song et al. 2023). Different from H_2O_2 and PS, PI is more easily activated to produce ROS and reactive iodine species because of the longer I—O bonds and weaker steric hindrance in its structure (Sun et al. 2023b; He et al. 2022). It is difficult to independently use PI to oxidize and remove organic compounds owing to its moderate oxidation capacity $\left(E^{\circ}(IO_4^-/IO_3^-) = +1.60\ V\right)$ (Zhuo et al. 2024; Xiang et al. 2023). Some activation methods, such as ultraviolet (UV), alkaline, freezing, and catalysts, have been utilized to activate PI to generate ROS for the removal of organic contaminants (Du et al. 2020a). At present, ABC can serve as effective catalysts to promote the activation of PI in agriculture wastewater treatment (Table 11.1). For example, the water hyacinth stems-derived biochar was employed to activate PI for degrading a typical antibiotic, and 100% TC (10 mg/L) could be effectively removed within 30 minutes in this system (Duan et al. 2024).

Both metal-free ABC and metal/ABC composites have been applied in the treatment of agriculture wastewater via the activation of PI. Radical pathways and nonradical pathways are involved in the degradation of refractory organic contaminants individually or jointly. In radical pathways, the key role of hydroxyl radical (•OH) and superoxide radical ($O_2^{\bullet -}$) have been proved via quenching experiments and electron spin resonance (ESR) analysis (Fang et al. 2022; Dai et al. 2023a). Fang et al. (2022) successfully synthesized the Mn_xO_y@BC to

Table 11.1 Representative reaction systems based on agriculture-derived biochar in agricultural wastewater treatment.

Types	Contaminants[a]	Catalysts[b]	Contaminants concentration (mg/L)	Oxidants	Removal efficiency	Dominated mechanisms[c]	References
H_2O_2-based AOPs	2,4-D	BC (0.1 g/L) + pyrite (1.5 g/L)	100	10 mM	> 95% (350 min)	•OH	Zhu et al. (2020)
	2,4-D	NSB (2.0 g/L) + Fe (III)	20	3 mM	100% (120 min)	—	Lai et al. (2023)
	2,4-DCP	PEG-nZVI@BC (0.8 g/L)	50	4 mM	92.94% (—)	•OH	Yu et al. (2023b)
	SMZ	FM700 (1 g/L)	2.8	3 mM	100% (120 min)	•OH	Zhang et al. (2020)
	SMZ	SBC600 (0.5 g/L)	10	10 mM	~100% (240 min)	•OH	Deng et al. (2020)
	SDZ	NSB (2.0 g/L) + Fe (III)	20	3 mM	100% (120 min)	•OH	Lai et al. (2023)
	MNZ	MBC (0.2 g/L)	20	5 mM	> 97.4% (90 min)	•OH	Yi et al. (2021)
	MNZ	SMBC (1 g/L)	40	5 mM	100% (< 120 min)	•OH	Yi et al. (2019)
	TC	NBGC-600 (1.5 g/L)	50	—	~100% (660 min)	•OH	Huang et al. (2023a)
	TC	PM500 (0.5 g/L)	30	5 mM	100% (240 min)	•OH	Huang et al. (2019)
	CIP	CuCo-LDH/BC (0.2 g/L)	10	20 mM	84.7% (90 min)	$^1O_2$ and $O_2^{\bullet-}$	Li et al. (2022a)

(Continued)

Table 11.1 (Continued)

Types	Contaminants[a]	Catalysts[b]	Contaminants concentration (mg/L)	Oxidants	Removal efficiency	Dominated mechanisms[c]	References
	CIP	FeMn/BC (0.2 g/L) + HA	10	68 mg/L	92.8% (90 min)	•OH and $O_2^{\bullet-}$	Li et al. (2021)
	CIP	BC/nZVI (0.4 g/L)	100	40 mM	61.9% (60 min)	•OH	Mao et al. (2019)
	CIP	FeS_2/BC (1.5 g/L)	30	5 mM	96.8% (20 min)	•OH	Sang et al. (2022)
	OFX	BC600 (0.3 g/L) + GT (0.2 g/L)	20	2 mM	> 94% (240 min)	•OH	Liu et al. (2020b)
	17β-estradiol	Fe/HC (0.5 g/L)	2.7	5 mM	~100% (300 min)	•OH and $O_2^{\bullet-}$	Yu et al. (2019)
PS-based AOPs	ATZ	BC (1.88 g/L)	7.2	PDS (7.7 mM)	99.8% (120 min)	$SO_4^{\bullet-}$	El-Bestawy et al. (2023)
	ATZ	Fe-Mn BC (2.5 g/L)	10	PDS (2 mM)	96.7% (120 min)	•OH and $SO_4^{\bullet-}$	Liang et al. (2024b)
	ATZ	CoBC500 (0.1 g/L)	2.2	PMS (1 mM)	> 99% (6 min)	$SO_4^{\bullet-}$	Liu et al. (2020a)
	2,4-DCP	KBC800 (0.1 g/L)	20	PMS (1 mM)	100% (15 min)	—	Zhao et al. (2022)
	2,4-DCP	NBC (0.1 g/L)	100	PMS (2 mM)	100% (60 min)	$^1O_2$ and $O_2^{\bullet-}$	Yan et al. (2023)
	Thiamethoxam	NSBC (0.3 g/L)	10	PDS (1 mM)	72.31% (60 min)	$^1O_2$, •OH, and $SO_4^{\bullet-}$	Wang et al. (2023f)

Quinclorac	γ-Fe_2O_{3-x}@BC (0.3 g/L)	10	PMS (0.8 mM)	100% (30 min)	•OH and $SO_4^{\bullet-}$	Zhong et al. (2022)
Thiacloprid	$MnFe_2O_4$-BC (0.2 g/L)	20	PMS (0.65 mM)	92.6% (45 min)	—	Chen et al. (2023)
Imidacloprid	Fe@PB9 (0.67 g/L)	10	PDS (5 mM)	~100% (360 min)	•OH and $SO_4^{\bullet-}$	Wang et al. (2022)
γ-HCH	nZVI@P-BC (1.25 g/L)	1	PDS (4 mM)	92.6% (10 min)	•OH, $^1O_2$, and ET	Huang et al. (2023b)
17β-estradiol	nZVI/PGBC (0.1 g/L)	3	PDS (1.68 mM)	100% (45 min)	•OH and $SO_4^{\bullet-}$	Ding et al. (2021)
Dimethomorph	NBC (0.4 g/L)	5	PMS (2 mM)	~90% (150 min)	$^1O_2$	Yu et al. (2023a)
SMX	PBC (0.1 g/L)	50.7	PDS (1 mM)	> 95% (90 min)	$^1O_2$	Sun et al. (2021)
SMX	WGBC (0.1 g/L)	10	PDS (0.1 mM)	98% (120 min)	$^1O_2$ and $O_2^{\bullet-}$	Du et al. (2020b)
SMX	NBC (0.1 g/L)	10	PMS (0.5 mM)	95% (60 min)	$^1O_2$ and ET	Wang and Chen (2022)
SMX	FeCo-NBC (0.1 g/L)	10	PMS (0.4 mM)	~100% (30 min)	•OH, $SO_4^{\bullet-}$, HVMs	Wang and Wang (2023)
SDZ	SG650 (1.8 g/L)	9.1	PDS (9.1 mM)	94.4% (240 min)	ET	Feng et al. (2022)
MNZ	NZVI/CuO/BC-3 (0.5 g/L)	10	PMS (1 mM)	100% (6 min)	$^1O_2$ and $SO_4^{\bullet-}$	Xu et al. (2024)

(Continued)

Table 11.1 (Continued)

Types	Contaminants[a]	Catalysts[b]	Contaminants concentration (mg/L)	Oxidants	Removal efficiency	Dominated mechanisms[c]	References
	TC	NBBC (0.2 g/L)	30	PDS (1 mM)	~80% (420 min)	ET	Dou et al. (2022)
	TC	PBC800-A (0.5 g/L)	100	PDS (2 mM)	100% (210 min)	$^1O_2$	Zhang et al. (2021a)
	TC	NiFeO/BC (1 g/L)	10	PDS (8 mM)	76.72% (240 min)	•OH and $SO_4^{\bullet-}$	Huang et al. (2022)
	NOR	CBC (0.8 g/L)	10	PDS (3.76 mM)	94.21% (300 min)	•OH and $SO_4^{\bullet-}$	Wang et al. (2019)
	NOR	Co-BC (1.0 g/L)	20	PDS (10 mM)	97.66% (180 min)	$^1O_2$	Yi et al. (2024)
	CEX	Fe_2O_3@LBC (0.4 g/L)	10	PDS (0.42 mM)	73.9% (200 min)	•OH and $SO_4^{\bullet-}$	Song et al. (2021)
	CAP	Co_3O_4-BC (0.2 g/L)	30	PMS (10 mM)	~97.6% (10 min)	$SO_4^{\bullet-}$	Xu et al. (2020)
	FLO	LaMBC (0.5 g/L)	20	PDS (5 mM)	99.5% (120 min)	•OH and $O_2^{\bullet-}$	Peng et al. (2024)
	LFH	CFO@EPBC (0.2 g/L)	10	PMS (0.2 mM)	~80% (30 min)	$SO_4^{\bullet-}$	Wang et al. (2023d)
PI-based AOPs	SMZ	MKBC (0.5 g/L)	10	2 mM	100% (100 min)	IR and $^1O_2$	Elmitwalli et al. (2024)

SMZ	PTBC (0.5 g/L)	10	2 mM	84% (120 min)	—	Elmitwalli et al. (2024)
SDZ	$WSBC_{350}$ (0.3 g/L)	5	0.5 mM	100% (50 min)	$^1O_2$ and $O_2^{\bullet-}$	Dai et al. (2023a)
SDZ	$WSBC_{800}$ (0.5 g/L)	5	2 mM	92.4% (20 h)	ET	Dai et al. (2023b)
TC	WHSBC-700 (0.2 g/L)	10	0.5 mM	100% (30 min)	ET	Duan et al. (2024)
OTC	Mn_xO_y@CSBC (0.3 g/L)	20	0.25 mM	97.5% (30 min)	$^1O_2$, •OH, and $O_2^{\bullet-}$	Fang et al. (2022)

a. 2,4-DCP, 2,4-dichlorophenol; MNZ, Metronidazole; CIP, Ciprofloxacin; 2,4-D, 2,4-dichlorophenolic acid; OFX, Ofloxacin; SMZ, Sulfamethazine; SDZ, Sulfadiazine; TC, Tetracycline; OTC, Oxytetracycline; ATZ, Atrazine; γ-HCH, Gamma-hexachlorocyclohexane; NOR, Norfloxacin; CEX, Cephalexin; CAP, Chloramphenicols; FLO, Florfenicol; LFH, Lomefloxacin hydrochloride; SMX, Sulfamethoxazole.

b. BC, biochar; PEG-nZVI@BC, PEGylated nanoscale zero-valent iron supported by biochar; MBC, Magnetic biochar; FM700, Fe-impregnated maize straw-derived biochar at 700 °C; BC/nZVI, Biochar-supported nanoscale zero-valent iron; NSB, N,S-co-doped biochar; CuCo-LDH/BC, CuCo layered double hydroxide by compositing with biochar; NBGC-600, N-doped biochar geopolymer composite at 600 °C; PM500, Pig-manure biochar; GT, Goethite; Fe/HC, Magnetic iron (hydr)oxides immobilized hydrochar composite; HA, Hydroxylamine; SBC600, Swine manure biochar pyrolyzed at 600 °C; KBC800, Potassium hydroxide activated biochar pyrolyzed at 800 °C; NBC, Nitrogen-doped biochar; NSBC, Nitrogen and sulfur co-doped porous biochar; Fe@PB9, Fe-based biochar pyrolyzed at 900 °C; nZVI@P-BC, P-doped biochar loaded with nano zero-valent iron (nZVI) composite; nZVI/PGBC, Nanoscale zero-valent iron-loaded porous graphitized biochar; PBC, Porous biochar; WGBC, A novel and environmentally friendly graphitized BC; SG650, Pyrolyzing sorghum stalks at 650 °C; NBBC, N/B co-doped biochar; PBC800-A, Post-deashing graphitized biochar; CBC, Corn stalk biochar; LaMBC, Lanthanum-doped magnetic biochar; CFO@EPBC, $CoFe_2O_4$-loaded biochar prepared from the marine biomass Enteromorpha prolifera; MKBC, Mulukhiyah-derived biochar; PTBC, Potato-derived biochar; $WSBC_{350}$, Wheat straw-derived biochar at 350 °C; WHSBC-700, Water hyacinth stems-derived biochar at 700 °C; Mn_xO_y@CSBC, Manganese oxide/corn straw-derived biochar composites; $WSBC_{800}$, Wheat straw-derived biochar at 800 °C; SMBC, Sugarcane bagasse-derived magnetic biochar; FeS_2/BC, Pyrite modified rice straw biochar; FeMn/BC, MnOx-Fe_3O_4/biochar composite; Fe-Mn BC, Iron (Fe)-manganese (Mn) bimetallic biochar composites; CoBC500, Cobalt impregnated biochar prepared at 500 °C; $MnFe_2O_4$-BC, $MnFe_2O_4$ modified biochar; FeCo-NBC, Bimetallic and nitrogen co-doped biochar; γ-Fe_2O_{3-x}@BC, γ-Fe_2O_{3-x} modified biochar catalysts; NZVI/CuO/BC-3, Zero-valent bimetallic shell/core iron/copper/biochar composite; Fe_2O_3@LBC, Fe_2O_3 modified loofah biochar; NiFeO/BC, Biochar-supported iron-containing bimetallic oxides; Co_3O_4-BC, Biochar-supported Co_3O_4 composite.

c. ET, Electron transfer; HVMs, High-valent metal-oxo species; IR, Iodate radicals.

activate PI for the removal of oxytetracycline (OTC). And almost 98% OTC could be removed, and the removal rate of total organic carbon (TOC) reached at 75% under the combined effect of •OH, $O_2^{\bullet-}$, and $^1O_2$. Different from H_2O_2-based AOPs and PS-based AOPs, reactive iodine species is a unique active species in PI-based AOPs, which could also contribute to organic pollutants degradation in agricultural wastewater. For instance, iodate radicals have a great influence on the degradation of SMZ in the mulukhiyah-derived BC/PI system (Elmitwalli et al. 2024). Common reactive iodine species mainly include $IO_3^{\bullet}$, $IO_4^{\bullet}$, HOI, I_2, I^-, and IO_3^- (Zong et al. 2021). Among these active species, $IO_3^{\bullet}$ and $IO_4^{\bullet}$ possess strong oxidation capacity and are often of concern (Xiang et al. 2023). And the quenching experiments can be used to reveal the contribution of reactive iodine species. It is worth mentioning that some iodine species (e.g., HOI and I_2) may serve as precursors and react with organic matter to form toxic iodinated disinfection by-products (I-DBPs) (Song et al. 2023), which deserves more attention in the treatment of agriculture wastewater. In nonradical pathways, electron-transfer regime and $^1O_2$ could contribute to the elimination of organic contaminants in ABC/PI systems (Table 11.1). Dai et al. (2023b) found that the removal of SDZ was dominated by electron-transfer regime, in which BC was the bridge to promote electron transfer from organic pollutants to PI. And the formation of the high potential BC-PI* complexes with strong electrophilicity is a key step in the electron-transfer regime. Besides, the galvanic oxidation process (GOP) and electrochemical characterizations are often used to verify the crucial role of electron-transfer regime (Duan et al. 2024; Dai et al. 2023b). In summary, although ABC can effectively activate PI to degrade refractory organics present in agricultural wastewater, the current research is not extensive, and there are still many problems that need to be solved (e.g., the potential threat on the formation of I-DBPs and the influence of pH value on the ROS generation). And more research could focus on these aspects.

11.5 O_3-based Catalysis Processes

Ozone (O_3) is a potent and versatile oxidant widely used in water treatment, air purification, and disinfection due to its high oxidation potential ($E_0 = 2.07$ V) (Lim et al. 2022). It effectively decomposes various contaminants, such as pathogens, pharmaceuticals, pesticides, and refractory pollutants (Nawrocki et al. 2010). Despite its impressive oxidation power, ozone has limitations when applied in wastewater treatment processes. One of the main drawbacks is its low solubility in water, which hinders its reactivity. Furthermore, while ozone can break down many organic compounds, its ability to fully mineralize these compounds into harmless byproducts is limited, especially in the case of more complex water environment. Heterogeneous catalytic ozonation enhances ozone utilization by activating it on catalyst surfaces, accelerating ROS generation (Fijołek et al. 2024). Traditional metal oxides or carbon catalysts are difficult to ensure their own structural stability in the ozone oxidation environment with strong oxidizing effect, thus limiting their long-term use in catalytic ozone oxidation. Biochar, produced from biomass pyrolysis, is a sustainable and cost-effective catalyst characterized by its porous structure and numerous surface functional groups. Biochar-catalyzed ozone oxidation has shown considerable potential in improving the breakdown of persistent organic compounds in agricultural wastewater (Wang et al. 2023b). Ozone can be catalytically decomposed by biochar into ROS, which mainly include hydroxyl radicals ($^{\bullet}OH$), singlet oxygen ($^1O_2$), superoxide radicals ($O_2^{\bullet-}$), etc., and their activities are often high enough to further oxidize the target. The $^{\bullet}OH$ is the most prevalent, with a standard redox potential of 2.80 V, surpassing ozone by 1.35 times. It exhibits rapid reaction kinetics, with rate constants typically between 10^8 and $10^9 M^{-1} s^{-1}$, enabling it to effectively degrade a wide range of organic pollutants in agricultural wastewater. In addition to its catalytic properties, biochar offers several other advantages in the treatment of agricultural wastewater. One of the primary benefits of biochar as a catalyst is its high surface area, which provides ample adsorption sites for pollutants. This property allows biochar to not only degrade organic contaminants through catalytic ozonation but also adsorb a range of pollutants, improving the overall removal efficiency. Furthermore, biochar is highly stable and resistant to chemical degradation, which means it can be reused multiple times without significant loss of activity. Research has demonstrated the enduring stability and reusability of biochar catalysts in catalytic ozonation, with certain catalysts preserving high degradation efficiencies across multiple reaction cycles (Lv 2025; Noman et al. 2025). Biochar offers cost-effectiveness and environmental sustainability, making it advantageous over catalytic materials like metal oxides and precious metals.

11.5.1 Performance of Biochar-catalyzed O_3 Oxidation Technology

The performance of biochar catalysts in the ozonation of agricultural wastewater pollutants has been extensively investigated in recent years. Biochar can improve catalytic ozonation and enhance the removal of pollutants like antibiotics, pesticides, pharmaceuticals, and dyes commonly present in agricultural wastewater (Table 11.2). For antibiotics like TC, in the study by Zhao et al. (2024), nitrogen-doped biochar (HSBC) was produced by co-pyrolyzing sludge with water hyacinth (Zhao et al. 2024). HSBC achieved 93% removal in 130 minutes under optimized conditions (0.1 g/L catalyst, 25 mg/L TC, pH 7–9). Metal-free biochar outperforms metal-based catalysts (e.g., Mn-Fe-Ce oxides) by avoiding secondary pollution from metal leaching. In real printing/dyeing wastewater, HSBC-O_3

Table 11.2 Representative ozone reaction systems based on agriculture-derived biochar in agricultural wastewater treatment.

Catalysts[a] (dose)	Contaminants[b]	Contaminants concentration	Ozone dose	Removal efficiency	Dominated mechanisms[c]	References
SBE-N_2-650 (1 g/L)	Benzoic acid	100 mg/L	30 mg/L	86% (40 min)	$^{\bullet}OH$, $O_2^{\bullet -}$, and $^1O_2$	Liu et al. (2020c)
HS-based biochar (0.5 g/L)	Ketoprofen	5 mg/L	0.5 L/min	99.9% (5 min)	$^{\bullet}OH$, $O_2^{\bullet -}$, and $^1O_2$	Li et al. (2022a)
MnO_x/biochar (20 mg/L)	ATZ	5 μM	2.5 mg/L	83% (30 min)	$^{\bullet}OH$	Tian et al. (2021)
FeO_x/biochar (20 mg/L)	ATZ	5 μM	2.5 mg/L	100% (30 min)	$^{\bullet}OH$	Tian et al. (2021)
Fe/BC (5 g/L)	MB	200 mg/L	0.3 g/L	96% (60 min)	$^{\bullet}OH$	Babar et al. (2022)
OHBC (20 mg/L)	2,4-D	1.0 mg/L	1.0 mg/L	89% (10 min)	$^{\bullet}OH$	Zhu et al. (2022)
N-doped biochar (40 mg/L)	ATZ	5 μM	3.62 mg/min	74.4% (15 min)	$^{\bullet}OH$	Cheng et al. (2023)
KBB (0.03 g/L)	Imazapic	5 μM	3 mg/L	97% (30 min)	$^{\bullet}OH$, $O_2^{\bullet -}$, and $^1O_2$	Wang et al. (2023a)
Fe-Mn-C (0.5 g/L)	Ibuprofen	50 mg/L	4.93 mg/min	> 95% (9 min)	$^{\bullet}OH$ and $O_2^{\bullet -}$	Xu et al. (2023)
Mn-5WT (0.05 g/L)	TMP	5 μM	3 mg/L	95.3% (30 min)	$^{\bullet}OH$	He et al. (2023)
HPB700 (0.1 g/L)	1,4-dioxane	25 mg/L	13.2 mg/L	97.8% (30 min)	$^{\bullet}OH$	Yang et al. (2023)
soybean biochar (0.03 g/L)	ATZ	5 μM	3 mg/L	88.2% (30 min)	$^{\bullet}OH$, $O_2^{\bullet -}$, and $^1O_2$	Dong et al. (2023)
Zn@SBC-CD (400 mg/L)	MB	200 mg/L	0.04 g/L	100% (10 min)	$O_2^{\bullet -}$	Yang et al. (2024)

BRBC (1 g/L)	CIP	80 mg/L	15 mg/L	~94% (10 min)	$^{\bullet}OH$, $O_2^{\bullet -}$, and $^1O_2$	Noman e t al. (2024)
2Co-NPC (150 mg/L)	Ibuprofen	10 mg/L	4.5 mg/L	99.9% (10 min)	$^{*}O$, $^{\bullet}OH$, and ET	Wei et al. (2024)
Co_9S_8/MgO-NPC (60 mg/L)	Ibuprofen	10/mg L	4.5 mg/L	100% (30 min)	$^{*}O$ and $^{\bullet}OH$	Wei et al. (2025)

a. SBE-N_2-650, spent bleaching earth calcined at 650 °C under N_2; MnO_2/BioC, piggery biogas residue biochar (BioC) to support MnO_2; HS-based biochar, peanut shell (HS)-based biochar; Fe/BC, Fe-loaded biochar; OHBC, NaOH-pretreated rice straw biochar; KBB, base-activated carbonized kelp biomass; Fe-Mn-C, biochar co-doped with iron and manganese; Mn-5WT, MnO_2-loaded biochar synthesized from waste tea leaves; HPB700, pyrolyzing waste coffee grounds at 700 °C; Zn@SBC-CD, cow dung-doped sludge biochar; BRBC, biogas residue biochar; 2Co-NPC, pyrolysis of nitrogen and phosphorus-rich *Chlorella pyrenoidosa* waste combined with cobalt nitrate; Co_9S_8/MgO-NPC, a one-step pyrolysis technique employing $Co(NO_3)_2 \cdot 6H_2O$, $MgSO_4 \cdot 7H_2O$, and N, P-enriched *Chlorella pyrenoidosa* (CP).

b. 2,4-D, 2,4-dichlorophenoxyacetic acid; MB, methylene blue; TMP, 2,3,5-trimethylpyrazine; ATZ, atrazine; CIP, ciprofloxacin.

c. ET, Electron transfer.

reduced COD from 9900 to 5105 mg/L while increasing biodegradability (BOD_5/COD ratio sevenfold). Such performance surpasses conventional methods like activated carbon (AC) adsorption (50–70% removal) and membrane filtration (high energy costs). For pesticides like atrazine (ATZ), MnO_x/biochar and FeO_x/biochar achieved 83 and 100% degradation in 30 minutes (pH 7.0, 2.5 mg/L O_3), respectively, far exceeding raw biochar (58.6%) and pure ozone (48%) (Tian et al. 2021). The improved performance is linked to a significant increase in Lewis acid sites, which are 3.5 times higher for MnO_x/biochar and 4.1 times higher for FeO_x/biochar, facilitating O_3 adsorption and decomposition (Tian et al. 2021). Zhu et al. (2023) used rice straw as biomass raw material and iron nitrate as iron source and obtained Fe_3C/BC composites by high-temperature pyrolysis in N_2 protective atmosphere. The degradation kinetics of 2,4-D revealed that the O_3-Fe_3C/BC catalytic system enhanced the degradation rate by 2.84 times compared to the ozone-only system, achieving 90% removal within 3 minutes. Additionally, the ozone utilization rate in the catalytic system was 1.55 times greater than in the ozone-only system. After 110 minutes of continuous aeration, the 2,4-D removal rate remained high, with iron ion leaching stabilizing at 80 μg/L. XRD analysis confirmed that the catalyst retained its Fe_3C and graphitic carbon crystal structure. For other pollutants, research by Moussavi and Khosravi (2012) showed that pistachio hull biochar (PHB) notably improved the efficiency of ozonation for Reactive Red 198 (RR198). At pH 10 and 2 g/L PHB, the COP achieved 98% decolorization within 60 minutes, compared to 44% with commercial AC under identical conditions. The pseudo-first-order rate constant for PHB (6.66 min^{-1}) exceeded that of AC (0.86 min^{-1}) by a factor of 7.7. Fe/Mn co-doped biochar (Fe–Mn–C) degraded 95% of ibuprofen (50 mg/L) and achieved 80.5% mineralization in 9 minutes at neutral pH (Xu et al. 2023). Comparatively, sole ozonation removed only 36% IBP, while MnO_2 and Fe_3O_4 catalysts achieved 50.7% and 70.2% removal, respectively. The reaction rate constant (k_{obs}) for Fe-Mn-C/O_3 was 0.316 min^{-1}, which is 6.58 times greater than that of sole ozonation and 2.3–4.1 times higher than other catalysts. Additionally, Fe-Mn-C maintained >92% efficiency after three reuse cycles with negligible metal leaching (Fe: 0.12 mg/L; Mn: 0.02 mg/L), highlighting its stability. These metrics highlight biochar's advantages over metal oxides and AC in catalytic ozonation.

The catalytic performance of biochar can be influenced by the type of feedstock used, the modification of biochar with heteroatoms or metal catalysts, and the operational conditions under the ozonation process. The surface of biochar contains a variety of functional groups, including hydroxyl groups, carboxyl groups, and phenolic groups, which contribute to its reactivity. Modifying biochar with metal oxides (Tian et al. 2021) or heteroatoms like nitrogen (N) (Cheng et al. 2023), sulfur (S) (Wang et al. 2023a), and phosphorus (P) (Wei et al. 2024) can enhance its catalytic activity. These modifications improve the biochar's ability to activate

O_3, leading to more efficient degradation of organic pollutants. Another important factor influencing the performance of biochar catalysts in ozonation is the presence of metal species (Tian et al. 2021; Xu et al. 2023; Wei et al. 2024), which can further enhance ozone activation. Biochar catalysts doped with metals like Co (Wei et al. 2024), Mn (Tian et al. 2021), (He et al. 2023), or Fe (Tian et al. 2021; Xu et al. 2023) enhance electron transfer and stimulate the production of ROS. Cobalt-doped biochar catalysts have been shown to improve the breakdown of ozone into *O and •OH (Wei et al. 2024). Catalytic ozonation efficiency is affected by reaction conditions, such as the catalyst dose, ozone concentration, pH, coexisting anions, organic matter, and temperature. He et al. (2023) has shown that increasing the ozone dosage improves the degradation efficiency, but the optimal ozone concentration varies depending on the type of pollutant and the catalyst used. In addition, the pH of the wastewater can affect the speciation of ozone and the formation of ROS. Biochar catalysts have been found to perform well across a wide range of pH values, making them suitable for treating complex wastewater matrices with varying compositions.

The practical application of biochar-catalyzed ozonation in agricultural wastewater treatment also requires consideration of the catalyst's reusability and stability. Several studies have investigated the long-term stability of biochar catalysts in ozonation process (Xu et al. 2023; Yang et al. 2024). For example, Noman et al. (2024) demonstrated that biogas residue biochar (BRBC) modified with glycerol maintained its catalytic activity over five consecutive cycles, achieving removal efficiencies of 92.45–94.41% for CIP. However, challenges remain in terms of minimizing catalyst deactivation over time, particularly due to the accumulation of pollutants or the leaching of metals from metal-doped biochar catalysts. Research is ongoing to improve the long-term stability of these catalysts through surface modifications or the development of more robust biochar materials.

11.5.2 Mechanisms of O_3 Oxidation Catalyzed by Biochar

In catalytic ozonation, biochar-based catalysts boost ozone's oxidation capacity by facilitating the production of ROS, crucial for degrading organic pollutants in wastewater. These ROS include hydroxyl radicals (•OH), superoxide radicals ($O_2^{\bullet-}$), singlet oxygen ($^1O_2$), and atomic oxygen *O (Wang and Chen 2020). The production and roles of these active species in the biochar-catalyzed ozonation process are fundamental to understanding how this technology can efficiently degrade recalcitrant contaminants in agricultural wastewater. Atomic oxygen (*O) is a less commonly discussed but reactive species generated during ozone decomposition. The formation of *O can occur when ozone is decomposed by metal-doped biochar or by biochar with oxygen vacancies (Wei et al. 2024; Wei et al. 2025). The atomic oxygen can then participate in oxidation reactions with organic molecules, leading to their degradation. The significance of *O is crucial in the degradation of pollutants like ibuprofen that necessitate electron transfer for decomposition (Wei et al. 2024). In summary, the production of active species like •OH, $O_2^{\bullet -}$, $^1O_2$, and *O is essential in biochar-catalyzed ozonation. These species have unique functions in breaking down organic pollutants, and their interaction improves the ozonation process's overall efficiency. The ability of biochar to generate and control the formation of these reactive species depends on its surface properties, including the presence of metal centers, nitrogen functional groups, and oxygen vacancies. Enhancing biochar catalysts to boost active species production will further improve catalytic ozonation efficiency in treating agricultural wastewater pollutants.

The catalytic mechanisms of biochar-catalyzed ozone oxidation involve complex interactions between ozone, biochar, and organic contaminants. These mechanisms are crucial for understanding how biochar enhances the degradation of pollutants and for optimizing the process. The O_3 oxidation mechanism catalyzed by biochar is categorized into free radical and non-free radical oxidation types (Yu et al. 2020). This involves the backbone structure (Cheng et al. 2023) (encompassing atomic configuration and electron distribution), oxygen-containing functional groups (Cheng et al. 2023; He et al. 2023; Liu et al. 2020c; Zhu et al. 2022; Yang et al. 2023) (e.g., basic groups —COOH, —OH, —C=O), structural defects (Yang et al. 2024), graphitic structures (Yang et al. 2024; Oh and Nguyen 2022), and heteroatoms (Wang et al. 2023a; Wei et al. 2024) (e.g., N, P, and S), and metal species (Xu et al. 2023; He et al. 2023), which are the potential active sites for catalyzing O_3 decomposition. Upon adsorption of O_3 on biochar surfaces, the active sites can either decompose O_3 to generate ROS or form oxidizing surface complexes with O_3. The process of O_3 adsorption to these active sites is influenced by the spatial structure of the biochar. Biochar's three-dimensional structure enhances its specific surface area, improving its capacity for O_3 adsorption and decomposition. Moreover, the isolated π-electrons in the material can improve the adsorption and decomposition of O_3 on its surface via electrophilic interaction (Xu et al. 2023; Cheng et al. 2023). From the perspective of electron transfer and reaction participation, the off-domain π-electrons have higher mobility and reactivity, so they have a stronger ability to participate in the O_3 reaction, and it is easy to realize the catalytic decomposition of O_3. In addition, the higher the graphitization degree of the biochar material, the better the electrical conductivity and the stronger the catalytic activity.

Biochar-derived catalysts exhibit multifaceted active sites and mechanisms in ozonation processes, driven by tailored heteroatom doping, metal incorporation, and structural defects. Nitrogen-functionalized sites, particularly pyrrolic-N and graphitic-N, play pivotal roles in ozone activation. For instance, pyrrolic-N in Enteromorpha prolifera-based biochar (EpC_2-650) demonstrated strong O_3 adsorption (adsorption energy = −2.9 eV) via orbital hybridization, where the 2p orbitals of N and O synergistically lowered the energy barrier for *O formation (Zhu et al. 2025). This process enabled electron transfer from biochar to O_3, producing ROS like •OH and $O_2^{\bullet-}$. Wei et al.'s study offered valuable insights into the active sites and mechanisms of biochar-based catalysts made from N and P-rich *Chlorella pyrenoidosa* (CP) waste combined with cobalt oxide (Wei et al. 2024). The electron-rich Co(II) centers and electron-deficient NPC matrix create a coupled system through covalent bonds (C—O—Co and C—N—Co), facilitating efficient ozone decomposition and ROS generation. The primary active sites identified include the electron-rich Co(II) centers, which directly adsorb and decompose ozone into adsorbed atomic oxygen (*O) via the reaction in Eq. (11.5) (Wei et al. 2024). Subsequently, *O reacts with protons to yield •OH, a dominant radical species (Eq. 11.6) (Wei et al. 2024). The NPC matrix aids in electron transfer from organic pollutants through polarized C=N, C=O, and P groups to Co(III), thereby regenerating the active Co(II) state and maintaining catalytic cycles (Eq. 11.7) (Wei et al. 2024). This redox cycling minimizes Co(III) accumulation and enhances catalyst stability.

$$\text{C—O(N)—Co(II)} + O_3 \rightarrow \text{C—O(N)—Co(II)—O*} + {*O_2} \tag{11.5}$$

$$\text{C—O(N)—Co(II)—O*} + H^+ \rightarrow \text{C—O(N)—Co(III)} + {}^{\bullet}OH \tag{11.6}$$

$$\text{C—O(N)—Co(III)} + e^-(\text{from NPC}) \rightarrow \text{C—O(N)—Co(II)} \tag{11.7}$$

Biochar catalytic performance is heavily dependent on the synergistic interplay between structural features and surface functionalities. In the study by Yang et al. (2024), a novel $ZnCl_2$-activated biochar composite derived from sludge and cow dung (Zn@SBC-CD) demonstrated exceptional catalytic ozonation efficiency attributed to its high graphitization degree, abundant defects, and uniform Zn distribution, which collectively optimized electron transfer and active site exposure. The catalytic mechanism involved both radical and non-radical pathways, primarily driven by $O_2^{\bullet-}$ with $^1O_2$ and •OH playing minor roles. The generation of $O_2^{\bullet-}$ was linked to the electron transfer from graphitic carbon and oxygen functionalities to ozone. Concurrently, non-radical pathways involving surface-bound $^1O_2$ were attributed to the interaction between ozone and carbonyl groups (C=O) on the biochar surface. The exceptional catalytic performance of biochar in ozonation systems arises from synergistic interactions among multiple active sites. In the study by Dong et al. (2023), soybean-derived biochar (SB) exhibited superior catalytic activity compared to soybean shell (SBS) and soy straw (SS) biochar, primarily attributed to its unique active sites: pyridine nitrogen (pyridine-N), carbonyl groups (C=O), delocalized π-electrons, and Lewis acid sites. Iron-copper

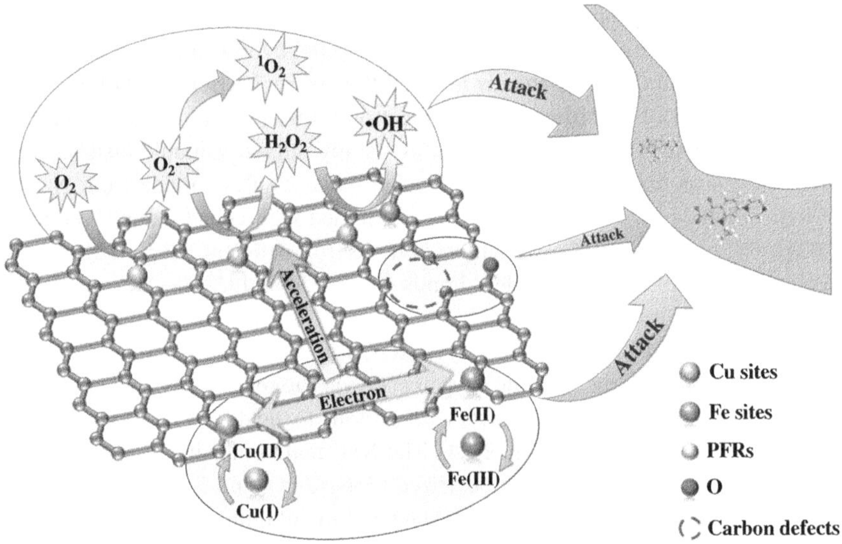

Figure 11.3 Proposed mechanism of CIP degradation by Fe/Cu-BC. *Source:* Luo et al. 2024/The Royal Society of Chemistry.

modified biochar (Fe/Cu-BC) was synthesized at low temperature (250 °C) to activate dissolved oxygen (DO) for efficient degradation of CIP without external energy and oxidants; the results suggested that the synergistic catalysis of Fe and Cu was crucial for the reaction, which improved the electron donor capacity of Cu and promoted the redox cycle of Fe (Figure 11.3) (Luo et al. 2024).

11.5.3 Outlook

Biochar-catalyzed ozonation is a promising AOP for treating agricultural wastewater with organic pollutants. However, despite its potential, there remain several challenges that hinder its widespread implementation, especially on a large scale. These challenges include issues related to catalyst deactivation, catalyst preparation and modification, stability, cost-effectiveness, and scalability. Addressing these challenges will require further optimization of biochar properties, improved reaction conditions, and innovative approaches for large-scale applications.

1. Regeneration and Reusability of Catalysts: Regeneration techniques that minimize catalyst deactivation will be crucial for the long-term viability of biochar-catalyzed ozonation. Research into cleaning methods, surface regeneration, self-healing, and reactivation of metal species could help improve the catalyst's lifespan and reduce operational costs.

2. Assessment of Environmental and Health Impacts: This includes evaluating the potential for the leaching of hazardous byproducts from biochar catalysts and ensuring that the degradation products of pollutants are non-toxic. Investigating the toxicity and long-term impacts of ozonation byproducts on ecosystems and human health is essential for the safe large-scale application of biochar-catalyzed ozonation.
3. Artificial intelligence (AI)-Informed Solutions in Biochar-Ozone Catalytic Systems: AI and machine learning (ML) offer transformative solutions. Predictive models, such as random forests or neural networks, can correlate feedstock properties with biochar characteristics by training on datasets spanning diverse agricultural residues. For instance, ML algorithms trained on pore size distributions, surface area, X-ray photon spectroscopy (XPS), and Fourier transform infrared spectroscopy (FTIR) data can predict optimal pyrolysis parameters to maximize surface area or oxygen-containing groups. Furthermore, reinforcement learning frameworks enable dynamic adjustment of ozone injection rates based on real-time wastewater quality sensors, enhancing oxidant utilization.
4. Big Data and Digital Twins in Scaling Catalytic Ozonation Processes: The integration of big data analytics and digital twin technology is reshaping the scalability of biochar-ozone systems. Practical application requires handling nonlinear interactions between wastewater composition, catalyst deactivation, and process economics. Legacy systems often lack granular data on long-term catalyst stability or byproduct formation, hindering lifecycle assessments. For example, temporal graph neural networks (GNNs) can predict catalyst fouling trends by analyzing time-series data on pore blockage and active site oxidation. Concurrently, digital twins—virtual replicas of physical systems—simulate scenarios like fluctuating agricultural wastewater inflows or seasonal feedstock variability.
5. Scale-up and Industrial Applications: Finally, research on scaling up biochar-catalyzed ozonation systems for industrial and agricultural applications is needed. This includes designing reactors that maximize the contact between biochar catalysts and contaminants, as well as developing more efficient and cost-effective ways to produce and apply biochar at large scales.

Future research must address the dual goals of technical efficacy and environmental sustainability. Autonomous experimentation platforms, combining robotic pyrolysis systems and high-throughput characterization, could accelerate biochar catalyst discovery. Emerging opportunities lie in coupling biochar-ozone systems with complementary technologies. For instance, hybrid models integrating catalytic ozonation with AI-driven membrane filtration or anaerobic digestion could achieve closed-loop resource recovery. The synergy between agricultural waste-derived biochar catalysts and AI-driven technologies holds immense promise for sustainable wastewater treatment. Biochar-catalyzed ozonation presents a promising solution for the treatment of agricultural wastewater, offering high efficiency, low cost, and environmental sustainability. Biochar-catalyzed ozonation has the potential to significantly enhance sustainable agricultural wastewater management and eliminate emerging contaminants harmful to human health and the environment.

11.6 PAA-based Catalysis Processes

Peracetic acid ($CH_3C(O)OOH$, PAA) is an organic peroxyacid with a high redox potential (1.06–1.96 V) and fast oxidation kinetics, effective against various pathogens and organic pollutants in water treatment (Wang et al. 2023c). It is a mixture of acetic acid and H_2O_2. Compared to traditional oxidants like H_2O_2 and PMS, PAA has a lower O—O bond energy of PAA (~159 kJ/mol vs. 213 kJ/mol in H_2O_2 and 317 kJ/mol in PMS), enabling more efficient radical generation with reduced external energy input (Gasim et al. 2023). Another major attraction is that the decomposition byproducts of PAA—mainly acetic acid, oxygen, and water—tend not to form toxic chlorinated organics, thus lowering the risk of secondary pollution. However, PAA has certain limitations in practical applications. While its naturally generated ROS are highly oxidative, the degradation efficiency of PAA is often slow in real-world wastewater conditions. Additionally, factors like water quality and pH affect its activity (Miao et al. 2024). Therefore, activation methods, such as UV light, heat, or metal catalysts, including iron and cobalt, are often employed to enhance its oxidative capacity. Yet, these traditional methods can result in secondary pollution or catalyst leaching issues. Thus, finding an efficient, green, and renewable catalyst to activate PAA has become an important research focus in the water treatment field. Biochar derived from agricultural waste is an effective catalyst for activating PAA to remove contaminants like antibiotics, pesticides, and pathogens from agricultural wastewater. Biochar serves as a catalyst for the activation of PAA, increasing its efficacy in pollutant degradation. Biochar's surface is abundant in functional groups like hydroxyl, carboxyl, and carbonyl groups, enhancing PAA adsorption and activation. The porous structure and high surface area of biochar improve its catalytic properties by offering more active sites for reactions. In particular, the conversion of lignocellulosic residues into engineered biochar, as well as the functionalization of biochar with transition metals (such as Co, Fe, and their oxides) has broadened the potential for highly efficient and eco-friendly treatments. These composites maintain the benefits of the carbon matrix (adsorption sites, electron shuttling functionalities) while reducing the leaching of metals (Li et al. 2024a). In addition, doping the biochar with nitrogen or sulfur can shift electron density to the carbon skeleton, favoring the electron transfer route essential for efficient radical generation (Miao et al. 2024). In sum, the synergy between PAA's strong oxidizing nature and biochar's surface functionalities or embedded atoms sets the stage for enhanced agricultural wastewater treatment.

11.6.1 Performance of Biochar-catalyzed PAA Oxidation Technology

Biochar-enhanced PAA systems effectively eliminate pollutants from agricultural wastewater. Studies have shown that biochar-catalyzed PAA systems can effectively remove antibiotics, pesticides, and other pollutants from agricultural wastewater (Table 11.3). As demonstrated in the research conducted by Dong et al. (2022), $CoFe_2O_4$@BC catalyst was synthesized by combining $CoFe_2O_4$ with biochar derived from lignin to activate PAA (Dong et al. 2022). This system achieved 100% removal of 10 mg/L SMX in just 60 minutes under optimal conditions, including a PAA concentration of 825 μM and a $CoFe_2O_4$@BC dosage of 100 mg/L. The study concluded that SMX degradation was enhanced through the formation of acetyl(per)oxyl groups ($CH_3C(O)O^\bullet$ and $CH_3C(O)OO^\bullet$) facilitated by the redox cycling of $CoFe_2O_4$@BC. These radicals were identified as the main reactive species responsible for the breakdown of SMX. Pesticides are essential for modern agriculture but significantly threaten ecosystems and human health due to their persistence and potential for bioaccumulation. Agrochemicals, such as organophosphates, organochlorines, and neonicotinoids, often contaminate agricultural wastewater at concentrations ranging from ng/L to mg/L levels. Liang et al. (2024a) developed a defect-rich biochar-supported cobalt-iron alloy/oxide heterojunction catalyst (CoFe@DBC) for PAA activation to degrade organochlorine pesticides in agricultural water. The catalyst was prepared through a single-step pyrolysis process utilizing rice husk biomass as the carbon source. Biomass was co-impregnated with $Co(NO_3)_2{\cdot}6H_2O$ and $Fe(NO_3)_3{\cdot}9H_2O$, then pyrolyzed at 500–800 °C under a nitrogen atmosphere. The CoFe@DBC2.0-700/PAA system removed 95.7% of 2,4-dichlorophenoxyacetic acid (2,4-D) in 60 minutes, exhibiting a kinetic constant (k_{obs}) of 0.0260 min^{-1}, which is 1.7 times greater than that of Nano CoFe/PAA ($k_{obs} = 0.0153\,min^{-1}$). The system demonstrated broad pH adaptability, maintaining >90% efficiency at pH 4–11. Magnetic recovery tests confirmed >84.7% efficiency after five cycles, with minimal metal leaching (Co: 0.346 mg/L; Fe: 0.146 mg/L). The system's effectiveness in treating 7000 mL of real water over 116 hours in a fixed-bed reactor highlights its practical application for agricultural wastewater remediation.

Pollutant removal efficiency and PAA activation are significantly affected by biochar type, contaminant nature, operational conditions (e.g., pH, PAA concentration), and environmental factors (e.g., ionic species). Several factors

Table 11.3 Representative PAA reaction systems based on agriculture-derived biochar in agricultural wastewater treatment.

Catalysts[a] (dose)	Contaminants[b]	Contaminants concentration	PAA dose	Removal efficiency	Dominated mechanisms[c]	References
$CoFe_2O_4$@BC (100 mg/L)	SMX	10 mg/L	825 μM	95.8% (60 min)	$CH_3C(O)O^•$ and $CH_3C(O)OO^•$	Dong et al. (2022)
Fe-biochar (0.3 g/L)	Acid orange	0.143 mM	1.144 mM	99.9% (25 min)	$CH_3C(O)O^•$ and $CH_3C(O)OO^•$	Shi et al. (2022)
PS-SDBC (0.3 g/L)	4-CP	5 mg/L	1.8 mM	100% (60 min)	$CH_3C(O)O^•$ and $CH_3C(O)OO^•$	Wu et al. (2022)
ABC (0.1 g/L)	SMZ	5 mg/L	5 mg/L	72.8% (100 min)	$^1O_2$	Zhang et al. (2022)
$Co(OH)_2$/BC (0.1 g/L)	TCH	10 μmol/L	100 μmol/L	100% (7 min)	$CH_3C(O)O^•$ and $CH_3C(O)OO^•$ and $^•OH$, $^1O_2$	Sun et al. (2023a)
PAC (0.02 g/L)	SMX	10 μM	100 μM	87% (150 min)	$^1O_2$ and ET	Wang et al. (2023e)
CFB-2 (1 g/L)	TCH	10 mg/L	0.5 mM	~100% (5 min)	$^1O_2$	Li et al. (2023b)
BMCBC (0.6 g/L)	SDZ	10 mg/L	2 mM	93% (60 min)	$CH_3C(O)O^•$, $CH_3C(O)OO^•$ and $^•OH$	Zeng et al. (2023)
Fe-SBC (0.6 g/L)	SMX	10 μM	0.5 mM	92% (30 min)	$CH_3C(O)OO^•$, $O_2^{•-}$, and $^1O_2$	Kong et al. (2023)
CBC900 (0.2 g/L)	SMX	10 μM	2 mM	~100% (30 min)	$R—O^•$	Hou et al. (2023)
Co-BC-850 (50 mg/L)	CBZ	5 mg/L	1 mM	99.9% (20 min)	$R—O^•$, $^1O_2$, and Co(IV)	Zhang et al. (2024)
SDBC (1.6 g/L)	EFA	80 μM	3 mM	99.98 % (30 min)	$CH_3C(O)O^•$ and $CH_3C(O)OO^•$ and $^1O_2$	Zhou et al. (2024)

BC-SR (1 g/L)	Triclosan	10 mg/L	1 mM	90% (120 min)	$O_2^{\bullet-}$ and $^1O_2$	Tong et al. (2024)
$BN_3Z_{0.5}BC$ (1.0 g/L)	SDZ	10 mg/L	1.0 mM	94.8% (5 min)	$O_2^{\bullet-}$, $^1O_2$ and ET	Li et al. (2024b)
FeN-BC (0.2 g/L)	Naproxen	10 μM	0.1 mM	> 96.8% (10 min)	$^1O_2$	Ning et al. (2024)
ZIF-67@KSB3 (0.2 g/L)	Acetaminophen	0.02 mM	0.4 mM	94.3% (30 min)	R—O$^\bullet$, $O_2^{\bullet-}$, and $^1O_2$	Nguyen et al. (2024b)
BC-$CoFe_2O_4$ (0.3 g/L)	CBZ	4 mg/L	0.8 mM	100% (30 min)	$CH_3C(O)OO^\bullet$	Xue et al. (2024)
CN900 (0.15 g/L)	Acetaminophen	0.02 mM	0.4 mM	98.6% (60 min)	$O_2^{\bullet-}$	Nguyen et al. (2024a)
PSBC (0.2 g/L)	Acetaminophen	0.1 mM	3 mM	90.4% (20 min)	ET	Wu et al. (2025)
CBC (0.1 g/L)	TC	10 μM	50 μM	97.7% (30 min)	R—O$^\bullet$ and $^1O_2$	Li et al. (2025)

a. $CoFe_2O_4$@BC, Lignin-derived biochar supporting $CoFe_2O_4$; PS-SDBC, Pyrolysis of primary and secondary sludge mixture; ABC, nut shells pyrolyzed at 900 °C. $Co(OH)_2$/BC, $Co(OH)_2$ nanoparticles on biochar; PAC, powdered activated carbon; CFB-2, biochar composite from corn straw and sludge with $CoFe_2O_4$; BMCBC, ball-milled magnetic corn straw biochar; Fe-SBC, iron-modified sludge biochar; CBC900, chicken manure biochar prepared at 900 °C; Co-BC-850, peanut shell biochar with Co nanoparticles; SDBC, sludge biochar; BC-SR, biochar from saccharification residue; $BN_3Z_{0.5}BC$, corn straw as the feedstock to synthesize biochar (BC) loaded with cobalt-zeolitic imidazolate framework nanoparticles and boron nitride quantum dots. FeN-BC, co-doping waste biomass with Fe and N; ZIF-67@KSB3, ZIF-67 doped onto sodium bicarbonate-modified biochar derived from kumquat peels; BC-$CoFe_2O_4$, magnetic biochar (BC)-ferrospinel $CoFe_2O_4$ nanocomposites; CN900, chestnut shell-derived biochar; PSBC, peanut shell-derived biochar; CBC, Corn straw-derived biochar.

b. SMX, sulfamethoxazole; 4-CP, p-chlorophenol; SMZ, sulfamethazine; TCH, tetracycline hydrochloride; SDZ, sulfadiazine; EFA, enrofloxacin; CBZ, carbamazepine; TC, tetracycline.

c. ET, Electron transfer; R—O$^\bullet$,organic radicals.

significantly influence the performance of biochar-catalyzed PAA systems in agricultural wastewater treatment. pH is crucial because it influences PAA activation and ROS formation. Studies have demonstrated that PAA activation is pH-dependent, with the highest efficiency observed under acidic to neutral conditions. The biogas residue biochar (BRBC)-PAA system demonstrated optimal SMX degradation under acidic conditions, achieving a 92% removal rate within 120 minutes (Chen et al. 2025). This is because PAA, being an acidic oxidant, is more reactive in lower pH environments, facilitating the generation of •OH and other ROS. In contrast, when the biochar is doped with transition metals such as cobalt and iron, the pH range for efficient PAA activation can be broadened. Liang et al. demonstrated that the CoFe alloy/oxide system performs exceptionally well across a broad pH spectrum from acidic to alkaline environments. The bimetallic CoFe catalyst's catalytic properties facilitate the production of $^1O_2$ and high-valent metal oxides ($Mn^{IV}=O$) even in neutral to alkaline conditions. The metal doping in biochar helps stabilize the PAA and modulate the active sites, thereby improving the system's resilience to changes in water chemistry.

Moreover, biochar's resistance to interference from other components in agricultural wastewater, such as organic matter and salts, makes it a robust catalyst. It has been observed that biochar maintains its catalytic activity even in complex, multi-contaminant systems, where conventional catalysts may degrade or lose efficiency. Environmental factors like pH, temperature, and ionic strength significantly impact pollutant removal efficiency. Bicarbonate ions were found to inhibit the catalytic process, while chloride had minimal effect (Dong et al. 2022; Wang et al. 2023e). As demonstrated in the provided studies, chloride ions (Cl^-) exhibit concentration-dependent effects. In the Fe-SBC/PAA system, low Cl^- concentrations (1–10 mM) marginally reduced SMX degradation (from 92 to 88%), while 100 mM Cl^- enhanced SMX removal to 99% within 15 minutes (Kong et al. 2023). In the BC-$CoFe_2O_4$/PAA system, low Cl^- concentrations (0.01 M) slightly hindered carbamazepine (CBZ) degradation because of competitive adsorption, whereas higher Cl^- concentrations (0.05 M) had minimal impact, possibly due to compensatory $Cl^•/Cl_2^-$ formation (Xue et al. 2024). Humic acid (HA), a model natural organic matter (NOM), universally suppressed degradation. Background dissolved organic matter (DOM) or humic substances can act as radical scavengers or can block catalyst active sites (Zhang et al. 2022). In Fe-SBC/PAA, the presence of 50 mg/L HA reduced SMX removal from 92 to 79% by quenching •OH and R—O• radicals ($k = 2.5 \times 10^4 - 5.8 \times 10^4\,M^{-1}s^{-1}$) and obstructing active sites through adsorption (Kong et al. 2023). Similarly, in BC-$CoFe_2O_4$/PAA, 50 mg/L HA reduced CBZ degradation by 26% due to HA's dual role as a radical scavenger ($k \approx 10^4$ L/mg/s) and surface passivator (Xue et al. 2024). However, In PSBC (peanut shell-derived biochar)/PAA, HA slightly promoted acetaminophen

(ACP) degradation by participating in PAA activation (Wu et al. 2025). Electron transfer-dominated systems (e.g., PSBC/PAA) are more resilient to anion interference, while radical-driven systems (e.g., CBC/PAA) are susceptible to scavenging and site competition (Wu et al. 2025; Li et al. 2025).

11.6.2 Mechanisms of Biochar-catalyzed PAA Oxidation Technology

Pollutant degradation in biochar-activated PAA systems is facilitated by reactive species produced during activation. These species are classified into two primary categories: free radicals and non-radical species. Both contribute to the degradation of organic pollutants, but their relative importance depends on the biochar's characteristics and the activation conditions. The activation of PAA leads to the cleavage of its O—O bond, generating various ROS, including $^{\bullet}OH$, acetoxyl radicals ($CH_3C(O)O^{\bullet}$), acetylperoxy radicals ($CH_3C(O)OO^{\bullet}$), $O_2^{\bullet -}$, and $^1O_2$. $CH_3C(O)O^{\bullet}$ and $CH_3C(O)OO^{\bullet}$ are critical reactive species generated during PAA activation. These highly reactive radicals are crucial in degrading organic pollutants, particularly those with aromatic rings or readily oxidizable functional groups. Studies have identified $CH_3C(O)O^{\bullet}$ and $CH_3C(O)OO^{\bullet}$ as the primary reactive species responsible for degrading pollutants like acid orange (Shi et al. 2022), SMX (Dong et al. 2022; Hou et al. 2023), 4-CP (Wu et al. 2022), and CBZ (Xue et al. 2024). The ability of biochar to generate acetoxyl radicals is particularly enhanced when biochar is doped with metals such as cobalt or iron. For instance, in cobalt-doped biochar systems (e.g., magnetic biochar-ferrospinel $CoFe_2O_4$ nanocomposites), cobalt ions assist in the electron transfer process, promoting the formation of acetoxyl radicals and enhancing the overall degradation efficiency (Xue et al. 2024).

In addition, singlet oxygen($^1O_2$) is a non-radical species that has been identified in several biochar-PAA systems as a major contributor to pollutant degradation (Zhang et al. 2022; Li et al. 2023b; Ning et al. 2024. Wang et al. (2023e) elucidated the non-radical mechanisms of PAA activation by powdered activated carbon (PAC) for SMX degradation, emphasizing $^1O_2$ generation and electron transfer pathways (ETPs). The generation of $^1O_2$ was attributed to the involvement of PFRs and electron-donating groups such as C—OH on the PAC surface. The proposed mechanism involves the transformation of C—OH groups to C=O, which facilitates the generation of $^1O_2$ via the following reactions (Eqs. 11.8–11.10) (Wang et al. 2023e):

$$PAC{-}OH + CH_3C(O)OOH \rightarrow PAC{-}OO^{\bullet} + CH_3C(O)OH + H^+ \quad (11.8)$$

$$PAC{-}OO^{\bullet} + CH_3C(O)OOH \rightarrow PAC{=}O + CH_3C(O)OH + {}^1O_2 \quad (11.9)$$

$$PFRs{-}e^- + 2CH_3C(O)OOH \rightarrow 2CH_3C(O)OH + {}^1O_2 \quad (11.10)$$

The electron transfer process was another non-radical pathway identified in the PAC/PAA system (Wang et al. 2023e). The study utilized electrochemical methods, such as open circuit potential (OCP), linear sweep voltammetry (LSV), and chronoamperometric measurements along with in situ Raman spectroscopy to investigate the ETP. The results indicated that PAA was adsorbed on the PAC surface, forming a reactive complex (PAC-PAA*). The complex facilitated electron transfer from SMX (electron donor) to PAA (electron acceptor), with PAC serving as the electron-transfer mediator. The ETP was evidenced by the immediate increase in OCP upon the addition of PAA and the subsequent decrease when SMX was introduced, indicating electron transfer between PAC-PAA* and SMX. The PSBC/PAA system operates via a nonradical electron transfer mechanism, distinct from conventional radical-dominated pathways (Wu et al. 2025). Quenching experiments with methanol (MeOH) and TEMP confirm minimal contributions from $^{\bullet}OH$, $CH_3C(O)O^{\bullet}$, or $^1O_2$. Electron paramagnetic resonance (EPR) spectra further exclude radical adducts. Electrochemical analyses indicate the formation of a metastable PSBC-PAA complex evidenced by O—O bond elongation (Raman shift from 872.7 to 863.1 cm^{-1}) and FTIR blue shifts (1102–1093 cm^{-1}). OCP measurements show a potential rise to +0.54 V upon PAA adsorption on PSBC700, followed by a drop upon ACP addition, indicative of direct electron transfer from ACP to the activated complex. Galvanic oxidation experiments corroborate this mechanism, with 15.8% ACP removal occurring via electron flow across a proton exchange membrane. This pathway avoids radical scavenging by water matrices, ensuring selective and efficient degradation.

Optimizing AOPs for degrading organic pollutants crucially involves regulating and transforming reactive species, especially through the interaction of radical and non-radical pathways. (Deng et al. 2024) explored pH-driven shifts in reactive sites and oxidant roles using Fe/Co@BC, a bimetallic biochar composite, to activate PAA. Under acidic conditions (pH 3.0–5.0), Fe^0 corrosion released Fe^{2+}, which reacted with H_2O_2 (a PAA decomposition byproduct) to generate hydroxyl radicals ($^{\bullet}OH$) via Fenton-like reactions. As the pH rose to neutral or alkaline levels (7.0–8.2), the reactive sites transitioned to Co species. Here, Co^{2+} facilitated the activation of PAA, generating organic radicals ($R—O^{\bullet}$, such as $CH_3C(O)OO^{\bullet}$) through single-electron transfer, which accounted for 94% of SMT removal. Concurrently, Fe(IV) species formed through two-electron oxidation of Fe^{2+} by PAA.

A key component in comprehending biochar-catalyzed PAA systems is identifying the active sites and catalytic pathways that enable bond cleavage and radical formation. Active sites on the biochar surface are essential for PAA adsorption and activation, resulting in the production of ROS. Biochar surfaces exhibit unique characteristics influenced by pyrolysis temperature, feedstock composition, and post-treatments like ball milling, acid washing, and metal impregnation. These features include oxygen-containing functional groups (Wang et al. 2023e) (e.g., —OH, —COOH, —C=O), graphitic domains (Wang et al. 2023e), defects (such as nitrogen–oxygen vacancies) (Li et al. 2023b; Hou et al. 2023: Li et al. 2024b), PFRs, (Tong et al. 2024) and π electrons (Zhou et al. 2024). Each of these structural characteristics contributes to catalytic performance in PAA-based AOPs. Surface hydroxyl groups on peanut shell-derived biochar (PSBC) are the main active sites for PAA activation. XPS analysis reveals a strong correlation $\left(R^2 = 0.922\right)$ between k_{obs} and C—O content (Wu et al. 2025). Carboxyl groups (—COOH) are crucial in PAA activation by facilitating electron transfer and aiding in the cleavage of the O—O bond in PAA (Hou et al. 2023). These groups facilitate the production of ROS, including $CH_3C(O)O^{\bullet}$ and $^{\bullet}OH$, which play a role in pollutant degradation. Carbonyl groups (C=O) on biochar surfaces can activate PAA by facilitating electron transfer processes that cleave the O–O bond in PAA (Li et al. 2025; Nguyen et al. 2024a). The C=O group can act as a key active site, mediating electron transfer processes essential for PAA activation. Both experimental and theoretical studies indicate that the electron-rich C=O group, due to its lone pair electrons, can donate electrons to PAA, promoting O—O bond cleavage and the formation of reactive species like $CH_3C(O)O^{\bullet}$ (Li et al. 2025). Biochar often contains structural defects that are created during the pyrolysis process. These defects, such as vacancies or disordered regions (Hou et al. 2023), provide additional sites for electron transfer and increase the material's reactivity. Surface defects on biochar can promote radical generation including $CH_3C(O)O^{\bullet}$, by facilitating electron transfer necessary for the homolytic cleavage of the O—O bond in PAA. PFRs formed during pyrolysis can also play a vital role by facilitating electron transfer in repetitive cycles; they act as redox mediators between adsorbed oxygen (or PAA) and the biochar matrix (Wu et al. 2022). Recent advancement in sludge-derived biochar catalysts have unveiled PFRs as pivotal active sites for PAA activation, offering a sustainable strategy for wastewater remediation. Wu et al. (2022) showed that biochar derived from the co-pyrolysis of primary sludge (rich in inorganic metals) and secondary sludge (abundant in organics), referred to as mixed sludge-derived biochar (PS-SDBC), is more effective in PAA activation than biochar produced from either sludge alone (P-SDBC or S-SDBC). Notably, the PS-SDBC/PAA system exhibits a self-sustaining catalytic cycle: Organic radicals (R—O$^{\bullet}$) species regenerate PFRs by replenishing electron-deficient sites on the biochar surface. This regeneration is facilitated by redox-active metals (e.g., Fe^{2+}/Fe^{3+} transitions) and graphitic nitrogen, which stabilize unpaired electrons via π-conjugation. The system effectively operates across a wide pH range (3–11), addressing the alkaline instability of traditional PAA systems ($pK_a = 8.2$). Although not an active site in the conventional sense, the porosity, surface area, and porous structure of biochar are essential for offering sufficient space to adsorb PAA molecules and pollutants for degradation (Ning et al. 2024). A superior pore structure allows for more active sites to be available for the activation of PAA (Zeng et al. 2023). This results in a more efficient degradation process, as the increased surface area facilitates better interaction between PAA, biochar, and the pollutants.

Doping biochar with transition metals like Co (Zhang et al. 2022), Fe (Shi et al. 2022), and Mn is another effective method to enhance its catalytic performance. These metals create extra active sites that promote electron transfer, thereby enhancing PAA activation and ROS generation. Cobalt ions (Co^{2+}/Co^{3+}) serve as the primary catalytic centers, driving the redox cycle that activates PAA (Wu et al. 2022). Specifically, Co^{2+} reacts with PAA ($CH_3C(O)OOH$) to generate Co^{3+} and acetoxyl radicals ($CH_3C(O)O^{\bullet}$), while Co^{3+} is subsequently reduced back to Co^{2+} by PAA, producing acetylperoxy radicals ($CH_3C(O)OO^{\bullet}$) (Eqs. 11.11 and 11.12) (Dong et al. 2022):

$$\equiv Co^{2+} + CH_3C(O)OOH \rightarrow \equiv Co^{3+} + CH_3C(O)O^{\bullet} + OH^{-} \tag{11.11}$$

$$\equiv Co^{3+} + CH_3C(O)OOH \leftrightarrow \equiv Co^{2+} + CH_3C(O)OO^{\bullet} + H^{+} \tag{11.12}$$

In addition, metal-loaded biochar can create oxygen vacancies, which boost $^1O_2$ production via interactions with adsorbed oxygen or with radical intermediates. Collectively, these enhancements underscore that metal doping or co-pyrolysis with metal-rich wastes can drastically improve both the rate and extent of PAA activation. Taken together, the catalytic mechanisms of PAA activation by biochar pivot around an interplay of structural (surface area, pore structure, defects), compositional (metal doping, functional groups), and electronical (persistent radicals, electron density) attributes. Understanding these elements allows researchers to tailor the production and modification of biochar-based catalysts to optimize efficiency and mitigate potential challenges such as metal leaching or undesired byproduct formation.

The oxidative removal of pollutants in biochar-catalyzed PAA systems involves several mechanisms, including oxidation, coupling, and polymerization reactions. The key mechanism behind the removal of pollutants, especially antibiotics, is oxidative degradation, where ROS such as $^{\bullet}OH$ and $^1O_2$ attack the chemical bonds of pollutants, leading to their breakdown into smaller, non-toxic byproducts. This process is particularly effective for organic pollutants like antibiotics and pesticides. One critical aspect of pollutant removal is the oxidative coupling and polymerization of pollutant molecules. This polymerization can effectively immobilize the pollutants, rendering them less toxic and easier to remove. For example, ACP degradation in a PSBC-PAA system was found to involve the oxidative coupling of the aromatic ring, leading to the formation of insoluble polymeric compounds that were removed through filtration or adsorption onto the biochar surface (Wu et al. 2025). LC-MS identifies dimer (P323), trimer (P472), and tetramer (P623) intermediates, generated through phenoxyl radical coupling. Sequential reactions involve deacetylation, dehydrogenation, and deamination, yielding smaller fragments like P227 and P198. The proposed pathway avoids mineralization, favoring polymerization—a unique trait of nonradical systems. This mechanism minimizes toxic byproducts but raises concerns about residual polymer toxicity. Compared to radical-based systems (e.g., •OH-mediated cleavage), electron transfer offers selective oxidation, aligning with the inertness of the PSBC/PAA system to background anions and NOM.

For antibiotic molecules, in sulfonamides oxidation, radicals ($CH_3C(O)O^•$ or •OH) attack the aromatic ring or the S—N linkage, leading to ring hydroxylation, ring opening, or formation of sulfonic byproducts. Further steps can yield smaller fragments, eventually mineralizing into CO_2, H_2O, and inorganic ions (Zeng et al. 2024). Tetracyclines, on the other hand, often experience hydroxylation at ring positions, followed by cleavage of the ring system. In many cases, $^1O_2$ selectively oxidizes certain double bonds or electron-rich regions in the TC scaffold, producing intermediate species that are more amenable to further radical oxidation (Li et al. 2025; He et al. 2025). For pesticides, the pathways can be similarly diverse. Triazine herbicides, for instance, typically undergo dealkylation and ring cleavage under the attack of •OH or R—O•. Organophosphates may be converted to organophosphate acids through cleavage of P—O or P—S bonds, subsequently forming less hazardous phosphate derivatives. LC-MS and DFT identified three degradation pathways for 2,4-D (Liang et al. 2024a): (1) hydroxylation/decarboxylation (P1→P2), (2) electrophilic ring-opening (P5→P8), and (3) dechlorination (P9→P12). Fukui index analysis highlighted attacks on C—Cl bonds ($f_0 = 0.196$) and benzene rings (HOMO localized). Toxicity assays confirmed detoxification, with intermediates (P11–P15) showing low ecotoxicity.

In summary, pollutant removal in PAA/biochar systems occurs via radical and non-radical pathways, driven by the synergy between various ROS (•OH, R—O•, $^1O_2$) and adsorption or electron-transfer mechanisms. Antibiotics, pesticides, and pathogens all display susceptibility to these oxidative routes, though the exact degradation pattern hinges on both pollutant structure and reaction conditions. Key considerations include the impact of background ions, potential byproducts, and scaling challenges for field applications.

11.6.3 Outlook

The utilization of agricultural waste-derived biochar catalysts for PAA activation presents a promising strategy for sustainable wastewater treatment. However, critical challenges persist. First, the heterogeneous composition of agricultural waste (e.g., lignocellulosic variability, ash content) introduces unpredictability in biochar properties, such as surface functionality, porosity, and catalytic active sites. Second, the complex reaction mechanisms of PAA activation—involving radical and non-radical pathways—require precise control of catalyst-electron transfer properties, which are poorly understood at the atomic scale. Third, scalability issues arise due to energy-intensive pyrolysis conditions and insufficient lifetime of biochar catalysts under repetitive oxidative stress. These gaps underscore the need for advanced computational tools to decode structure–activity relationships and streamline process design.

1. Optimization of Biochar Catalyst Design and Synthesis: A critical challenge in utilizing agricultural waste-derived biochar catalysts for PAA activation lies in achieving consistent physicochemical properties (e.g., porosity, surface functional groups, and metal doping) due to the heterogeneous nature of agricultural feedstocks. Traditional trial-and-error synthesis methods are time-consuming and fail to account for complex interactions between pyrolysis parameters (temperature, heating rate) and feedstock composition. ML offers a solution by enabling predictive modeling of biochar characteristics through regression algorithms trained on datasets integrating feedstock properties (e.g., lignin content, ash composition) and process variables. For instance, neural networks can optimize co-pyrolysis of mixed agricultural residues to enhance active site density. Future research should prioritize developing open-access databases of biochar synthesis parameters and performance metrics, coupled with explainable AI frameworks to decode feature importance. Hybrid approaches combining ML with molecular dynamics simulations could further elucidate structure–activity relationships at atomic scales, addressing current limitations in mechanistic interpretation.
2. Ambiguous Reaction Mechanisms and Pathway Control: Contradictory reports persist regarding dominant PAA activation mechanisms (radical vs. non-radical), exacerbated by inadequate in situ detection methods and variable water matrices. Advanced characterization techniques (in situ FTIR, isotope labeling) coupled with probe compounds have clarified context-dependent mechanism predominance. Critical knowledge gaps remain in correlating biochar's defect density/metallic impurities with electron transfer efficiency. Future work must develop standardized diagnostic protocols and engineer biochar heterostructures to directionally enhance specific degradation pathways for priority pollutants.
3. Scalability, Environmental Impact, and Practical Implementation: Despite promising laboratory-scale results, the practical deployment of biochar-derived catalysts for PAA-based wastewater treatment faces challenges in scalability,

cost-effectiveness, and environmental impact. Industrial-scale production methods should be optimized to ensure uniform biochar quality and cost efficiency. The development of circular economy approaches, such as utilizing agricultural residues as feedstocks, can enhance sustainability. Future research should concentrate on assessing long-term environmental impacts, conducting life-cycle analyses, and implementing pilot-scale demonstration projects to validate the feasibility of biochar catalysts in practical applications. The integration of biochar-based catalytic systems with existing wastewater treatment infrastructures should be explored to enhance operational efficiency and sustainability.

The integration of AI, materials genomics, and autonomous systems is revolutionizing agricultural waste valorization for PAA-based wastewater treatment. Finally, from the standpoint of agricultural waste management, using biochar derived from crop residues or other biomass resources confers a dual environmental benefit: it valorizes waste streams while enabling the decontamination of agricultural wastewater. Given that many agricultural waste streams are locally abundant and inexpensive, the scale-up feasibility of PAA/biochar processes appears promising if design and operational parameters are carefully optimized.

References

Babar, M., Munir, H.M.S., Nawaz, A. et al. (2022). Comparative study of ozonation and ozonation catalyzed by Fe-loaded biochar as catalyst to remove methylene blue from aqueous solution. *Chemosphere* 307: 135738.

Bokare, A.D. and Choi, W. (2014). Review of iron-free Fenton-like systems for activating H_2O_2 in advanced oxidation processes. *Journal of Hazardous Materials* 275: 121–135.

Chen, N., Huang, Y., Hou, X. et al. (2017). Photochemistry of hydrochar: reactive oxygen species generation and sulfadimidine degradation. *Environmental Science & Technology* 51 (19): 11278–11287.

Chen, X., Li, Y., Wu, L. et al. (2025). Mechanistic insights into activation of peracetic acid by sludge biogas residue biochar for efficient sulfamethoxazole degradation in aqueous solution. *Bioresource Technology* 418: 131857.

Chen, X., Oh, W.-D., and Lim, T.-T. (2018). Graphene-and CNTs-based carbocatalysts in persulfates activation: material design and catalytic mechanisms. *Chemical Engineering Journal* 354: 941–976.

Chen, X.-L., Li, H., Lai, L. et al. (2023). Peroxymonosulfate activation using $MnFe_2O_4$ modified biochar for organic pollutants degradation: performance and mechanisms. *Separation and Purification Technology* 308: 122886.

Cheng, Y., Wang, B., Yan, P. et al. (2023). In-situ formation of surface reactive oxygen species on defective sites over N-doped biochar in catalytic ozonation. *Chemical Engineering Journal* 454: 140232.

Dai, J., Cai, T., Li, X. et al. (2023a). Revisit the selectivity of metal-free biochar activated periodate for the oxidation of emerging contaminants. *Chemical Engineering Journal* 476: 146795.

Dai, J., Wang, Z., Zhu, A. et al. (2023b). Elimination of sulfadiazine and its metabolite from hydrolyzed urine by biochar activated periodate process: lower environmental risk compared with $CoFe_2O_4$/peroxymonosulfate process. *Chemical Engineering Journal* 472: 145080.

Deng, J., Dong, H., Zhang, S. et al. (2024). Insights into the pH-dependent mechanism of peracetic acid activation by biochar-supported zero-valent iron/cobalt bimetallic nanoparticles: the shift of reactive sites and the dual role of hydrogen peroxide. *Journal of Hazardous Materials* 476: 135207.

Deng, R., Luo, H., Huang, D. et al. (2020). Biochar-mediated Fenton-like reaction for the degradation of sulfamethazine: role of environmentally persistent free radicals. *Chemosphere* 255: 126975.

Ding, J., Xu, W., Liu, S. et al. (2021). Activation of persulfate by nanoscale zero-valent iron loaded porous graphitized biochar for the removal of 17β-estradiol: synthesis, performance and mechanism. *Journal of Colloid and Interface Science* 588: 776–786.

Dong, J., Xu, W., Liu, S. et al. (2022). Lignin-derived biochar to support $CoFe_2O_4$: effective activation of peracetic acid for sulfamethoxazole degradation. *Chemical Engineering Journal* 430: 132868.

Dong, S., Shen, X., Guo, Q. (2023). Valorization of soybean plant wastes in preparation of N-doped biochar for catalytic ozonation of organic contaminants: atrazine degradation performance and mechanistic considerations. *Chemical Engineering Journal* 472: 145153.

Dou, J., Cheng, J., Lu, Z. et al. (2022). Biochar co-doped with nitrogen and boron switching the free radical based peroxydisulfate activation into the electron-transfer dominated nonradical process. *Applied Catalysis B: Environmental* 301: 120832.

Du, J., Xiao, G., Xi, Y. et al. (2020a). Periodate activation with manganese oxides for sulfanilamide degradation. *Water Research* 169: 115278.

Du, L., Xu, W., Liu, S. et al. (2020b). Activation of persulfate by graphitized biochar for sulfamethoxazole removal: the roles of graphitic carbon structure and carbonyl group. *Journal of Colloid and Interface Science* 577: 419–430. https://doi.org/10.1016/j.jcis.2020.05.096.

Duan, W., Zhang, M., and Zhou, R. (2024). Efficient degradation of antibiotic wastewater by biochar derived from water hyacinth stems via periodate activation:

pyridinic N and carbon structures improved the electron transfer process. *Water Science & Technology* 89 (1): 212–224.
El-Bestawy, E.A., Gaber, M., Shokry, H., et al. (2023). Effective degradation of atrazine by spinach-derived biochar via persulfate activation system: process optimization, mechanism, degradation pathway and application in real wastewater. *Environmental Research* 229: 115987.
Elmitwalli, T., Fouad, M., Mossad, M. et al. (2024). Periodate activation by mulukhiyah stalks and potato peels-derived biochars for the efficient degradation of sulfamethazine. *Journal of Environmental Chemical Engineering* 12 (2): 112101.
Fang, G., Li, J., Zhang, C. et al. (2022). Periodate activated by manganese oxide/biochar composites for antibiotic degradation in aqueous system: combined effects of active manganese species and biochar. *Environmental Pollution* 300: 118939.
Fang, Q., Liu, N., Gu, Y. et al. (2024) Mechanism exploration in tetracycline degradation by Ni-Fe layered double hydroxide-biochar/peroxymonosulfate system: nonradical-dominated oxidation process. *Separation and Purification Technology* 335: 126122. https://doi.org/10.1016/j.seppur.2023.126122.
Fang, Q., Yang, H., Ye, S. et al. (2023). Generation and identification of $^1O_2$ in catalysts/peroxymonosulfate systems for water purification. *Water Research* 245: 120614. https://doi.org/10.1016/j.watres.2023.120614.
Fang, Q., Ye, S., Yang, H. et al. (2021). Application of layered double hydroxide-biochar composites in wastewater treatment: recent trends, modification strategies, and outlook. *Journal of Hazardous Materials* 420: 126569.
Feng, Z., Zhou, B., Yuan, R. et al. (2022). Biochar derived from different crop straws as persulfate activator for the degradation of sulfadiazine: influence of biomass types and systemic cause analysis. *Chemical Engineering Journal* 440: 135669.
Fijołek, L., Świetlik, J., and Frankowski, M. (2024). Ozonation and catalytic ozonation-Sources of error. What do we need to know? *Journal of Environmental Management* 370: 123031.
Gasim, M.F., Bao, Y., Elgarahy, A.M. et al. (2023). Peracetic acid activation using heterogeneous catalysts for environmental decontamination: a review. *Catalysis Communications* 180: 106702.
Gasim, M.F., Choong, Z.-Y., Koo, P.-L. et al. (2022). Application of biochar as functional material for remediation of organic pollutants in water: an overview. *Catalysts* 12 (2): 210.
Geng, A., Xu, L., Gan, L., et al. (2020). Using wood flour waste to produce biochar as the support to enhance the visible-light photocatalytic performance of BiOBr for organic and inorganic contaminants removal. *Chemosphere* 250: 126291.
Ghanbari, F. and Moradi, M. (2017). Application of peroxymonosulfate and its activation methods for degradation of environmental organic pollutants. *Chemical Engineering Journal* 310: 41–62.

Grisales-Cifuentes, C.M., Serna-Galvis, E.A., Acelas, N. et al. (2022). Biochar from palm fiber wastes as an activator of different oxidants for the elimination of pharmaceuticals from diverse classes in aqueous samples. *Journal of Environmental Management* 323: 116148.

He, C., Yao, Z., Zhang, A. et al. (2025). The synergistic effect of peracetic acid and nitrogen doped biochar prepared at pyrolysis temperature of 1000°C on tetracycline degradation: performance and mechanism. *Colloids and Surfaces A: Physicochemical and Engineering Aspects* 704: 135438.

He, L., Yang, C., Ding, J. et al. (2022). N-doped carbonaceous catalyst activating periodate for micropollutant removal: significant role of electron transfer. *Applied Catalysis B: Environmental* 303: 120880.

He, Y., Chen, Y., Li, J. et al. (2023). Efficient degradation of 2, 3, 5-trimethylpyrazine by catalytic ozonation over MnO_x supported on biochar derived from waste tea leaves. *Chemical Engineering Journal* 464: 142525.

Hou, J., Xu, J., Tang, R. et al. (2023). Activation of peracetic acid by biochar for sulfamethoxazole degradation: revealing roles of the active sites. *Journal of the Taiwan Institute of Chemical Engineers* 152: 105184.

Huang, C., Qin, F., Zhang, C. et al. (2022). Effects of heterogeneous metals on the generation of persistent free radicals as critical redox sites in iron-containing biochar for persulfate activation. *ACS ES&T Water* 3 (2): 298–310.

Huang, D., Luo, H., Zhang, C. et al. (2019). Nonnegligible role of biomass types and its compositions on the formation of persistent free radicals in biochar: insight into the influences on Fenton-like process. *Chemical Engineering Journal* 361: 353–363.

Huang, J., Wang, M., Luo, S. et al. (2023a). In situ preparation of highly graphitized N-doped biochar geopolymer composites for efficient catalytic degradation of tetracycline in water by H_2O_2. *Environmental Research* 219: 115166.

Huang, P., Zhang, P., Wang, C. et al. (2023b). P-doped biochar regulates nZVI nanocracks formation for superefficient persulfate activation. *Journal of Hazardous Materials* 450: 130999.

Kong, D., He, L., Shen, S. et al. (2023). Unveiling the mechanisms of peracetic acid activation by iron-rich sludge biochar for sulfamethoxazole degradation with wide adaptability. *Journal of Environmental Management* 347: 119119.

Kumar, A., Kumar, A., Sharma, G. et al. (2017) Sustainable nano-hybrids of magnetic biochar supported g-C_3N_4/$FeVO_4$ for solar powered degradation of noxious pollutants-synergism of adsorption, photocatalysis & photo-ozonation. *Journal of Cleaner Production* 165: 431–451.

Kumar, A., Kumar, A., Sharma, G. et al. (2018). Biochar-templated g-C_3N_4/$Bi_2O_2CO_3$/$CoFe_2O_4$ nano-assembly for visible and solar assisted photo-degradation of paraquat, nitrophenol reduction and CO_2 conversion. *Chemical Engineering Journal* 339: 393–410.

Kumar, A., Sharma, G., Naushad, M. et al. (2019). Visible photodegradation of ibuprofen and 2, 4-D in simulated waste water using sustainable metal free-hybrids based on carbon nitride and biochar. *Journal of environmental management* 231: 1164–1175.

Kumar, M., Ambika, S., Hassani, A. et al. (2023). Waste to catalyst: role of agricultural waste in water and wastewater treatment. *Science of the Total Environment* 858: 159762.

Lai, M., Li, J., Li, H. et al. (2023). S-codoped biochar outperformed N-doped biochar on co-activation of H_2O_2 with trace dissolved Fe(III) for enhanced oxidation of organic pollutants. *Environmental Pollution* 334: 122208.

Lee, J., Von Gunten, U., and Kim, J.-H. (2020). Persulfate-based advanced oxidation: critical assessment of opportunities and roadblocks. *Environmental Science & Technology* 54 (6): 3064–3081.

Li, H., Liu, S., Qiu, S. et al. (2022a). Catalytic ozonation oxidation of ketoprofen by peanut shell-based biochar: effects of the pyrolysis temperatures. *Environmental Technology* 43 (6): 848–860.

Li, H., Qiu, R., Tang, Y. et al. (2023a). Formation of Fe(IV) over a wide pH range via iron-carbon composite-catalyzed persulfate activation. *Chemical Engineering Journal* 461. https://doi.org/10.1016/j.cej.2023.141951.

Li, L., Cheng, M., Qin, L. et al. (2022a). Enhancing hydrogen peroxide activation of CuCo layered double hydroxide by compositing with biochar: performance and mechanism. *Science of the Total Environment* 828: 154188.

Li, L., Liu, S., Cheng, M. et al. (2021). Improving the fenton-like catalytic performance of MnO_x-Fe_3O_4/biochar using reducing agents: a comparative study. *Journal of Hazardous Materials* 406: 124333.

Li, L., Wang, X., Fu, H. et al. (2020). Dissolved black carbon facilitates photoreduction of Hg(II) to Hg(0) and reduces mercury uptake by lettuce (*Lactuca sativa* L.). *Environmental Science and Technology* 54 (18): 11137–11145. https://doi.org/10.1021/acs.est.0c01132.

Li, L., Zhao, J., Zhao, X. et al. (2025). Insight into the mechanism of peracetic acid activation by corn straw-derived biochar as efficient green activator mediating electron transfer: crucial role of carbonyl functional group. *Separation and Purification Technology* 354: 129005.

Li, S., Dai, C., Duan, Y. et al. (2024a). Non-radical pathways in peracetic acid-based micropollutant degradation: a comprehensive review of mechanisms, detection methods, and promising applications. *Separation and Purification Technology* 330: 125240.

Li, S., Liu, Y., Zheng, H. et al. (2023b). Biochar loaded with $CoFe_2O_4$ enhances the formation of high-valent Fe(IV) and Co(IV) and oxygen vacancy in the peracetic acid activation system for enhanced antibiotic degradation. *Bioresource Technology* 387: 129536.

Li, S., Liu, Y., Zheng, H. et al. (2024b). Mechanism of biochar composite ($BN_3Z_{0.5}BC$) activated peracetic acid for efficient antibiotic degradation: synergistic effect between free radicals and non-free radicals. *Bioresource Technology* 397: 130452.

Liang, X., Li, X., Zhao, Y. et al. (2024a). Efficient activation of peroxyacetic acid by cobalt-iron alloy/oxide heterojunctions anchored in defect-rich biochar for pesticide degradation in water: unravelling the radical-unradical mechanism. *Chemical Engineering Journal* 500: 156810.

Liang, Y., Tao, R., Zhao, B. et al. (2024b). Roles of iron and manganese in bimetallic biochar composites for efficient persulfate activation and atrazine removal. *Biochar* 6 (1): 41.

Lim, S., Shi, J.L., von Gunten, U. et al. (2022). Ozonation of organic compounds in water and wastewater: a critical review. *Water Research* 213: 118053.

Liu, B., Guo, W., Wang, H. et al. (2020a). Activation of peroxymonosulfate by cobalt-impregnated biochar for atrazine degradation: the pivotal roles of persistent free radicals and ecotoxicity assessment. *Journal of Hazardous Materials* 398: 122768.

Liu, G., Zhang, Y., Yu, H. et al. (2020b). Acceleration of goethite-catalyzed Fenton-like oxidation of ofloxacin by biochar. *Journal of Hazardous Materials* 397: 122783.

Liu, L., Li, Y., Yoza, B.A. et al. (2020c). A char-clay composite catalyst derived from spent bleaching earth for efficient ozonation of recalcitrants in water. *Science of the Total Environment* 699: 134395.

Liu, Y., Shi, S., and Wang, Y. (2021). Removal of pollutants from gas streams using Fenton (-like)-based oxidation systems: a review. *Journal of Hazardous Materials* 416: 125927.

Lu, Y., Cai, Y., Zhang, S. et al. (2022). Application of biochar-based photocatalysts for adsorption-(photo) degradation/reduction of environmental contaminants: mechanism, challenges and perspective. *Biochar* 4 (1): 45.

Luo, Y., Li, H., Yang, H. et al. (2024). Critical role of dissolved oxygen and iron–copper synergy in dual-metal/char catalyst systems. *Environmental Science: Nano* 11(5): 2091–2102. https://doi.org/10.1039/D3EN00911D.

Luo, Y., Zheng, A., Li, J. et al. (2023). Integrated adsorption and photodegradation of tetracycline by bismuth oxycarbonate/biochar nanocomposites. *Chemical Engineering Journal* 457: 141228.

Lv, M. (2025). Resource utilization of iron sludge: synthesis of high performance FeN internal doped biochar to catalyze the ozonation degradation of tetracycline. *Journal of Water Process Engineering* 70: 106758.

Lyu, H., Zhang, Q., and Shen, B. (2020). Application of biochar and its composites in catalysis. *Chemosphere* 240, 124842.

Mao, Q., Zhou, Y., Yang, Y. et al. (2019). Experimental and theoretical aspects of biochar-supported nanoscale zero-valent iron activating H_2O_2 for ciprofloxacin removal from aqueous solution. *Journal of Hazardous Materials* 380: 120848.

Miao, F., Ren, W., Zhou, H. et al. (2024). Carbon-based peracetic acid activation towards advanced water purification. *Applied Catalysis B: Environment and Energy* 363: 124807.

Moussavi, G. and Khosravi, R. (2012). Preparation and characterization of a biochar from pistachio hull biomass and its catalytic potential for ozonation of water recalcitrant contaminants. *Bioresource Technology* 119: 66–71.

Nawrocki, J. and Kasprzyk-Hordern, B. (2010). The efficiency and mechanisms of catalytic ozonation. *Applied Catalysis B: Environmental* 99 (1–2), 27–42.

Nguyen, T.-B., Chen, C.-W., Chen, W.-H. et al. (2024a). Boosting acetaminophen degradation in water by peracetic acid activation: a novel approach using chestnut shell-derived biochar at varied pyrolysis temperatures. *Environmental Research* 252: 119143.

Nguyen, T.-B., Chen, C.-W., Chen, W.-H. et al. (2024b). Kumquat peel-derived biochar to support zeolitic imidazole framework-67 (ZIF-67) for enhancing peracetic acid activation to remove acetaminophen from aqueous solution. *Environmental Pollution* 350: 123970.

Nidheesh, P., Gopinath, A., Ranjith, N. et al. (2021). Potential role of biochar in advanced oxidation processes: a sustainable approach. *Chemical Engineering Journal* 405: 126582.

Ning, R., Dong, Y., Yang, S.-R. et al. (2024). Fe-N co-doped biochar derived from biomass waste triggers peracetic acid activation for efficient water decontamination. *Journal of Hazardous Materials* 470: 134139.

Noman, M., Yu, G., Awugichew, D.T. et al. (2024). Synthesis of surficial-modified green biochar catalyst generated by biogas residue biochar and potential application for catalytic ozonation degradation of ciprofloxacin. *Environmental Research* 257: 119314.

Noman, M., Yu, G., Elimian, E.A. et al. (2025). One step hydrothermal synthesis of green iron-modified catalyst by food waste digestate biogas residue for ozonation of ciprofloxacin: characterization, degradation mechanism, DFT study and reactive oxygen species role. *Journal of Environmental Chemical Engineering* 13 (1): 115236.

Oh, S.-Y. and Nguyen, T.-H.A. (2022). Ozonation of phenol in the presence of biochar and carbonaceous materials: the effect of surface functional groups and graphitic structure on the formation of reactive oxygen species. *Journal of Environmental Chemical Engineering* 10 (2): 107386.

Peng, Y., Xue, C., Luo, J., et al. (2024). Lanthanum-doped magnetic biochar activating persulfate in the degradation of florfenicol. *Science of The Total Environment* 916: 170312.

Ren, W., Cheng, C., Shao, P. et al. (2021). Origins of electron-transfer regime in persulfate-based nonradical oxidation processes. *Environmental Science & Technology* 56 (1): 78–97.

Ren, W., Nie, G., Zhou, P. et al. (2020). The intrinsic nature of persulfate activation and N-doping in carbocatalysis. *Environmental Science & Technology* 54 (10): 6438–6447.

Sang, F., Yin, Z., Wang, W. et al. (2022). Degradation of ciprofloxacin using heterogeneous Fenton catalysts derived from natural pyrite and rice straw biochar. *Journal of Cleaner Production* 378: 134459.

Shen, X., Zheng, T., Yang, J. et al. (2020). Removal of Cr(VI) from acid wastewater by BC/$ZnFe_2O_4$ magnetic nanocomposite via the synergy of absorption-photocatalysis. *ChemCatChem* 12 (16): 4121–4131.

Shi, C., Wang, Y., Zhang, K. et al. (2022). Fe-biochar as a safe and efficient catalyst to activate peracetic acid for the removal of the acid orange dye from water. *Chemosphere* 307: 135686.

Song, H., Li, Q., Ye, Y. et al. (2021). Degradation of cephalexin by persulfate activated with magnetic loofah biochar: performance and mechanism. *Separation and Purification Technology* 272: 118971.

Song, T., Gao, Y., Yang, J. et al. (2023). Recent advances in periodate activated by carbon and carbon-doped materials in removing recalcitrant organics in water. *Journal of Water Process Engineering* 56: 104506.

Subramaniam, M.N., Wu, Z., Goh, P.S. et al. (2023). The state-of-the-art development of biochar based photocatalyst for removal of various organic pollutants in wastewater. *Journal of Cleaner Production* 429: 139487.

Sun, C., Chen, T., Huang, Q. et al. (2021) Selective production of singlet oxygen from zinc-etching hierarchically porous biochar for sulfamethoxazole degradation. *Environmental Pollution* 290: 117991.

Sun, F., Yang, X., Shao, F. et al. (2023). In-situ construction of $Co(OH)_2$ nanoparticles decorated biochar for highly efficient degradation of tetracycline hydrochloride via peracetic acid activation. *Chinese Chemical Letters* 34 (9): 108563.

Sun, P., Sun, P., Fang, Y. et al. (2023). Enhanced degradation of bisphenol S in water using a combination process of electrochemical oxidation with periodate activation. *Journal of Water Process Engineering* 55: 104203.

Tian, S.-Q., Qi, J.-Y., Wang, Y.-P. et al. (2021). Heterogeneous catalytic ozonation of atrazine with Mn-loaded and Fe-loaded biochar. *Water Research* 193: 116860.

Tong, W.K., Dai, C., Jia, C. et al. (2024). Eco-friendly and stable triclosan removal from groundwater using peroxyacetic acid activated with biochar produced from saccharification residues. *Chemical Engineering Journal* 481: 148422.

Wang, B., Li, Y.-N., and Wang, L. (2019). Metal-free activation of persulfates by corn stalk biochar for the degradation of antibiotic norfloxacin: activation factors and degradation mechanism. *Chemosphere* 237: 124454.

Wang, D., Dong, S., Fu, S. et al. (2023a). Catalytic ozonation for imazapic degradation over kelp-derived biochar: promotional role of N- and S-based active sites. *Science of The Total Environment* 860: 160473.

Wang, J. and Chen, H. (2020). Catalytic ozonation for water and wastewater treatment: recent advances and perspective. *Science of the Total Environment* 704: 135249.

Wang, J. and Wang, S. (2018). Activation of persulfate (PS) and peroxymonosulfate (PMS) and application for the degradation of emerging contaminants. *Chemical Engineering Journal* 334: 1502–1517.

Wang, J., Yuan, S., Dai, X. et al. (2023b). Application, mechanism and prospects of Fe-based/Fe-biochar catalysts in heterogenous ozonation process: a review. *Chemosphere* 319: 138018.

Wang, L., Wei, J., Li, Y. et al. (2023c). A state-of-the-art review on heterogeneous catalysts-mediated activation of peracetic acid for micropollutants degradation: classification of reaction pathways, mechanisms, influencing factors and DFT calculation. *Chemical Engineering Journal* 477: 147051.

Wang, S. and Wang, J. (2023). Bimetallic and nitrogen co-doped biochar for peroxymonosulfate (PMS) activation to degrade emerging contaminants. *Separation and Purification Technology* 307: 122807.

Wang, W. and Chen, M. (2022). Catalytic degradation of sulfamethoxazole by peroxymonosulfate activation system composed of nitrogen-doped biochar from pomelo peel: important roles of defects and nitrogen, and detoxification of intermediates. *Journal of Colloid and Interface Science* 613: 57–70.

Wang, W., Zhang, J., Chen, T. et al. (2020). Preparation of TiO_2-modified biochar and its characteristics of photo-catalysis degradation for enrofloxacin. *Scientific Reports* 10 (1): 6588.

Wang, X., Zhang, C., Zhang, Y. et al. (2023d). Activation of peroxymonosulfate by an Enteromorpha prolifera derived biochar supported $CoFe_2O_4$ catalyst for highly efficient lomefloxacin hydrochloride degradation under a wide pH range. *Separation and Purification Technology* 316: 123846.

Wang, X., Zhang, P., Wang, C. et al. (2022). Metal-rich hyperaccumulator-derived biochar as an efficient persulfate activator: role of intrinsic metals (Fe, Mn, and Zn) in regulating characteristics, performance and reaction mechanisms. *Journal of Hazardous Materials* 424: 127225.

Wang, Z., Chen, Z., Li, Q. et al. (2023e). Non-radical activation of peracetic acid by powdered activated carbon for the degradation of sulfamethoxazole. *Environmental Science & Technology* 57 (28): 10478–10488.

Wang, Z., Meng, X., Yang, Y. et al. (2023f). Persulfate activation with nitrogen and sulfur co-doped porous biochar to remove thiamethoxam: a combination of radicals and nonradicals. *Journal of Environmental Chemical Engineering* 11 (5): 110738.

Wei, X., Zhang, H., and Cao, M. (2024). Electron transfer via CoO coupled to NPC derived from N, P-rich *Chlorella pyrenoidosa* enhanced radical-nonradical process in catalytic ozonation for efficient ibuprofen removal. *Chemical Engineering Journal* 500: 157327.

Wei, X., Zhang, H., and Cao, M. (2025). Ozone catalyzation modulated by oxygen vacancies over Co/Mg loading on biochar derived from *Chlorella pyrenoidosa* for wastewater purification. *Chemical Engineering Journal* 503: 158493.

Wu, F., Yue, X., Xu, Y. et al. (2025). Peanut-shell-derived biochar mediated peracetic acid activation for the elimination of acetaminophen via an electron-transfer regime. *Separation and Purification Technology* 353: 128150.

Wu, L., Li, Z., Cheng, P. et al. (2022). Efficient activation of peracetic acid by mixed sludge derived biochar: critical role of persistent free radicals. *Water Research* 223: 119013.

Xiang, L., Almatrafi, E., Yang, H. et al. (2023). Coupled carbon structure and iron species for multiple periodate-based oxidation reaction. *Chemical Engineering Journal* 455: 140560.

Xiao, S., Cheng, M., Zhong, H. et al. (2020a). Iron-mediated activation of persulfate and peroxymonosulfate in both homogeneous and heterogeneous ways: a review. *Chemical Engineering Journal* 384: 123265.

Xiao, Y., Lyu, H., Tang, J. et al. (2020b). Effects of ball milling on the photochemistry of biochar: enrofloxacin degradation and possible mechanisms. *Chemical Engineering Journal* 384: 123311.

Xu, H., Zhang, Y., Li, J. et al. (2020). Heterogeneous activation of peroxymonosulfate by a biochar-supported Co_3O_4 composite for efficient degradation of chloramphenicols. *Environmental Pollution* 257: 113610.

Xu, L., He, Z., Wei, X. et al. (2023). Facile-prepared Fe/Mn co-doped biochar is an efficient catalyst for mediating the degradation of aqueous ibuprofen via catalytic ozonation. *Chemical Engineering Journal* 461: 142028.

Xu, W., Liang, J., Li, J et al. (2024). Biochar encapsulated metal nanoflowers for high efficient degradation of metronidazole via peroxymonosulfate activation. *Separation and Purification Technology* 328: 125081.

Xue, Y., Kamali, M., Kakavandi, B. et al. (2024). Activation of peracetic acid by a magnetic biochar-ferrospinel AFe_2O_4 (A = Cu, Co, or Mn) nanocomposite for the degradation of carbamazepine-A comparative and mechanistic study. *Chemical Engineering Journal* 490: 151932.

Yan, J., Chen, C., Sun, H. et al. (2023). Mechanism of nitrogen-doped biochar activated peroxymonosulfate for degradation of 2, 4-dichlorophenol. *Environmental Science and Pollution Research* 30 (13): 37475–37486.

Yan, Y., Wang, W., Peng, Y. et al. (2021). Heterogeneous NiS/NiSe/3D porous biochar for as removal from water by interface engineering-induced nickel lattice distortion. *Science of the Total Environment* 776: 145874.

Yang, H., Ye, S., Zeng, Z. et al. (2020). Utilization of biochar for resource recovery from water: a review. *Chemical Engineering Journal* 397: 125502.

Yang, J., Li, Y., Yang, Z. et al. (2023). Hydrogen peroxide as a key intermediate for hydroxyl radical generation during catalytic ozonation of biochar: mechanistic insights into the evolution and contribution of radicals. *Separation and Purification Technology* 324: 124525.

Yang, Y., Ma, K., Cui, Y. et al. (2024). Novel cow dung-doped sludge biochar as an efficient ozone catalyst: synergy between graphitic structure and defects induces free radical pathways. *Environmental Research* 251: 118747.

Ye, S., Yan, M., Tan, X. et al. (2019) Facile assembled biochar-based nanocomposite with improved graphitization for efficient photocatalytic activity driven by visible light. *Applied Catalysis B: Environmental* 250: 78–88.

Ye, S., Zeng, G., Tan, X. et al. (2020). Nitrogen-doped biochar fiber with graphitization from *Boehmeria nivea* for promoted peroxymonosulfate activation and non-radical degradation pathways with enhancing electron transfer. *Applied Catalysis B: Environmental* 269: 118850. https://doi.org/10.1016/j.apcatb.2020.118850.

Yi, Y., Fu, Y., Wang, Y. et al. (2024). Persulfate oxidation of norfloxacin by cobalt doped water hyacinth biochar composite: the key role of cobalt and singlet oxygen. *Journal of Water Process Engineering* 59: 104967.

Yi, Y., Luo, J., and Fang, Z. (2021). Magnetic biochar derived from *Eichhornia crassipes* for highly efficient Fenton-like degradation of antibiotics: mechanism and contributions. *Journal of Environmental Chemical Engineering* 9 (5): 106258.

Yi, Y., Tu, G., Zhao, D. et al. (2019). Pyrolysis of different biomass pre-impregnated with steel pickling waste liquor to prepare magnetic biochars and their use for the degradation of metronidazole. *Bioresource Technology* 289: 121613.

Yu, B., Man, Y., Wang, P. et al. (2023a). Catalytic degradation of dimethomorph by nitrogen-doped rice husk biochar. *Ecotoxicology and Environmental Safety* 257: 114908.

Yu, G., Wang, Y., Cao, H. et al. (2020). Reactive oxygen species and catalytic active sites in heterogeneous catalytic ozonation for water purification. *Environmental Science & Technology* 54 (10): 5931–5946.

Yu, J., Zhang, X., Zhao, X. et al. (2023b). Heterogeneous Fenton oxidation of 2,4-dichlorophenol catalyzed by PEGylated nanoscale zero-valent iron supported by biochar. *Environmental Science and Pollution Research* 30 (14): 41333–41347.

Yu, J., Zhu, Z., Zhang, H. et al. (2019). Efficient removal of several estrogens in water by Fe-hydrochar composite and related interactive effect mechanism of H_2O_2 and iron with persistent free radicals from hydrochar of pinewood. *Science of the Total Environment* 658: 1013–1022.

Zeng, C., Ma, Y., Li, P. et al. (2024). Efficient degradation of sulfadiazine by UV-triggered electron transfer on oxalic acid-functionalized corn straw biochar for activating peroxyacetic acid: performance, mechanism, and theoretical calculation. *Bioresource Technology* 407: 131103.

Zeng, C., Qian, S., Ma, Y. et al. (2023). Ball milling enhanced magnetic corn straw biochar for peroxyacetic acid activation towards efficient degradation of sulfadiazine. *Industrial Crops and Products* 204: 117346.

Zhang, R., Zheng, X., Zhang, D. et al. (2021a). Insight into the roles of endogenous minerals in the activation of persulfate by graphitized biochar for tetracycline removal. *Science of the Total Environment* 768: 144281.

Zhang, W., Li, M., Lin, L. et al. (2024). Peroxyacetic acid activation by cobalt-doped peanut shell biochar for efficient CBZ degradation: role of organic radicals, singlet oxygen, and high-valent cobalt species. *Separation and Purification Technology* 330: 125592.

Zhang, X., Sun, P., Wei, K. et al. (2020). Enhanced H_2O_2 activation and sulfamethoxazole degradation by Fe-impregnated biochar. *Chemical Engineering Journal* 385: 123921.

Zhang, Y., Xu, M., Liu, X. et al. (2021b). Regulation of biochar mediated catalytic degradation of quinolone antibiotics: important role of environmentally persistent free radicals. *Bioresource Technology* 326: 124780.

Zhang, Z., Duan, Y., Dai, C. et al. (2022). Oxidation of sulfamethazine by peracetic acid activated with biochar: reactive oxygen species contribution and toxicity change. *Environmental Pollution* 313: 120170.

Zhao, H., Lv, M., Shang, X. et al. (2024). High performance of heterogeneous catalytic ozonation for tetracycline removal by a N-doped biochar derived from co-pyrolysis of sludge and water hyacinth. *Chemical Engineering and Processing-Process Intensification* 205: 110031.

Zhao, L., Zhang, H., Dai, Z. et al. (2022). Recycling chestnut shell for superior peroxymonosulfate activation in contaminants degradation via the synergistic radical/non-radical mechanisms. *Journal of Hazardous Materials* 430: 128471.

Zhong, M.-e., Tong, G., Sun, J. et al. (2022). Low-temperature reduction synthesis of $\gamma\text{-}Fe_2O_{3-x}$@ biochar catalysts and their combining with peroxymonosulfate for quinclorac degradation. *International Journal of Environmental Research and Public Health* 19 (24): 16790.

Zhou, X., Zhu, Y., Niu, Q. et al. (2021). New notion of biochar: a review on the mechanism of biochar applications in advanced oxidation processes. *Chemical Engineering Journal* 416: 129027.

Zhou, Z., Yang, Y., Xue, G. et al. (2024). Efficient degradation of fluoroquinolone antibiotics in the landfill leachate by sludge biochar activated peracetic acid: radical vs non-radical process. *Chemical Engineering Journal* 480: 148048.

Zhu, B., Li, C., Liu, J. et al. (2025). Adsorption promoted heterogeneous catalytic ozonation for removing emerging contaminations: the intrinsic correlations with biochar properties and complementary effect of adsorption and ozonation processes. *Separation and Purification Technology* 354: 128726.

Zhu, X., Li, J., Xie, B. et al. (2020). Accelerating effects of biochar for pyrite-catalyzed Fenton-like oxidation of herbicide 2, 4-D. *Chemical Engineering Journal* 391: 123605.

Zhu, X., Shen, J., Kang, J. et al. (2023). Surface atomic oxygen species mediated the in-situ formation of hydroxyl radicals on Fe_3C decorated biochar for enhancing catalytic ozonation. *Chemical Engineering Journal* 473: 145380.

Zhu, X., Wang, B., Kang, J. et al. (2022). Interfacial mechanism of the synergy of biochar adsorption and catalytic ozone micro-nano-bubbles for the removal of 2,4-dichlorophenoxyacetic acid in water. *Separation and Purification Technology* 299: 121777.

Zhu, Z., Fan, W., Liu, Z. et al. (2018) Fabrication of the metal-free biochar-based graphitic carbon nitride for improved 2-Mercaptobenzothiazole degradation activity. *Journal of Photochemistry and Photobiology A: Chemistry* 358: 284–293.

Zhuo, S.-N., Zhang, W., Ren, H.-Y. et al. (2024). Highly efficient degradation of acetaminophen via nano zero-valent iron biochar with periodate system at low temperature. *Bioresource Technology* 395: 130349.

Zong, Y., Shao, Y., Zeng, Y. et al. (2021). Enhanced oxidation of organic contaminants by iron(II)-activated periodate: the significance of high-valent iron–oxo species. *Environmental Science & Technology* 55 (11): 7634–7642.

12

Waste-derived Biochar for Efficient CO_2 Capture

Leichang Cao[1,2], Jieni Wang[1,2], Shuqin Zhang[1,2], Haodong Hou[1,2], Yuqing Sun[3], and Daniel C.W. Tsang[4]

[1] *College of Chemistry and Molecular Sciences, Henan University, Kaifeng, China*
[2] *Miami College, Henan University, Kaifeng, China*
[3] *School of Agriculture and Biotechnology, Sun Yat-sen University, Guangzhou, Guangdong, China*
[4] *Department of Civil and Environmental Engineering, The Hong Kong University of Science and Technology, Clear Water Bay, Hong Kong, China*

12.1 Introduction

Global climate change is one of the most pressing challenges for humanity, with CO_2 recognized as a leading greenhouse gas (Ahmed et al. 2022). Mitigating this crisis requires the development of efficient CO_2 capture and storage technologies. While conventional CO_2 capture methods have been widely studied and implemented over the past decades (Bai et al. 2023; Bernardo et al. 2021), growing attention has shifted toward physical adsorbents as a cleaner and more efficient alternative. Materials such as zeolites, hyper cross-linked resins, covalent organic frameworks (COFs), and metal-organic frameworks (MOFs) have demonstrated strong potential as CO_2 adsorbents due to their high surface areas and tunable pore structures. However, their practical application is often hindered by high production costs, significant energy demands, and scalability issues. These limitations underscore the urgent need for innovative, sustainable, and cost-effective CO_2 capture technologies.

Biomass-based porous carbon materials have emerged as a promising alternative for CO_2 capture. Notably, processes such as pyrolysis and hydrothermal treatment enable the conversion of biomass waste or by-products into high-performance porous carbons (Cheng et al. 2023). These methods not only utilize renewable feedstocks but also allow for precise control over the porosity and surface chemistry of the resulting materials. Compared with conventional

Waste-derived Biochar for Sustainable Rural Revitalization, First Edition. Edited by Yuqing Sun and Daniel C.W. Tsang.
Published 2026 by John Wiley & Sons, Inc.

carbon materials, biomass-derived carbons offer distinct advantages, including cost-effectiveness, renewability, tunable porous structures, and exceptional CO_2 adsorption performance (Sun et al. 2021; Meng et al. 2024). These attributes make them as a compelling option for sustainable CO_2 capture.

This chapter reviews the recent advancements in biomass-based carbon materials for CO_2 capture. It explores the sources and transformation methods of various biomass types, alongside the preparation processes for biomass-derived carbon materials (Guo et al. 2021). Subsequently, the discussion shifts to strategies aimed at enhancing CO_2 adsorption capacity and selectivity, encompassing physical and chemical modifications as well as surface functionalization (Javed et al. 2023). Additionally, it delves into the recent development of different functionalized biochar materials, highlighting the introduction of heteroatom doping and metal supporting in the context of CO_2 capture (Nagarajan et al. 2022; Sun et al. 2024), aiming to elucidate the properties and practical potential of these materials. As a sustainable, environmentally friendly, and cost-efficient solution, biomass-derived carbon materials hold considerable promise for reducing CO_2 emissions and mitigating global climate change. Their adaptability across diverse industrial applications underscores their pivotal role in advancing next-generation CO_2 capture technologies. The insights provided in this chapter are expected to help guide future research and industrial applications in this field (Zou et al. 2023; Zhang et al. 2022b).

12.2 Biomass-based Carbon Materials

12.2.1 Preparation Techniques for Biomass-based Carbon Materials

12.2.1.1 Hydrothermal Carbonization

Hydrothermal carbonization (HTC) is an emerging technique designed to transform biomass into carbon-rich solids (i.e., hydrochar) under high-temperature (180–250 °C) and moderate-pressure aqueous conditions (0.5–2 MPa) (Liu et al. 2019). It encompasses a series of chemical reactions, including hydrolysis, dehydration, condensation, and polymerization, which work in tandem to convert biomass into carbonaceous materials. Specifically, polysaccharides in biomass are degraded into monosaccharides, which subsequently dehydrate at high temperatures to form intermediates like furfural, eventually yielding carbonaceous solids (Zhang et al. 2020). HTC not only minimizes secondary pollution but is also compatible with diverse biomass feedstocks, such as lignin, cellulose, and agricultural residues, highlighting its versatility (Cao et al. 2019). Furthermore, HTC is characterized by its energy efficiency, with low energy requirements, making it an economically viable technology for biomass conversion (Sun et al. 2022).

The carbonaceous solid yield from HTC typically ranges from 30 to 70%, depending on factors such as biomass characteristics, pretreatment methods, operational temperature, residence time, reactor design, water-to-biomass ratio, and pH levels, etc. (Xiao et al. 2020). Hydrochar exhibits poor performance in CO_2 capture due to their low specific surface area even has abundant surface functional groups. Further treatments such as chemical activation, doping, and surface modification can significantly enhance the CO_2 capture performance of hydrochar. For instance, doping with elements such as nitrogen or phosphorus introduces additional active sites, thereby improving its CO_2 adsorption capacity (Cao et al. 2019).

12.2.1.2 Pyrolytic Carbonization

Pyrolytic carbonization, a thermochemical decomposition process, converts biomass into biochar, bio-oil, and syngas under an oxygen-free environment at temperatures ranging from 300 to 900 °C. During pyrolysis, the structural components of biomass, including lignocellulose, undergo depolymerization, forming volatile intermediates that subsequently recombine to yield carbon-rich biochar, bio-oil, and syngas (Figure 12.1a) (Fan et al. 2024). The chemical changes in pyrolytic

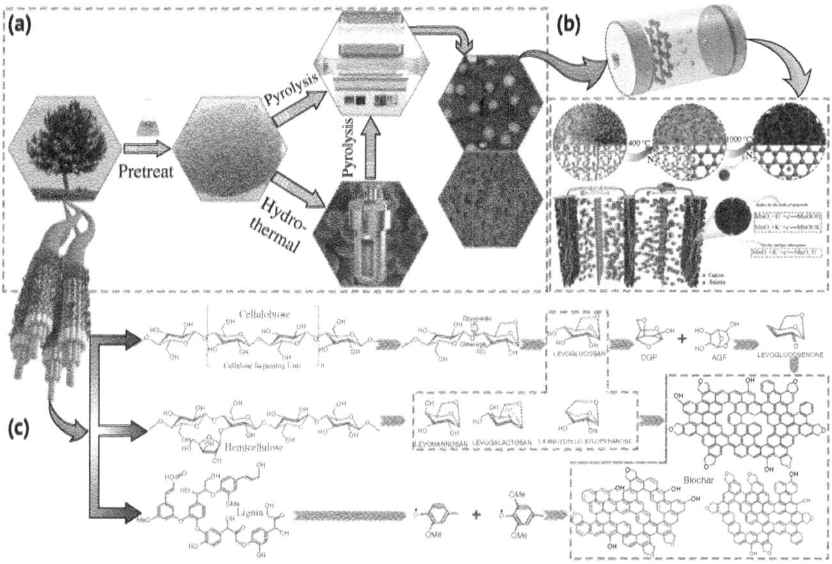

Figure 12.1 The preparation process of biomass-based carbon materials and the carbonization mechanism of cellulose, hemicellulose, cellulose, and lignin. (a) Preparation of biochar materials. (b) The application of biochar materials as functional materials. (c) The molecular degradation mechanism of biomass during the the biochar formation process. *Source:* Fan et al. 2024/with permission of Elsevier.

carbonization can be categorized into four stages: below 120 °C, evaporation of moisture in the biomass; between 220 and 315 °C, carbonization of hemicellulose; between 315 and 400 °C, carbonization and degradation of cellulose; above 400 °C, decomposition of lignin, accompanied by extensive carbonization (Wang et al. 2023b). These temperature-dependent stages critically influence the physicochemical properties of the resultant biochar, such as porosity, surface area, and functional group composition, making pyrolysis a flexible method for tailoring biochar characteristics to specific applications.

The yield and composition of pyrolytic products are influenced by multiple factors, including feedstock type, particle size, pyrolysis rate, and operating temperature (Patel et al. 2024). Pyrolysis processes can be broadly categorized into fast and slow pyrolysis based on heating rates. Fast pyrolysis, characterized by rapid heating, is optimized for bio-oil production, whereas slow pyrolysis, with gradual heating, is more conducive to higher biochar yields (Figure 12.1b). The physicochemical properties of biochar are strongly dependent on the composition of the biomass feedstock (Figure 12.1c). Critical determinants include the lignin content, cellulose decomposition behavior, and ash composition. Feedstocks with high lignin content typically produce biochar with superior yields, enhanced porosity, and larger specific surface areas due to the lignin's resistance to thermal decomposition and its structural complexity (Wang et al. 2023a).

By precisely controlling pyrolysis conditions, carbon materials with tailored pore distributions and properties can be synthesized (Zhu et al. 2020). For instance, rapid heating tends to generate three-dimensional carbon architectures, whereas slower heating promotes the formation of highly porous structures with extensive surface areas (Song et al. 2022). Additionally, the carbonization temperature is a decisive factor in influencing the degree of graphitization and nitrogen retention in the final carbon materials. High-temperature pyrolysis, typically above 700 °C, facilitates the transformation of amorphous carbon into ordered graphitic structures, enhancing the material's suitability for CO_2 adsorption. By carefully tuning pyrolysis parameters, such as heating rate and operating temperature, it is possible to optimize the composition, structure, and functionality of pyrolysis-derived products to meet specific application requirements (Zhu et al. 2023a). This adaptability highlights the potential of pyrolysis technologies in producing advanced carbon materials for CO_2 capture and other industrial uses.

12.2.2 Factors Influencing the Performance of Carbon Materials

12.2.2.1 Feedstock

Biomass-derived porous carbon materials have gained significant attention in adsorption applications due to their inherently rich micro- and mesoporous structures, diverse oxygen-containing surface functional groups (e.g., carboxyl, hydroxyl, carbonyl, alcohol, and lactone groups), and varying inorganic mineral components (He et al. 2022b). Compared to traditional petroleum-based coke, biomass-derived activated carbon exhibits highly tunable pore architectures and a higher abundance of oxygen-containing functional groups, offering distinct advantages such as high specific surface areas, well-developed porosity, and exceptional CO_2 adsorption capacities. Current research in biochar focus on developing materials that combine micro- and mesoporous architectures, nitrogen- or oxygen-enriched functional groups, and high carbon content (Yan et al. 2023).

Biomass feedstocks for carbon material production are diverse and widely available, including wood and its residues (Tian et al. 2021), agricultural by-products (Rao et al. 2019), livestock waste, and food scraps, etc. (Zhang et al. 2023). Widely utilized biomass raw materials comprise straw, nutshells, fruit peels, antibiotic fermentation residues, bacterial cellulose, egg whites, edible fungi, and even human hair, etc. (Zhao et al. 2024; Zhu et al. 2024; Huang et al. 2022; Zhou et al. 2021). Biomass precursors with high carbon content and low ash content are particularly advantageous for biochar production, as they contribute to enhanced porosity, specific surface areas, and functional group diversity in the resultant carbon materials. The wide availability and renewable nature of these feedstocks further underscore their potential for sustainable and scalable production of high-performance CO_2 adsorbents.

Based on their sources, biomass feedstocks can be broadly categorized into three groups: plant-based, animal-based, and microorganism-based (Zhang et al. 2023). Plant-based biomass, including roots, stems, branches, leaves, and nutshells, is rich in cellulose and was among the earliest raw materials utilized for preparing biomass-derived porous carbon. These feedstocks can be further divided into two major subcategories: agricultural waste (such as crop straw, which accounts for the largest proportion of plant-based biomass) and woody biomass (such as by-products from industrial pulping processes). Plant-based feedstocks often face limitations such as low thermal stability and low carbon yields due to the dominance of cellulose and hemicellulose. Feedstocks with higher lignin content, such as wood-based materials, are therefore more suitable for applications requiring higher carbon yields and thermal stability (Sasono et al. 2023). Animal-derived biomass includes bones, shells, skins, hair, horns, and hooves (Wang et al. 2016). These materials are rich in structural proteins like chitin and keratin, which form strong hydrogen bonds between molecules. As a result, they exhibit relatively high thermal stability and higher carbon yields compared to most plant-based feedstocks. These characteristics make animal-based biomass an excellent option for biochar production, particularly in applications requiring durable carbon materials. In recent years, microorganism-based biomass has garnered significant attention due to its rapid growth potential and resource abundance (Wang et al. 2016). Examples include bacterial cellulose and fungi, which are notable for their high porosity, exceptional tensile strength, and biodegradability. These properties render microorganism-based biomass particularly effective in biochar production, with promising potential for applications in advanced material synthesis and environmental remediation.

12.2.2.2 Temperature

Temperature is a critical factor influencing the structural and functional properties of biomass-based carbon materials. Precise temperature control during the preparation and application of these materials is essential to optimizing their adsorption efficiency (Li et al. 2024b).

At low temperatures (300–500 °C), carbon materials typically exhibit a higher proportion of amorphous regions, smaller pore sizes, and greater hydrophilicity (Hou et al. 2024). These characteristics can enhance interactions with polar substances but may limit the overall adsorption capacity for non-polar compounds. Additionally, the reduced thermal energy at low temperatures inhibits the formation of highly organized structures, resulting in less stability in the material's framework.

As the temperature rises (600–800 °C), the carbon material undergoes structural stabilization, accompanied by the development of multiple surface functional groups that enhance its cation exchange capacity. Simultaneously, higher temperatures promote the formation of larger pore sizes, increased specific surface areas, and greater pore volumes, which collectively boost the material's adsorption performance (Wang et al. 2024). Enhanced hydrophobicity at elevated temperatures further improves the material's suitability for non-polar contaminant adsorption, broadening its application scope.

At excessively high temperatures (>800 °C), the balance between pore formation and pore collapse is disrupted. The rate of pore collapse may exceed pore formation, leading to a reduction in specific surface area and diminished adsorption capacity. Furthermore, extreme temperatures can result in the loss of surface functional groups, compromising the material's ability to engage in chemical interactions (Petrović et al. 2024). Thus, maintaining an optimal temperature range during the synthesis and application of carbon materials is essential for achieving maximum adsorption performance.

12.2.2.3 Reaction Time

Reaction time is a critical parameter in the preparation of biomass-based carbon materials, as it directly affects the carbonization process, product properties, and energy efficiency (Acemioğlu 2019). Precise control of reaction time not only ensures the attainment of desired carbonization levels but also minimizes unnecessary energy consumption. In HTC, extending the reaction time typically increases the relative carbon content while reducing the oxygen content of the prepared hydrochar. These compositional changes directly affect the dispersion, aggregation, and particle size of the hydrochar (Sun et al. 2023a), which in turn influences its adsorption and catalytic performance (Chagas et al. 2020). However, excessively long reaction times may lead to particle aggregation or unwanted side reactions, underscoring the need for optimized time control to balance compositional and structural properties. For pyrolytic carbonization, the reaction time is equally critical. Studies indicate that shorter reaction times yield lower amounts of biochar due to insufficient thermal decomposition of the biomass (Li et al. 2020c). To achieve optimal carbonization, slower heating rates combined with extended gas retention times are recommended. These conditions facilitate the formation of stable carbon structures, enhance pore development, and improve adsorption capacities of the final product. Thus, optimizing reaction time in pyrolytic carbonization ensures better material performance while balancing yield and energy consumption.

12.2.2.4 Heating Rate

Heating rate, defined as the speed at which the temperature of biomass feedstock increases during carbonization or activation, is a critical parameter influencing the performance of biomass-based carbon materials. It directly impacts heat and mass transfer processes within biomass particles, affecting the structure and functionality of the resulting carbon materials (Leng et al. 2021).

At low heating rates, the reaction progresses gradually, allowing sufficient time for uniform heat distribution and carbonization. This controlled process often results in carbon materials with a more stable structure and enhanced thermal stability (Huang et al. 2022a). Additionally, the slower release of volatile components reduces structural stress, leading to a well-preserved pore architecture. Such characteristics make low-heating-rate carbon materials particularly suited for applications requiring durability and consistent adsorption performance. In contrast, high heating rates markedly enhance reaction kinetics, accelerating heat and mass transfer within biomass particles. Rapid heating promotes the formation of richer microporous structures and higher specific surface areas, significantly boosting the CO_2 adsorption performance of carbon materials (Wu et al. 2025). Furthermore, faster heating facilitates the swift removal of volatile substances, reducing the overall pyrolysis time and contributing to the development of additional pore networks (Amer et al. 2022). These advantages make high-heating-rate processes suitable for producing activated carbon materials optimized for CO_2 capture and similar applications. However, excessively high heating rates can have detrimental effects. For example, the rapid temperature rise may cause biochar particles to melt partially, resulting in smoother surfaces and reduced pore formation. Additionally, incomplete pyrolysis under such conditions can lead to the development of unstable or damaged pore structures, thereby compromising the material's adsorption capacity. These limitations highlight the importance of controlling heating rates to balance pore development and material stability. By carefully optimizing the heating rate, it is possible to produce carbon materials with well-developed microporous structures, high specific surface areas, and superior adsorption properties. Such materials can effectively meet the demands of CO_2 capture and other advanced adsorption applications.

12.3 Activation Methods for Carbon Materials

12.3.1 Physical Treatments

Activation is a critical step in developing the pore structure of carbon materials during the carbonization process. Physical activation usually involves two stages: carbonization and activation (He et al. 2022c). Organic precursors are first carbonized at temperatures ranging from 300 to 500 °C in an inert atmosphere. During this process, unstable chemical bonds are broken, and volatile substances (e.g., gases and tars) are released from the precursor, forming an initial porous structure. In the activation process, the biochar is activated at high temperatures between 600 and 900 °C using activating agents such as steam or CO_2. During this stage, the carbonized material undergoes partial gasification, further developing its pore structure and significantly increasing its specific surface area. Initially, tar-like products are removed, existing pores are widened, and new pores are formed. However, with prolonged activation, excessive gasification may reduce both the specific surface area and pore volume. By carefully controlling activation conditions, carbon materials with optimal pore structures and surface areas can be produced to meet specific application needs.

12.3.1.1 Steam Activation

Steam activation is a widely used physical activation method that employs steam, or a mixture of steam and CO_2, for high-temperature gasification of char, typically at temperatures exceeding 800 °C (Li and Guo 2025). Steam acts as a mild oxidizing agent (operating within the range of 800–950 °C), selectively reacting with carbon atoms in the char to form CO and/or a mixture of CO and H_2. This reaction enhances pore development by increasing the porosity of the carbon material. The degree of char combustion, or weight loss, during steam activation is influenced by parameters such as temperature, residence time, and steam concentration. The high reactivity and efficiency of steam activation make it a preferred method for enhancing the pore structure of carbon materials. By carefully controlling activation parameters, this method can produce materials with well-developed pore networks.

12.3.1.2 Air Activation

Thermal Air Activation (TAA) is an effective method for enhancing the pore structure and specific surface area of biochar. This process uses air as an oxidizing agent at controlled temperatures to selectively burn carbon and optimize porosity. In the initial carbonization stage, feedstocks were carbonized at temperatures ranging from 300 to 800 °C, grouped into low (300–400 °C), medium (500–600 °C), and high (700–800 °C) temperature categories (Zeng et al. 2024). Higher activation temperatures can significantly enhance surface area development within optimal time frames. TAA also exhibited notable capabilities in enhancing microporosity and generating new pores. Activation at 500 °C generally achieve superior porosity characteristics compared to the 400 °C experiments. However, prolonged activation times introduced challenges, such as pore wall disruption and the merging of micropores into larger pores, which reduced specific surface area. This highlights a critical trend also observed in steam activation: porosity increases initially with greater combustion but decreases beyond optimal activation times.

TAA is a versatile and efficient method for producing carbon materials with well-developed pore structures. By balancing carbonization and activation parameters, such as temperature and time, TAA can achieve superior specific surface areas and pore volumes for applications in CO_2 capture.

12.3.1.3 CO_2 Activation

The core reaction of the CO_2 activation process is represented by "$CO_2 + C \rightarrow 2CO$." During this process, CO_2 molecules are chemically adsorbed onto active sites of free carbon, then decomposed on the carbon surface to form active carbon–oxygen complexes (C(O)) and carbon monoxide (CO) (Ogungbenro et al. 2018). Subsequently, the surface oxides migrate from the solid phase to the gas phase, removing carbon atoms from the carbon structure and generating new vacant active sites for the following chemical adsorption of CO_2 molecules. Under carbonization conditions of 800 °C, modified biochar exhibits strong CO_2 adsorption capacities of 141.14 mg/g (20 °C, 101 kPa) and 129.23 mg/g (25 °C, 101 kPa) (Shafawi et al. 2021; Dissanayake et al. 2020). The CO_2 activation process involves the following key reactions:

$$C + H_2O \rightarrow C(O) + H_2 \tag{12.1}$$

$$2C + H_2 \rightarrow 2C(H) \tag{12.2}$$

$$CO + H_2O \rightarrow CO_2 + H_2 \tag{12.3}$$

$$C + CO_2 \rightarrow 2CO \tag{12.4}$$

$$C + 2H_2 \rightarrow CH_4 \tag{12.5}$$

$$CH_4 + H_2O \rightarrow CO + 3H_2 \tag{12.6}$$

In this process, CO_2 adsorbed on the carbon surface forms surface oxides, which are subsequently displaced by active free carbon centers, the resulting oxides are released as CO. Simultaneously, the generated H_2 reacts with carbon on the carbon material's surface to form surface hydrogen complexes (C(H)). Additionally, the interaction between steam and carbon further enhances the activation process. The generated CO_2 and H_2 react with the carbon surface, thus activating the carbon surface, increasing porosity, and creating additional active sites. This dual mechanism of chemical reactions and structural modifications makes CO_2 activation a potent method for enhancing the porosity and functional active site density of biochar, significantly improving its performance in adsorption and other applications.

12.3.2 Chemical Treatments

12.3.2.1 Acidic Activation

Acidic activation is an effective method for chemically modifying carbon materials. This process typically involves treating biomass-derived porous carbon materials with acidic solutions to introduce oxygen-containing functional groups, optimize pore structures, and ultimately achieve efficient CO_2 capture (Horri et al. 2020). Research demonstrates that acidic activator represented by phosphoric acid (H_3PO_4) is an outstanding acidic activator. For instance, Wang et al. (2020) used phosphoric acid to activate montmorillonite-derived biochar, achieving a specific surface area of 1558 m^2/g. Similarly, Ismail et al. (2022) applied phosphoric acid at varying concentrations to bamboo-derived activated carbon, resulting in BET surface areas exceeding 1000 m^2/g. Under conditions of 25 °C and 1 bar, the maximum CO_2 adsorption capacity reached 1.45 mmol/g. These findings indicate that phosphoric acid activation is a feasible approach for converting bamboo waste and other biomass into highly efficient CO_2 adsorbents. In addition to phosphoric acid, other acidic activators such as sulfuric acid and nitric acid can also enhance carbon materials by increasing specific surface area, adjusting pore size distribution, and introducing new functional groups. However, with the continuous enhancement of environmental awareness, future research on acidic activation is expected to shift toward greener modifications. For example, the development of bio-based acids, recyclable acids, and other environmentally friendly activators not only improves material performance but also reduces environmental impacts, supporting achievements in sustainable development (Kamran and Park 2020).

12.3.2.2 Alkaline Activation

Alkaline activation, like acidic activation, facilitates pore formation and increases the specific surface area of carbon materials. However, alkaline activation could introduce additional basic groups onto the biochar surface, enhancing its alkalinity, which is particularly advantageous for the adsorption of acidic gases such as CO_2 (Wu and Bao 2023). Among the commonly used alkaline activators, potassium hydroxide (KOH) and sodium hydroxide (NaOH) are the most prevalent. When using KOH as an activator, the process typically involves thoroughly mixing KOH with carbon samples followed by thermal treatment. The KOH activation process is particularly effective for forming micropores, significantly enhancing the CO_2 adsorption capacity of biochar. Under high-temperature conditions, KOH reacts with carbon materials through the following chemical reactions:

$$6KOH + 2C \rightarrow 2K + 3H_2 + 2K_2CO_3 \tag{12.7}$$

$$K_2CO_3 \rightarrow K_2O + CO_2 \tag{12.8}$$

$$K_2O + C \rightarrow 2K + CO \tag{12.9}$$

These reactions not only aid in generating more micropores but also promote the escape of gases, further expanding the pore structure. Additionally, the decomposition products of KOH, such as metallic potassium, can embed into the micropores of the carbon material, enhancing its basic characteristics and adsorption capacity. Alkaline activation using KOH is highly effective for applications that require materials with enhanced CO_2 adsorption properties. The ability to manipulate pore structures and introduce functional groups makes KOH an optimal choice for biochar modification. Compared to KOH, NaOH exhibits similar activation effects but is less corrosive to biochar. This results in materials with slightly smaller microporous surface areas. However, NaOH is safer to handle, making it suitable for processing biochar for certain specific applications (Zhang et al. 2024a). At temperatures above 700 °C, potassium (K) integrates into the carbon structure, further expanding porosity and enhancing the alkalinity of biochar (Amer et al. 2022). For example, Li et al. (2022) used KOH to activate biochar derived from a combination of pinewood and sludge. This process increased the specific surface area of the biochar from an initial 31.6 m^2/g to an impressive 2623 m^2/g, with a maximum CO_2 adsorption capacity of 177.1 mg/g. Similarly, NaOH as an activator produces effects akin to KOH but with lower corrosiveness, resulting in biochar with relatively smaller microporous surface areas (Guo et al. 2022). Ji et al. (2023) prepared biochar from coconut shells and activated it using both KOH and NaOH. The KOH-activated biochar achieved a specific surface area of 1701.39 m^2/g and a CO_2 adsorption capacity of 2.11 mmol/g, significantly outperforming the NaOH-activated biochar. In contrast to KOH and NaOH, which improve CO_2 adsorption primarily by etching the carbon structure to form pores, ammonia gas (NH_3) represents a different class of alkaline activators. NH_3 enhances the interaction with CO_2 by introducing nitrogen-containing functional groups onto the carbon surface (Zhou et al. 2022). Common nitrogen-containing functional groups include pyrrolic-N (N-5), pyridinic-N (N-6), oxidized-N (N-O), and graphitic-N (N-Q). These groups increase the basicity of the carbon material, substantially boosting its adsorption capacity for acidic gases like CO_2 (Xie et al. 2022).

12.3.2.3 Salt Activation

Salts play a crucial role in the activation of carbon materials by regulating porosity and surface chemical characteristics, thereby significantly enhancing the biochar's CO_2 adsorption capacity. Different types of salts as activators yield varying effects. Due to its versatility, cost-effectiveness, and ability to improve the CO_2 adsorption capacity of carbon materials, salt activation has garnered significant attention (Shen 2023). Salt activation is generally performed using two main approaches: Impregnation Method, the biochar or precursor is soaked in a salt solution. Direct Mixing Method, the salt is evenly mixed with the precursor before thermal activation. The remarkable effects of salt activation are attributed to its ability to control hierarchical pore structures, adjust surface functional groups, and introduce alkali metal elements during the activation process (Zhu et al. 2023b). For instance, Zhu et al. (2023b) used cotton lint as a feedstock and potassium citrate ($K_3C_6H_5O_7$) as the activator to successfully produce biochar with excellent CO_2 adsorption capabilities. Their study revealed that the activated biochar achieved a maximum CO_2 adsorption capacity of 4.19 mmol/g, demonstrating the potential of salt activation in optimizing the performance of biochar for CO_2 capture.

12.3.2.4 Heteroatom Doping

Heteroatom doping involves introducing non-carbon atoms (such as nitrogen, sulfur, or phosphorus) into biomass-derived carbon materials to improve their adsorption performance and chemical activity (Luo et al. 2021). Doping elements can effectively alter the electronic structure and surface chemical properties of carbon materials, thereby significantly enhancing their CO_2 capture capability.

Nitrogen Doping: This is achieved by reacting nitrogen sources (e.g., urea, ammonia) with biochar at high temperatures. Nitrogen doping forms various nitrogen-containing functional groups, such as pyridinic nitrogen and graphitic nitrogen (Luo et al. 2021). These groups increase the basicity and electron density of carbon materials, enhancing their CO_2 adsorption capacity (He et al. 2022a). Sulfur Doping: Sulfur doping uses sulfur-containing compounds (e.g., thiourea, sulfuric acid) to introduce sulfur atoms into carbon materials. Sulfur can form functional groups such as thioethers and thiophenes (Li et al. 2021). These functional groups enhance the chemical activity of carbon materials, increase adsorption sites, and strengthen their CO_2 adsorption performance (Matsagar et al. 2021). Phosphorus

Doping: Phosphorus doping is performed by treating carbon materials with phosphoric acid or phosphate salts. This introduces phosphorus-containing functional groups, such as phosphate esters and phosphoryl groups, onto the carbon surface (Li et al. 2020b). Phosphorus doping enhances the surface polarity and chemical activity of the carbon material, further improving its CO_2 capture efficiency (Yang et al. 2020). Heteroatom doping provides a powerful strategy for optimizing the performance of biomass-derived carbon materials. Its flexibility and efficiency make it a promising approach for CO_2 capture and other functional applications, offering broad prospects for advancing carbon materials in environmental and industrial contexts.

12.3.2.5 Metal Impregnation

Metal impregnation is a modification method where metal ions (e.g., potassium, sodium, calcium, iron) are introduced into biomass-derived carbon materials through impregnation techniques (Li et al. 2020a). These metal ions form metal oxides or metal particles within the carbon material's pores, significantly enhancing its surface alkalinity and reactivity, thereby improving its CO_2 adsorption efficiency. Research has shown that metals such as potassium, sodium, calcium, and iron effectively enhance the adsorption performance of carbon materials. Potassium and sodium can increase the surface alkaline sites of carbon materials, facilitating chemical adsorption reactions with CO_2 and thus boosting adsorption efficiency (Zhang et al. 2022a). Calcium-based materials, known for their high alkalinity and reactivity, are widely applied in CO_2 capture. Under high-temperature conditions, calcium reacts with CO_2 to form stable calcium carbonate ($CaCO_3$), significantly enhancing capture capacity (Wei et al. 2018). Iron impregnation forms zero-valent iron or iron oxides, which not only improve the pore structure of carbon materials but also enhance CO_2 adsorption through chemical interactions (Sekhon and Park 2021). Metal impregnation serves as an effective modification approach, providing critical support for the application of biomass-derived carbon materials in CO_2 capture. By reasonably selecting metal types and optimizing impregnation conditions, the performance and adsorption efficiency of carbon materials can be further enhanced (Tang et al. 2019; Hao et al. 2021).

12.4 The Recent Advances of Functionalized Biochar Materials for CO_2 Capture

Currently, researchers are exploring new preparation methods and functionalization strategies to further improve the adsorption performance, cyclic stability, and environmental compatibility of functionalized biochar materials. These studies provide critical technical support for large-scale CO_2 capture and reduction,

Table 12.1 The reported work on the application of biomass derived porous carbon in CO_2 capture.

Carbon precursor	CO_2 uptake (mmol/g)		CO_2/N_2 Selectivity (1 bar)	Stability (adsorption capacity after 5th runs)	References
	25 °C	0 °C			
Cigarette butts	4.93	6.83	23.4	95.33%	Liu et al. (2025)
Rice husk	2.1	2.9	3.4	—	He et al. (2021)
Pineapple	4.25	5.32	18.2	87.4%	Zhu et al. (2019)
Coconut shells	3.59	5.31	—	—	Bai et al. (2024)
Chitosan	3.26	5.73	12.5	94.9%	Chen et al. (2023)
Pomelo peel and ZIF-8	5.51	9.01	—	85.30%	Gai et al. (2021)
Cupuassu shell	4.4	7.8	—	—	Cruz et al. 2022)
Peanut shells	3.76	5.7	—	95.74%	Xu et al. (2024)
Birch	4.22	6.00	—	—	Serafin et al. (2023)
Corn cobs	6.43	4.57	—	—	Serafin et al. (2023)
Antibiotic fermentation residues	4.38	6.4	32	96%	Liu et al. (2023)
Chitosan foam	3.59	—	90.3	98.61%	Zhang et al. (2024b)

offering sustainable solutions for achieving carbon neutrality (Table 12.1). The superior performance of functionalized biochar materials can be attributed to the following factors: (1) high specific surface area and developed porous structure: These properties provide ample space for the adsorption and diffusion of CO_2 molecules; (2) introduction of surface functional groups: Polar groups such as carboxyl (–COOH), hydroxyl (–OH), and amino ($-NH_2$) significantly enhance the interaction between the material and CO_2 molecules; (3) enhanced active sites: Doping with metals or non-metal elements (e.g., nitrogen, sulfur, and phosphorus) increases the number of active sites on the surface, improving the capture of acidic gases such as CO_2.

12.4.1 Mechanisms for CO_2 Capture

CO_2 capture commonly employs adsorption methods, a technique that selectively enriches CO_2 molecules on the surface of specific materials via physical or chemical adsorption. Adsorption relies on the affinity of adsorbents for CO_2, enabling efficient separation and recovery of CO_2 molecules (Wei et al. 2018). Depending on the adsorption mechanism, adsorbents can be regenerated through various processes to release the captured CO_2, facilitating their cyclic use. Physical adsorption depends on weak van der Waals forces (Zhang et al. 2022a). This process typically does not involve the formation of chemical bonds, allowing for rapid adsorption and desorption cycles. It is most effective under low-temperature conditions for CO_2 capture. Chemical adsorption involves the formation of strong bonds between specific functional groups on the adsorbent surface and CO_2 molecules (Zhang et al. 2022a). This process is temperature-sensitive and is ideal for high-selectivity and high-capacity CO_2 capture. Many solid adsorbents for CO_2 capture leverage a combination of physical and chemical adsorption mechanisms. CO_2 molecules interact with the adsorbent surface via van der Waals forces, enabling initial adherence. Functional groups on the adsorbent surface interact chemically with CO_2 molecules, forming stronger bonds that enhance adsorption capacity and selectivity (Sekhon and Park 2021). The synergistic effect of these dual mechanisms makes the CO_2 adsorption process more intricate while significantly improving efficiency and selectivity.

12.4.1.1 Physical Adsorption

Physical adsorption is a process in which adsorbate molecules are bound to the surface of an adsorbent through physical forces such as dipole–dipole interactions, electrostatic effects, nonpolar forces, hydrophobic interactions, and van der Waals forces (Ho et al. 2014). The binding energy for this process typically ranges from 8 to 41 kJ/mol, indicating relatively weak interactions. Due to these weaker forces, physical adsorption exhibits certain limitations in stability and selectivity, and its performance is highly sensitive to changes in environmental temperature and pressure. Despite these limitations, physical adsorption offers significant advantages in scenarios requiring frequent regeneration. Since the desorption process is relatively simple—often achieved through degassing or pressure reduction—this method is well-suited for high-throughput applications. For instance, physical adsorption is widely used in the preliminary separation of CO_2 from industrial waste gases or flue gases, where cost-effectiveness and high processing capacity are prioritized.

12.4.1.2 Chemical Adsorption

Chemical adsorption involves the formation of strong chemical bonds, such as covalent, ionic, or metallic bonds, between the adsorbate and the adsorbent surface (Goel et al. 2021). The binding energy for chemical adsorption is significantly higher than that of physical adsorption, typically ranging from 60 to 418 kJ/mol. This strong bonding imparts greater stability and selectivity to chemical adsorption

(Su et al. 2022). However, the strong binding forces in chemical adsorption make the regeneration process more complex, often requiring higher energy inputs or chemical reactions to release the adsorbed molecules. This limitation makes chemical adsorption less suitable for applications requiring frequent regeneration. Nevertheless, chemical adsorption excels in scenarios where permanent fixation or highly efficient capture of target molecules is required, such as for CO_2 sequestration or capturing specific hazardous gases. By leveraging the complementary mechanisms of physical and chemical adsorption, appropriate adsorption methods can be chosen based on specific application needs, enabling efficient and cost-effective CO_2 capture solutions.

12.4.2 Carbon Materials for CO_2 Capture

In this section, four types of carbon materials commonly used for CO_2 capture will be discussed: raw carbon materials (non-activated carbon), activated carbon materials, functionalized carbon materials, and heteroatom-doped carbon materials. These materials have distinct characteristics in terms of pore structure, surface chemistry, and adsorption performance, providing multiple pathways to achieve efficient CO_2 capture.

12.4.2.1 Raw Carbon Materials (Non-activated Carbon)

Raw carbon materials are primarily derived from biomass waste, coal, coke, and other carbon-rich precursors. These materials are typically prepared through thermochemical processes like high-temperature carbonization. While raw carbon materials exhibit some capacity for CO_2 adsorption, their performance is limited due to a lack of surface functional groups and relatively simple pore structures, which are dominated by macropores and mesopores with fewer micropores (Goel et al. 2021). This limitation makes them less suitable for high-efficiency CO_2 capture. Huang et al. (2022b) investigated the application of biochar prepared from sludge and *Leucaena leucocephala* wood at varying ratios through microwave co-pyrolysis for CO_2 capture. The study revealed that the CO_2 adsorption capacity of biochar was significantly influenced by the mixing ratio and microwave power. Especially, biochar derived from *Leucaena* wood demonstrated the highest CO_2 adsorption capacities at microwave powers of 200 W (61.66 mg/g) and 250 W (61.97 mg/g), respectively, which are significantly higher than the absorption capacity of pure sludge biochar (13.47 and 13.26 mg/g). Additionally, the study indicated that adjusting the sludge-to-wood mixing ratio (e.g., 50 sludge and 50% wood) can achieve a balance between CO_2 adsorption performance and biochar yield. Biochar prepared with this optimal ratio exhibited strong durability during 20 adsorption–desorption cycles, with CO_2 adsorption capacity decreasing by only 2–10%. These findings highlight that by precisely controlling carbonization process parameters, the adsorption performance of biochar can be optimized. However, raw carbon materials still require further functionalization or activation to meet the demands of high-efficiency CO_2 capture. Igalavithana et al. (2020) compared the performance of biochar derived from pine sawdust and paper mill sludge for CO_2 capture. Their results demonstrated that biochar prepared from pine sawdust (PS550 and PSS550) significantly outperformed biochar derived from paper mill sludge (PMS300 and PMS600) in terms of CO_2 adsorption capacity. Pine sawdust-derived biochar exhibited a higher specific surface area (581.74 m^2/g) and an optimized micropore structure, with micropore diameters concentrated in the range of 0.3–0.8 nm, which is ideal for CO_2 physical adsorption. Additionally, steam activation of pine sawdust biochar (PSS550) increased the CO_2/N_2 selectivity from 18.28 to 26.70. This improvement was attributed to an increase in micropore volume and a higher proportion of oxygen-containing functional groups (e.g., hydroxyl and carbonyl groups) introduced during the activation process. These changes enhanced the electrostatic interactions and hydrogen bonding between the biochar and CO_2 molecules. In contrast, biochar derived from paper mill sludge exhibited lower CO_2 adsorption capacity. This was mainly due to its lower specific surface area and the presence of acidic surface functional groups (e.g., C—OH and O=C—O), which inhibited effective CO_2 adsorption. Although high-temperature pyrolysis (PMS600) increased the micropore volume and slightly improved adsorption performance, it is still short in the capabilities of pine sawdust-derived biochar. The study concluded that pine sawdust biochar, with its superior pore structure and surface chemistry, is a cost-effective and efficient material for CO_2 adsorption. Steam activation further enhanced its performance and selectivity, making it a promising candidate for industrial CO_2 capture applications. These findings underscore the importance of feedstock selection and activation treatment in optimizing biochar performance, suggesting avenues for exploring its large-scale application potential in the future.

12.4.2.2 Activated Carbon

Activated carbon exhibits outstanding performance in CO_2 capture due to its large specific surface area and well-developed porous structure. The activation process removes residual volatile substances and further expands existing pores, optimizing micropore distribution (Hidayu and Muda 2016). As a result, activated carbon achieves enhanced specific surface area and adsorption activity. Gong and Bao (2023) investigated N,O co-doped porous carbon prepared from alkali lignin through pre-carbonization and chemical activation using KOH for CO_2 capture. The study revealed that the pre-carbonization temperature, the KOH-to-carbon ratio, and the final carbonization temperature significantly affected the material's pore structure and adsorption performance. Under optimal conditions (pre-carbonization at 450 °C, KOH/carbon ratio of 2, and final carbonization at 600 °C), the activated carbon displayed exceptional CO_2 adsorption capacities of 5.82 mmol/g at 25 °C and 3.98 mmol/g at 0 °C (Figure 12.2a). The superior performance of the material was attributed to its high specific surface area (2050 m^2/g), rich microporous structure (Figures 12.2b, c), and presence of N and O functional groups, which enhanced the material's physical adsorption and chemical affinity for CO_2. The study also demonstrated that the activated carbon material exhibited more pronounced CO_2 adsorption at lower temperatures and excellent dynamic adsorption performance. Furthermore, it achieved impressive gas selectivity with a CO_2/N_2 ratio of 16.27 (Figure 12.2d), making it a promising candidate for selective CO_2 capture under industrial conditions. Converting waste lignin into high-performance CO_2 adsorbents provides valuable insights for the development of economical, efficient, and environmentally friendly biomass-based adsorbent materials. Lignin, with its complex structure and high carbon content, serves as an ideal precursor for carbon materials, enabling the development of sustainable and

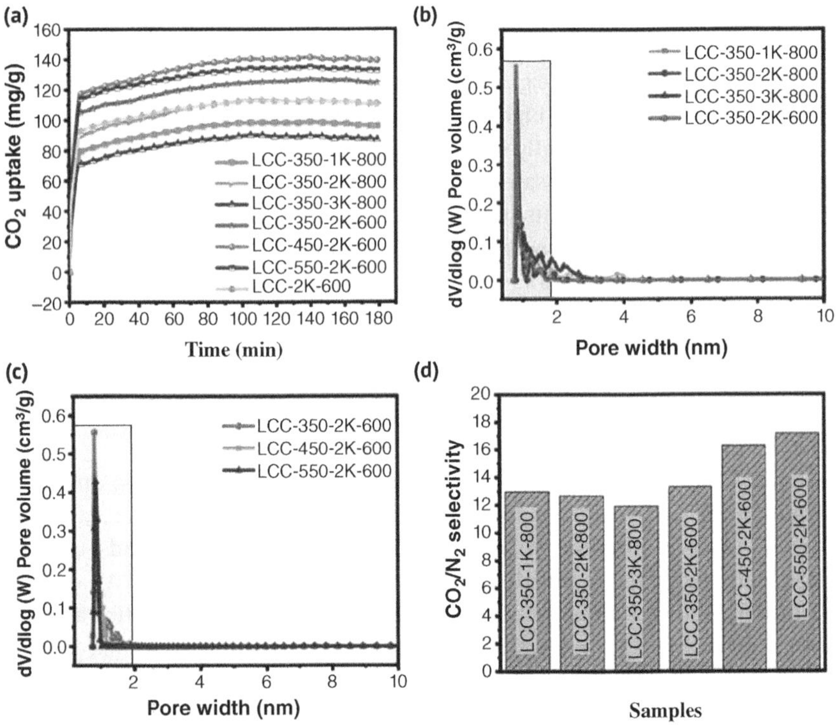

Figure 12.2 (a) CO_2 adsorption curve of porous carbon at 25 °C, (b, c) pore size distribution curve of porous carbon, and (d) IAST selectivity of the sample in CO_2/N_2 gas mixture (10:90, v/v) at 25 °C. *Source:* Gong and Bao 2023/with permission of Elsevier.

eco-friendly adsorbents. Detailed experimental work in the study described the preparation and structural characterization of lignin-derived porous carbon materials. Overall, this research offers a promising strategy for the effective utilization of lignin and the production of advanced carbon materials for CO_2 capture and storage. Ahmed et al. (2023) studied activated carbon (AC-W) derived from walnut shells, which exhibited exceptional performance in CO_2 adsorption. Their findings highlighted the decisive impact of preparation conditions on adsorption capacity. Activated carbon prepared with a 1 : 2 carbon-to-KOH ratio at 800 °C achieved a CO_2 adsorption capacity of 7.2 mmol/g under 1 atm, which increased to 18.2 mmol/g under 30 atm. These differences were attributed primarily to the specific surface area and active group characteristics of the activated carbon, which largely determine its adsorption efficiency. Optimization of preparation conditions can significantly enhance the adsorption performance of activated carbon, making it an effective solution for addressing CO_2 emissions. Activated carbon, with its excellent performance, shows immense potential as a high-efficiency adsorbent for capturing and processing CO_2 emissions from flue gases and industrial waste gases. However, some technical challenges remain for its practical application, such as further increasing adsorption capacity and selectivity.

Developing efficient regeneration and recycling processes for activated carbon. Addressing these issues will require ongoing efforts and technological innovations from researchers to unlock the full potential of activated carbon for industrial-scale CO_2 capture and environmental applications.

12.4.2.3 Heteroatom-doped Carbon Materials

The CO_2 adsorption performance of carbon materials is primarily influenced by pore size, porosity, and nitrogen content. Among these, nitrogen content is a critical factor for improving adsorption capacity. Studies have shown that heteroatom doping (e.g., nitrogen, sulfur, phosphorus) significantly enhances the adsorption performance of carbon materials by introducing new active sites and optimizing pore structures, thereby increasing their potential for CO_2 capture (Cui et al. 2022). The nitrogen-doped porous carbons with excellent microporous structures generally have enhanced CO_2 adsorption performance (Ge et al. 2022). Samples with higher nitrogen content demonstrate significantly improved CO_2 selectivity, particularly under low-concentration conditions, while their affinity for N_2 remains much lower than that for CO_2 (Burrow et al. 2022). Yuan et al. (2022) employed a two-step pyrolysis method using corn stalks as the carbon source, melamine as the nitrogen source, phytic acid as the phosphorus source, and K_2CO_3 as the activator to prepare N, P co-doped porous carbon (Figure 12.3). The optimal sample, NPCPB-600-3, achieved a CO_2 adsorption capacity of 3.11 mmol/g at 25 °C and 1 bar, significantly higher than the undoped material (0.75 mmol/g).

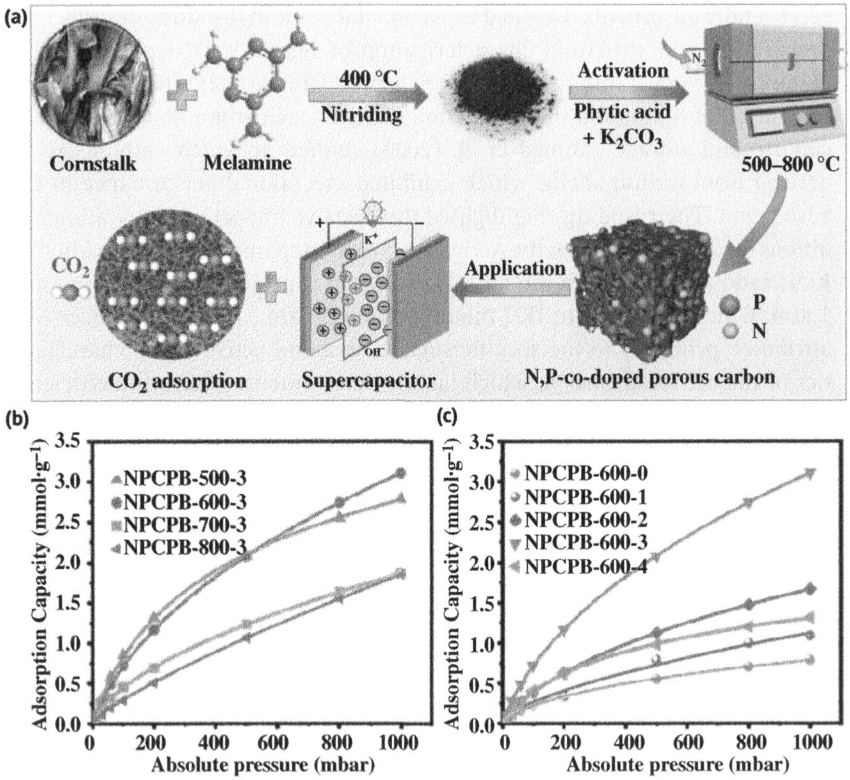

Figure 12.3 (a) Schematic diagram of two-step pyrolysis of N and P co-doped porous biochar prepared from corn stover, (b, c) CO_2 isotherm adsorption curves of NPCPB-T-3 (T = 500, 600, 700, 800 °C) and NPCPB-600-n (n = 0, 1, 2, 3, 4). *Source:* Yuan et al. 2022/with permission of Elsevier.

This improvement in CO_2 adsorption performance was attributed to the formation of abundant basic active sites, such as pyridinic-N and pyrrolic-N, as well as a high specific surface area (1136.31 m^2/g) and micropore volume (0.36 cm^3/g). Key features of heteroatom-doped carbon materials: enhanced basicity—doping with nitrogen, phosphorus, or other heteroatoms introduces basic functional groups that increase CO_2 adsorption sites. Nitrogen doping can also significantly improve CO_2 selectivity, particularly under low-pressure or low-concentration conditions. Sun et al. (2023b) investigated the physicochemical interactions between nitrogen and sulfur in carbon materials. Using HTC, they prepared heteroatom-doped carbon materials with urea, sodium thiosulfate, and thiourea as additives. The study found that nitrogen and sulfur doping independently improved the CO_2 adsorption performance of biochar by 11.9 and 8.5%, respectively. However, co-doping did not show a significant improvement, possibly due to partial competition between nitrogen and sulfur's mechanisms under certain conditions. Heteroatom doping enhances CO_2 adsorption capacity by introducing abundant active sites (e.g., pyridinic-N, pyrrolic-N) and optimizing pore structures. These modifications enable stronger interactions with CO_2 molecules while improving the material's selectivity and reactivity.

12.4.2.4 Metal-loaded Carbon Materials

Metal salt loading is an effective modification strategy to enhance the CO_2 adsorption capacity of biochar. Using wet impregnation methods to introduce metal salts (e.g., magnesium salts and lithium salts) into carbon materials can significantly increase surface alkalinity and chemical adsorption capabilities, thus improving CO_2 adsorption performance (Kim et al. 2023). Zubbri et al. (2020) investigated the impact of magnesium salt modification on the CO_2 adsorption performance of biochar. The study employed a wet impregnation method to introduce magnesium nitrate, magnesium sulfate, magnesium chloride, and magnesium acetate into biochar, followed by pyrolysis at 900 °C under a nitrogen atmosphere to produce metal-loaded carbon materials. The results showed that compared to raw biochar (68.74 mg/g), biochar modified with magnesium nitrate exhibited an enhanced adsorption capacity of 76.80 mg/g. This improvement was primarily attributed to the incorporation of magnesium oxides, which significantly increased the alkalinity and surface chemical properties of the carbon material, thereby promoting chemical adsorption of CO_2. The modification effects of magnesium salts on CO_2 adsorption performance followed the order: magnesium nitrate > magnesium sulfate > magnesium chloride > magnesium acetate.

Li et al. (2024a) examined the impact of molten LiCl modification on the CO_2 adsorption performance of rice husk biochar. Their study revealed that biochar treated with molten LiCl and pyrolyzed (e.g., MRH700R7.5) achieved a CO_2 adsorption capacity of 2.70 mmol/g, representing a 36% improvement compared to unmodified biochar (1.98 mmol/g). This enhancement was primarily attributed to the etching effect of LiCl, which significantly increased the biochar's specific surface area, chemical adsorption oxygen content, and carbon defects, thereby optimizing its pore structure and surface functional groups. While some metal-modified carbon materials show significant CO_2 adsorption improvements, they often fall short of the ideal range for industrial applications (3–4 mmol/g). However, a too high salt/biomass ratio can bring additional operating costs and equipment corrosion risk. By optimizing pore structure, introducing active metal oxides, and adjusting surface chemical properties, metal loading provides an important technical means to enhance the CO_2 trapping capacity of biochar. While metal-modified carbon materials exhibit significantly enhanced adsorption performance, their industrial application remains limited due to these challenges.

Future efforts should focus on optimizing metal loading techniques to minimize costs and maximize adsorption efficiency, enabling scalable and practical solutions for CO_2 capture.

12.5 Conclusion and Outlook

Biomass-derived porous carbon, sourced from renewable materials such as agricultural waste, forestry residues, and food processing by-products, is a low-cost and environmentally friendly material for CO_2 adsorption. Significant advancements have been made in this research field, showcasing the potential of biomass-based carbon materials for sustainable CO_2 capture applications. At present, the primary methods for preparing biomass-based carbon materials include HTC and pyrolytic carbonization. Studies suggest that combining hydrothermal and pyrolytic processes can generate microporous structures with higher stability and larger specific surface areas than standalone pyrolysis. This enhances CO_2 adsorption performance and CO_2/N_2 selectivity. Biomass-derived porous carbon can be modified using physical activation (e.g., CO_2 or steam) and chemical activation (e.g., acidic or alkaline treatments, heteroatom doping, and metal impregnation). The CO_2 adsorption performance of porous carbon is influenced by the specific surface area, pore size distribution (developed microporous structures), and surface chemical properties. The surface activity of carbon materials can be further enhanced by heteroatom doping, such as the introduction of nitrogen-containing functional groups or alkaline sites. The introduction of metal oxides not only increases the polarity of carbon materials, but also boosts CO_2 affinity and adsorption efficiency.

With advancements in materials science and engineering, biomass-derived porous carbon materials hold great promise for breakthroughs in CO_2 capture and multifunctional applications. Their potential for integration into systems addressing climate change and resource sustainability makes them a vital area of research. By tackling the challenges of scalability, cost, and multifunctionality, these materials could provide critical technological support for achieving global sustainability and mitigating the impacts of climate change. Their large-scale production faces significant challenges. Future research must address the following key issues:

1. Scaling Up and Cost Control: Developing low-cost, efficient preparation processes to reduce reliance on expensive chemicals and energy-intensive steps. Optimizing the production techniques to make the scaling-up process economically viable.

2. Material Consistency and Performance Optimization: Ensuring the structural and performance uniformity of carbon materials during large-scale production. Improving process-control to maintain key properties such as specific surface area, pore structure, and adsorption performance.
3. Multifunctional Integrated Design: Combining pore structure optimization with surface functionalization to enhance CO_2 selectivity and adsorption capacity. Exploring integrating CO_2 capture with catalytic conversion, energy storage, or other functionalities to develop multifunctional materials to improve overall system efficiency and economic feasibility.

References

Acemioğlu, B. (2019). Removal of a reactive dye using NaOH-activated biochar prepared from peanut shell by pyrolysis process. *International Journal of Coal Preparation and Utilization* 42: 671–693.

Ahmed, A.S., Alsultan, M., Sabah, A.A. et al. (2023). Carbon dioxide adsorption by a high-surface-area activated charcoal. *Journal of Composites Science* 7: 050179.

Ahmed, A.T.A., Bathula, C., Soni, R. et al. (2022). Nanostructurally engineered TiO_2 embedded Mentha aquatica biowaste derived carbon for supercapacitor applications. *Chemosphere* 289: 133197.

Amer, N.M., Lahijani, P., Mohammadi, M. et al. (2022). Modification of biomass-derived biochar: a practical approach towards development of sustainable CO_2 adsorbent. *Biomass Conversion and Biorefinery* 14: 7401–7448.

Bai, J., Huang, J., Yu, Q. et al. (2023). Fabrication of coconut shell-derived porous carbons for CO_2 adsorption application. *Frontiers of Chemical Science and Engineering* 17: 1122–1130.

Bai, J., Shao, J., Yu, Q. et al. (2024). Sulfur-doped porous carbon adsorbent: a promising solution for effective and selective CO_2 capture. *Chemical Engineering Journal* 479: 147667.

Bernardo, M., Lapa, N., Fonseca, I. et al. (2021). Biomass valorization to produce porous carbons: applications in CO_2 capture and biogas upgrading to biomethane-a mini-review. *Frontiers in Energy Research* 9: 625188.

Burrow, J.N., Eichler, J.E., Wang, Y. et al. (2022). N-rich porous carbons with tunable affinity for CO_2 adsorption achieve size-sieving CO_2/N_2 selectivity in turbostratic interlayers. *Journal of Materials Chemistry A* 10: 24649–24661.

Cao, L., Yu, I.K.M., Cho, D.W. et al. (2019). Microwave-assisted low-temperature hydrothermal treatment of red seaweed (*Gracilaria lemaneiformis*) for production of levulinic acid and algae hydrochar. *Bioresource Technology* 273: 251–258.

Chagas, J.A.O., Crispim, G.O., Pinto, B.P. et al. (2020). Synthesis, characterization, and CO_2 uptake of adsorbents prepared by hydrothermal carbonization of chitosan. *ACS Omega* 5: 29520–29529.

Chen, J.X., Li, J.H., and Bao, A. (2023). B/N codoped porous carbon materials with rich microporous structure and their CO_2 adsorption properties. *Biomass Conversion and Biorefinery* 14: 04832–04833.

Cheng, J., Cheng, X., Wang, Z. et al. (2023). Multifunctional carbon aerogels from *Typha orientalis* for applications in adsorption: hydrogen storage, CO_2 capture and VOCs removal. *Energy* 263: 125984.

Cruz, O.F., Campello-Gómez, I., Casco, M.E. et al. (2022). Enhanced CO_2 capture by cupuassu shell-derived activated carbon with high microporous volume. *Carbon Letters* 33: 727–735.

Cui, H., Xu, J., Shi, J. et al. (2022). Oxamic acid potassium salt as a novel and bifunctional activator for the preparation of N-doped carbonaceous CO_2 adsorbents. *Journal of CO_2 Utilization* 64: 102198.

Dissanayake, P.D., You, S., Igalavithana, A.D. et al. (2020). Biochar-based adsorbents for carbon dioxide capture: a critical review. *Renewable and Sustainable Energy Reviews* 119: 109582.

Fan, Q., Song, C., and Fu, P. (2024). Advances in the improvement of the quality and efficiency of biomass-derived porous carbon: a comprehensive review on synthesis strategies and heteroatom doping effects. *Journal of Cleaner Production* 452: 142169.

Gai, F., Zhu, D., Wu, Y. et al. (2021). Tailored N-doped porous carbons via a MOF assembly process for high-performance CO_2 uptake. *Materials Advances* 2: 692–699.

Ge, K., He, Y., Cai, W. et al. (2022). Nitrogen-doped microporous carbon materials derived from DBU-modified carboxylic acid polymers for CO_2 capture. *Journal of Environmental Chemical Engineering* 10: 107849.

Goel, C., Mohan, S., and Dinesha, P. (2021). CO_2 capture by adsorption on biomass-derived activated char: a review. *Science of the Total Environment* 798: 149296.

Gong, L. and Bao, A. (2023). High-value utilization of lignin to prepare N,O-codoped porous carbon as a high-performance adsorbent for carbon dioxide capture. *Journal of CO_2 Utilization* 68: 102374.

Song, G., Wang, Y.., Huang, D.. et al. (2022). The heating rate and final temperature impacts on the coconut shell char structure characteristics during photo-thermal pyrolysis. *Journal of Analytical and Applied Pyrolysis* 167: 105695.

Guo, L., Lamb, K.J., and North, M. (2021). Recent developments in organocatalysed transformations of epoxides and carbon dioxide into cyclic carbonates. *Green Chemistry* 23: 77–118.

Guo, S., Li, Y., Wang, Y. et al. (2022). Recent advances in biochar-based adsorbents for CO_2 capture. *Carbon Capture Science & Technology* 4: 100059.

Hao, X., An, X., Patil, A.M. et al. (2021). Biomass-derived N-doped carbon for efficient electrocatalytic CO_2 reduction to CO and Zn-CO_2 batteries. *ACS Applied Materials & Interfaces* 13: 3738–3747.

He, J., Jin, Z., Gan, F. et al. (2022a). Liquefiable biomass-derived porous carbons and their applications in CO_2 capture and conversion. *Green Chemistry* 24: 3376–3415.

He, M., Xu, Z., Hou, D. et al. (2022b). Waste-derived biochar for water pollution control and sustainable development. *Nature Reviews Earth & Environment* 3: 444–460.

He, M., Zhu, X., Dutta, S. et al. (2022c). Catalytic co-hydrothermal carbonization of food waste digestate and yard waste for energy application and nutrient recovery. *Bioresource Technology* 344: 126395.

He, S., Chen, G., Xiao, H. et al. (2021). Facile preparation of N-doped activated carbon produced from rice husk for CO_2 capture. *Journal of Colloid and Interface Science* 582: 90–101.

Hidayu, A.R. and Muda, N. (2016). Preparation and characterization of impregnated activated carbon from palm kernel shell and coconut shell for CO_2 capture. *Procedia Engineering* 148: 106–113.

Ho, T.M., Howes, T., and Bhandari, B.R. (2014). Encapsulation of gases in powder solid matrices and their applications: a review. *Powder Technology* 259: 87–108.

Horri, N., Sanz-Pérez, E.S., Arencibia, A. et al. (2020). Effect of acid activation on the CO_2 adsorption capacity of montmorillonite. *Adsorption* 26: 793–811.

Hou, Y., Chen, Y., He, X. et al. (2024). Insights into the adsorption of CO_2, SO_2 and NO_x in flue gas by carbon materials: a critical review. *Chemical Engineering Journal* 490: 151424.

Huang, Y., Tang, Z., Zhou, S. et al. (2022a). Renewable waste biomass-derived carbon materials for energy storage. *Journal of Physics D: Applied Physics* 55: 313002.

Huang, Y.F., Chiueh, P.T., and Lo, S.L. (2022b). Carbon capture of biochar produced by microwave co-pyrolysis: adsorption capacity, kinetics, and benefits. *Environmental Science and Pollution Research* 30: 22211–22221.

Igalavithana, A.D., Choi, S.W., Shang, J. et al. (2020). Carbon dioxide capture in biochar produced from pine sawdust and paper mill sludge: effect of porous structure and surface chemistry. *Science of the Total Environment* 739: 139845.

Ismail, I.S., Rashidi, N.A., and Yusup, S. (2022). Production and characterization of bamboo-based activated carbon through single-step H_3PO_4 activation for CO_2 capture. *Environmental Science and Pollution Research* 29: 12434–12440.

Javed, M., Zahoor, M., Mazari, S.A. et al. (2023). An overview of effect of process parameters for removal of CO_2 using biomass-derived adsorbents. *Biomass Conversion and Biorefinery* 13: 4495–4513.

Ji, Y., Xie, R.Y., Wu, C. et al. (2023). Cost-effective carbon-based amine adsorbents for carbon capture: equilibrium, kinetics and selectivity. *Journal of CO_2 Utilization* 69: 102422.

Kamran, U. and Park, S.J. (2020). Tuning ratios of KOH and NaOH on acetic acid-mediated chitosan-based porous carbons for improving their textural features and CO_2 uptakes. *Journal of CO_2 Utilization* 40: 101212.

Kim, C., Talapaneni, S.N., and Dai, L. (2023). Porous carbon materials for CO_2 capture, storage and electrochemical conversion. *Materials Reports: Energy* 3: 100199.

Leng, L., Xiong, Q., Yang, L. et al. (2021). An overview on engineering the surface area and porosity of biochar. *Science of the Total Environment* 763: 144204.

Li, B., Li, M., Xie, X. et al. (2024a). Pyrolysis of rice husk in molten lithium chloride: biochar structure evolution and CO_2 adsorption. *Journal of the Energy Institute* 113: 101526.

Li, C. and Guo, X. (2025). Insight into the mechanism of H_2O promoted $CaCO_3$ decomposition in CO_2 atmosphere. *Separation and Purification Technology* 357: 130067.

Li, K., Niu, X., Zhang, D. et al. (2022). Renewable biochar derived from mixed sewage sludge and pine sawdust for carbon dioxide capture. *Environmental Pollution* 306: 119399.

Li, P., Zhu, Y., Wu, J. et al. (2024b). Amino-modified mesoporous carbon material for CO_2 adsorption in tunnel engineering: materials characterization and application prospects. *Case Studies in Construction Materials* 20: e02940.

Li, Q., Liu, S., Peng, W. et al. (2020a). Preparation of biomass-derived porous carbons by a facile method and application to CO_2 adsorption. *Journal of the Taiwan Institute of Chemical Engineers* 116: 128–136.

Li, Q., Lu, T., Wang, L. et al. (2021). Biomass based N-doped porous carbons as efficient CO_2 adsorbents and high-performance supercapacitor electrodes. *Separation and Purification Technology* 275: 119204.

Li, Y., Wang, S., Wang, B. et al. (2020b). Sustainable biomass glucose-derived porous carbon spheres with high nitrogen doping: as a promising adsorbent for $CO_2/CH_4/N_2$ adsorptive separation. *Nanomaterials* 10: 010174.

Li, Y., Xing, B., Ding, Y. et al. (2020c). A critical review of the production and advanced utilization of biochar via selective pyrolysis of lignocellulosic biomass. *Bioresource Technology* 312: 123614.

Liu, C., Wang, J., Zhang, S. et al. (2025). Ultramicropore-rich N-doped porous biochar from discarded cigarette butts for efficient CO_2 capture with ultra-high adsorption capacity and selectivity. *Separation and Purification Technology* 358: 130205.

Liu, P., Qin, S., Wang, J. et al. (2023). Effective CO_2 capture by in-situ nitrogen-doped nanoporous carbon derived from waste antibiotic fermentation residues. *Environmental Pollution* 333: 121972.

Liu, W.J., Jiang, H., and Yu, H.Q. (2019). Emerging applications of biochar-based materials for energy storage and conversion. *Energy & Environmental Science* 12: 1751–1779.

Luo, J.Y., Liu, B.G., Shi, R. et al. (2021). The effects of nitrogen functional groups and narrow micropore sizes on CO_2 adsorption onto N-doped biomass-based porous carbon under different pressure. *Microporous and Mesoporous Materials* 327: 111404.

Matsagar, B.M., Yang, R.X., Dutta, S. et al. (2021). Recent progress in the development of biomass-derived nitrogen-doped porous carbon. *Journal of Materials Chemistry A* 9: 3703–3728.

Meng, Y., Zhu, X., Zhang, Y. et al. (2024). Valorizing inherent resources from waste streams for in-situ CO_2 capture and sequestration in the steel industry. *Journal of Cleaner Production* 458: 142486.

Nagarajan, L., Kumaraguru, K., Saravanan, P. et al. (2022). Facile synthesis and characterization of microporous-structured activated carbon from agro waste materials and its application for CO_2 capture. *Environmental Technology* 43: 3983–3992.

Ogungbenro, A.E., Quang, D.V., Al-Ali, K.A. et al. (2018). Physical synthesis and characterization of activated carbon from date seeds for CO_2 capture. *Journal of Environmental Chemical Engineering* 6: 4245–4252.

Patel, H., Mohanty, A., and Misra, M. (2024). Post-combustion CO_2 capture using biomass based activated porous carbon: latest advances in synthesis protocol and economics. *Renewable and Sustainable Energy Reviews* 199: 114484.

Petrović, J., Ercegović, M., Simić, M. et al. (2024). Hydrothermal carbonization of waste biomass: a review of hydrochar preparation and environmental application. *Processes* 12: 207.

Rao, L., Liu, S., Wang, L. et al. (2019). N-doped porous carbons from low-temperature and single-step sodium amide activation of carbonized water chestnut shell with excellent CO_2 capture performance. *Chemical Engineering Journal* 359: 428–435.

Sasono, S.R.A., Rois, M.F., Widiyastuti, W. et al. (2023). Nanofiber-enrich dispersed activated carbon derived from coconut shell for supercapacitor material. *Results in Engineering* 18: 101070.

Sekhon, S.S. and Park, J.S. (2021). Biomass-derived N-doped porous carbon nanosheets for energy technologies. *Chemical Engineering Journal* 425: 129017.

Serafin, J., Kishibayev, K., Tokpayev, R. et al. (2023). Functional activated biocarbons based on biomass waste for CO_2 capture and heavy metal sorption. *ACS Omega* 8: 48191–48210.

Shafawi, A.N., Mohamed, A.R., Lahijani, P. et al. (2021). Recent advances in developing engineered biochar for CO_2 capture: an insight into the biochar modification approaches. *Journal of Environmental Chemical Engineering* 9: 106869.

Shen, Y. (2023). Molten salt-mediated carbon capture and conversion. *Fuel* 339: 127473.

Su, Y., Liu, S., and Gao, X. (2022). Impact of impure gas on CO_2 capture from flue gas using carbon nanotubes: a molecular simulation study. *Molecules* 27: 051627.

Sun, W., Bai, L., Chi, M. et al. (2023a). Study on the evolution pattern of the aromatics of lignin during hydrothermal carbonization. *Energies* 16: 1089.

Sun, Y., Jia, J., Liu, Z. et al. (2023b). Heteroatom-doped biochar devised from cellulose for CO_2 adsorption: a new vision on competitive behavior and interactions of N and S. *Biochar* 5: 002751.

Sun, Y., Xiong, X., He, M. et al. (2021). Roles of biochar-derived dissolved organic matter in soil amendment and environmental remediation: a critical review. *Chemical Engineering Journal* 424: 130387.

Sun, Y., Xu, Z., He, M. et al. (2024). Unlocking the solution-phase molecular transformation of biochar during intensive rainfall events: implications for the long-term carbon cycle under climate change. *Science of the Total Environment* 954: 176708.

Sun, Y.Q., Zhang, Q., Clark, J.H. et al. (2022). Tailoring wood waste biochar as a reusable microwave absorbent for pollutant removal: structure-property-performance relationship and iron-carbon interaction. *Bioresource Technology* 362: 127838.

Tang, F., Wang, L., and Liu, Y.N. (2019). Biomass-derived N-doped porous carbon: an efficient metal-free catalyst for methylation of amines with CO_2. *Green Chemistry* 21: 6252–6257.

Tian, Y., Zhang, F., Wang, J. et al. (2021). A review on solid acid catalysis for sustainable production of levulinic acid and levulinate esters from biomass derivatives. *Bioresource Technology* 342: 125977.

Wang, B., Li, Y., Gu, Z. et al. (2023a). Synthesis and design of biomass-derived heteroatom-doped hierarchical porous carbon systems for high-voltage supercapacitors. *Fuel Processing Technology* 247: 107776.

Wang, B., Yang, W., McKittrick, J. et al. (2016). Keratin: structure, mechanical properties, occurrence in biological organisms, and efforts at bioinspiration. *Progress in Materials Science* 76: 229–318.

Wang, H., Niu, Y., Xue, M. et al. (2024). Development of porous N/O hybrid carbon from spent tires for CO_2 capture in the framework of circular economy: parallel upgrading of pore and surface chemical structure. *Journal of Environmental Chemical Engineering* 12: 112113.

Wang, L., Wang, T., Hao, R. et al. (2023b). Synthesis and applications of biomass-derived porous carbon materials in energy utilization and environmental remediation. *Chemosphere* 339: 139635.

Wang, Q., Lv, Y., and Zhu, T. (2020). Silane coupling agent assisting dopamine-functionalized biomass porous carbons for enhanced adsorption of organic acids: effects of acid–alkali activation on microstructure. *Carbon Letters* 31: 29–37.

Wei, H., Chen, J., Fu, N. et al. (2018). Biomass-derived nitrogen-doped porous carbon with superior capacitive performance and high CO_2 capture capacity. *Electrochimica Acta* 266: 161–169.

Wu, R. and Bao, A. (2023). Preparation of cellulose carbon material from cow dung and its CO_2 adsorption performance. *Journal of CO_2 Utilization* 68: 102377.

Wu, W., Liu, J., Zhang, G. et al. (2025). Adsorption of CO_2 by sludge/bamboo shoot shell hybrid biochar prepared by a single-step K_2CO_3 activation. *Fuel* 381: 133555.

Xiao, K., Liu, H., Li, Y. et al. (2020). Excellent performance of porous carbon from urea-assisted hydrochar of orange peel for toluene and iodine adsorption. *Chemical Engineering Journal* 382: 122997.

Xie, W.H., Yao, X., Li, H. et al. (2022). Biomass-based n-rich porous carbon materials for CO_2 capture and in-situ conversion. *ChemSusChem* 15: 1004.

Xu, Y., Liu, Y., Zhan, W. et al. (2024). Enhancing CO_2 capture with K_2CO_3-activated carbon derived from peanut shell. *Biomass & Bioenergy* 183: 107148.

Yan, B., Zheng, J., Feng, L. et al. (2023). Pore engineering: structure-capacitance correlations for biomass-derived porous carbon materials. *Materials & Design* 229: 111904.

Yang, Z., Guo, X., Zhang, G. et al. (2020). One-pot synthesis of high N-doped porous carbons derived from a N-rich oil palm biomass residue in low temperature for CO_2 capture. *International Journal of Energy Research* 44: 4875–4887.

Yuan, X., Xiao, J., Yılmaz, M. et al. (2022). N, P co-doped porous biochar derived from cornstalk for high performance CO_2 adsorption and electrochemical energy storage. *Separation and Purification Technology* 299: 121719.

Zeng, F., Zhang, Y., Lv, Q. et al. (2024). Preparation of hierarchical porous carbon materials from bamboo shoot shells via air activation for high-performance supercapacitors. *International Journal of Electrochemical Science* 19: 100691.

Zhang, F., Wang, J., Tian, Y. et al. (2023). Effective removal of tetracycline antibiotics from water by magnetic functionalized biochar derived from rice waste. *Environmental Pollution* 330: 121681.

Zhang, J., Chen, Z., Liu, Y. et al. (2024a). Removal of emerging contaminants (ECS) from aqueous solutions by modified biochar: a review. *Chemical Engineering Journal* 479: 147615.

Zhang, P., O'Connor, D., Wang, Y. et al. (2020). A green biochar/iron oxide composite for methylene blue removal. *Journal of Hazardous Material* 384: 121286.

Zhang, X., Xu, Y., Zhang, G. et al. (2022a). Nitrogen-doped porous carbons derived from sustainable biomass via a facile post-treatment nitrogen doping strategy: efficient CO_2 capture and DRM. *International Journal of Hydrogen Energy* 47: 24388–24397.

Zhang, Y., Wei, Z., Liu, X. et al. (2022b). Synthesis of palm sheath derived-porous carbon for selective CO_2 adsorption. *RSC Advances* 12: 8592–8599.

Zhang, Z., Zou, W., Wang, H. et al. (2024b). Facile synthesis and biomimetic amine-functionalization of chitosan foam for CO_2 capture. *International Journal of Biological Macromolecules* 282: 136870.

Zhao, X., Wang, J., Zhu, G. et al. (2024). Efficient removal of high concentration dyes from water by functionalized in-situ N-doped porous biochar derived from waste antibiotic fermentation residue. *Chemosphere* 364: 143215.

Zhou, J., Zhang, S., Zhou, Y.N. et al. (2021). Biomass-derived carbon materials for high-performance supercapacitors: current status and perspective. *Electrochemical Energy Reviews* 4: 219–248.

Zhou, Y., Tan, P., He, Z. et al. (2022). CO_2 adsorption performance of nitrogen-doped porous carbon derived from licorice residue by hydrothermal treatment. *Fuel* 311: 122507.

Zhu, G., Wang, J., Zhao, X. et al. (2024). Effective removal of tetracycline antibiotics from water by in-situ nitrogen-doped porous biochar derived from waste antibiotic fermentation residues. *Journal of Environmental Chemical Engineering* 12: 114433.

Zhu, J., Yan, C., Zhang, X. et al. (2020). A sustainable platform of lignin: from bioresources to materials and their applications in rechargeable batteries and supercapacitors. *Progress in Energy and Combustion Science* 76: 100788.

Zhu, M., Cai, W., Verpoort, F. et al. (2019). Preparation of pineapple waste-derived porous carbons with enhanced CO_2 capture performance by hydrothermal carbonation-alkali metal oxalates assisted thermal activation process. *Chemical Engineering Research and Design* 146: 130–140.

Zhu, X., Luo, Z., Zhang, Q. et al. (2023a). Valorization of slow pyrolysis vapor from biomass waste: comparative study on pyrolysis characteristics, evolved gas evaluation, and adsorption effects. *Bioresource Technology* 386: 129543.

Zhu, X., Sun, M., Zhu, X. et al. (2023b). Molten salt shielded pyrolysis of biomass waste: development of hierarchical biochar, salt recovery, CO_2 adsorption. *Fuel* 334: 126565.

Zou, L., Kong, W., Peng, L. et al. (2023). First application of nitrogen-doped carbon nanosheets derived from lotus leaves as the electrode catalyst for Li-CO_2O_2 battery. *Catalysts* 13: 113030577.

Zubbri, N.A., Mohamed, A.R., Kamiuchi, N. et al. (2020). Enhancement of CO_2 adsorption on biochar sorbent modified by metal incorporation. *Environmental Science and Pollution Research* 27: 11809–11829.

13

Biomass Waste-derived Biochar as Graphitic Carbon for Agricultural Applications

Baojun Yi[1], Fang Huang[2], and Jiaqi Deng[1]*

[1] *College of Engineering, Huazhong Agricultural University, Wuhan, P. R. China*
[2] *School of Environment and Health, Jianghan University, Wuhan, P. R. China*

*Corresponding author: bjyi@mail.hzau.edu.cn

13.1 Literature Statistics Methodology

Biomass possesses the remarkable ability to regenerate in a short period and widely distributed, making it the largest renewable energy source currently available (Tursi 2019). Approximately, 550 billion tons of carbon in the form of biomass exists on earth. Plants contribute about 80% to this total—equivalent to roughly 450 billion tons of carbon (Bar-On et al. 2018). Over its entire lifecycle, biomass is regarded as a carbon-neutral energy option. The reason lies in the fact that the process of biomass formation involves the uptake of carbon dioxide from the atmosphere (Ni et al. 2006; Kang 2021). The processing and treatment of biomass generate a significant amount of biomass waste. This category of waste encompasses agricultural residues, forest waste, algal biomass, the organic components of sludge, along with cardboard, pulp, and slaughterhouse waste. It also includes food waste, municipal solid waste, and other assorted organic waste materials (Carpenter et al. 2014). Biomass holds several advantages over fossil fuel-derived carbon precursors in carbon material production. It's cost-effective, has a sustainable supply chain, and is widely and easily available. Biomass waste stands out as the most promising carbon source within the biomass spectrum, primarily due to its environmental benefits. Currently, a significant portion of biomass waste is either landfilled or incinerated. These disposal methods not only pollute the air—releasing unpleasant odors—and disrupt ecosystems but also consume valuable

Waste-derived Biochar for Sustainable Rural Revitalization, First Edition. Edited by Yuqing Sun and Daniel C.W. Tsang.
Published 2026 by John Wiley & Sons, Inc.

land resources. By contrast, using biomass waste as a carbon precursor can convert this seemingly worthless waste into high-value functional carbon materials (Ayele et al. 2021; Wang et al. 2018).

13.1.1 Graphitized Carbon

Thermochemical conversion enables the transformation of biomass waste into carbon-rich biochar. The formation mechanism primarily involves two stages: below 600 °C, lignocellulose decomposes to generate free radicals, and at 600–800 °C, these free radicals stabilize through condensation reactions to form carbonized structure (Shirley and Kammen 2015). At temperatures above 800 °C, biochar gradually undergoes graphitization. Under high temperatures (1000–1400 °C), it exhibits significant X-ray diffraction (XRD) characteristics (such as an enhanced 002 peak at $2\theta = 26.5°$) and sharper D and G peaks in Raman spectra, indicating a notable trend toward graphitization (Stephen et al. 2016; Lindgren 2016). Prolonging the pyrolysis time further enhances the degree of graphitization and reduces the content of heteroatoms (Senneca et al. 2018; Chen et al. 2021b). At ultra-high temperatures (>2500 °C), amorphous carbon precursors can be converted into crystalline graphite.

Carbon precursor materials fall into two main types: graphitized carbon, also called soft carbon, and non-graphitized carbon, known as hard carbon (Franklin and Randall 1997). In soft carbon, the microcrystalline structures are ordered and easily graphitized, whereas in hard carbon, microcrystals are cross-linked in random orientations, making it difficult to transform into crystalline graphite even at high temperatures. Research suggests that microporous carbon may possess a "fullerene-like" structure, composed of curved graphene sheets and non-hexagonal carbon rings (Harris 2013; Forse et al. 2015; Townsend et al. 1992; Terzyk et al. 2007). The microstructure of carbon materials significantly impacts the graphitization process and determines how their properties are characterized.

13.1.2 Literature Statistics

This section comprehensively reviews the research progress on graphitic carbon mentioned above, portraying the current research landscape in this field. Bibliometrics allows for a macroscopic examination of literature evolution, publications, research focal points, methodologies, and the distribution of authors in the research domain. It also captures the networks, structures, interactions, intersections, and evolutionary trends of the field, thereby facilitating the prediction of future developmental trajectories. To offer reliable recommendations for future research directions, we conducted a bibliometric analysis of biomass-derived graphitic carbon preparation technologies. The bibliometric data for studies on biomass-derived graphitic carbon technologies were retrieved from the Web of Science core databases on February 6, 2025. Search terms such as "graphitic carbon" and "biomass" were utilized, with a time span from 2018 to 2024. After preliminary screening and the elimination of duplicates, 2000 articles were selected for analysis based on relevance ranking. Data visualization was carried out using VOSviewer and bibliometric software, identifying over 500 keywords with a minimum occurrence threshold of 30. The visualization results are presented in Figure 13.1.

Keyword analysis reveals that, with the rapid development of carbon material science and green manufacturing technologies, biomass-based graphitic carbon preparation has emerged as a forefront research area. Biomass, a renewable and widely accessible carbon-rich resource, significantly contributes to the development of new energy technologies and high-value resource utilization. Traditional graphite production is highly dependent on fossil feedstocks and is associated with high energy consumption and pollution. In contrast, biomass-derived graphitic carbon technologies enable the efficient conversion of agricultural and forestry waste while reducing reliance on environmentally harmful production processes. An in-depth exploration of biomass graphitization mechanisms and structural regulation methods is crucial for the development of green, low-carbon advanced carbon material synthesis. Biomass-derived graphitic carbon preparation—through processes such as pyrolysis, catalytic carbonization, etc.—transforms cellulose, lignin, and other components into carbon materials with ordered lattice structures. The graphitic carbon obtained exhibits outstanding conductivity and stability when applied in lithium-ion batteries, supercapacitors, and catalyst supports. Evidently, it holds substantial potential. These trends mirror the dual requirements of clean energy transition and high-end material innovation. Future research should prioritize clarifying the structure–property relationships among biomass precursors, structural modulation, and performance optimization. Overcoming technical challenges such as high graphitization temperatures and complex processes will be essential for promoting biomass-based graphitic carbon from laboratory-scale research to large-scale industrial production.

13.2 Biomass Waste Feedstocks Suitable for the Preparation of Graphitic Carbon

Biomass waste broadly falls into two categories: plant biomass and animal biomass. Formed primarily via photosynthesis, plant biomass chiefly consists of macromolecules like cellulose, hemicellulose, and lignin, along with small molecules such as glucose and fructose (Zhang et al. 2022a). In contrast, animal biomass comes from animal waste or residues. It's mainly made up of proteins, minerals, and

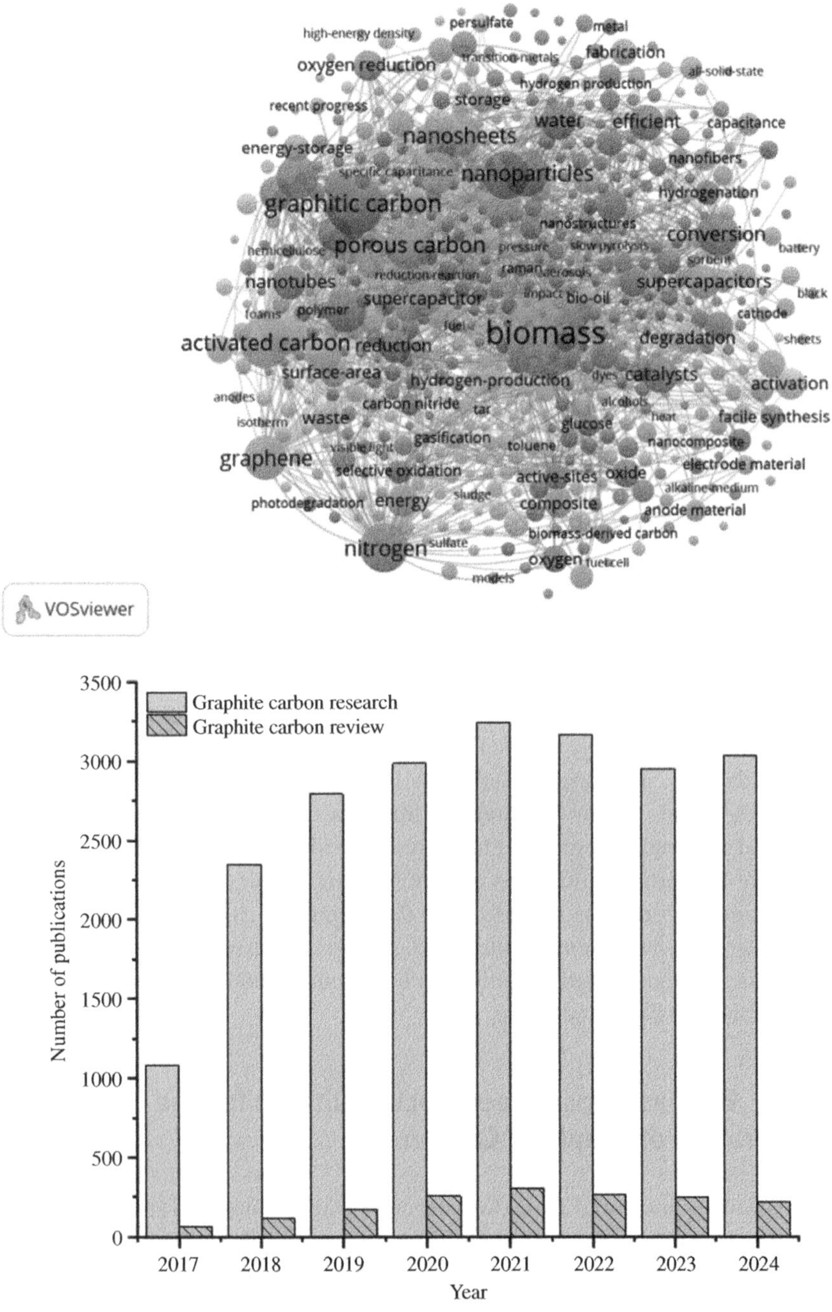

Figure 13.1 Visualization of literature statistics on graphitic carbon.

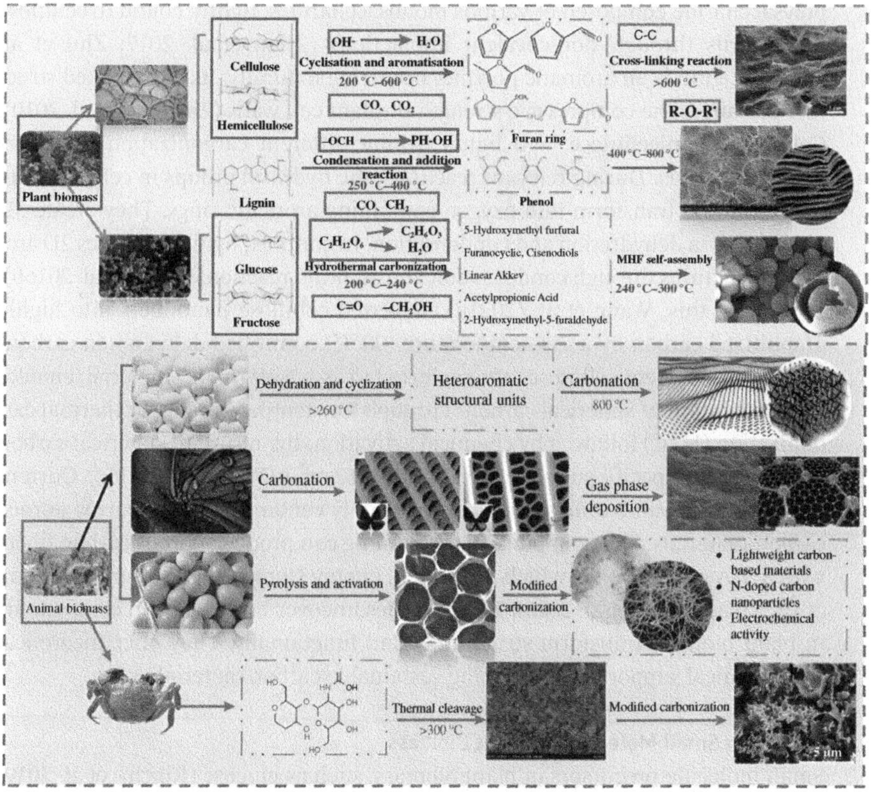

Figure 13.2 Types of biomass feedstocks for the preparation of graphitic carbon. *Source:* He et al. 2021b/with permission of American Chemical Society; Gao et al. 2018b/with permission of American Chemical Society; Vinodh et al. 2021/with permission of Elsevier.

polysaccharides including chitosan and chitin (Borghei et al. 2018). Figure 13.2 depicts the origins of plant and animal biomass, the two main types of biomass.

13.2.1 Plant Biomass

13.2.1.1 Macromolecules in Plant Biomass

Plant biomass primarily contains cellulose, hemicellulose, and lignin (Yang et al. 2007). These components jointly construct the structural framework of plant cell walls. Cellulose is composed of (1,4)-D-glucopyranose unit chains. Its polymerization degree, varying between 100 and 40,000, depends on the biomass type and extraction method (McKendry 2002; Costes et al. 2017). It has both crystalline and amorphous parts (Song et al. 2018). Hemicellulose is a heterogeneous polysaccharide composed of various monosaccharides, tightly bound to cellulose microfibrils through non-covalent interactions (Zhou et al. 2017; Zhu et al. 2016b). Lignin is an aromatic polymer occurring naturally. Its cross-linked structure enhances the compressive strength of plant cell walls (Zakzeski et al. 2010). Cellulose, hemicellulose, and lignin exhibit different carbonization behaviors during pyrolysis. During relevant reactions, the hydroxyl groups in cellulose and hemicellulose transform into oxygen-containing aromatic rings. They create 3D structures via dehydration and condensation. In contrast, lignin generates 2D aromatic structures through condensation and addition reactions (Deng et al. 2016b). In view of this, Wang et al. (2021a) converted cellulose nanofibers into highly graphitized carbon by a low-temperature (90 °C) controlled dehydration carbonation method. Hemicellulose, often referred to as a "soft tissue" material, enables the preparation of spherical carbon. Through low-temperature hydrothermal carbonization (HTC) followed by chemical activation, the resultant spherical carbon exhibits a high specific surface area (SSA) of 3062 m^2/g (Wang et al. 2015c). Current efforts in high-value utilization of lignin mainly center around preparing porous carbon materials. For example, electrospinning can produce porous carbon fibers with a SSA of 1880 m^2/g, which are used in energy storage applications (You et al. 2016; Ribadeneyra et al. 2020). These studies uncover how the main components in plant biomass transform structurally and functionally. They offer theoretical and technical support for developing sustainable carbon materials.

13.2.1.2 Small Molecules in Plant Biomass

Small molecule precursors in plant biomass, such as glucose (Ribeiro et al. 2019) and fructose (Mahyari et al. 2014), have simple structures and can be used to synthesize carbon materials with controllable morphologies under mild conditions through HTC. During hydrolysis, polysaccharides derived from starch and cellulose can be decomposed into monosaccharides, providing raw materials for carbon synthesis. Studies have shown that different hydrothermal regulation methods enable precise design of carbon structures from the micrometer to molecular scale, synthesizing various specific morphologies of carbon materials, including hollow bowls and mesoporous nanowires (Wang et al. 2014; Chen et al. 2017a; Xie et al. 2018; Chen et al. 2018; Xie et al. 2021).

As the basic structural unit of cellulose and starch, glucose first dehydrates to form 5-hydroxymethylfurfural (HMF) during HTC, which further polymerizes to form carbon spheres. Their structures can evolve into low-graphitized carbon materials through heat treatment (Lourvanij and Rorrer 1993; Antal et al. 1990; Cui et al. 2006). Zhang et al. (2014) developed a low-cost, large-scale method for synthesizing graphene using glucose and $FeCl_3$ as a template and catalyst, resulting in graphene sheets with excellent properties. Moreover, carbon materials derived from glucose serve as catalytic supports. Ribeiro et al. (2019) optimized the dispersion of ruthenium metal particles using their high microporosity and low acidity for the production of cellulose to sorbitol.

Fructose, another important small molecule precursor, undergoes HTC at lower temperatures (130 °C) (Mahyari et al. 2014), making it suitable for preparing carbon supports for metal nanoparticles. Heckel et al. (2008) synthesized a composite material of silver nanoparticles encapsulated in fructose-derived carbon shells via HTC, where the carbon shells improved particle stability and reduced damping effects. Fructose-derived carbon supports can also be functionalized to enhance performance, such as thiol functionalization improving the stability of Au nanoparticles (Mahyari et al. 2014). Furthermore, Kurniawan et al. (2022) used fructose to prepare graphene quantum dots (GQDs), demonstrating their broad application potential in environmental probes. These studies highlight the efficiency and diverse application prospects of small molecule plant biomass in carbon material preparation.

13.2.2 Animal Biomass

13.2.2.1 Protein-based Biomass

Proteins, rich in nitrogen (average content of 16% (Sharma et al. 2015)), are crucial precursors for in situ synthesizing nitrogen-doped carbon materials. These nitrogen-doped materials show better electrochemical performance than pure carbon materials (Wang et al. 2012). Recently, biomass rich in proteins, including milk (Chen et al. 2020b), silk fibroin (He et al. 2021a), and eggs (Ma et al. 2017; Li et al. 2013) has seen extensive use in preparing carbon materials and optimizing their performance.

Silk fibroin, with its unique β-sheet crystal structure, is an ideal precursor for high-performance carbon materials. Studies have shown that silk fibroin decomposes at temperatures above 260 °C, forming heteroaromatic structural units through intermolecular dehydration and cyclization reactions. Subsequent high-temperature heat treatment yields carbon materials with sp^2-hybridization characteristics (He et al. 2021; Cho et al. 2015). These carbon materials exhibit a 3D layered structure, where the carbon stacking in the (002) layer is closely related to the molecular stacking of the β-sheet structure in silk fibroin (Cho et al. 2015). Additionally, carbonizing silk fabric in an inert atmosphere produces small but structurally intact 3D porous carbon materials (Wang et al. 2016), which demonstrate excellent performance in multilayer graphene and nitrogen doping, particularly in supercapacitors (Sahu et al. 2015). Subsequent research indicates that silk fibroin, when combined with templates like $FeCl_3$ and $ZnCl_2$, forms porous carbon nanosheets post-carbonization. These nanosheets feature numerous Fe-N_x-C active sites and SSA of 2105 m_2/g (Wang et al. 2019c). Fe ions play a crucial role in promoting active site generation and graphitization, while

Zn ions introduce additional pores through activation. The template method improves the control over the structure and chemistry of silk fibroin-derived carbon materials. This makes these materials very promising for electrocatalytic applications, including the oxygen reduction and oxygen evolution reactions (Wang et al. 2019b). High-temperature graphitization of silk fibroin improves the conductivity and mechanical properties of carbon materials (Wang et al. 2017b). These studies offer fresh perspectives and approaches to the design and application of nitrogen-doped carbon materials.

The butterfly wings composed of glycoproteins have a unique 3D structure, which can maintain the original hierarchical structure after carbonization and be transformed into high-performance carbon materials (Gao et al. 2018). In addition, after the butterfly wings treated with metal salts are carbonized, a graphene–graphite film can be formed, which exhibits excellent oxygen reduction electrocatalytic performance and has broad application prospects (Li et al. 2019). Keratin serves as the primary constituent of chicken feathers and contains C, N, and O elements. By undergoing low-temperature pyrolysis, chicken feathers can be converted into heteroatom-doped (N, O) carbon-based materials with a distinct graphite layer structure (Singh et al. 2018).

Human hair is mainly composed of α-keratin rich in nitrogen and sulfur. As a precursor of carbon materials, through an improved carbonization process, hierarchical and porous heteroatom-doped carbon nanosheets can be synthesized and applied to high-performance energy storage devices (Chaudhari et al. 2014). Similarly, cashmere and wool are rich in keratin and amino acids. After chemical activation and carbonization treatment, N, O, S triple-doped porous carbon materials can be prepared, showing excellent electrochemical properties (Ren et al. 2019; Zhou et al. 2015).

Due to its high nitrogen content and special β-sheet crystal structure, spider silk can be transformed into N, S-doped carbon fibers or carbon nanospheres after pyrolysis treatment, showing significant advantages in the fields of energy storage and bioimaging (Ruan et al. 2014; Zhou et al. 2016).

Eggs and their derivatives are common nitrogen-containing biomass precursors. After pyrolysis and activation treatment of the eggshell membrane, a 3D macroporous carbon membrane with good electrical conductivity (EC) can be formed (Li et al. 2012). By introducing carbon nanotubes (CNTs) or metal oxide nanostructures, lightweight carbon-based materials suitable for the fields of energy storage and catalysis can be prepared (Geng et al. 2015). In addition, after mixing egg white with inorganic salts and carbonizing it, N-doped carbon nanoparticles embedded with metal oxides can be obtained, which exhibit excellent performance in energy storage devices (Gao et al. 2016). The multiple nitrogen functional groups in egg-derived carbon materials further enhance their electrochemical activity (Li et al. 2014b).

13.2.2.2 Chitin-based Biomass

Chitin ranks second abundant biopolymer on Earth. It's commonly present in crustacean shells and fungal cell walls. Composed of *N*-acetylglucosamine units joined by β-1,4-glycosidic bonds, chitin molecules can have molecular weights in the millions. Because of strong intermolecular hydrogen bonds, it's insoluble in water and alkaline solutions (Ramos Berger et al. 2018). During pyrolysis, chitin starts decomposing at 300 °C. Carbonization leads to a mass loss of up to 78.3%. With the temperature rises, the nitrogen content in the resulting carbon material gradually drops (Gao et al. 2015).

Chitin-derived carbon materials can be obtained through pyrolysis, with crab shells and shrimp shells being common sources. During carbonization, $CaCO_3$ acts as a template that can be removed to yield carbon materials with porous structures (Shao et al. 2017). Carbonized shrimp shells yield carbon materials featuring large SSA and total pore volume (TPV). Their porous structure boosts electrochemical performance (Mondal et al. 2017). Additionally, other chitin sources, such as cuttlebones, can be used to synthesize carbon nanosheets with high levels of nitrogen and oxygen doping (Guo et al. 2018).

Through pyrolysis and further carbonization, chitin-derived materials can be transformed into porous carbon nanofibers or carbon nanocomposites, which exhibit excellent performance in supercapacitors and lithium-ion batteries (Nogi et al. 2010; Liu et al. 2021a). Additionally, chitin solutions are applied in restructuring processes to fabricate carbon materials with various structures like carbon dots and aerogels. These carbon materials show extensive application potential (Duan et al. 2016; Hao et al. 2018; Briscoe et al. 2015).

Chitosan, a natural alkaline polysaccharide derived from chitin through deacetylation, exhibits high thermal stability and adsorption properties (Vinodh et al. 2021; Croisier and Jérôme 2013). Chitosan decomposes at lower temperatures than chitin and can be used to synthesize various carbon materials, widely applied in heavy metal ion adsorption, sodium-ion batteries, and conductive aerogels (Ba et al. 2018; Duan et al. 2019; Tian et al. 2021).

13.3 Graphitization and Carbonization Processes of Waste Biomass and Characteristics of Graphitic Carbon

The structure of biomass-derived carbon materials is frequently restricted by the inherent structure of their precursors. This makes it difficult to achieve precise control. Recently, multiple strategies have been adopted to optimize the structures of biomass-derived carbon materials across 0–3D configurations. This expansion has broadened their range of applications. 0D materials exhibit photoluminescence properties, making them suitable for bioimaging and nanomedicine (Ayiloor Rajesh et al. 2022; Thangaraj et al. 2021). 1D carbon-based materials possess benefits for energy storage devices. Their continuous electron transport pathways and large ion-accessible surface areas contribute to this advantage (Ouyang et al. 2017; Tran et al. 2009). 2D sheet-shaped carbon materials make excellent electrode frameworks for lithium-ion batteries (Gao et al. 2018a). 3D carbon materials exhibit outstanding performance in gas adsorption, water treatment, and electromagnetic shielding. Their high SSA and interconnected porous structures account for this enhanced performance (Liu et al. 2019; Hirst et al. 2018; Zhou et al. 2019). Carbonization strategies can be classified into "structure-inheriting carbonization" and "reconstructive carbonization." In structure-inheriting carbonization, the macro/microstructure of the biomass is preserved during conversion, making it suitable for constructing 1D carbon tubes (Wang et al. 2021). Conversely, reconstructive carbonization breaks down the initial structure of biomass. Tube furnaces offer a broad temperature range and adjustable heating procedures for structure-inheriting carbonization (Blankenship 2021). Low-temperature HTC is suitable for synthesizing nanoscale 0D biochar at 150–350 °C (Hu et al. 2010; Jain et al. 2016). Laser or Joule heating can produce highly graphitized biochar in a short time (Sha et al. 2017; Muñoz-Hernández et al. 2018). Catalytic graphitization promotes graphitization at lower temperatures. Template methods use strong acids or bases to remove templates, yielding ordered porous structures (Deng et al. 2016a). The final structure of carbon materials is significantly affected by several factors. These include the carbonization temperature, heating rate, reaction atmosphere, choice of catalysts, and templates (Chen et al. 2017b). Figure 13.3 shows the technology roadmap for graphite carbon preparation. The following sections will detail each preparation technique.

13.3.1 Hydrothermal Carbonization

HTC effectively prepares carbonaceous materials from high-moisture biomass. By mimicking natural coalification, it transforms waste biomass into high-value carbon products (Sharma et al. 2020). HTC involves heat treatment of biomass-water mixtures at high temperatures (180–350 °C) and pressures, promoting reactions such as dehydration, fragment polymerization, and intermolecular dehydration carbonization. The structure, composition, and morphology of carbon materials are significantly shaped by reaction conditions. These include temperature, residence time, and the selection of precursors (Lu et al. 2012). During HTC, different biomass precursors exhibit distinct transformation characteristics. For example, carbon spheres derived from monosaccharides are more regular and require shorter reaction times compared to those from polysaccharides and natural biomass (Titirici et al. 2008). Increasing temperature significantly reduces the oxygen content in hydrochar

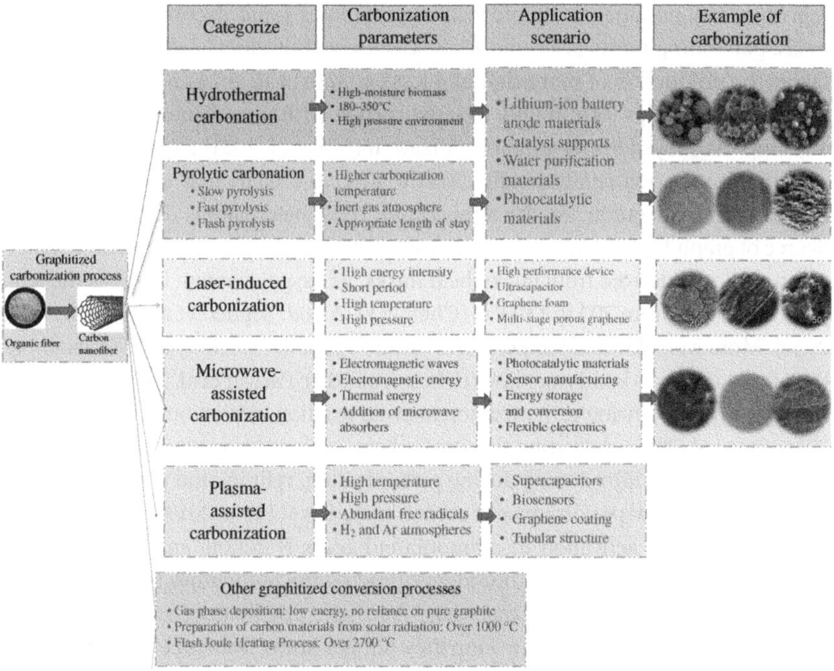

Figure 13.3 Technology roadmap for graphitic carbon preparation.

(Zhang et al. 2020a), while extending residence time helps form microspheres or carbon microspheres, enhancing porosity and SSA (Sun and Li 2004; Romero-Anaya et al. 2014; Nizamuddin et al. 2017). Additionally, precursors with higher lignin content tend to yield more hydrochar with larger particle sizes (Yang et al. 2007; Chen et al. 2021a; Thomsen et al. 2008). Chemical additives can alter the structure and composition of carbon materials. For instance, MnO_2/carbon sphere composites exhibit larger SSA, facilitating electron and ion transport (Wang et al. 2015a), while Fe^{2+} and Fe_2O_3 catalyze carbonization reactions and regulate product morphology (Cui et al. 2006). Template approaches, like the use of silica templates, can regulate the morphology of carbon materials. This allows for the production of hollow carbon spheres (Joo et al. 2008; Liu et al. 2011). These studies demonstrate that optimizing HTC conditions allows precise control over carbon material structures, enhancing their performance.

13.3.2 Pyrolysis Carbonization

Pyrolysis carbonization is a prevalent way to produce biochar. Adjusting carbonization parameters enables customization of the structure and properties of carbon materials. Temperature, heating rate, residence time, and reaction atmosphere represent key parameters. Rapid heating promotes rapid decomposition of biomass, generating more gases and fewer carbonized products, while slow heating favors the production of more char and less tar (Kan et al. 2016). Studies show that slow pyrolysis helps form highly ordered carbon structures (Guerrero et al. 2005) and improves crystallinity and aromatization (Liu et al. 2021b; Yuan et al. 2020).

Carbonization temperature can directly affect the elemental composition and degree of graphitization. As the temperature rises, heteroatoms decrease, and the degree of graphitization improves (Frank et al. 2014; Chen et al. 2020a). For instance, when carbon is derived from buckwheat husks, its oxygen content drops as the temperature climbs from 700 to 1300 °C (Chen et al. 2020d; Septien et al. 2014; Lei et al. 2021). Moreover, elevated temperatures affect functional group types. For example, the number of C=O bonds increases (Chen et al. 2020; Pariyar et al. 2020). Inert gases are typically used during carbonization to avoid reactions with biomass (Cheng et al. 2016; Ou et al. 2015). Adjusting the atmosphere, such as using ammonia, can introduce nitrogen or sulfur doping during carbonization, refining the pore architecture and electrochemical performance of carbon materials (Li et al. 2019; Luo et al. 2014). Studies show that activation in ammonia enhances N-doping and improves pore characteristics (Li et al. 2019b). Residence time significantly impacts the pore characteristics of activated carbon materials. Pore size, distribution, and SSA of these materials can vary based on how long they remain in the reaction environment. An appropriate residence time helps achieve optimal pore structures, while excessively long residence times may lead to pore shrinkage (Lua 2020).

13.3.3 Laser-induced Carbonization

Biomass can be efficiently transformed into graphene through laser-induced carbonization. By generating extremely high temperatures in a short period, it can rapidly create high temperature and pressure in a local area, promoting the rapid pyrolysis of biomass and the formation of graphene. When irradiated by lasers, hemicellulose and cellulose tend to decompose, creating more defects. Meanwhile, aromatic lignin generates a low-degree graphene structure (Ye et al. 2017; Han et al. 2018; Mahmood et al. 2019). The formation of high-viscosity amorphous carbon from lignin after laser treatment indicates that the cross-linked structure is crucial for the conversion process (Mahmood et al. 2019). Lignin-derived porous graphene was crafted by Mahmood et al. (2019) via laser writing. They regulated the structural defects using laser power and obtained hierarchical porous carbon materials with excellent EC. Graphene can be formed with the laser's high energy intensity, eliminating the need for metal catalysts and carbon gases (Kurra et al. 2019). High temperatures and quick heating during laser-induced carbonization may lead to the combustion of biomass precursors. Therefore, pre-treatment and special laser equipment are crucial for biomass carbonization. Ye et al. (2017)

converted pine wood into hierarchical porous graphene through laser-induced transformation. The obtained graphene has good EC and mechanical properties, making it suitable for the manufacture of high-performance devices. Chyan et al. (2018) used a multi-pulse laser scribing strategy to convert biomass such as coconut and cork into laser-induced graphene (LIG). The obtained materials can be applied in supercapacitors. In addition, Zang et al. (2018) fabricated micro-supercapacitors with 3D LIG electrodes. These electrodes exhibited excellent EC and SSA. By laser-writing a nickel/sucrose blend in an H_2 atmosphere, Zhang et al. (2018a) successfully synthesized graphene foams (GFs). This material has excellent structural and electrochemical properties.

13.3.4 Microwave-assisted Carbonization

As an emerging approach, microwave-assisted pyrolysis (MAP) enables biomass conversion to carbon materials. It utilizes electromagnetic waves in the frequency range of 300 MHz to 300 GHz to directly transfer heat, converting electromagnetic energy into thermal energy through dielectric heating, thereby accelerating the pyrolysis process and reducing processing time (Arpia et al. 2021; Liew et al. 2019). MAP not only effectively removes moisture but also allows control over processing parameters such as irradiation time and energy density (Luque et al. 2012; Tseng et al. 2013). Although MAP has certain limitations regarding raw materials and often requires additional microwave absorbers (Haeldermanse et al. 2020), it enables faster biomass pyrolysis compared to traditional carbonization methods and can adjust the structure and composition of carbon materials through doping. Guiotoku et al. (2009) first explored microwave-assisted HTC of biomass, while Elaigwu and Greenway (2016) compared the carbonization efficiency of microwave-assisted HTC on *Prosopis africana* shells. Studies have shown that microwave irradiation exhibits higher efficiency in decomposing and carbonizing biomass. Microwave irradiation rapidly hydrolyzes cellulose from bamboo, cotton, and starch into reducing sugars. This simultaneously activates cellulose molecules, further accelerating hydrolysis (Wu et al. 2010). Wang et al. (2008) synthesized carbon nanostructures with high SSA using MAP of Fe/conductive polymer/biomass composites.

MAP not only effectively promotes biomass conversion but also modifies carbon-based materials, as carbon materials are typically excellent microwave absorbers (Nizamuddin et al. 2018). Villota et al. (2019) prepared porous carbon derived from waste cocoa pod husks via microwave-assisted activation, finding that H_3PO_4 was more effective than KOH in promoting pore development. Kaewtrakulchai et al. (2020) applied MAP in tandem with KOH activation to derive nanoporous carbon from oil palm male flowers. As a result of 6 minutes of irradiation, the SSA surged to 991 m^2/g, with more notable pore development.

13.3.5 Plasma-assisted Carbonization

As a conversion approach, plasma technology is used to turn biomass into carbon materials like graphene and CNTs (Zhang et al. 2021a). Zhang et al. (2021a) and Wang et al. (2015d) devised the microwave plasma irradiation (MPI) approach for nanocarbon production from rice husks. The resultant products included graphene, CNTs, and graphitized CNTs (g-CNTs). Temperature and pressure during the MPI process are critical for pyrolysis and the generation of graphitic carbon. Seo et al. (2015) used Ar/H_2 plasma to prepare graphene from honeycomb materials for supercapacitors and biosensors. The electric field within the plasma sheath spurred the vertical growth of graphene. Shah et al. (2018) found that as exposure time lengthens, monolayer graphene forms. Usually, this monolayer graphene is exfoliated from multilayer graphene. Plasma irradiation within H_2 and Ar atmospheres supplies high temperature and pressure. Moreover, it releases a wealth of radicals, ions, and molecules. These decompose biomass rapidly, generating graphene building units. These units can form uniform graphene layers via chemical vapor deposition (CVD) methods (Seo et al. 2015; Shah et al. 2018). Shah et al. (2018) discovered that even though chemical bonds break during 750 °C pyrolysis, the formed macromolecules don't promote graphene growth. Post plasma activation, carbon atoms and radicals settle on a copper substrate. Meanwhile, hydrogen radicals etch the substrate, spurring graphene nucleation and growth. Wang et al. (2015) demonstrated that temperature and pressure during plasma processes are crucial for graphene formation, with higher temperatures favoring the formation of graphene layers from precursors such as CH_4, CO, CO_2, and bio-oil on nickel surfaces. Additionally, nickel clusters generated by plasma irradiation act as templates and catalysts during graphitization, while high-pressure plasma helps graphene layers wrap into tubular structures.

13.3.6 Other Graphitization Conversion Processes

CVD synthesizes carbon nanomaterials with low energy input, eliminating the need for pure graphite. In CVD, gaseous hydrocarbons flow over a substrate coated with metal catalysts and decompose, with carbon atoms growing from catalyst particles through mechanisms such as floating growth or base growth to form carbon nanostructures (Ruan et al. 2011). CVD efficiently generates high-purity carbon nanomaterials. By tuning experimental parameters like reaction temperature, catalyst type, and substrate chemistry, it enables control over the final product's structure. As a result, CVD is extensively applied in industrial carbon nanomaterial manufacturing. Ruan et al. (2011) devised an affordable CVD approach. They grew graphene at 1050 °C in an H_2/Ar atmosphere, using copper foil as the substrate and low-value biomass waste as the carbon source. This method efficiently utilizes waste as a raw material to produce high quality graphene. Other studies have shown that carbon sources from food, insects, and waste can also be successfully converted into high-quality graphene via CVD (Saha and Dutta 2022).

In addition, Hu et al. (2022) put forward a technique to prepare carbon materials using concentrated solar radiation. This method uses focused sunlight to convert biomass into graphene within a short period of time, with the temperature exceeding 1000 °C, enabling rapid conversion. This approach applies to diverse biomass raw materials like banana peels, cantaloupe peels, and orange peels. It also enables the design of graphene morphology.

On the other hand, Luong et al. (2020) created a flash Joule heating approach. With this, they converted affordable carbon sources such as biochar and biological waste to graphene. This method generates a high temperature exceeding 2700 °C through Joule heating and converts amorphous carbon into turbostratic graphene within 100 ms. Molecular dynamics simulation analysis indicates biomass density and temperature are crucial in the graphene graphitization process. Biomass with a low density structure facilitates graphene formation. Additionally, elevated processing temperatures boost the development of a layered structure.

13.4 Optimization Methods for Biomass Waste-derived Graphitic Carbon

The structure of carbon materials is significantly influenced by their precursors. Many biomass precursors generally feature underdeveloped pore structures, low SSA, and inferior conductivity (Weldekidan et al. 2022). Additionally, traditional carbonization processes require temperatures as high as 3000 °C to achieve highly graphitized carbon materials, which are complex, costly, and require precise control of operating parameters. Therefore, to meet the diverse performance requirements of carbon materials, researchers have proposed various methods to optimize their structure and properties. In this context, structural regulation of carbon materials has become a research hotspot. Current optimization strategies include: (1) improving pore structure and SSA through activation processes; (2) using templating agents to control material morphology and pore distribution; (3) introducing catalysts to enhance graphitization and conductivity. These approaches significantly improve the SSA, conductivity, mechanical properties, and thermal stability of carbon materials, meeting the needs of various applications such as environmental governance and modification, supercapacitors, energy storage devices, and catalysis. Figure 13.4 summarizes various graphite carbon activation methods.

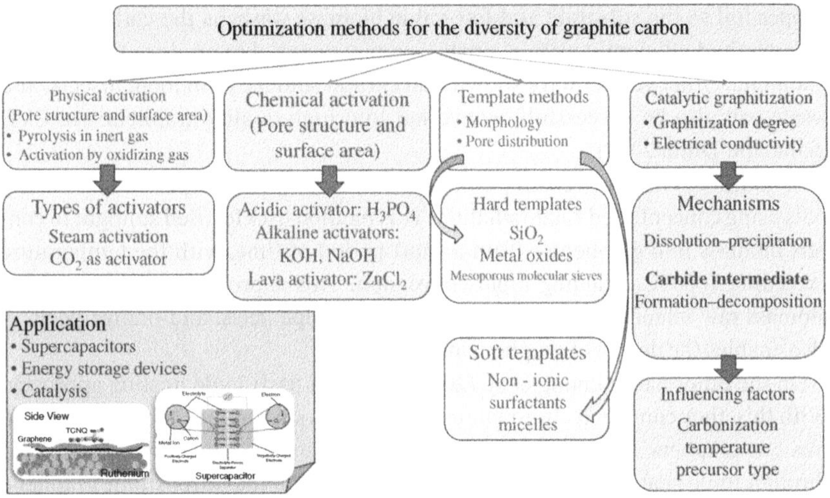

Figure 13.4 Summary of activation methods on graphitic carbon.

13.4.1 Physical Activation

In physical activation, biomass undergoes a two-stage reaction. The first stage involves pyrolysis in an inert gas, partially carbonizing the biomass and removing non-carbon elements (Sajjadi et al. 2019). During the second stage, activation takes place using oxidizing gases such as steam, CO_2, or air mixtures. The gas reacts with the carbon structure, eliminating pore-clogging substances and expanding open pores. Oxidation generates CO_2, leading to pore formation (Ioannidou and Zabaniotou 2007). Chen et al. (2015) activated carbonized logs in air at 350 °C to produce porous carbon materials with large SSA, while Li et al. (2012) pyrolyzed chicken manure in argon and activated it in air at 300 °C for 2 hours to obtain high-performance porous carbon materials.

Steam activation is used as an environmentally friendly and cost-effective way of oxidation through milder oxidation, which usually requires temperature higher than 800 °C. The steam activation process can be carried out in a variety of ways. Prakash Kumar et al. (2006) reported that increasing steam activation temperature enhances the SSA of rubberwood sawdust-derived carbon materials, while Wang et al. (2015b) synthesized low-cost spherical carbon materials with large pore volumes, demonstrating effective pore structure adjustment. Steam flow rate and activation time significantly affect the pore structure of carbon materials (Wei and Li 2021).

CO_2 is a commonly used activator with a slower etching rate, facilitating operation and control, especially at high temperatures (Taer et al. 2010; Guo and

Lua 2002). Guo et al. (2009) found that high temperatures and prolonged activation time promote mesopore formation, while high flow rates accelerate reactions and influence pore structure. CO_2 activation preferentially forms mesoporous carbon, whereas steam tends to generate microporous carbon (Chang et al. 2000). Yang's team synthesized carbon fibers with highly interconnected pore structures using dual activation with CO_2 and KOH (Wang et al. 2020).

Although physical activation causes less reactor corrosion, lower environmental pollution, and simpler operation compared to chemical activation, it requires higher temperatures and longer times, increasing energy consumption and production costs. Therefore, improving activation methods, increasing reaction rates, and shortening activation times are key to enhancing application efficiency (Li et al. 2022b).

13.4.2 Chemical Activation

Chemical activation plays a vital role in preparing biomass-derived carbon materials. It effectively controls the pore structure and can achieve a large SSA, ranging from 1000 to 4000 m^2/g (Yin et al. 2020; Sun et al. 2019). Activated carbon is typically prepared through one-step direct activation or two-step carbonization and activation processes. In chemical activation, parameters like activator type, the ratio of precursor to activator, activation temperature, and time can be manipulated to optimize the pore structure—SSA, pore size, pore volume, and pore geometry—along with surface functional groups and the graphitization degree. Chen et al. (2020e) and Tyagi et al. (2020) pointed out that chemical activation serves as the main approach to produce activated carbon.

In this process, the carbon precursor is fully impregnated with the activator, and porous carbon materials are obtained by carbonizing in an inert atmosphere. As a result of reactions like dehydration, cross-linking, or condensation between the activator and precursor, carbon atoms are etched. The detached carbon atoms facilitate the formation of a porous structure by separating from the carbon skeleton. Gases such as CO_2, CO, and H_2O further drive the formation of pores. Because activators exhibit high activity, elevated carbonization temperatures are conducive to pore generation (Xu et al. 2022). However, excessively high activation temperatures or activator usage may lead to overly large pores, carbon skeleton collapse, reduced SSA, and poor pore size distribution, affecting capacitive performance Wei et al. (2011). Common chemical activators include alkaline substances, acidic substances, and molten salts.

13.4.2.1 Alkaline Activators

KOH is a commonly used alkaline activator that significantly increases the SSA of activated carbon, up to 4000 m^2/g (Zhang et al. 2022c). The KOH activation mechanism involves chemical activation, physical activation, and the synergistic effect of metal K insertion (Wang and Kaskel 2012). KOH and its derivatives chemically etch the carbon skeleton, releasing gases to promote pore formation, while metallic K vapor embeds into the carbon lattice at high temperatures, expanding the lattice and forming a highly porous structure (Wang et al 2017a). Sun et al. (2020) further showed that inserting metal K broadens the scope of graphitic carbon. This promotes the rearrangement of carbon atoms, leading to the formation of graphitic structures (Xu et al. 2021). Muramatsu et al. (2014) identified that both molten and gaseous KOH show preference in etching amorphous carbon. This etching process generates carbon materials with smooth edges, and simultaneously increases their crystallinity. Han et al. (2020) discovered that KOH has a greater tendency to react with C—N bonds rather than C—C bonds, resulting in a drop in the nitrogen content of the carbon materials after activation. As temperature climbs and the proportion of KOH increases, the SSA and TPV of carbon materials typically rise until they hit a peak (Yu et al. 2018; Dong et al. 2018) But, overly high temperatures and an excessive amount of KOH can cause pores to collapse. As a result, the SSA and TPV decline (Wang et al. 2014; Hou et al. 2017; Xie et al. 2016; Ding et al. 2015; Manyala et al. 2016). Yu et al. (2017) pointed out that the SSA and TPV of activated carbon originate from corn stover are affected by the KOH content. Wang et al. (2014) obtained pig skin-derived activated carbon (PSAC) at 800 °C, which exhibited a large SSA and TPV. Therefore, the selection of appropriate KOH activation parameters needs to be adjusted according to different types of biomass.

Additionally, NaOH and $KHCO_3$ function as activators to improve the pore structure. Wu et al. (2022) synthesized microwave-adsorbing materials from coconut shells using NaOH activation, achieving a large SSA and pore size. Shi et al. (2022) used $KHCO_3$ to synthesize porous carbon spheres with a SSA of 3576 m^2/g.

13.4.2.2 Acidic Activators

Commonly, H_3PO_4 is employed for synthesizing microporous carbon materials, as it introduces pore structures and phosphorus functional groups at high temperatures, forming porous carbon materials rich in phosphorus functional groups (Qu et al. 2015). As an acidic activator, H_3PO_4 is widely chosen for its ease of recovery, ability to promote mesopore formation, and ability to increase carbon yield, TPV, and SSA. Investigations demonstrate that by activating with H_3PO_4 at low temperatures, a high carbon yield can be achieved, along with lower emissions and a wealth of surface functional groups, including esters, hydroxyls, and phosphorus-containing moieties (Neme et al. 2022). Multiple processes make up the H_3PO_4 activation mechanism. Biomass undergoes hydrolysis, dehydration, cross-linking, aromatization, and during this, pores are formed (Zhu et al. 2016a; Niu et al. 2021). Ionization leads biomass like cellulose to swell into a colloidal state. Meanwhile, hydrolysis and oxidation reactions generate small-molecule saccharides, both polysaccharides and monosaccharides. This contributes to the creation of interlocking graphitic microcrystals and the formation of pores. During activation, H_3PO_4 cross-links with polymers, generating phosphate esters and pyrophosphates, further promoting porous structure formation (Lin et al. 2020). Moreover, washing can remove excess H_3PO_4 and inorganic components. This enhances the pore structure and adsorption capacity of carbon materials.

H_3PO_4-activated carbon materials exhibit excellent performance in electrochemical energy storage and adsorption applications. Hulicova-Jurcakova et al. (2009) found that carbon materials activated by H_3PO_4 at 800 °C showed high and stable performance in electrochemical energy storage devices. To reduce macropore formation, H_3PO_4 is often combined with other methods, such as CO_2 physical activation, to further increase skeleton density and form narrow pore size distributions (Prauchner et al. 2016). Mohammad et al. (2022) used H_3PO_4 to activate water hyacinth, obtaining porous materials with a SSA of 1421 m^2/g at 600 °C. Zhang et al. (2021b) investigated how varying activation temperatures impacted the pore structure of porous carbon materials. They determined that an activation temperature of 650 °C helped preserve a favorable porous structure.

13.4.2.3 Molten Salt Activators

$ZnCl_2$ molten salt activation is widely used as a chemical activation method to produce porous carbon materials. This method can effectively drive the formation of the pore structure. Compared to KOH, $ZnCl_2$ activation is milder and promotes biomass dehydration and aromatic condensation, forming open pores (Sevilla and Mokaya 2014). At high temperatures (>263 °C), $ZnCl_2$ can also expand carbon layers and produce exfoliation effects (Hayashi et al. 2000). Additionally, $ZnCl_2$ catalyzes biomass dehydration and promotes aromatization reactions, avoiding tar formation at high temperatures (Hayashi et al. 2000). In the 450–600 °C temperature interval, $ZnCl_2$ leads to an augmentation in the SSA of carbon materials (Wang et al. 2022b).

The low boiling point and volatility of $ZnCl_2$ facilitate mesopore formation, while aromatic condensation and $ZnCl_2$ templating during activation can result in activated carbon materials with mesopore proportions exceeding 90% (Huang et al. 2009). Liu et al. (2022b) found that $ZnCl_2$ activation significantly improves biomass structure, synthesizing mesoporous carbon materials suitable for fuel cells and catalysts. The mass ratio between the activator and biomass significantly impacts the surface properties of carbon materials. For example, a 1 : 2 mass ratio of $ZnCl_2$ to cotton yields the highest SSA and TPV (Ma et al. 2015). Additionally, $ZnCl_2$-activated carbon materials are widely used for removing heavy metals and organic pollutants (Ma et al. 2015). Luo et al.(2022b) managed to synthesize bio-based adsorbents. These adsorbents are abundant in adsorption sites and demonstrate remarkable adsorption capacities for Cr(VI) and methylene blue (MB).

Despite its advantages, the high cost and release of hydrogen chloride limit the large-scale application of $ZnCl_2$ activation. Therefore, composite molten salt systems are often used for activation (Wang et al. 2019c). Gopalakrishnan et al. (2020) added thiourea to a NaCl/KCl molten salt system, successfully synthesizing nitrogen-sulfur co-doped activated carbon for supercapacitors, demonstrating excellent charge transfer performance. However, excessively high activator-to-biomass ratios and corrosive activation processes remain bottlenecks for cost and yield improvement.

13.4.3 Templating Methods

Templating is a mature method for preparing carbon materials, using templates of different types, structures, sizes, and shapes to control and regulate material pore structure. The fundamental principle involves impregnating a template with a biomass precursor. After that, carbonization and template removal are carried out to acquire carbon materials featuring uniform pores. Depending on their properties, templating methods are classified into soft-template and hard-template approaches.

Soft templates are typically composed of organic molecules, such as non-ionic surfactants and micelles, which form framework structures through interactions with precursors. Successful soft templating requires precise selection of precursors and templates, with hydrogen bonding and electrostatic interactions ensuring pore structure stability. Common soft templating methods include evaporation-induced self-assembly (EISA) and aqueous self-assembly. EISA forms ordered mesoporous structures through solvent evaporation but often suffers from pore collapse due to volume shrinkage (Deng et al. 2016). In contrast, aqueous self-assembly is more environmentally friendly, efficiently generating highly ordered structures and suitable for large-scale production (Zhang et al. 2021c).

The hard templating process requires filling templates with biomass precursors either physically or chemically. Once filled, carbonization is carried out, and subsequently, the templates are eliminated to create porous carbon. Common hard templates include mesoporous molecular sieves, SiO_2, and metal oxides, which are widely used for preparing porous carbon due to their ordered pore structures (Nelson et al. 2016; Li et al. 2019c). Hard templating provides precise spatial control, enabling ordered micro- and 3D structures in carbon materials. Unlike soft templates, hard templates are typically non-degradable and require post-treatment such as acid washing to remove residues. In summary, templating is an effective method for preparing ordered porous carbon materials, with different templates offering varying advantages and applicability.

In recent years, researchers have explored not only hard and soft templating for preparing multifunctional carbon materials but also proposed combining both strategies to achieve higher stability in hierarchical carbon nanomaterials. Haynes et al. (2020) utilized a combination of carbon black hard templates and Pluronics P123 and F127 soft templates. This enabled the successful synthesis of silica microcapsules with varying pore sizes. The synthesized microcapsules exhibited outstanding catalytic performance. Additionally, bio-templating techniques provide new pathways for preparing high-performance functional carbon materials. Chen et al. (2019) prepared 3D hierarchical porous carbon materials using rice straw cellulose as both a template and carbon source. When these materials were used in supercapacitors, they demonstrated low resistivity and high specific capacitance. Recently, new methods such as melamine, ice, and protein templating (Hong et al. 2020; Liang et al. 2022; Ma et al. 2021) have also been successfully applied to prepare functional carbon materials.

13.4.4 Catalytic Graphitization

In general, carbonization above 1000 °C causes carbon atoms to rearrange, forming conjugated sp^2-C atoms and promoting the evolution of graphitic structures. According to (Zhang et al. 2018b), biomass-derived carbon mostly appears in the form of disordered hard carbon with multiple structural and morphological features. This makes the conversion of such carbon into perfect graphite a challenging task, even at 2500 °C. Besides, carbonization at elevated temperatures (over 1000 °C) not only consumes a large amount of energy but also causes pore collapse. This leads to underdeveloped pores and a small SSA. Catalytic graphitization is a simple and reliable method to convert amorphous carbon into graphitic carbon at temperatures not exceeding 1000 °C (Lee et al. 2009). Transition metal elements (Sun et al. 2022; Sevilla and Fuertes 2006) reduce the activation energy for carbon phase transformation. Moreover, metal particles can serve in the role of hard templates. As such, they create additional pores (Gang et al. 2021). This is why catalytic graphitization has extensive applications in manufacturing functional carbon materials.

13.4.4.1 Factors Influencing Catalytic Graphitization

Carbonization temperature is a key factor affecting catalytic graphitization. As temperatures rise, more graphitic carbon forms, increasing the ratio of graphitic to amorphous carbon (Sevilla and Fuertes 2010; Li et al. 2020a). The intensity ratio of the G and D bands (I_G/I_D) is proportional to graphitization degree (Cao et al. 2022; Gao et al. 2022). As carbonization temperature increases, the I_G/I_D value increases, following the trend 1.4 (1000 °C) < 1.6 (1100 °C) < 3.4 (1200 °C), indicating enhanced conductivity. Sun et al. (2014) synthesized N-doped porous graphitic carbon from glucose, using $Ni(NO_3)_2$ and melamine. They discovered that higher temperatures enhance conductivity and regulate graphitic and porous structures. As the temperature rose from 800 to 900 °C, the SSA increased from 649 to 1027 m^2/g. But when the temperature further increased to 1000 °C, the carbon skeleton degraded, decreasing the SSA to 579 m^2/g.

Typically, carbon precursors significantly influence catalytic graphitization. Most biomass-derived graphitic carbon materials have higher I_G/I_D values (>1) compared to activated carbon. Under identical graphitization conditions, graphitic carbon derived directly from biomass precursors shows a higher graphitization degree compared to that derived from activated carbon precursors (Liu et al. 2013). This is primarily due to two reasons: (1) Carbon atom rearrangement easily occurs during biomass dehydration, dehydrogenation, and aromatization reactions. Yet, activated carbon already has a glassy amorphous structure composed of non-hexagonal rings, which may resist graphitization and result in insignificant conductivity improvement (Guo et al. 2012). (2) Uniform distribution of metal catalysts is hard to achieve when simply mixing activated carbon with them. (Sevilla et al. 2008) found that catalysts dispersed at the molecular level in carbon precursors improve catalytic efficiency. Yu et al. (2016) showed metal ions chelating with biomass oxygen-containing groups improve catalyst distribution, promoting graphitization.

13.4.4.2 Catalytic Graphitization Mechanisms

Two widely accepted mechanisms for catalytic graphitization are (Sevilla and Fuertes 2010): (1) The dissolution–precipitation mechanism involves amorphous carbon dissolving into the catalyst until saturation, followed by its precipitation as graphitic carbon. (2) Under the carbide intermediate formation–decomposition mechanism, amorphous carbon is dissolved by carbide droplets and then precipitates as ordered graphitic structures.

Fu's group (Wang et al. 2013) detected the formation of cementite (Fe_3C) during the carbonization of corn stover–Fe but not in corn stover-Co or corn stover -Ni samples. They also determined that the enthalpy of carbide formation follows the sequence Fe–C < Co–C < Ni–C. This makes Fe_3C easier to form at low carbonization temperatures (Wang et al. 2010). In the carbonization process, metallic Fe reacts with amorphous carbon to form Fe_3C, which further decomposes into graphitic carbon and metallic Fe (Boonprachai et al. 2022; Xia et al. 2020; Kim et al. 2017). Fe-based catalysts operate via the carbide intermediate formation–decomposition mechanism. In contrast, Ni- and Co-based graphitization likely follows the dissolution–precipitation mechanism.

13.4.4.3 Activation-catalytic Graphitization

Traditionally, porous carbon contains a significant amount of amorphous and disordered carbon. It readily forms during activation but collapses at extremely high temperatures. High graphitization means the carbon structure has numerous conjugated sp^2 carbon atoms. These structures form at high temperatures yet can be etched by activators. Researchers have made substantial efforts to address the trade-off between pore structure and graphitization. They aim to develop effective methods for preparing carbon materials with both a high SSA and good conductivity (graphitization). As a result, integrating activation and graphitization has emerged as a viable approach to endow carbon materials with a high SSA and good conductivity.

13.4.4.3.1 Single-function Activators and Catalysts

Functional graphitic carbon is prepared through a two-step carbonization process: activation and graphitization. Li et al. (2022a) prepared PGC-Ni-Mo by activating sawdust with KOH, then graphitizing it with bimetallic Ni-Mo catalysts at 750 °C. Compared to carbon-based materials prepared by single activation (PC), PGC-Ni-Mo showed a 9.4% increase in graphitization but a decrease in SSA from 1847.3 (PC) to 1366.54 m^2/g (PGC-Ni-Mo). Thus, activation followed by graphitization slightly improves conductivity but sacrifices pore structure (Wang et al. 2022a). First, graphitization occurs, followed by activation. Additional activation can boost the SSA of graphitic carbon. However, it still fails to supply enough active sites for energy storage (Li et al. 2022). Both approaches are flawed. Significantly improving both the SSA and electronic conductivity of carbon materials remains a formidable challenge.

To tackle this issue, researchers devised a straightforward and cost-efficient one-step carbonization approach for concurrent activation–graphitization. This method involves introducing activators and graphitization catalysts into the carbon precursor at once. Specifically, KOH chemical activation paired with $Ni(NO_3)_2$ catalytic graphitization was employed to produce honeycomb-like functional graphitic carbon from bio-oil (Wang et al. 2020b). The synthesized carbon material demonstrated a SSA of 1302 m^2/g, high EC, and outstanding specific capacitance, rate performance, and cycling stability in 6 M KOH electrolyte. Similar methods have been applied to synthesize various carbon materials from coconut shells (Destyorini et al. 2022), rice wine (Wu et al. 2019), dried kelp (Tan et al. 2022a), and others. Wang et al. (2021b) suggested that controlling the $KOH/FeCl_2$ ratio optimizes the graphitic/amorphous structure. However, when KOH and transition metal salts are combined, insoluble precipitates often form in aqueous solutions. This makes it hard to evenly distribute KOH and graphitization catalysts within the carbon matrix. Therefore, activators such as $K_2C_2O_4$ and $ZnCl_2$ are commonly used instead of KOH for activation–graphitization. Singh et al. (2022) used $K_2C_2O_4$ as an activator and $K_2C_2O_4$ as a graphitization catalyst. A mixed solution of $K_2C_2O_4$ and $K_2C_2O_4$ was evenly distributed over pistachio shell precursors. This ensured well-developed porosity and uniformly structured graphitic carbon (Kumar and Kim 2018; Moniruzzaman et al. 2022; Anil Kumar and Kim 2018; Kumar et al. 2022), where K_2CO_3 provides activation effects and Fe phases contribute to graphitization.

Molten salt activator $ZnCl_2$ is often paired with graphitization catalyst $FeCl_3$ to prepare carbon-based functional materials. Sun et al. (2013) employed this combination to synthesize porous graphene. In the carbonization process, Fe transforms amorphous carbon into graphitic carbon by the carbide intermediate formation–decomposition mechanism. Meanwhile, $ZnCl_2$ activates the carbon skeleton to create numerous pores. The chelation of Fe^{2+}/Zn^{2+} with the functional groups in biomass promotes the uniform distribution of graphitic structures and pores. Perondi et al. (2021) used $FeCl_3$ and $ZnCl_2$ to convert cellulose into graphene carbon. Higher $ZnCl_2$/biomass ratios result in higher graphitization and total pore volume. Other combinations of activators and graphitization catalysts have been developed, such as KOH/$FeCl_3$ (Wu et al. 2019), KOH/K_2FeO_4 (Tan et al. 2022a), KOH/$FeCl_2$ (Wang et al. 2021). Most carbon material samples exhibit large SSA (>1000 m^2/g) and high graphitization ($I_G/I_D > 1.0$). Therefore, activation–graphitization is an effective method for simultaneously achieving porous structures and good conductivity.

13.4.4.3.2 Dual-function Activation-catalysts

Adding two different chemical reagents to biomass precursors is complex and time-consuming in practice. Common activators such as $ZnCl_2$ and KOH are corrosive and toxic, corroding equipment and polluting the environment. Using one chemical reagent to activate and graphitize biomass precursors is more efficient, quicker, and less polluting (Sun et al. 2022).

K_2FeO_4 serves as an efficient and gentle activation–graphitization agent. It enables concurrent activation and graphitization during one-step carbonization. Pan's group (Gong et al. 2020) successfully synthesized hierarchical carbon materials from chitosan aerogel coated with K_2FeO_4. The synthesized carbon material featured 3D hierarchical porosity, a large accessible SSA, short ion-transport pathways, and numerous electroactive sites. Moreover, its localized graphitic structures guaranteed high EC, thereby promoting electron transfer. K_2FeO_4 has also been applied to various biomass sources, such as sodium lignosulfonate (Wang et al. 2021c), bamboo (Gong et al. 2017), peanut shells (Zhang et al. 2019b), dandelion flower stems (Tan et al. 2022b), and wood (Liu et al. 2022). Pan's group (Gong et al. 2017) systematically reviewed the mechanism of K_2FeO_4, primarily based on its reaction with H_2O to generate KOH and $Fe(OH)_3$. KOH acts as an effective activator, etching the carbon matrix to generate pores, while $Fe(OH)_3$ transforms into Fe_2O_3, which is then reduced to Fe_3O_4 or even Fe (Gong et al. 2017). Subsequently, Fe promotes graphitic carbon development via intermediate phases (Fe_3C).

Yang's group (Liu et al. 2022a) prepared carbon materials from a K_2FeO_4/wood mixture. The research team synthesized carbon materials with distinct chemical reagents for comparison: GPC (K_2FeO_4), GC ($FeCl_3$), and PC (KOH). GPC showed the sharpest and strongest graphitic carbon diffraction peak and the highest I_G/I_D value of 1.10, confirming K_2FeO_4's excellent graphitization effect. Pore structure curves revealed GPC's hierarchical porosity, demonstrating K_2FeO_4's efficient activation.

Fu's group (Wang et al. 2013) reported that $K_4Fe(CN)_6$, based on Fe's self-templating and catalytic graphitization effects, successfully converted corn stover into porous graphitic carbon at high temperatures. The group (Zhang et al. 2017) analyzed the impact of $K_4Fe(CN)_6$ on the microstructure of willow catkin-derived carbon materials. At carbonization temperatures of 500 and 600 °C, $K_4Fe(CN)_6$ thermally decomposed, generating K_2CO_3 and Fe_3C through a series of reactions, which promoted porous structure and graphitic carbon formation. BET analysis of various willow catkin carbon microtubes synthesized with $K_4Fe(CN)_6$ (PGCMT), $NiCl_2$ (GCMT-Ni), $FeCl_2$ (GCMT-Fe), and no catalyst (CMT) showed that PGCMT had the largest SSA (1066.6 m^2/g) and the most micropores and mesopores, indicating $K_4Fe(CN)_6$'s effective activation. PGCMT also exhibited significantly improved graphitization ($I_G/I_D = 1.22$), proving $K_4Fe(CN)_6$'s effective activation–graphitization function.

Another green and effective activation–graphitization agent, $K_3[Fe(C_2O_4)_3]$, has also been explored (Zhang et al. 2019c; Zhang et al. 2021). At high temperatures, $K_3[Fe(C_2O_4)_3]$ thermally decomposes into $K_2C_2O_4$ and FeC_2O_4 (Zhang et al. 2020b). $K_2C_2O_4$ is an effective activator with low corrosiveness. The impressive activation–graphitization effect of $K_3[Fe(C_2O_4)_3]$ is primarily based on its decomposition products, $K_2C_2O_4$ and Fe. Unlike K_2FeO_4 and $K_4Fe(CN)_6$, whose derivatives (KOH or KCN) are corrosive or toxic, $K_3[Fe(C_2O_4)_3]$ exhibits more environmentally friendly properties and facilitates a mild activation–graphitization process. In light of this analysis, the activation and graphitization effects of K_2FeO_4, $K_4Fe(CN)_6$, and $K_3[Fe(C_2O_4)_3]$ depend on the types of K and Fe phases. Considering the fixed K/Fe ratios in these compounds, the SSA and graphitization degree of carbon-based materials can be regulated by adjusting the biomass/catalyst mass ratio and carbonization temperature.

13.5 Removal of Organic Pollutants from Water and Soil by Waste-derived Graphitic Carbon

Industrial and agricultural activities are contaminating water and soil globally with toxic heavy metal ions, organic dyes, and other pollutants (Hoang et al. 2022). Biomass-sourced porous carbon materials exhibit large SSA, pore volume, customizable surface features, mechanical strength, and chemical/thermal stability. Therefore, they're promising for remediating water and soil pollutants (Benzigar et al. 2018). The adsorption capabilities of porous carbon materials hinge on their pore structures and surface chemistries. Micropores, sized closely to the molecular dimensions of organic pollutants, demonstrate substantial adsorption capacities. In contrast, mesopores and macropores create channels, enabling organics to reach the internal carbon surfaces (Rossner et al. 2009). Carbon surfaces featuring π-electrons can establish π–π interactions with organic pollutants having benzene rings or C=C bonds. This indicates that a high degree of graphitization enhances the adsorption performance (Chen et al. 2007). Moreover, N and O surface functional groups on carbon materials can form hydrogen bonds or Lewis acid–base interactions with organics having —OH or —COOH groups (Yang et al. 2008). Biomass-derived graphitic carbon materials with minimal defects demonstrate efficient electron-accepting and micro-transport capabilities, rendering them suitable for metal-free catalysis. Incorporating heteroatom additives or leveraging inherent element in biomass precursors is an effective strategy for synthesizing heteroatom-doped graphitic carbon materials. These doped materials enhance the catalytic degradation of pollutants.

Recently, researchers have carried out wide studies to assess carbon materials' use in eliminating diverse organic pollutants from water and soil (Zhang et al. 2013). Target organic pollutants include cationic aromatic dyes (Cao et al. 2011; Inyang and Dickenson 2015; Mondal et al. 2016; Jung et al. 2015; Xu et al. 2011; Yu et al. 2009), agrochemicals, antibiotics/pharmaceuticals, and industrial chemicals including polycyclic aromatic hydrocarbons (PAHs) (Adeel et al. 2017; Brebu and Vasile 2010; Li et al. 2014a). Figure 13.5 illustrates the applications of graphite carbon across various fields, which will be detailed below.

13.5.1 Adsorption and Catalytic Removal of Water Pollutants by Waste-derived Graphitic Carbon

MB, a common cationic dye in aqueous solutions, frequently serves as the target pollutant in adsorption research. Li et al. (2020b) fabricated a cellulose-derived graphene aerogel for MB adsorption. This aerogel exhibited an outstanding MB adsorption capacity of 610.85 mg/g. It outperformed porous carbon derived from bean shells (469.00 mg/g) (Gupta et al. 2019), straw-derived activated carbon (386.00 mg/g) (Khalil et al. 2017), and bamboo-derived biochar (5.80 mg/g) (Huff et al. 2014). When it comes to MB adsorption, unlike the adsorption of gases, metals, and non-metal ions, the process is mainly driven by π–π interactions between the graphene structure and the aromatic structure of MB. Electrostatic interactions play a relatively minor role (Huff et al. 2014; Fang et al. 2020).

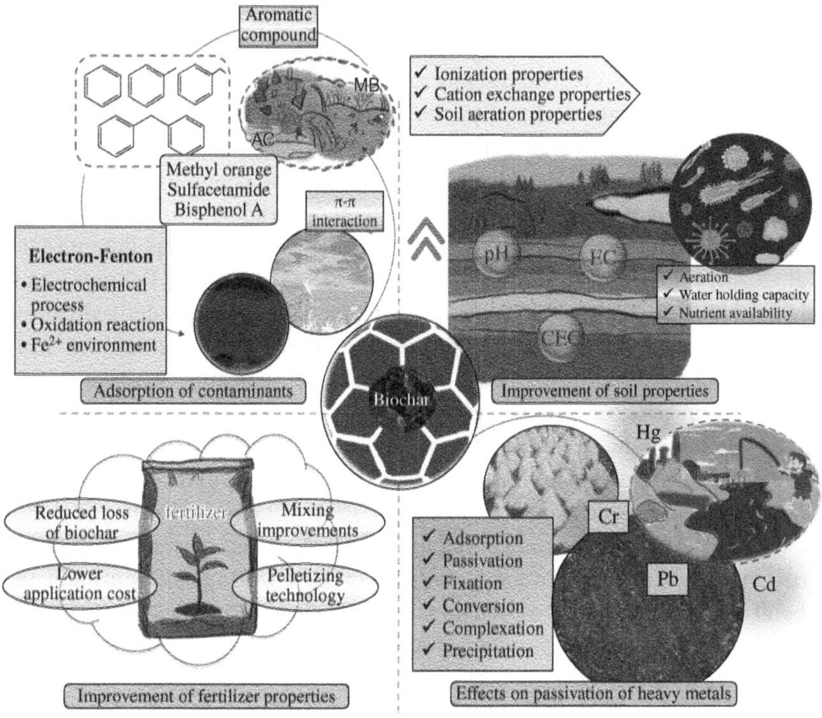

Figure 13.5 Applications of graphite carbon in various fields.

Electron-Fenton (EF) ranks among the popular water treatment technologies. It oxidizes organic pollutants via electrochemical reactions to purify water (Rezgui et al. 2018). In the presence of Fe^{2+}, organic pollutants were strongly oxidatively degraded by in situ electrogenerated H_2O_2 in the presence of two-electron ORR and hydroxyl radicals (—OH) (Ganiyu et al. 2019). Carbon materials featuring a large SSA and high EC are applied as cathodes in EF processes. Ganiyu et al. (2021) engineered sucrose-derived carbon foam, characterized by a SSA of 160 $m^2\ g^{-1}$. The synthesized carbon foam was then deployed in the EF system for the degradation of sulfamethoxazole. Using the carbon foam cathode, the in situ production of H_2O_2 reached 7 mg/L at 60 mA, and a 0.5 mM synthetic sulfamethoxazole solution was completely degraded within 4 hours, with chemical oxygen demand (COD) removal. Doping N- and O-containing functional groups into the carbon skeleton is one strategy for preparing EF cathodes. Zhu et al. (2021) fabricated nitrogen-doped porous carbon materials using *Artemisia argyi*. These materials were employed as cathodes in the EF system for the degradation of MB. The porous carbon cathodes exhibited a nitrogen content of 4.0% and a substantial SSA of 1302 m_2/g. Notably, they used 3D printing technology to fabricate a flexible triboelectric nanogenerator to power the EF system. Throughout the EF treatment, UV-Vis spectroscopy revealed that the characteristic absorption peak of MB at 664 nm declined as the reaction time elapsed. Within 58 minutes, the degradation efficiency of MB reached 98.1%. Moreover, the degradation efficiency demonstrated a positive correlation with the content of C—O—C and —COOH functional groups present in the carbon material.

Bisphenol A (BPA) represents a classic type of environmental pollutant (Fang et al. 2020), sulfacetamide (SAM) (Chen et al. 2020f), and methyl orange (Amir Faiz et al. 2020), can also be catalytically degraded by biomass-derived carbon materials in persulfate activation processes. Carbon materials sourced from biomass—such as graphene sheets—are capable of activating peroxydisulfate. This activation generates sulfate radicals, which facilitate the degradation of BPA. The degradation rate of BPA is strongly influenced by the pH of the solution. Specifically, as the pH value rises, the degradation rate of BPA tends to decline (Fang et al. 2020). Additionally, Chen et al. (2020f) fabricated nitrogen-doped carbon nanosheets derived from chitosan. These nanosheets were used to activate peroxymonosulfate (PMS) for the degradation of SAM. During PMS activation, the N-bond configuration underwent a transformation from graphitic N to pyrrolic N. This change suggested that graphitic N serves as the key reactive functional group for PMS activation. In the advanced oxidation process, non-radical oxidation, with singlet oxygen ($^1O_2$) as the main oxidizing species, was the predominant catalytic mechanism. Apart from pH, the degradation of SAM was also affected by the inhibitory effects of nitrate ions (NO_3^-), chloride ions (Cl^-), and humic acid.

13.5.2 Stabilization and Degradation of Organic Pollutants in Soil by Waste-derived Graphitic Carbon

Regarding organic pollutants in the soil, specifically antibiotics, biochar's high adsorption capacity hinders their transfer. Antibiotics are adsorbed strongly by carbon-based substances, causing weak desorption and thus further confining their mobility in the soil environment. Liu et al. (2017) investigated how straw biochar influenced the migration behavior of sulfonamide antibiotics. In comparison with soil that had not been treated, the soil treated with biochar retained 12–20% more sulfonamide antibiotics. This was attributed to the biochar's excellent physical and chemical properties. Moreover, in the soil amended with biochar, the migration of these antibiotics was more restricted.

Apart from their function in eliminating antibiotics, numerous researchers have delved into the utilization of carbon-based materials for the removal of pesticides (Kamarulazam et al. 2022). You et al. (2021) discovered in a lettuce pot experiment that wood biochar had the ability to selectively absorb metalaxyl, which is a fungicide, and it also had an impact on changing the soil bacterial communities. They further showed that biochar enhanced the population density of pesticide-degrading bacteria, specifically *Methalophilus*. Additionally, it decreased the absorption and accumulation of metalaxyl as well as its metabolites in lettuce. This discovery offers a novel approach for efficiently eliminating chiral pesticide pollution. Applying 1–5% wheat residue and cow manure-derived biochar to soil reduced three herbicides in contaminated soil by nearly 65–88% (Cao et al. 2011; Zhang et al. 2005).

Research has also been conducted on how biochar reduces the bioavailability of organic pollutants in the soil and the subsequent uptake of these pollutants by microorganisms and plants (Yu et al. 2009; Kookana 2010). Researchers explored the impact of adding 0.5–1.0% biochar on the depletion and toxicity reduction of pesticides like pentachlorobenzene (PeCB), carbofuran, and insecticides such as chlorpyrifos and fipronil in plants. Results showed that the dominant adsorption process involved partitioning, followed by diffusion into both the non-carbonized and carbonized components of biochar. Although non-ionic in nature, electrostatic attraction also played a part in removing these contaminants (Yu et al. 2009; Kookana 2010). The removal of contaminants from biochar-treated soils is around 40–50%, but lower than in untreated soils. However, pesticide removal occurred more effectively in soils without biochar amendments. In the absence of biochar, pesticides remained unbound. This unbound form was easily absorbed by plants, making them highly toxic. Only a fraction of these pesticides was reduced through biodegradation and sequestration by soil microorganisms. In contrast, in biochar-amended soils, the contaminant is largely immobilized on the surface of the biochar, reducing its availability to the plant and hence its toxicity to the plant (Yu et al. 2009; Kookana 2010). However, pesticide use in biochar-amended soils increased with increasing doses of biochar (Zhang et al. 2005). Irrespective of how long chlorpyrifos and carbofuran persisted in the soil, adding 1% biochar to the soil reduced the amount these pesticides taken up by plants. Compared to non-amended soils, the uptake decreased by 10–25%. Migration of chlorpyrifos, carbofuran, and fipronil in contaminated soil was reduced by about 32–51% when pyrolyzed cotton straw and wood chip biochar were mixed with soil at 0.1–1.0% at temperatures above 400 °C (Yu et al. 2009; Yang et al. 2010). Therefore, carbon-based materials exhibit excellent performance in removing soil organic pollutants and reducing their bioavailability.

13.6 Improvement of Soil Properties by Waste-derived Graphitic Carbon

Soil serves as the main medium within terrestrial ecosystems. It's an essential carrier for the survival of plants, animals, and microorganisms. However, excessive use of pesticides, fertilizers, and long-term lack of soil restoration in agricultural activities can disrupt soil ecosystems, leading to soil degradation and affecting crop growth. Thus, the restoration of degraded soil and enabling its safe reuse have emerged as prominent subjects in agronomy, ecology, and environmental science. Research indicates that incorporating biochar can markedly boost crop yields and soil fertility. As such, biochar serves as an effective means of enhancing soil quality (Khosravi et al. 2022).

Biochar improves soil quality by enhancing various chemical properties, such as pH, EC, cation exchange capacity (CEC), and organic carbon content, thereby aiding nutrient retention in soil. The alkaline nature of biochar stems from its carbonization process, with biochar produced at lower temperatures (<400 °C) having lower ash content and pH compared to that produced at higher temperatures (>400 °C) (Lehmann et al. 2011). Countless research works have evidenced that biochar can efficiently enhance the soil's pH value and decrease the degree of soil acidification (Agegnehu et al. 2016; Yang et al. 2016). Moreover, biochar remarkably enhances the soil's EC, which in turn stimulates the uptake of nutrients by plants. This effect is particularly pronounced when biochar is produced at elevated temperatures (Wang et al. 2022c).

The CEC of biochar is also significantly enhanced, depending on the material type and production conditions. Under certain mixing ratios, CEC can increase by more than 1.4 times (Muche et al. 2022). Furthermore, carbon-rich biochar can improve the organic matter content of nutrient-poor soils, thereby enhancing soil quality (Ippolito et al. 2016; Zhang et al. 2022b). In biochar derived from corn stover and wood chips, increased carbon content aids in the remediation of saline–alkali soils (Wang et al. 2022c).

Biochar not only improves soil chemical properties but also promotes the healthy development of soil microbial communities. Studies have indicated that biochar creates an optimal living environment for soil microorganisms. By enhancing soil aeration, water retention capacity, and nutrient availability, it helps to reshape the microbial community structure. As a result, soil fertility is enhanced, and crop yields are increased (Aziz et al. 2020). Additionally, biochar addition can enhance soil enzyme activity and improve the living environment of soil microorganisms Martins Filho 2021). Certain studies have likewise demonstrated that biochar has the potential to be integrated with other carbon-based materials, such as graphene. This combination can further enhance the efficiency of nitrogen management in agricultural soils and effectively minimize nitrate loss (Li et al. 2024; Xiang et al. 2024). In conclusion, biochar holds great promise in enhancing soil quality, augmenting its fertility, and fostering the sustainable development of agriculture. As such, it serves as an efficient means for soil remediation.

13.7 Improvement of Fertilizer Properties by Waste-derived Graphitic Carbon

Carbon-based fertilizers, as an environmentally friendly solution, have become a key component of the circular economy. Biochar-based fertilizers are produced through various methods, including mixing biochar with fertilizers or soil amendments or using granulation technology to achieve slow-release functionality, improving nutrient control and slow-release effects. Studies have shown that mixing biochar with fertilizers effectively reduces field operations and increases efficiency. Biochar derived from orchard pruning residues mixed with pig manure and ammonium sulfate significantly reduced N_2O emissions when applied to Japanese pear trees (Oo et al. 2018). Hirst (2018) showed that when biochar and NPK fertilizers were jointly applied during rice cultivation, the crop yield rose from 2.1 to 3 tons per hectare. Liao et al. (2020) discovered that a novel fertilizer integrating biochar and urea significantly improved rapeseed yield by 17% and nitrogen use efficiency by 59% (Liao et al. 2020).

In fertilizer production, granulation is a common method that effectively reduces biochar loss and lowers soil application costs. Especially during heavy rain or wind, fine biochar particles can lead to resource loss and human irritation. Granulation technology can slow nutrient release by increasing diffusion resistance, optimizing its application. Depending on the characteristics of the raw materials, biochar-based fertilizers are typically granulated in two ways: (1) impregnating biochar with other nutrients before granulation; (2) mixing biochar with fertilizer particles and granulating via disc granulation or mechanical extrusion.

In some experiments, biochar–manure mixed granules exhibited good adsorption of NH_4^+-N and promoted plant growth (Shin and Park 2018). The performance of granular fertilizers depends on biomass characteristics and operating conditions. For example, pressing and steam treatment can optimize the mechanical properties and nutrient release of granules. The binder-free biochar granules have poor mechanical properties, thus organic binders are often used to improve granule strength, though their poor degradability may pose environmental issues. Therefore, biochar granules with binders still face challenges in cost and preparation complexity. In practice, optimizing the formulation and production process of granular fertilizers can effectively enhance sustained nutrient release, improve soil moisture regulation, and increase crop yield. Granular fertilizers prepared by mixing biochar with diammonium phosphate, potassium chloride, and bentonite increased corn yield by 46.2% in field trials (Gwenzi et al. 2018). These results demonstrate the environmental benefits and agricultural potential of biochar-based fertilizers.

13.8 Immobilization of Heavy Metals by Waste-derived Graphitic Carbon

With the rapid progress of industrialization and urbanization, heavy metal pollution in the environment has become increasingly severe. Heavy metals like mercury (Hg), cadmium (Cd), chromium (Cr), and lead (Pb) are widely encountered. They are extremely toxic, remaining persistent in the environment. Moreover, they have a strong tendency to bio-accumulate. Since environmental and health laws place severe constraints on the release of these elements, it's essential to create highly efficient treatment methods. Moreover, heavy metal ion pollution accelerates soil degradation. These elements readily accumulate in the soil, resist degradation, and are toxic. When heavy metals enter the food chain or water systems, they present substantial risks to ecosystems and human well-being.

Graphitic carbon materials feature stable aromatic structures, a wealth of oxygen-containing functional groups, and other functional components. This unique composition endows them with the potential to adsorb and immobilize heavy metals. Graphitic carbon-based materials can adsorb and fix heavy metals in soil. Additionally, they can convert the active states of heavy metals into stable ones, decreasing their mobility and bioavailability, which effectively facilitates soil remediation. Luo et al. (2022a) prepared P-doped biomass-derived carbon materials from cow manure and straw, effectively immobilizing Pb and Cd, reducing their mobility in soil, enhancing soil fertility, and lowering pollution risks. Zhang et al. (2019a) used sewage sludge to produce biochar for treating Hg-contaminated soil, significantly reducing methylmercury (MeHg) and total mercury (T_{Hg}) accumulation in rice by 73.4 and 81.9%, respectively.

Moreover, research findings indicate that biochar can effectively adsorb and immobilize heavy metal ions like Pb^{2+}, Zn^{2+}, Cu^{2+}, and CrO_4^{2-}. This occurs via multiple mechanisms, including electrostatic interactions, complexation reactions, ion-exchange processes, and precipitation (Qiu et al. 2021). Qiu et al. (2021) discovered that biochar made from rice and wheat straw notably increased the adsorption of Pb^{2+}. This was achieved by raising the CEC and pH level of the soil. Functional groups on the biochar surface, like carboxyl (—COOH), hydroxyl (—OH), and amino ($-NH_2$), play a crucial role in heavy metal immobilization. They boost the biochar's attraction to these heavy metals (Nguyen et al. 2019). Through ion exchange mechanisms, biochar can effectively remove heavy metals from soil (Gholizadeh and Hu 2021). Ca^{2+} in biochar can exchange with Pb^{2+} due to their similar ionic radii, enhancing Pb^{2+} immobilization (Chen et al. 2020c). Additionally, components such as carbonates and phosphates introduced by biochar can form insoluble precipitates with heavy metal ions, further stabilizing their presence in soil (Gholizadeh and Hu 2021; Bousdra et al. 2023).

In summary, graphitic carbon-based materials demonstrate significant potential in immobilizing heavy metals, reducing their bioavailability, and enhancing soil remediation through adsorption, ion exchange, complexation, and precipitation mechanisms.

13.9 Conclusions and Prospects

In summary, preparing graphitic carbon materials using biomass waste provides substantial advantages in agriculture. Different components in plant and animal biomass exert distinct influences on the structure of carbon materials during the thermal conversion process. This variety provides numerous choices for synthesizing graphitic carbon with customized properties. Regarding preparation technologies, multiple methods such as hydrothermal carbonization and pyrolytic carbonization possess unique features. The structure and properties of carbon materials can be precisely adjusted by changing carbonization conditions and incorporating different additives. Simultaneously, optimization strategies including physical activation, chemical activation, templating methods, and catalytic graphitization can effectively enhance the pore structure and EC of carbon materials for the diverse requirements of agricultural applications. Biomass-derived graphitic carbon shows great potential in various practical applications. For instance, it effectively removes organic pollutants from water and soil. Additionally, it improves soil properties, optimizes fertilizer performance, and immobilizes heavy metals. It not only enables the efficient adsorption and catalytic degradation of organic pollutants but also enhances soil fertility and microbial activity, improves the slow-release of fertilizers, and reduces the bioavailability of heavy metals. These functions play a crucial role in promoting sustainable agricultural development. Nevertheless, this field is currently confronted with numerous challenges. The high-temperature graphitization process is intricate because of energy intensive and large-scale production process, which restricts the translation of biomass derived graphitic carbon in industrial applications. Moreover, in depth research on the long-term interaction mechanisms between biomass derived graphitic carbon and soil, crops, as well as its potential impacts on the ecological environment, will provide a more robust theoretical foundation and practical guidance for its extensive application in the agricultural sector, thereby propelling the green development of agriculture to new heights.

References

Adeel, M., Song, X., Wang, Y. et al. (2017). Environmental impact of estrogens on human, animal and plant life: a critical review. *Environment International* 99: 107–119. https://doi.org/10.1016/j.envint.2016.12.010.

Agegnehu, G., Nelson, P.N., and Bird, M.I. (2016). Crop yield, plant nutrient uptake and soil physicochemical properties under organic soil amendments and nitrogen fertilization on Nitisols. *Soil and Tillage Research* 160: 1–13. https://doi.org/10.1016/j.still.2016.02.003.

Amir Faiz, M.S., Che Azurahanim, C.A., Yazid, Y. et al. (2020). Preparation and characterization of graphene oxide from tea waste and it's photocatalytic application of TiO_2/graphene nanocomposite. *Materials Research Express* 7: 015613. https://doi.org/10.1088/2053-1591/ab689d.

Anil Kumar, Y. and Kim, H.-J. (2018). Preparation and electrochemical performance of $NiCo_2O_4$@$NiCo_2O_4$ composite nanoplates for high performance supercapacitor applications. *New Journal of Chemistry* 42: 19971–19978. https://doi.org/10.1039/C8NJ05401K.

Antal, M.J., Mok, W.S.L., Richards, G.N. et al. (1990). Mechanism of formation of 5-(hydroxymethyl)-2-furaldehyde from d-fructose and sucrose. *Carbohydrate Research* 199: 91–109. https://doi.org/10.1016/0008-6215(90)84096-D.

Arpia, A.A., Chen, W.-H., Lam, S.S. et al. (2021). Sustainable biofuel and bioenergy production from biomass waste residues using microwave-assisted heating: a comprehensive review. *Chemical Engineering Journal* 403: 126233. https://doi.org/10.1016/j.cej.2020.126233.

Ayele, M., Limeneh, D.Y., Tesfaye, T. et al. (2021). Review on utilization routes of the leather industry biomass. *Advances in Materials Science and Engineering* 2021: 1503524. https://doi.org/10.1155/2021/1503524.

Ayiloor Rajesh, G., John, V.L., Pookunnath Santhosh, A. et al. (2022). Carbon dots from natural sources for biomedical applications. *Particle & Particle Systems Characterization* 39: 2200017. https://doi.org/10.1002/ppsc.202200017.

Aziz, S., Yaseen, L., Jamal, A. et al. (2020). Fabrication of biochar from organic wastes and its effect on wheat growth and soil microflora. *Polish Journal of Environmental Studies* 29: 1069–1076. https://doi.org/10.15244/pjoes/99825.

Ba, Y., Pan, W., Pi, S. et al. (2018). Nitrogen-doped hierarchical porous carbon derived from a chitosan/polyethylene glycol blend for high performance supercapacitors. *RSC Advances* 8: 7072–7079. https://doi.org/10.1039/C8RA00016F.

Bar-On, Y.M., Phillips R., and Milo, R. (2018). The biomass distribution on Earth. *Proceedings of the National Academy of Sciences* 115: 6506–6511. https://doi.org/10.1073/pnas.1711842115.

Benzigar, M.R., Talapaneni, S.N., Joseph, S. et al. (2018). Recent advances in functionalized micro and mesoporous carbon materials: synthesis and applications. *Chemical Society Reviews* 47: 2680–2721. https://doi.org/10.1039/C7CS00787F.

Blankenship, L.S. (2021). Correction: cigarette butt-derived carbons have ultra-high surface area and unprecedented hydrogen storage capacity. *Energy & Environmental Science* 14: 3633–3633. https://doi.org/10.1039/D1EE90031E.

Boonprachai, R., Autthawong, T., Namsar, O. et al. (2022). Natural porous carbon derived from popped rice as anode materials for lithium-ion batteries. *Crystals* 12 (2): 223. https://doi.org/10.3390/cryst12020223.

Borghei, M., Lehtonen, J., Liu, L. et al. (2018). Advanced biomass-derived electrocatalysts for the oxygen reduction reaction. *Advanced Materials* 30: 1703691. https://doi.org/10.1002/adma.201703691.

Bousdra, T., Papadimou, S.G., and Golia, E.E. (2023). The use of biochar in the remediation of Pb, Cd, and Cu-contaminated soils. the impact of biochar feedstock and preparation conditions on its remediation capacity. *Land* 12 (2): 383. https://doi.org/10.3390/land12020383.

Brebu, M. and Vasile, C. (2010). Thermal degradation of lignin—a review. *Cellulose Chemistry and Technology* 44: 353–363. https://webofscience.clarivate.cn/wos/alldb/full-record/WOS:000287385000006.

Briscoe, J., Marinovic, A., Sevilla, M. et al. (2015). Biomass-derived carbon quantum dot sensitizers for solid-state nanostructured solar cells. *Angewandte Chemie International Edition* 54: 4463–4468. https://doi.org/10.1002/anie.201409290.

Cao, M., Liu, Y., Sun, K. et al. (2022). Coupling Fe_3C nanoparticles and N-doping on wood-derived carbon to construct reversible cathode for Zn–air batteries. *Small* 18: 2202014. https://doi.org/10.1002/smll.202202014.

Cao, X., Ma, L., Liang, Y. et al. (2011). Simultaneous immobilization of lead and atrazine in contaminated soils using dairy-manure biochar. *Environmental Science & Technology* 45: 4884–4889. https://doi.org/10.1021/es103752u.

Carpenter, D., Westover, T.L., Czernik, S. et al. (2014). Biomass feedstocks for renewable fuel production: a review of the impacts of feedstock and pretreatment on the yield and product distribution of fast pyrolysis bio-oils and vapors. *Green Chemistry* 16: 384–406. https://doi.org/10.1039/C3GC41631C.

Chang, C.-F., Chang, C.-Y., and Tsai, W.-T. (2000). Effects of burn-off and activation temperature on preparation of activated carbon from corn cob agrowaste by CO_2 and steam. *Journal of Colloid and Interface Science* 232: 45–49. https://doi.org/10.1006/jcis.2000.7171.

Chaudhari, K.N., Song, M.Y., Yu, J.-S. (2014). Transforming hair into heteroatom-doped carbon with high surface area. *Small* 10: 2625–2636. https://doi.org/10.1002/smll.201303831.

Chen, C., Huang, Y., Zhu, Y. et al. (2020a). Nonignorable influence of oxygen in hard carbon for sodium ion storage. *ACS Sustainable Chemistry & Engineering* 8: 1497–1506. https://doi.org/10.1021/acssuschemeng.9b05948.

Chen, C., Li, X., Deng, J. et al. (2018). Shape engineering of biomass-derived nanoparticles from hollow spheres to bowls through solvent-induced buckling. *Chemistry Sustainability Energy Materials* 11: 2540–2546. https://doi.org/10.1002/cssc.201801215.

Chen, C., Wang, H., Han, C. et al. (2017a). Asymmetric flasklike hollow carbonaceous nanoparticles fabricated by the synergistic interaction between soft template and biomass. *Journal of the American Chemical Society* 139: 2657–2663. https://doi.org/10.1021/jacs.6b10841.

Chen, C., Zhang, Y., Li, Y. et al. (2017b). Highly conductive, lightweight, low-tortuosity carbon frameworks as ultrathick 3D current collectors, *Advanced Energy Materials* 7: 1700595. https://doi.org/10.1002/aenm.201700595.

Chen, H., Lu, X., Wang, H. et al. (2020b). Controllable fabrication of nitrogen-doped porous nanocarbons for high-performance supercapacitors via supramolecular modulation strategy. *Journal of Energy Chemistry* 49: 348–357. https://doi.org/10.1016/j.jechem.2020.02.043.

Chen, H., Yuan, X., Xiong, T. et al. (2020c). Biochar facilitated hydroxyapatite/calcium silicate hydrate for remediation of heavy metals contaminated soils. *Water, Air, & Soil Pollution* 231: 66. https://doi.org/10.1007/s11270-020-4425-1.

Chen, L., Ji, T., Brisbin, L. et al. (2015). Hierarchical porous and high surface area tubular carbon as dye adsorbent and capacitor electrode. *ACS Applied Materials & Interfaces* 7: 12230–12237. https://doi.org/10.1021/acsami.5b02697.

Chen, L., Lv, A., Guo, F. et al. (2020d). Solid–liquid coexisting $LiNO_3$ electrolyte for extremely stable lithium metal anodes on a bare Cu foil. *ACS Sustainable Chemistry & Engineering* 8: 706–713. https://doi.org/10.1021/acssuschemeng.9b06624.

Chen, T., Ye, T., Zhu, J. et al. (2021a). Small-sized biomass-derived hydrothermal carbon with enriched oxygen groups quickens benzene hydroxylation to phenol with dioxygen. *Applied Catalysis A: General* 626: 118356. https://doi.org/10.1016/j.apcata.2021.118356.

Chen, W., Duan, L., and Zhu, D. (2007). Adsorption of polar and nonpolar organic chemicals to carbon nanotubes. *Environmental Science & Technology* 41: 8295–8300. https://doi.org/10.1021/es071230h.

Chen, W. Gong, M. Li, K. et al. (2020e). Insight into KOH activation mechanism during biomass pyrolysis: chemical reactions between O-containing groups and KOH. *Applied Energy* 278: 115730. https://doi.org/10.1016/j.apenergy.2020.115730.

Chen, X., Oh, W.-D., Zhang, P.-H. et al. (2020f). Surface construction of nitrogen-doped chitosan-derived carbon nanosheets with hierarchically porous structure for enhanced sulfacetamide degradation via peroxymonosulfate activation:

maneuverable porosity and active sites. *Chemical Engineering Journal* 382: 122908. https://doi.org/10.1016/j.cej.2019.122908.

Chen, Y., Zou, K., Xin, D. et al. (2021b). Polysulfide filter and dendrite inhibitor: highly graphitized wood framework inhibits polysulfide shuttle and lithium dendrites in Li–S batteries. *Advanced Functional Materials* 31. https://doi.org/10.1002/adfm.202102458.

Chen, Z., Wang, X., Ding, Z. et al. (2019). Biomass-based hierarchical porous carbon for supercapacitors: effect of aqueous and organic electrolytes on the electrochemical performance. *ChemSusChem* 12: 5099–5110. https://doi.org/10.1002/cssc.201902218.

Cheng, P., Li, T., Yu, H. et al. (2016). Biomass-derived carbon fiber aerogel as a binder-free electrode for high-rate supercapacitors. *The Journal of Physical Chemistry C* 120: 2079–2086. https://doi.org/10.1021/acs.jpcc.5b11280.

Cho, S.Y., Yun, Y.S., Lee, S. et al. (2015). Carbonization of a stable β-sheet-rich silk protein into a pseudographitic pyroprotein. *Nature Communications* 6: 7145. https://doi.org/10.1038/ncomms8145.

Chyan, Y., Ye, R., Li, Y. et al. (2018). Laser-induced graphene by multiple lasing: toward electronics on cloth, paper, and food. *ACS Nano* 12: 2176–2183. https://doi.org/10.1021/acsnano.7b08539.

Costes, L., Laoutid, F., Brohez, S. et al. (2017). Bio-based flame retardants: when nature meets fire protection. *Materials Science and Engineering: R: Reports* 117: 1–25. https://doi.org/10.1016/j.mser.2017.04.001.

Croisier, F. and Jérôme, C. (2013). Chitosan-based biomaterials for tissue engineering. *European Polymer Journal* 49: 780–792. https://doi.org/10.1016/j.eurpolymj.2012.12.009.

Cui, X., Antonietti, M., and Yu, S.-H. (2006). Structural effects of iron oxide nanoparticles and iron ions on the hydrothermal carbonization of starch and rice carbohydrates. *Small* 2: 756–759. https://doi.org/10.1002/smll.200600047.

Deng, C., Ding, F., Li, X. et al. (2016a). Templated-preparation of a three-dimensional molybdenum phosphide sponge as a high performance electrode for hydrogen evolution. *Journal of Materials Chemistry A* 4: 59–66. https://doi.org/10.1039/C5TA05453B.

Deng, J., Xiong, T., Wang, H. et al. (2016b). Effects of cellulose, hemicellulose, and lignin on the structure and morphology of porous carbons, *ACS Sustainable Chemistry & Engineering* 4: 3750–3756. https://doi.org/10.1021/acssuschemeng.6b00388.

Destyorini, F., Amalia, W.C., Irmawati, Y. et al. (2022). High graphitic carbon derived from coconut coir waste by promoting potassium hydroxide in the catalytic graphitization process for lithium-ion battery anodes. *Energy & Fuels* 36: 5444–5455. https://doi.org/10.1021/acs.energyfuels.2c00632.

Ding, J., Wang, H., Li, Z. et al. (2015). Peanut shell hybrid sodium ion capacitor with extreme energy–power rivals lithium ion capacitors. *Energy & Environmental Science* 8: 941–955. https://doi.org/10.1039/C4EE02986K.

Dong, S., He, X., Zhang, H. et al. (2018). Surface modification of biomass-derived hard carbon by grafting porous carbon nanosheets for high-performance supercapacitors. *Journal of Materials Chemistry A* 6: 15954–15960. https://doi.org/10.1039/C8TA04080J.

Duan, B., Gao, X., Yao, X. et al. (2016). Unique elastic N-doped carbon nanofibrous microspheres with hierarchical porosity derived from renewable chitin for high rate supercapacitors. *Nano Energy* 27: 482–491. https://doi.org/10.1016/j.nanoen.2016.07.034.

Duan, J., Deng, S., Wu, W. et al. (2019). Chitosan derived carbon matrix encapsulated CuP_2 nanoparticles for sodium-ion storage. *ACS Applied Materials & Interfaces* 11: 12415–12420. https://doi.org/10.1021/acsami.8b21213.

Elaigwu, S.E. and Greenway, G.M. (2016). Microwave-assisted and conventional hydrothermal carbonization of lignocellulosic waste material: comparison of the chemical and structural properties of the hydrochars. *Journal of Analytical and Applied Pyrolysis* 118: 1–8. https://doi.org/10.1016/j.jaap.2015.12.013.

Fang, Z., Gao, Y., Bolan, N. et al. (2020). Conversion of biological solid waste to graphene-containing biochar for water remediation: a critical review. *Chemical Engineering Journal* 390: 124611. https://doi.org/10.1016/j.cej.2020.124611.

Forse, A.C., Merlet, C., Allan, P.K. et al. (2015). New insights into the structure of nanoporous carbons from NMR, Raman, and pair distribution function analysis. *Chemistry of Materials* 27: 6848–6857. https://doi.org/10.1021/acs.chemmater.5b03216.

Frank, E., Steudle, L.M., Ingildeev, D. et al. (2014). Carbon fibers: precursor systems, processing, structure, and properties. *Angewandte Chemie International Edition* 53: 5262–5298. https://doi.org/10.1002/anie.201306129.

Franklin, R.E. and Randall, J.T. (1997). Crystallite growth in graphitizing and non-graphitizing carbons. *Proceedings of the Royal Society of London. Series A. Mathematical and Physical Sciences* 209: 196–218. https://doi.org/10.1098/rspa.1951.0197.

Gang, B., Zhang, F., Li, X. et al. (2021). A ulva lactuca-derived porous carbon for high-performance electrode materials in supercapacitor: synergistic effect of porous structure and graphitization degree. *Journal of Energy Storage* 33: 102132. https://doi.org/10.1016/j.est.2020.102132.

Ganiyu, S.O., de Araújo, M.J.G., de Araújo Costa, E.C.T. et al. (2021). Design of highly efficient porous carbon foam cathode for electro-Fenton degradation of antimicrobial sulfanilamide. *Applied Catalysis B: Environmental* 283: 119652. https://doi.org/10.1016/j.apcatb.2020.119652.

Ganiyu, S.O., Oturan, N., Raffy, S. et al. (2019). Efficiency of plasma elaborated sub-stoichiometric titanium oxide (Ti_4O_7) ceramic electrode for advanced electrochemical degradation of paracetamol in different electrolyte media. *Separation and Purification Technology* 208: 142–152. https://doi.org/10.1016/j.seppur.2018.03.076.

Gao, F., Geng, C., Xiao, N. et al. (2018a). Hierarchical porous carbon sheets derived from biomass containing an activation agent and in-built template for lithium ion batteries. *Carbon* 139: 1085–1092. https://doi.org/10.1016/j.carbon.2018.08.010.

Gao, F., Liang, L., Mao, L. et al. (2022). Achievement of excellent microwave absorption on the Co/C-based composite from coal hydrogasification residue. *Diamond and Related Materials.* 129: 109374. https://doi.org/10.1016/j.diamond.2022.109374.

Gao, L., Gan, W., Qiu, Z. et al. (2018b). Biomorphic carbon-doped TiO_2 for photocatalytic gas sensing with continuous detection of persistent volatile organic compounds. *ACS Applied Nano Materials* 1: 1766–1775. https://doi.org/10.1021/acsanm.8b00209.

Gao, S., Yan, Y., and Chen, G. (2016). External water-free approach toward TiO_2 nanoparticles embedded in biomass-derived nitrogen-doped carbon. *ACS Sustainable Chemistry & Engineering* 4: 844–850. https://doi.org/10.1021/acssuschemeng.5b00904.

Gao, Y., Chen, X., Zhang, J. et al. (2015). Chitin-derived mesoporous, nitrogen-containing carbon for heavy-metal removal and styrene epoxidation. ChemPlusChem 80: 1556–1564. https://doi.org/10.1002/cplu.201500293.

Geng, J., Wu, H., Al-Enizi, A.M. et al. (2015). Freestanding eggshell membrane-based electrodes for high-performance supercapacitors and oxygen evolution reaction. *Nanoscale* 7: 14378–14384. https://doi.org/10.1039/C5NR04603C.

Gholizadeh, M. and Hu, X. (2021). Removal of heavy metals from soil with biochar composite: a critical review of the mechanism. *Journal of Environmental Chemical Engineering* 9: 105830. https://doi.org/10.1016/j.jece.2021.105830.

Gong, Y., Li, D., Fu, Q. et al. (2020). Nitrogen self-doped porous carbon for high-performance supercapacitors. *ACS Applied Energy Materials* 3: 1585–1592. https://doi.org/10.1021/acsaem.9b02077.

Gong, Y., Li, D., Luo, C. et al. (2017). Highly porous graphitic biomass carbon as advanced electrode materials for supercapacitors. *Green Chemistry.* 19: 4132–4140. https://doi.org/10.1039/C7GC01681F.

Gopalakrishnan, A., Raju, T.D., and Badhulika, S., (2020). Green synthesis of nitrogen, sulfur-co-doped worm-like hierarchical porous carbon derived from ginger for outstanding supercapacitor performance. *Carbon* 168: 209–219. https://doi.org/10.1016/j.carbon.2020.07.017.

Guerrero, M., Ruiz, M.P., Alzueta, M.U. et al. (2005). Pyrolysis of eucalyptus at different heating rates: studies of char characterization and oxidative reactivity. *Journal of Analytical and Applied Pyrolysis* 74: 307–314. https://doi.org/10.1016/j.jaap.2004.12.008.

Guiotoku, M., Rambo, C.R., Hansel, F.A. et al. (2009). Microwave-assisted hydrothermal carbonization of lignocellulosic materials. *Materials Letters* 63: 2707–2709. https://doi.org/10.1016/j.matlet.2009.09.049.

Guo, J. and Lua, A.C. (2002). Characterization of adsorbent prepared from oil-palm shell by CO_2 activation for removal of gaseous pollutants. *Materials Letters* 55: 334–339. https://doi.org/10.1016/S0167-577X(02)00388-9.

Guo, J., Morris, J.R., Ihm, Y. et al. (2012). Topological defects: origin of nanopores and enhanced adsorption performance in nanoporous carbon. *Small* 8: 3283–3288. https://doi.org/10.1002/smll.201200894.

Guo, S., Peng, J., Li, W. et al. (2009). Effects of CO_2 activation on porous structures of coconut shell-based activated carbons. *Applied Surface Science* 255: 8443–8449. https://doi.org/10.1016/j.apsusc.2009.05.150.

Guo, Y., Liu, W., Wu, R. et al. (2018). Marine-biomass-derived porous carbon sheets with a tunable N-doping content for superior sodium-ion storage. *ACS Applied Materials & Interfaces* 10: 38376–38386. https://doi.org/10.1021/acsami.8b14304.

Gupta, K., Gupta, D., and Khatri, O.P. (2019). Graphene-like porous carbon nanostructure from Bengal gram bean husk and its application for fast and efficient adsorption of organic dyes. *Applied Surface Science* 476: 647–657. https://doi.org/10.1016/j.apsusc.2019.01.138.

Gwenzi, W., Nyambishi, T.J., Chaukura, N. et al. (2018). Synthesis and nutrient release patterns of a biochar-based N–P–K slow-release fertilizer. *International Journal of Environmental Science and Technology* 15: 405–414. https://doi.org/10.1007/s13762-017-1399-7.

Haeldermans, T., Campion, L., Kuppens, T. et al. (2020). A comparative techno-economic assessment of biochar production from different residue streams using conventional and microwave pyrolysis. *Bioresource Technology* 318: 124083. https://doi.org/10.1016/j.biortech.2020.124083.

Han, P., Cheng, M., Luo, D. et al. (2020). Selective etching of C-N bonds for preparation of porous carbon with ultrahigh specific surface area and superior capacitive performance. *Energy Storage Materials* 24: 486–494. https://doi.org/10.1016/j.ensm.2019.07.009.

Han, X., Ye, R., Chyan, Y. et al. (2018). Laser-induced graphene from wood impregnated with metal salts and use in electrocatalysis. *ACS Applied Nano Materials* 1: 5053–5061. https://doi.org/10.1021/acsanm.8b01163.

Hao, R., Lan, H., Kuang, C. et al. (2018). Superior potassium storage in chitin-derived natural nitrogen-doped carbon nanofibers. *Carbon* 128: 224–230. https://doi.org/10.1016/j.carbon.2017.11.064.

Harris, P.J.F. (2013). Fullerene-like models for microporous carbon. *Journal of Materials Science* 48: 565–577. https://doi.org/10.1007/s10853-012-6788-1.

Hayashi, J.I., Kazehaya, A., Muroyama, K. et al. (2000). Preparation of activated carbon from lignin by chemical activation. *Carbon* 38: 1873–1878. https://doi.org/10.1016/S0008-6223(00)00027-0.

Haynes, T., Bougnouch, O., Dubois, V. et al. (2020). Preparation of mesoporous silica nanocapsules with a high specific surface area by hard and soft dual templating approach: application to biomass valorization catalysis. *Microporous and Mesoporous Materials* 306: 110400. https://doi.org/10.1016/j.micromeso.2020.110400.

He, H. Zhang, Y., Wang, P. et al. (2021a). Preparation of sponge-cake-like N-doped porous carbon materials derived from silk fibroin by chemical activation. *Microporous and Mesoporous Materials* 317: 110998. https://doi.org/10.1016/j.micromeso.2021.110998.

He, H., Zhang, Y., Zhang, W. et al. (2021b). Dual metal-loaded porous carbon materials derived from silk fibroin as bifunctional electrocatalysts for hydrogen evolution reaction and oxygen evolution reaction. *ACS Applied Materials & Interfaces* 13: 30678–30692. https://doi.org/10.1021/acsami.1c07058.

Heckel, J.C., Farhan, F.F., and Chumanov, G. (2008). The effect of fructose derived carbon shells on the plasmon resonance and stability of silver nanoparticles. *Colloid and Polymer Science* 286: 1545–1552. https://doi.org/10.1007/s00396-008-1929-4.

Hirst, E.A., Taylor, A., and Mokaya, R. (2018). A simple flash carbonization route for conversion of biomass to porous carbons with high CO_2 storage capacity. *Journal of Materials Chemistry A* 6: 12393–12403. https://doi.org/10.1039/C8TA04409K.

Hirst, G. (2018). Practical application of biochar fertilizer for use in rice cultivation. *'One Ton per Rai I' experiment in eastern Thailand, 2018 International Conference and Utility Exhibition on Green Energy for Sustainable Development (ICUE)*, pp. 1–14. https://doi.org/10.23919/ICUE-GESD.2018.8635691.

Hoang, A.T., Nižetić, S., Cheng, C.K. et al. (2022). Heavy metal removal by biomass-derived carbon nanotubes as a greener environmental remediation: a comprehensive review. *Chemosphere* 287: 131959. https://doi.org/10.1016/j.chemosphere.2021.131959.

Hong, X., Wang, R., Li, S. et al. (2020). Hydrophilic macroporous SnO_2/rGO composite prepared by melamine template for high efficient photocatalyst. *Journal of Alloys and Compounds* 816: 152550. https://doi.org/10.1016/j.jallcom.2019.152550.

Hou, J., Jiang, K., Wei, R. et al. (2017). Popcorn-derived porous carbon flakes with an ultrahigh specific surface area for superior performance supercapacitors. *ACS Applied Materials & Interfaces* 9: 30626–30634. https://doi.org/10.1021/acsami.7b07746.

Hu, B., Wang, K., Wu, L. et al. (2010). Engineering carbon materials from the hydrothermal carbonization process of biomass. *Advanced Materials* 22: 813–828. https://doi.org/10.1002/adma.200902812.

Hu, X.-H., Zhang, R., Wu, Z. et al. (2022). Concentrated solar induced graphene. *ACS Omega* 7: 27263–27271. https://doi.org/10.1021/acsomega.2c02159.

Huang, Y.a., Hu, S., Zuo, S. et al. (2009). Mesoporous carbon materials prepared from carbohydrates with a metal chloride template. *Journal of Materials Chemistry* 19: 7759–7764. https://doi.org/10.1039/B911011A.

Huff, M.D., Kumar, S., Lee, J.W. (2014). Comparative analysis of pinewood, peanut shell, and bamboo biomass derived biochars produced via hydrothermal conversion and pyrolysis. *Journal of Environmental Management* 146: 303–308. https://doi.org/10.1016/j.jenvman.2014.07.016.

Hulicova-Jurcakova, D., Puziy, A.M., Poddubnaya, O.I. et al. (2009). Highly stable performance of supercapacitors from phosphorus-enriched carbons. *Journal of the American Chemical Society* 131: 5026–5027. https://doi.org/10.1021/ja809265m.

Inyang, M. and Dickenson, E. (2015). The potential role of biochar in the removal of organic and microbial contaminants from potable and reuse water: a review. *Chemosphere* 134: 232–240. https://doi.org/10.1016/j.chemosphere.2015.03.072.

Ioannidou, O. and Zabaniotou, A. (2007). Agricultural residues as precursors for activated carbon production—a review. *Renewable and Sustainable Energy Reviews* 11: 1966–2005. https://doi.org/10.1016/j.rser.2006.03.013.

Ippolito, J.A., Ducey, T.F., Cantrell, K.B. et al. (2016). Designer, acidic biochar influences calcareous soil characteristics. *Chemosphere* 142: 184–191. https://doi.org/10.1016/j.chemosphere.2015.05.092.

Jain, A., Balasubramanian, R., and Srinivasan, M.P., (2016). Hydrothermal conversion of biomass waste to activated carbon with high porosity: A review. *Chemical Engineering Journal* 283: 789–805. https://doi.org/10.1016/j.cej.2015.08.014.

Joo, J.B., Kim, P., Kim, W. et al. (2008). Simple preparation of hollow carbon sphere via templating method. *Current Applied Physics* 8: 814–817. https://doi.org/10.1016/j.cap.2007.04.038.

Jung, C., Oh, J., and Yoon, Y. (2015). Removal of acetaminophen and naproxen by combined coagulation and adsorption using biochar: influence of combined sewer overflow components. *Environmental Science and Pollution Research* 22: 10058–10069. https://doi.org/10.1007/s11356-015-4191-6.

Kaewtrakulchai, N., Faungnawakij, K., and Eiad-Ua, A. (2020). Parametric study on microwave-assisted pyrolysis combined KOH activation of oil palm male flowers derived nanoporous carbons. Materials 13 (12): 2876. https://doi.org/10.3390/ma13122876.

Kamarulazam, F., Bashir, S., Subramaniam, R. et al. (2022). Review of peak shaving features of the power box. *Energy Technology* 10: 2101055. https://doi.org/10.1002/ente.202101055.

Kan, T., Strezov, V., and Evans, T.J. (2016). Lignocellulosic biomass pyrolysis: a review of product properties and effects of pyrolysis parameters. *Renewable and Sustainable Energy Reviews* 57: 1126–1140. https://doi.org/10.1016/j.rser.2015.12.185.

Kang, K., Klinghoffer N.B., ElGhamrawy, I. et al. (2021). Thermochemical conversion of agroforestry biomass and solid waste using decentralized and mobile systems for renewable energy and products. *Renewable and Sustainable Energy Reviews* 149: 111372. https://doi.org/10.1016/j.rser.2021.111372.

Khalil, K.M.S., Allam, O.A.S., Khairy, M. et al. (2017). High surface area nanostructured activated carbons derived from sustainable sorghum stalk. *Journal of Molecular Liquids* 247: 386–396. https://doi.org/10.1016/j.molliq.2017.09.090.

Khosravi, A., Zheng, H., Liu, Q. et al. (2022). Production and characterization of hydrochars and their application in soil improvement and environmental remediation. *Chemical Engineering Journal* 430: 133142. https://doi.org/10.1016/j.cej.2021.133142.

Kim, D.-W., Kil, H.-S., Kim, J. et al. (2017). Highly graphitized carbon from non-graphitizable raw material and its formation mechanism based on domain theory. *Carbon* 121: 301–308. https://doi.org/10.1016/j.carbon.2017.05.086.

Kookana, R.S. (2010). The role of biochar in modifying the environmental fate, bioavailability, and efficacy of pesticides in soils: a review. *Soil Research* 48: 627–637. https://doi.org/10.1071/SR10007.

Kumar, Y.A., Al-Asbahi, B.A., Pallavolu, M.R. et al. (2022). Multiple structural defects in poor crystalline nickel-doped tungsten disulfide nanorods remarkably enhance supercapacitive performance. *International Journal of Energy Research* 46: 14227–14239. https://doi.org/10.1002/er.8137.

Kumar, Y.A. and Kim, H.-J. (2018). Effect of time on a hierarchical corn skeleton-like composite of CoO@ZnO as capacitive electrode material for high specific performance supercapacitors. *Energies* 11 (12): 3285. https://doi.org/10.3390/en11123285.

Kurniawan, D., Sharma, N., Rahardja, M.R. et al. (2022). Chiang, plasma nanoengineering of bioresource-derived graphene quantum dots as ultrasensitive environmental nanoprobes. *ACS Applied Materials & Interfaces* 14: 52289–52300. https://doi.org/10.1021/acsami.2c15251.

Kurra, N., Jiang, Q., Nayak, P. et al. (2019). Laser-derived graphene: a three-dimensional printed graphene electrode and its emerging applications. *Nano Today* 24: 81–102. https://doi.org/10.1016/j.nantod.2018.12.003.

Lee, K.T., Ji, X., Rault, M. et al. (2009). Simple synthesis of graphitic ordered mesoporous carbon materials by a solid-state method using metal phthalocyanines. *Angewandte Chemie International Edition* 48: 5661–5665. https://doi.org/10.1002/anie.200806208.

Lehmann, J., Rillig, M.C., Thies, J. et al. (2011). Biochar effects on soil biota—a review. *Soil Biology and Biochemistry* 43: 1812–1836. https://doi.org/10.1016/j.soilbio.2011.04.022.

Lei, L., Pan, F., Lindbråthen, A. et al. (2021). Carbon hollow fiber membranes for a molecular sieve with precise-cutoff ultramicropores for superior hydrogen separation. *Nature Communications* 12: 268. https://doi.org/10.1038/s41467-020-20628-9.

Li, G., Mao, L., Zhu, B. et al. (2020a). Novel ceramic-based microwave absorbents derived from gangue. *Journal of Materials Chemistry C* 8: 14238–14245. https://doi.org/10.1039/D0TC03633A.

Li, H., Zhang, L., Li, L. et al. (2019a). Two-in-one solution using insect wings to produce graphene-graphite films for efficient electrocatalysis. *Nano Research* 12: 33–39. https://doi.org/10.1007/s12274-018-2172-z.

Li, J., Wang, Q., Zheng, L. et al. (2020b). A novel graphene aerogel synthesized from cellulose with high performance for removing MB in water. *Journal of Materials Science & Technology* 41: 68–75. https://doi.org/10.1016/j.jmst.2019.09.019.

Li, J., Zhang, Z., Wang, Z. et al. (2022a). Low temperature graphitization and electrochemical properties of porous carbon catalyzed with bimetal Ni-Mo. *Diamond and Related Materials* 123: 108862. https://doi.org/10.1016/j.diamond.2022.108862.

Li, K., Chen, W., Yang, H. et al. (2019b). Mechanism of biomass activation and ammonia modification for nitrogen-doped porous carbon materials. *Bioresource Technology* 280: 260–268. https://doi.org/10.1016/j.biortech.2019.02.039.

Li, P., Wan, K., Chen, H. et al. (2022b). Long, value-added products from catalytic pyrolysis of lignocellulosic biomass and waste plastics over biochar-based catalyst: a state-of-the-art review. *Catalysts* 12 (9): 1067. https://doi.org/10.3390/catal12091067.

Li, Q., Xiong, Z., Xiang, P. et al. (2024). Effects of uranium mining on soil bacterial communities and functions in the Qinghai-Tibet plateau. *Chemosphere* 347: 140715. https://doi.org/10.1016/j.chemosphere.2023.140715.

Li, X., Guan, B.Y., Gao, S. et al. (2019c). A general dual-templating approach to biomass-derived hierarchically porous heteroatom-doped carbon materials for enhanced electrocatalytic oxygen reduction. *Energy & Environmental Science* 12: 648–655. https://doi.org/10.1039/C8EE02779J.

Li, Y., Shao, J., Wang, X. et al. (2014a). Characterization of modified biochars derived from bamboo pyrolysis and their utilization for target component (furfural) adsorption. *Energy & Fuels* 28: 5119–5127. https://doi.org/10.1021/ef500725c.

Li, Z., Xu, Z., Tan, X. et al. (2013). Mesoporous nitrogen-rich carbons derived from protein for ultra-high capacity battery anodes and supercapacitors. *Energy & Environmental Science* 6: 871–878. https://doi.org/10.1039/C2EE23599D.

Li, Z., Xu, Z., Wang, H. et al. (2014b). Colossal pseudocapacitance in a high functionality–high surface area carbon anode doubles the energy of an asymmetric supercapacitor. *Energy & Environmental Science* 7: 1708–1718. https://doi.org/10.1039/C3EE43979H.

Li, Z., Zhang, L., Amirkhiz, B.S. et al. (2012). Carbonized chicken eggshell membranes with 3D architectures as high-performance electrode materials for supercapacitors. *Advanced Energy Materials* 2: 431–437. https://doi.org/10.1002/aenm.201100548.

Liang, B., Chen, L., Lv, J. et al. (2022). Seaweed-like nitrogen-doped porous carbon superstructures via an ultrasonic atomization ice template as high-performance electrodes in supercapacitors. *ACS Applied Energy Materials* 5: 6163–6173. https://doi.org/10.1021/acsaem.2c00568.

Liao, J., Liu, X., Hu, A. et al. (2020). Effects of biochar-based controlled release nitrogen fertilizer on nitrogen-use efficiency of oilseed rape (*Brassica napus* L.). *Scientific Reports* 10: 11063. https://doi.org/10.1038/s41598-020-67528-y.

Liew, R.K., Chai, C., Yek, P.N.Y. et al. (2019). Innovative production of highly porous carbon for industrial effluent remediation via microwave vacuum pyrolysis plus sodium-potassium hydroxide mixture activation. *Journal of Cleaner Production* 208: 1436–1445. https://doi.org/10.1016/j.jclepro.2018.10.214.

Lin, G., Wang, Q., Yang, X. et al. (2020). Preparation of phosphorus-doped porous carbon for high performance supercapacitors by one-step carbonization. *RSC Advances* 10: 17768–17776. https://doi.org/10.1039/D0RA02398A.

Lindgren, R. (2016). Gasification of biomass as an alternative energy source to fuel wood and charcoal-a feasibility study in Malawi.

Liu, D., Han, Z., Ma, J. et al. (2021a). Dual-confined SiO_x encapsulated in PVA derived carbon layer and chitin derived N-doped carbon nanosheets for high-performance lithium storage. *Chemical Engineering Journal* 420: 129754. https://doi.org/10.1016/j.cej.2021.129754.

Liu, H., Su, S., Wang, H. et al. (2022a). sustainable one-step strategy for highly graphitized capacitive carbons with hierarchical micro–meso–macro porosity. *Nanoscale Advances* 4: 1394–1407. https://doi.org/10.1039/D1NA00856K.

Liu, R. Mahurin, S.M., Li, C. et al. (2011). Dopamine as a carbon source: the controlled synthesis of hollow carbon spheres and yolk-structured carbon nanocomposites. *Angewandte Chemie International Edition* 50: 6799–6802. https://doi.org/10.1002/anie.201102070.

Liu, W., Yang, J., Liu, S. et al. (2022b). Zinc chloride-activated micro-mesoporous carbon material derived from energy grass and its electrocatalytic performance toward oxygen reduction reaction. *Journal of Energy Storage* 46: 103697. https://doi.org/10.1016/j.est.2021.103697.

Liu, X., Ma, C., Li, J. et al. (2019). Biomass-derived robust three-dimensional porous carbon for high volumetric performance supercapacitors. *Journal of Power Sources* 412: 1–9. https://doi.org/10.1016/j.jpowsour.2018.11.032.

Liu, Y., Liu, Q., Gu, J. et al. (2013). Highly porous graphitic materials prepared by catalytic graphitization. *Carbon* 64: 132–140. https://doi.org/10.1016/j.carbon.2013.07.044.

Liu, Y., Zhang, Y., Zhang, Y.-Q. et al. (2021b). Investigation on the structural feature and gasification reactivity of bio-char derived from energy crop *Fuel* 289: 119904. https://doi.org/10.1016/j.fuel.2020.119904.

Liu, Z., Han, Y., Jing, M. et al. (2017). Sorption and transport of sulfonamides in soils amended with wheat straw-derived biochar: effects of water pH, coexistence copper ion, and dissolved organic matter. *Journal of Soils and Sediments* 17: 771–779. https://doi.org/10.1007/s11368-015-1319-8.

Lourvanij, K. and Rorrer, G.L. (1993). Reactions of aqueous glucose solutions over solid-acid Y-zeolite catalyst at 110–160 °C. *Industrial & Engineering Chemistry Research* 32: 11–19. https://doi.org/10.1021/ie00013a002.

Lu, X., Jordan, B., and Berge, N.D. (2012). Thermal conversion of municipal solid waste via hydrothermal carbonization: comparison of carbonization products to products from current waste management techniques. *Waste Management* 32: 1353–1365. https://doi.org/10.1016/j.wasman.2012.02.012.

Lua, A.C. (2020). A detailed study of pyrolysis conditions on the production of steam-activated carbon derived from oil-palm shell and its application in phenol adsorption. *Biomass Conversion and Biorefinery* 10: 523–533. https://doi.org/10.1007/s13399-019-00447-9.

Luo, W., Wang, B., Heron, C.G. et al. (2014). Pyrolysis of cellulose under ammonia leads to nitrogen-doped nanoporous carbon generated through methane formation. *Nano Letters* 14: 2225–2229. https://doi.org/10.1021/nl500859p.

Luo, Y., Li, Z., Xu, H. et al. (2022a). Development of phosphorus composite biochar for simultaneous enhanced carbon sink and heavy metal immobilization in soil. *Science of The Total Environment* 831: 154845. https://doi.org/10.1016/j.scitotenv.2022.154845.

Luo, Y., Zeng, L., Zhao, Y. et al. (2022b). Roles of $ZnCl_2$ and $FeCl_3$ in preparing high performance corn stover-based carbon materials for efficient removal of Cr (VI) from wastewater. *Journal of Water Process Engineering* 47: 102743. https://doi.org/10.1016/j.jwpe.2022.102743.

Luong, D.X., Bets, K.V., Algozeeb, W.A. et al. (2020). Gram-scale bottom-up flash graphene synthesis. *Nature* 577: 647–651. https://doi.org/10.1038/s41586-020-1938-0.

Luque, R., Menéndez, J.A., Arenillas, A. et al. (2021). Microwave-assisted pyrolysis of biomass feedstocks: the way forward? *Energy & Environmental Science* 5: 5481–5488 https://doi.org/10.1039/C1EE02450G.

Ma, G., Guo, D., Sun, K. et al. (2015). Cotton-based porous activated carbon with a large specific surface area as an electrode material for high-performance supercapacitors. *RSC Advances* 5: 64704–64710. https://doi.org/10.1039/C5RA11179J.

Ma, H. Li, C. Zhang, M. et al. (2017). Graphene oxide induced hydrothermal carbonization of egg proteins for high-performance supercapacitors. *Journal of Materials Chemistry A* 5: 17040–17047. https://doi.org/10.1039/C7TA04771A.

Ma, T., Liao, L., Zhang, X. et al. (2021). Hierarchical pores from microscale to macroscale boost ultrahigh lithium intercalation pseudocapacitance of biomass carbon. *Journal of Energy Storage* 33: 102068. https://doi.org/10.1016/j.est.2020.102068.

Mahmood, F., Zhang, C., Xie, Y. et al. (2019). Transforming lignin into porous graphene via direct laser writing for solid-state supercapacitors. *RSC Advances* 9: 22713–22720. https://doi.org/10.1039/C9RA04073K.

Mahyari, M., Shaabani, A., Behbahani, M. et al. (2014). Thiol-functionalized fructose-derived nanoporous carbon as a support for gold nanoparticles and its application for aerobic oxidation of alcohols in water. *Applied Organometallic Chemistry* 28: 576–583. https://doi.org/10.1002/aoc.3159.

Manyala, N., Bello, A., Barzegar, F. et al. (2016). Coniferous pine biomass: a novel insight into sustainable carbon materials for supercapacitors electrode. *Materials Chemistry and Physics* 182: 139–147. https://doi.org/10.1016/j.matchemphys.2016.07.015.

Martins Filho, A.P., Medeiros, E.V.D., Lima, J.R.S. et al. (2021). Impact of coffee biochar on carbon, microbial biomass and enzyme activities of a sandy soil cultivated with bean. *Anais da Academia Brasileira de Ciências* 93. https://doi.org/10.1590/0001-3765202120200096.

McKendry, P. (2002). Energy production from biomass (part 1): overview of biomass. *Bioresource Technology* 83: 37–46. https://doi.org/10.1016/S0960-8524(01)00118-3.

Mohammad, A.H., Radovic, I., Ivanović, M. et al. (2022). Adsorption of metformin on activated carbon produced from the water hyacinth biowaste using H_3PO_4 as a chemical activator. *Sustainability* 14 (18): 11144. https://doi.org/10.3390/su141811144.

Mondal, A.K., Kretschmer, K., Zhao, Y. et al. (2017). Naturally nitrogen doped porous carbon derived from waste shrimp shells for high-performance lithium ion batteries and supercapacitors. *Microporous and Mesoporous Materials* 246: 72–80. https://doi.org/10.1016/j.micromeso.2017.03.019.

Mondal, S., Bobde, K., Aikat, K. et al. (2016). Biosorptive uptake of ibuprofen by steam activated biochar derived from mung bean husk: equilibrium, kinetics, thermodynamics, modeling and eco-toxicological studies. *Journal of Environmental Management* 182: 581–594. https://doi.org/10.1016/j.jenvman.2016.08.018.

Moniruzzaman, M., Anil Kumar, Y., Pallavolu, M.R. et al. (2022). Two-dimensional core-shell structure of cobalt-soped@MnO_2 nanosheets grown on nickel foam as a binder-free battery-type electrode for supercapacitor application. *Nanomaterials* 12 (18): 3187. https://doi.org/10.3390/nano12183187.

Muche, M., Molla, E., Mohammed, S. et al. (2022). Evaluating slow pyrolysis of parthenium hysterophorus biochar: perspectives to acidic soil amelioration and growth of selected wheat (*Triticum aestivum*) varieties. *The Scientific World Journal* 8181742. https://doi.org/10.1155/2022/8181742.

Muñoz-Hernández, A. and Diaz, G. (2018). Modeling of thermal runaway of carbonaceous materials: graphite, biochar, and wood. *AIP Advances* 8: 095312. https://doi.org/10.1063/1.5038125.

Muramatsu, H., Kim, Y.A. Yang, K.-S. et al. (2014). Rice husk-derived graphene with nano-sized domains and clean edges. *Small* 10: 2766–2770. https://doi.org/10.1002/smll.201400017.

Nelson, K.M., Mahurin, S.M., Mayes, R.T. et al. (2016). Preparation and CO_2 adsorption properties of soft-templated mesoporous carbons derived from chestnut tannin precursors. *Microporous and Mesoporous Materials* 222: 94–103. https://doi.org/10.1016/j.micromeso.2015.09.050.

Neme, I., Gonfa, G., and Masi, C. (2022). Activated carbon from biomass precursors using phosphoric acid: a review. *Heliyon* 8. https://doi.org/10.1016/j.heliyon.2022.e11940.

Nguyen, V.-T., Nguyen, T.-B., Chen, C.-W. et al. (2019). Influence of pyrolysis temperature on polycyclic aromatic hydrocarbons production and tetracycline adsorption behavior of biochar derived from spent coffee ground. *Bioresource Technology* 284: 197–203. https://doi.org/10.1016/j.biortech.2019.03.096.

Ni, M., Leung, D.Y.C., Leung, M.K.H. et al. (2006). An overview of hydrogen production from biomass. *Fuel Processing Technology* 87: 461–472. https://doi.org/10.1016/j.fuproc.2005.11.003.

Niu, H., Jin, H., Sun, Q. et al. (2021). Activation of biochars by waste phosphoric acids: an integrated disposal route of waste acids and solid waste. *ACS Sustainable Chemistry & Engineering* 9: 16403–16414. https://doi.org/10.1021/acssuschemeng.1c06326.

Nizamuddin, S., Baloch, H.A., Griffin, G.J. et al. (2017). An overview of effect of process parameters on hydrothermal carbonization of biomass. *Renewable and Sustainable Energy Reviews* 73: 1289–1299. https://doi.org/10.1016/j.rser.2016.12.122.

Nizamuddin, S., Baloch, H.A., Siddiqui, M.T.H. et al. (2018). An overview of microwave hydrothermal carbonization and microwave pyrolysis of biomass. *Reviews in Environmental Science and Bio/Technology* 17: 813–837. https://doi.org/10.1007/s11157-018-9476-z.

Nogi, M., Kurosaki, F., Yano, H. et al. (2010). Preparation of nanofibrillar carbon from chitin nanofibers. *Carbohydrate Polymers* 81: 919–924. https://doi.org/10.1016/j.carbpol.2010.04.006.

Oo, Z.A., Gonai, T., Sudo, S. et al. (2018). Surface application of fertilizers and residue biochar on N_2O emission from Japanese pear orchard soil. *Plant, Soil and Environment* 64: 597–604. https://doi.org/10.17221/114/2018-PSE.

Ou, J., Zhang, Y., Chen, L. et al. (2015). Nitrogen-rich porous carbon derived from biomass as a high performance anode material for lithium ion batteries. *Journal of Materials Chemistry A* 3: 6534–6541. https://doi.org/10.1039/C4TA06614F.

Ouyang, T., Cheng, K., Yang, F. et al. (2017). From biomass with irregular structures to 1D carbon nanobelts: a stripping and cutting strategy to fabricate high performance supercapacitor materials. *Journal of Materials Chemistry A* 5: 14551–14561. https://doi.org/10.1039/C7TA02412F.

Pariyar, P., Kumari, K., Jain, M.K. et al. (2020). Evaluation of change in biochar properties derived from different feedstock and pyrolysis temperature for environmental and agricultural application. *Science of the Total Environment* 713: 136433. https://doi.org/10.1016/j.scitotenv.2019.136433.

Perondi, D., Bassanesi, G.R., Manera, C. et al. (2021). From cellulose to graphene-like porous carbon nanosheets. *Microporous and Mesoporous Materials* 323: 111217. https://doi.org/10.1016/j.micromeso.2021.111217.

Prakash Kumar, B.G., Shivakamy, K., Miranda, L.R. et al. (2006). Preparation of steam activated carbon from rubberwood sawdust (*Hevea brasiliensis*) and its adsorption kinetics. *Journal of Hazardous Materials* 136: 922–929. https://doi.org/10.1016/j.jhazmat.2006.01.037.

Prauchner, M.J., Sapag, K., and Rodríguez-Reinoso, F. (2016). Tailoring biomass-based activated carbon for CH_4 storage by combining chemical activation with H_3PO_4 or $ZnCl_2$ and physical activation with CO_2. *Carbon* 110: 138–147. https://doi.org/10.1016/j.carbon.2016.08.092.

Qiu, X., Zhou, G., Wang, H. et al. (2021). The behavior of antibiotic-resistance genes and their relationships with the bacterial community and heavy metals during sewage sludge composting. *Ecotoxicology and Environmental Safety* 216: 112190. https://doi.org/10.1016/j.ecoenv.2021.112190.

Qu, J., Geng, C., S. Lv, S. et al. (2015). Nitrogen, oxygen and phosphorus decorated porous carbons derived from shrimp shells for supercapacitors. *Electrochimica Acta* 176: 982–988. https://doi.org/10.1016/j.electacta.2015.07.094.

Ramos Berger, L.R., Montenegro Stamford, T.C., de Oliveira, K.Á.R. et al. (2018). Chitosan produced from Mucorales fungi using agroindustrial by-products and its efficacy to inhibit *Colletotrichum* species. *International Journal of Biological Macromolecules* 108: 635–641. https://doi.org/10.1016/j.ijbiomac.2017.11.178.

Ren, J., Zhou, Y., Wu, H. et al. (2019). Sulfur-encapsulated in heteroatom-doped hierarchical porous carbon derived from goat hair for high performance lithium–sulfur batteries. *Journal of Energy Chemistry* 30: 121–131. https://doi.org/10.1016/j.jechem.2018.01.015.

Rezgui, S., Amrane, A., Fourcade, F. et al. (2018). Electro-Fenton catalyzed with magnetic chitosan beads for the removal of Chlordimeform insecticide. *Applied Catalysis B: Environmental* 226: 346–359. https://doi.org/10.1016/j.apcatb.2017.12.061.

Ribadeneyra, M.C., Grogan, L., Au, H. et al. (2020). Lignin-derived electrospun freestanding carbons as alternative electrodes for redox flow batteries. *Carbon* 157: 847–856. https://doi.org/10.1016/j.carbon.2019.11.015.

Ribeiro, L.S., Rey-Raap, N., Figueiredo, J.L. et al. (2019). Glucose-based carbon materials as supports for the efficient catalytic transformation of cellulose directly to ethylene glycol. *Cellulose* 26: 7337–7353. https://doi.org/10.1007/s10570-019-02583-x.

Romero-Anaya, A.J., Ouzzine, M., Lillo-Ródenas, M.A. et al. (2014). Spherical carbons: synthesis, characterization and activation processes. *Carbon* 68: 296–307. https://doi.org/10.1016/j.carbon.2013.11.006.

Rossner, A., Snyder, S.A., and Knappe, D.R.U. (2009). Removal of emerging contaminants of concern by alternative adsorbents. *Water Research* 43: 3787–3796. https://doi.org/10.1016/j.watres.2009.06.009.

Ruan, G., Sun, Z., Peng, Z. et al. (2011). Growth of graphene from food, insects, and waste. *ACS Nano* 5: 7601–7607. https://doi.org/10.1021/nn202625c.

Ruan, S., Zhu, B., Zhang, H. et al. (2014). A simple one-step method for preparation of fluorescent carbon nanospheres and the potential application in cell organelles imaging. *Journal of Colloid and Interface Science* 422: 25–29. https://doi.org/10.1016/j.jcis.2014.02.006.

Saha, J.K. and Dutta, A. (2022). A review of graphene: material synthesis from biomass sources. *Waste and Biomass Valorization* 13: 1385–1429. https://doi.org/10.1007/s12649-021-01577-w.

Sahu, V., Grover, S., Tulachan, B. et al. (2015). Heavily nitrogen doped, graphene supercapacitor from silk cocoon. *Electrochimica Acta* 160: 244–253. https://doi.org/10.1016/j.electacta.2015.02.019.

Sajjadi, B., Chen, W.-Y., and Egiebor, N.O. (2019). A comprehensive review on physical activation of biochar for energy and environmental applications. *Reviews in Chemical Engineering* 35: 735–776. https://doi.org/10.1515/revce-2017-0113.

Senneca, O., Apicella, B., Russo, C. et al. (2018). Pyrolysis and thermal annealing of coal and biomass in CO_2-rich atmospheres. *Energy & Fuels* 32. https://doi.org/10.1021/acs.energyfuels.8b02417.

Seo, D.H., Pineda, S., Yick, S. et al. (2015). Plasma-enabled sustainable elemental lifecycles: honeycomb-derived graphenes for next-generation biosensors and supercapacitors. *Green Chemistry* 17: 2164–2171. https://doi.org/10.1039/C4GC02135E.

Septien, S., Valin, S., Peyrot, M. et al. (2014). Characterization of char and soot from millimetric wood particles pyrolysis in a drop tube reactor between 800°C and 1400°C. *Fuel* 121: 216–224. https://doi.org/10.1016/j.fuel.2013.12.026.

Sevilla, M. and Fuertes, A.B. (2006). Catalytic graphitization of templated mesoporous carbons. *Carbon* 44: 468–474. https://doi.org/10.1016/j.carbon.2005.08.019.

Sevilla, M., and Fuertes, A.B. (2010). Graphitic carbon nanostructures from cellulose. *Chemical Physics Letters* 490: 63–68. https://doi.org/10.1016/j.cplett.2010.03.011.

Sevilla, M. and Mokaya, R. (2014). Energy storage applications of activated carbons: supercapacitors and hydrogen storage. *Energy & Environmental Science* 7: 1250–1280. https://doi.org/10.1039/C3EE43525C.

Sevilla, M., Salinas Martínez-de Lecea, C., Valdés-Solís, T. et al. (2008). Solid-phase synthesis of graphitic carbon nanostructures from iron and cobalt gluconates and their utilization as electrocatalyst supports. *Physical Chemistry Chemical Physics* 10: 1433–1442. https://doi.org/10.1039/B714924G.

Sha, J., Li, Y., Villegas Salvatierra, R. et al. (2017). Three-dimensional printed graphene foams. *ACS Nano* 11: 6860–6867. https://doi.org/10.1021/acsnano.7b01987.

Shah, J., Lopez-Mercado, J., Carreon, M.G. et al. (2018). Plasma synthesis of graphene from mango peel. *ACS Omega* 3: 455–463. https://doi.org/10.1021/acsomega.7b01825.

Shao, H., Ai, F., Wang, W. et al. (2017). Crab shell-derived nitrogen-doped micro-/mesoporous carbon as an effective separator coating for high energy lithium–sulfur batteries. *Journal of Materials Chemistry A* 5: 19892–19900. https://doi.org/10.1039/C7TA05192A.

Sharma, H.B., Sarmah, A.K., and Dubey, B. (2020). Hydrothermal carbonization of renewable waste biomass for solid biofuel production: a discussion on process mechanism, the influence of process parameters, environmental performance and fuel properties of hydrochar. *Renewable and Sustainable Energy Reviews* 123: 109761. https://doi.org/10.1016/j.rser.2020.109761.

Sharma, S., Thind, S.S., and Kaur, A. (2015). In vitro meat production system: Why and how?. *Journal of Food Science and Technology* 52: 7599–7607. https://doi.org/10.1007/s13197-015-1972-3.

Shi, J., Cui, H., Xu, J. et al. (2022). Carbon spheres synthesized from $KHCO_3$ activation of glucose derived hydrochar with excellent CO_2 capture capabilities at both low and high pressures. *Separation and Purification Technology* 294: 121193. https://doi.org/10.1016/j.seppur.2022.121193.

Shin, J. and Park, S. (2018). Optimization of blended biochar pellet by the use of nutrient releasing model. *Applied Sciences* 8(11): 2274. https://doi.org/10.3390/app8112274.

Shirley, R. and Kammen, D. (2015). Energy planning and development in Malaysian Borneo: assessing the benefits of distributed technologies versus large scale energy mega-projects. *Energy Strategy Reviews* 8: 15–29. https://doi.org/10.1016/j.esr.2015.07.001.

Singh, G., Lee, J., Bahadur, R. et al. (2022). Highly graphitized porous biocarbon nanosheets with tunable Micro-Meso interfaces and enhanced layer spacing for CO_2 capture and LIBs. *Chemical Engineering Journal* 433: 134464. https://doi.org/10.1016/j.cej.2021.134464.

Singh, S.K., Prakash, H., Akhtar, M.J. et al. (2018). Lightweight and high-performance microwave absorbing heteroatom-doped carbon derived from chicken feather fibers. *ACS Sustainable Chemistry & Engineering* 6: 5381–5393. https://doi.org/10.1021/acssuschemeng.8b00183.

Song, J., Chen, C., Zhu, S. et al. (2018). Processing bulk natural wood into a high-performance structural material. *Nature* 554: 224–228. https://doi.org/10.1038/nature25476.

Stephen, J.D., Mabee, W.E., Pribowo, A. et al. (2016). Biomass for residential and commercial heating in a remote Canadian aboriginal community. *Renewable Energy* 86: 563–575. https://doi.org/10.1016/j.renene.2015.08.048.

Sun, F., Wu, D., Gao, J. et al. (2020). Graphitic porous carbon with multiple structural merits for high-performance organic supercapacitor. *Journal of Power Sources* 477: 228759. https://doi.org/10.1016/j.jpowsour.2020.228759.

Sun, L., Gong, Y., Li, D. et al. (2022). Biomass-derived porous carbon materials: synthesis, designing, and applications for supercapacitors. *Green Chemistry* 24: 3864–3894. https://doi.org/10.1039/d2gc00099g.

Sun, L., Tian, C., Fu, Y. et al. (2014). Nitrogen-doped porous graphitic carbon as an excellent electrode material for advanced supercapacitors. *Chemistry–A European Journal* 20: 564–574. https://doi.org/10.1002/chem.201303345.

Sun, L., Tian, C., Li, M. et al. (2013). From coconut shell to porous graphene-like nanosheets for high-power supercapacitors. *Journal of Materials Chemistry A* 1: 6462–6470. https://doi.org/10.1039/C3TA10897J.

Sun, N., Li, Z., Zhang, X. et al. (2019). Hierarchical porous carbon materials derived from kelp for superior capacitive applications. *ACS Sustainable Chemistry & Engineering*. https://doi.org/10.1021/acssuschemeng.9b00635.s001.

Sun, X. and Li, Y. (2004). Colloidal carbon spheres and their core/shell structures with noble-metal nanoparticles. *Angewandte Chemie International Edition* 43: 597–601. https://doi.org/10.1002/anie.200352386.

Taer, E., Deraman, M., Talib, I.A. et al. (2010). Physical, electrochemical and supercapacitive properties of activated carbon pellets from pre-carbonized rubber wood sawdust by CO_2 activation. *Current Applied Physics*. 10: 1071–1075. https://doi.org/10.1016/j.cap.2009.12.044.

Tan, Y., Ren, J., Li, X. et al. (2022a). Highly graphitic porous carbon prepared via K2FeO4-assisted KOH activation for supercapacitors. *New Journal of Chemistry* 46: 14338–14345. https://doi.org/10.1039/D2NJ02150A.

Tan, Y., Xu, Z., He, L. et al. (2022b). Three-dimensional high graphitic porous biomass carbon from dandelion flower activated by K_2FeO_4 for supercapacitor electrode. *Journal of Energy Storage* 52: 104889. https://doi.org/10.1016/j.est.2022.104889.

Terzyk, A.P., Furmaniak, S., Harris, P.J.F. et al. (2007). How realistic is the pore size distribution calculated from adsorption isotherms if activated carbon is composed of fullerene-like fragments? *Physical Chemistry Chemical Physics* 9: 5919–5927. https://doi.org/10.1039/b710552e.

Thangaraj, B., Solomon, P.R., Chuangchote, S. et al. (2021). Biomass-derived carbon quantum dots–a review. Part 1: preparation and characterization. *ChemBioEng Reviews* 8: 265–301. https://doi.org/10.1002/cben.202000029.

Thomsen, M.H., Thygesen, A., and Thomsen, A.B. (2008). Hydrothermal treatment of wheat straw at pilot plant scale using a three-step reactor system aiming at high hemicellulose recovery, high cellulose digestibility and low lignin hydrolysis. *Bioresource Technology* 99: 4221–4228. https://doi.org/10.1016/j.biortech.2007.08.054.

Tian, Y., Estevez, D., Wei, H. et al. (2021). Chitosan-derived carbon aerogels with multiscale features for efficient microwave absorption. *Chemical Engineering Journal* 421: 129781. https://doi.org/10.1016/j.cej.2021.129781.

Titirici, M.-M., Antonietti, M., and Baccile, N. (2008). Hydrothermal carbon from biomass: a comparison of the local structure from poly- to monosaccharides and pentoses/hexoses. *Green Chemistry* 10: 1204–1212. https://doi.org/10.1039/B807009A.

Townsend, S.J., Lenosky, T.J., Muller, D.A. et al. (1992). Negatively curved graphitic sheet model of amorphous carbon. *Physical Review Letters* 69: 921–924. https://doi.org/10.1103/PhysRevLett.69.921.

Tran, H.D., Li, D., and Kaner, R.B. (2009). One-dimensional conducting polymer nanostructures: bulk synthesis and applications. *Advanced Materials* 21: 1487–1499. https://doi.org/10.1002/adma.200802289.

Tseng, K.-H., Shiao, Y.-F., Chang, R.-F. et al. (2013). Optimization of microwave-based heating of cellulosic biomass using Taguchi method. *Materials* 6: 3404–3419. https://doi.org/10.3390/ma6083404.

Tursi, A. (2019). A review on biomass: importance, chemistry, classification, and conversion. *Biofuel Research Journal* 6: 962–979. https://doi.org/10.18331/brj2019.6.2.3.

Tyagi, A., Banerjee, S., Singh, S. et al. (2020). Biowaste derived activated carbon electrocatalyst for oxygen reduction reaction: effect of chemical activation. *International Journal of Hydrogen Energy* 45: 16930–16943. https://doi.org/10.1016/j.ijhydene.2019.06.195.

Villota, S.M., Lei, H., Villota, E. et al. (2019). Microwave-assisted activation of waste cocoa pod husk by H_3PO_4 and KOH—comparative insight into textural properties and pore development. *ACS Omega* 4: 7088–7095. https://doi.org/10.1021/acsomega.8b03514.

Vinodh, R., Sasikumar, Y., Kim, H.-J. et al. (2021). Chitin and chitosan based biopolymer derived electrode materials for supercapacitor applications: a critical review. *Journal of Industrial and Engineering Chemistry* 104: 155–171. https://doi.org/10.1016/j.jiec.2021.08.019.

Wang, C., Chen, W., Xia, K. et al. (2019a). Silk-derived 2D porous carbon nanosheets with atomically-dispersed Fe-N-C sites for highly efficient oxygen reaction catalysts. *Small* 15: 1804966. https://doi.org/10.1002/smll.201804966.

Wang, C., Li, X., Gao, E. et al. (2016). Wearable strain sensors: carbonized silk fabric for ultrastretchable, highly sensitive, and wearable strain sensors (Adv. Mater. 31/2016). *Advanced Materials* 28: 6639–6639. https://doi.org/10.1002/adma.201670215.

Wang, C., Ma, D., and Bao, X. (2008). Transformation of biomass into porous graphitic carbon nanostructures by microwave irradiation. *The Journal of Physical Chemistry C* 112: 17596–17602. https://doi.org/10.1021/jp805113y.

Wang, C., Xie, N.-H., Zhang, Y. et al. (2019b). Silk-derived highly active oxygen electrocatalysts for flexible and rechargeable Zn–Air batteries. *Chemistry of Materials* 31: 1023–1029. https://doi.org/10.1021/acs.chemmater.8b04572.

Wang, D., Nai, J., Xu, L. et al. (2019c). A potassium formate activation strategy for the synthesis of ultrathin graphene-like porous carbon nanosheets for advanced supercapacitor applications. *ACS Sustainable Chemistry & Engineering* 7: 18901–18911. https://doi.org/10.1021/acssuschemeng.9b04209.

Wang, D.-C., Yu, H.-Y., Qi, D. et al. (2021a). Confined chemical transitions for direct extraction of conductive cellulose nanofibers with graphitized carbon shell at low temperature and pressure. *Journal of the American Chemical Society* 143: 11620–11630. https://doi.org/10.1021/jacs.1c04710.

Wang, D.-W., Li, F., Yin, L.-C. et al. (2012). Nitrogen-doped carbon monolith for alkaline supercapacitors and understanding nitrogen-induced redox transitions. *Chemistry–A European Journal* 18: 5345–5351. https://doi.org/10.1002/chem.201102806.

Wang, G., Xu, H., Lu, L. et al. (2015a). One-step synthesis of mesoporous MnO_2/carbon sphere composites for asymmetric electrochemical capacitors. *Journal of Materials Chemistry A* 3: 1127–1132. https://doi.org/10.1039/C4TA03096F.

Wang, H., Li, X., Peng, J. et al. (2021b). Control of the interface graphitized/amorphous carbon of biomass-derived carbon microspheres for symmetric supercapacitors. *Nanoscale Advances* 3: 4858–4865. https://doi.org/10.1039/D1NA00262G.

Wang, J. and Kaskel, S. (2012). KOH activation of carbon-based materials for energy storage. *Journal of Materials Chemistry* 22: 23710–23725. https://doi.org/10.1039/C2JM34066F.

Wang, J. Nie, P., Ding, B. et al. (2017a). Biomass derived carbon for energy storage devices. *Journal of Materials Chemistry A* 5: 2411–2428. https://doi.org/10.1039/C6TA08742F.

Wang, L., Mu, G., Tian, C. et al. (2013). Porous graphitic carbon nanosheets derived from cornstalk biomass for advanced supercapacitors. *ChemSusChem* 6: 880–889. https://doi.org/10.1002/cssc.201200990.

Wang, L., Tian, C., Wang, H. et al. (2010). Mass production of graphene via an in situ self-generating template route and its promoted activity as electrocatalytic support for methanol electroxidization. *The Journal of Physical Chemistry C* 114: 8727–8733. https://doi.org/10.1021/jp911292p.

Wang, M., Zhang, H., Wang, Q. et al. (2015b). Steam-etched spherical carbon/sulfur composite with high sulfur capacity and long cycle life for Li/S battery application. *ACS Applied Materials & Interfaces* 7: 3590–3599. https://doi.org/10.1021/am5075562.

Wang, N., Zhang, G., Guan, T. et al. (2022a). Microphase separation engineering toward 3D porous carbon assembled from nanosheets for flexible all-solid-state supercapacitors. *ACS Applied Materials & Interfaces* 14: 13250–13260. https://doi.org/10.1021/acsami.1c23624.

Wang, Q., Jian, M., Wang C. et al. (2017b). Carbonized silk nanofiber membrane for transparent and sensitive electronic skin. *Advanced Functional Materials* 27: 1605657. https://doi.org/10.1002/adfm.201605657.

Wang, Q., Liu, F., Jin, Z. et al. (2020a). Hierarchically divacancy defect building dual-activated porous carbon fibers for high-performance energy-storage devices. *Advanced Functional Materials* 30: 2002580. https://doi.org/10.1002/adfm.202002580.

Wang, Q., Qin, B., Li, H.-X. et al. (2020b). Honeycomb-like carbon with tunable pore size from bio-oil for supercapacitor. *Microporous and Mesoporous Materials* 309: 110551. https://doi.org/10.1016/j.micromeso.2020.110551.

Wang, S., Han, C. Wang, J. et al. (2014). Controlled synthesis of ordered mesoporous carbohydrate-derived carbons with flower-like structure and N-doping by self-transformation. *Chemistry of Materials* 26: 6872–6877. https://doi.org/10.1021/cm503669v.

Wang, T., Hu, S., Wu, D. et al. (2021c). Boosting the capacity of biomass-based supercapacitors using carbon materials of wood derivatives and redox molecules from plants. *Journal of Materials Chemistry A* 9: 11839–11852. https://doi.org/10.1039/D1TA01542G.

Wang, Y., Yang, R., Li, M. et al. (2015c). Hydrothermal preparation of highly porous carbon spheres from hemp (*Cannabis sativa* L.) stem hemicellulose for use in energy-related applications. *Industrial Crops and Products* 65: 216–226. https://doi.org/10.1016/j.indcrop.2014.12.008.

Wang, Y., Yang, R., Wei, Y. et al. (2014). Preparation of novel pigskin-derived carbon sheets and their low-temperature activation-induced high capacitive performance. *RSC Advances* 4: 45318–45324. https://doi.org/10.1039/C4RA08055F.

Wang, Y., Zhang, M., Shen, X. et al. (2021d). Biomass-derived carbon materials: controllable preparation and versatile applications. *Small* 17: 2008079. https://doi.org/10.1002/smll.202008079.

Wang, Y., Zhuang, Y., Wang, S. et al. (2022b). Preparation and characterization of porous palygorskite/carbon composites through zinc chloride activation for wastewater treatment. *Clays and Clay Minerals* 70: 450–459. https://doi.org/10.1007/s42860-022-00187-4.

Wang, Z., Ogata, H., Morimoto, S. et al. (2015d). Nanocarbons from rice husk by microwave plasma irradiation: from graphene and carbon nanotubes to graphenated carbon nanotube hybrids. *Carbon* 94: 479–484. https://doi.org/10.1016/j.carbon.2015.07.037.

Wang, Z., Pan, X., Kuang, S. et al. (2022c). Amelioration of coastal salt-affected soils with biochar, acid modified biochar and wood vinegar: enhanced nutrient availability and bacterial community modulation, *International Journal of Environmental Research and Public Health* 19 (12):7282. https://doi.org/10.3390/ijerph19127282.

Wang, Z., Shen, D., Wu, C. et al. (2018). State-of-the-art on the production and application of carbon nanomaterials from biomass. *Green Chemistry* 20: 5031–5057. https://doi.org/10.1039/c8gc01748d.

Wei, L., Sevilla, M., Fuertes, A.B. et al. (2011). Hydrothermal carbonization of abundant renewable natural organic chemicals for high-performance supercapacitor electrodes. *Advanced Energy Materials* 1: 356–361. https://doi.org/10.1002/aenm.201100019.

Wei, X. and Li, T. (2021). Wooden activated carbon production for dioxin removal via a two-step process of carbonization coupled with steam activation from biomass wastes. *ACS Omega* 6: 5607–5618. https://doi.org/10.1021/acsomega.0c06032.

Weldekidan, H., Mohanty, A.K., and Misra, M. (2022). Upcycling of plastic wastes and biomass for sustainable graphitic carbon production: a Critical Review. *ACS Environmental Au* 2: 510–522. https://doi.org/10.1021/acsenvironau.2c00029.

Wu, D., Cheng, J., Wang, T. et al. (2019). A novel porous N- and S-self-doped carbon derived from Chinese rice wine lees as high-performance electrode materials in a supercapacitor. *ACS Sustainable Chemistry & Engineering* 7: 12138–12147. https://doi.org/10.1021/acssuschemeng.9b01225.

Wu, L., Zhang, X., and Si, Y. (2022). Polydopamine functionalized superhydrophilic coconut shells biomass carbon for selective cationic dye methylene blue adsorption. *Materials Chemistry and Physics* 279: 125767. https://doi.org/10.1016/j.matchemphys.2022.125767.

Wu, Y., Fu, Z. Yin, D. et al. (2010). Microwave-assisted hydrolysis of crystalline cellulose catalyzed by biomass char sulfonic acids. *Green Chemistry* 12: 696–700. https://doi.org/10.1039/B917807D.

Xia, S., Cai, N., Wu, J. (2020). Synthesis and formation mechanism of biomass-based mesoporous graphitic carbon. *Fuel Processing Technology* 209: 106543. https://doi.org/10.1016/j.fuproc.2020.106543.

Xiang, P., Liao, W., Xiong, Z. et al. (2024). Effects of polystyrene microplastics on the agronomic traits and rhizosphere soil microbial community of highland barley. *Science of the Total Environment* 907: 167986. https://doi.org/10.1016/j.scitotenv.2023.167986.

Xie, L., Sun, G., Su, F. et al. (2016). Hierarchical porous carbon microtubes derived from willow catkins for supercapacitor applications. *Journal of Materials Chemistry A* 4: 1637–1646. https://doi.org/10.1039/C5TA09043A.

Xie, L., Wang, H., Chen, C. et al. (2018), Cooperative assembly of asymmetric carbonaceous bivalve-like superstructures from multiple building blocks. *Research*: 5807980. https://doi.org/10.1155/2018/5807980.

Xie, L., Zhou, S., Liu, J. et al. (2021). Sequential superassembly of nanofiber arrays to carbonaceous ordered mesoporous nanowires and their heterostructure membranes for osmotic energy conversion. *Journal of the American Chemical Society* 143: 6922–6932. https://doi.org/10.1021/jacs.1c00547.

Xu, H., Zhang, Y., and Wang, L. (2021). Hierarchical porous biomass-derived carbon framework with ultrahigh surface area for outstanding capacitance supercapacitor. *Renewable Energy* 179: 1826–1835. https://doi.org/10.1016/j.renene.2021.08.008.

Xu, Q., Si, Y., He, R. et al. (2022). Silk-waste-derived porous carbon for fast electric heating under safe voltage. *ACS Applied Materials & Interfaces* 14: 6005–6015. https://doi.org/10.1021/acsami.1c21874.

Xu, R.-k., Xiao, S.-c., Yuan, J.-h. et al. (2011). Adsorption of methyl violet from aqueous solutions by the biochars derived from crop residues. *Bioresource Technology* 102: 10293–10298. https://doi.org/10.1016/j.biortech.2011.08.089.

Yang, H., Yan, R., Chen, H. et al. (2007). Characteristics of hemicellulose, cellulose and lignin pyrolysis. *Fuel* 86: 1781–1788. https://doi.org/10.1016/j.fuel.2006.12.013.

Yang, K., Wu, W., Jing, Q. et al. (2008). Aqueous adsorption of aniline, phenol, and their substitutes by multi-walled carbon nanotubes. *Environmental Science & Technology* 42: 7931–7936. https://doi.org/10.1021/es801463v.

Yang, X., Xu, G., Yu, H. et al. (2016). Preparation of ferric-activated sludge-based adsorbent from biological sludge for tetracycline removal. *Bioresource Technology* 211: 566–573. https://doi.org/10.1016/j.biortech.2016.03.140.

Yang, X.-B., Ying, G.-G., Peng, P.-Aet al. (2010). Influence of biochars on plant uptake and dissipation of two pesticides in an agricultural soil. *Journal of Agricultural and Food Chemistry* 58: 7915–7921. https://doi.org/10.1021/jf1011352.

Ye, R., Chyan, Y., Zhang, J. et al. (2017). Laser-induced graphene formation on wood. *Advanced Materials* 29: 1702211. https://doi.org/10.1002/adma.201702211.

Yin, J., Zhang, W., Alhebshi, N.A. et al. (2020). Synthesis strategies of porous carbon for supercapacitor applications. *Small Methods* 4: 1900853. https://doi.org/10.1002/smtd.201900853.

You, X., Duan, J., Koda, K. et al. (2016). Preparation of electric double layer capacitors (EDLCs) from two types of electrospun lignin fibers. *Holzforschung* 70: 661–671. https://doi.org/10.1515/hf-2015-0175.

You, X., Suo, F., Yin, S. et al. (2021). Biochar decreased enantioselective uptake of chiral pesticide metalaxyl by lettuce and shifted bacterial community in agricultural soil. *Journal of Hazardous Materials* 417: 126047. https://doi.org/10.1016/j.jhazmat.2021.126047.

Yu, H., Zhang, W., Li, T. et al. (2017). Capacitive performance of porous carbon nanosheets derived from biomass cornstalk. *RSC Advances* 7: 1067–1074. https://doi.org/10.1039/C6RA25899A.

Yu, P., Liang, Y., Dong, H. et al. (2018). Rational synthesis of highly porous carbon from waste bagasse for advanced supercapacitor application. *ACS Sustainable Chemistry & Engineering* 6: 15325–15332. https://doi.org/10.1021/acssuschemeng.8b03763.

Yu, X.-Y., Ying, G.-G., and Kookana, R.S. (2009). Reduced plant uptake of pesticides with biochar additions to soil. *Chemosphere* 76: 665–671. https://doi.org/10.1016/j.chemosphere.2009.04.001.

Yu, Z.-L., Xin, S., You, Y. et al. (2016). Ion-catalyzed synthesis of microporous hard carbon embedded with expanded nanographite for enhanced lithium/sodium storage. *Journal of the American Chemical Society* 138: 14915–14922. https://doi.org/10.1021/jacs.6b06673.

Yuan, T., He, W., Yin, G. et al. (2020). Comparison of bio-chars formation derived from fast and slow pyrolysis of walnut shell. *Fuel* 261: 116450. https://doi.org/10.1016/j.fuel.2019.116450.

Zakzeski, J., Bruijnincx, P.C.A., Jongerius, A.L. et al. (2010). The catalytic valorization of lignin for the production of renewable chemicals. *Chemical Reviews* 110: 3552–3599. https://doi.org/10.1021/cr900354u.

Zang, X., Shen, C., Chu, Y. et al. (2018). Laser-induced molybdenum carbide–graphene composites for 3D foldable paper electronics. *Advanced Materials* 30: 1800062. https://doi.org/10.1002/adma.201800062.

Zhang, B., Jiang, Y., Balasubramanian, R. et al. (2021a). Synthesis, formation mechanisms and applications of biomass-derived carbonaceous materials: a critical review. *Journal of Materials Chemistry A* 9: 24759–24802. https://doi.org/10.1039/D1TA06874A.

Zhang, B., Song, J., Yang, G. et al. (2014). Large-scale production of high-quality graphene using glucose and ferric chloride. *Chemical Science* 5: 4656–4660. https://doi.org/10.1039/C4SC01950D.

Zhang, G., Lei, B., Chen, S. et al. (2021b). Activated carbon adsorbents with micro-mesoporous structure derived from waste biomass by stepwise activation for toluene removal from air. *Journal of Environmental Chemical Engineering* 9: 105387. https://doi.org/10.1016/j.jece.2021.105387.

Zhang, G., Liu, X., Wang, L. et al. (2022a). Recent advances of biomass derived carbon-based materials for efficient electrochemical energy devices. *Journal of Materials Chemistry A* 10: 9277–9307. https://doi.org/10.1039/D2TA01442D.

Zhang, J., Wu, S., Xu, Z. et al. (2019a). The role of sewage sludge biochar in methylmercury formation and accumulation in rice. *Chemosphere* 218: 527–533. https://doi.org/10.1016/j.chemosphere.2018.11.090.

Zhang, K., Cheng, X., Dang, H. et al. (2013). Linking litter production, quality and decomposition to vegetation succession following agricultural abandonment. *Soil Biology and Biochemistry* 57: 803–813. https://doi.org/10.1016/j.soilbio.2012.08.005.

Zhang, P., Bing, X., Jiao, L. et al. (2022b). Amelioration effects of coastal saline-alkali soil by ball-milled red phosphorus-loaded biochar. *Chemical Engineering Journal* 431: 133904. https://doi.org/10.1016/j.cej.2021.133904.

Zhang, P., Sheng, G., Feng, Y. et al. (2005). Role of wheat-residue-derived char in the biodegradation of benzonitrile in soil: nutritional stimulation versus adsorptive inhibition. *Environmental Science & Technology* 39: 5442–5448. https://doi.org/10.1021/es0480670.

Zhang, P., Tan, X., Liu, S. et al. (2019b). Catalytic degradation of estrogen by persulfate activated with iron-doped graphitic biochar: process variables effects and matrix effects. *Chemical Engineering Journal* 378: 122141. https://doi.org/10.1016/j.cej.2019.122141.

Zhang, S., Zheng, M., Tang, Y. et al. (2022c). Understanding synthesis–structure–performance correlations of nanoarchitectured activated carbons for electrochemical applications and carbon capture. *Advanced Functional Materials* 32: 2204714. https://doi.org/10.1002/adfm.202204714.

Zhang, W., Cheng, R.-r., Bi, H.-h. et al. (2021c). A review of porous carbons produced by template methods for supercapacitor applications. *New Carbon Materials* 36: 69–81. https://doi.org/10.1016/S1872-5805(21)60005-7.

Zhang, W., Lei, Y. Ming, F. et al. (2018a). Lignin laser lithography: a direct-write method for fabricating 3D graphene electrodes for microsupercapacitors. *Advanced Energy Materials* 8: 1801840. https://doi.org/10.1002/aenm.201801840.

Zhang, X., Gao, B., Zhao, S. et al. (2020a). Optimization of a “coal-like” pelletization technique based on the sustainable biomass fuel of hydrothermal carbonization of wheat straw. *Journal of Cleaner Production* 242: 118426. https://doi.org/10.1016/j.jclepro.2019.118426.

Zhang, X., Li, H., Qin, B. et al. (2019c). Direct synthesis of porous graphitic carbon sheets grafted on carbon fibers for high-performance supercapacitors. *Journal of Materials Chemistry A* 7: 3298–3306. https://doi.org/10.1039/C8TA11844B.

Zhang, X., Liu, B., Yan, X. et al. (2020b). Design and structure optimization of 3D porous graphitic carbon nanosheets for high-performance supercapacitor. *Microporous and Mesoporous Materials* 309: 110580. https://doi.org/10.1016/j.micromeso.2020.110580.

Zhang, X., Zhang, K., Li, H. et al. (2017). Porous graphitic carbon microtubes derived from willow catkins as a substrate of MnO_2 for supercapacitors. *Journal of Power Sources* 344: 176–184. https://doi.org/10.1016/j.jpowsour.2017.01.107.

Zhang, X.-h., Gan, X.-y., Liu, B.-s. et al. (2021d). An interfacial self-assembly strategy to fabricate graphitic hollow porous carbon spheres for supercapacitor electrodes. *New Carbon Materials* 36: 594–605. https://doi.org/10.1016/S1872-5805(21)60062-8.

Zhang, Y., Li, X., Dong, P. et al. (2018b). Honeycomb-like hard carbon derived from pine pollen as high-performance anode material for sodium-ion batteries. *ACS Applied Materials & Interfaces* 10: 42796–42803. https://doi.org/10.1021/acsami.8b13160.

Zhou, L., Cao, H., Zhu, S. et al. (2015). Hierarchical micro-/mesoporous N- and O-enriched carbon derived from disposable cashmere: a competitive cost-effective material for high-performance electrochemical capacitors. *Green Chemistry* 17: 2373–2382. https://doi.org/10.1039/C4GC02032D.

Zhou, L., Fu, P., Cai, X. et al. (2016). Naturally derived carbon nanofibers as sustainable electrocatalysts for microbial energy harvesting: a new application of spider silk. *Applied Catalysis B: Environmental* 188: 31–38. https://doi.org/10.1016/j.apcatb.2016.01.063.

Zhou, P., Wan, J., Wang, X. et al. (2019). Three-dimensional hierarchical porous carbon cathode derived from waste tea leaves for the electrocatalytic degradation of phenol. *Langmuir* 35: 12914–12926. https://doi.org/10.1021/acs.langmuir.9b02017.

Zhou, X., Li, W., Mabon, R. et al. (2017). Corrigendum: a critical review on hemicellulose pyrolysis. *Energy Technology* 5: 216–216. https://doi.org/10.1002/ente.201600667.

Zhu, G.-z., Deng, X.-l., Hou, M. et al. (2016a). Comparative study on characterization and adsorption properties of activated carbons by phosphoric acid activation from corncob and its acid and alkaline hydrolysis residues. *Fuel Processing Technology* 144: 255–261. https://doi.org/10.1016/j.fuproc.2016.01.007.

Zhu, H., Luo, W., Ciesielski, P.N. et al. (2016b). Correction to wood-derived materials for green electronics, biological devices, and energy applications. *Chemical Reviews* 116: 12650–12650. https://doi.org/10.1021/acs.chemrev.6b00576.

Zhu, Y., Chen, C., Tian, M. et al. (2021). Self-powered electro-Fenton degradation system using oxygen-containing functional groups-rich biomass-derived carbon catalyst driven by 3D printed flexible triboelectric nanogenerator. *Nano Energy* 83: 105720. https://doi.org/10.1016/j.nanoen.2020.105720.

14

Waste-derived Biochar for Low-carbon Construction Materials in Rural Areas

Fulin Qu[1], Weijian Xu[2], Yizhe Wang[1], Yipu Guo[1], Su Yilin[3], and Daniel C.W. Tsang[2]*

[1] *Centre for Infrastructure Engineering and Safety, School of Civil and Environmental Engineering, The University of New South Wales, Kensington, NSW, Australia*
[2] *Department of Civil and Environmental Engineering, The Hong Kong University of Science and Technology, Clear Water Bay, Hong Kong, China*
[3] *Department of Civil and Environmental Engineering, The Hong Kong Polytechnic University, Kowloon, Hong Kong, China*

*Corresponding author: fulin.qu@unsw.edu.au

14.1 Introduction

The global demand for construction materials, especially cement and concrete, continues to grow due to increasing population and rapid urbanization (Qu et al. 2025). Cement production alone contributes ~2.8 billion tons of carbon dioxide emissions annually, accounting for nearly 8% of global emissions (Du et al. 2022b). This considerable carbon footprint results primarily from the calcination of limestone and the high thermal energy required by cement kilns (Du et al. 2022a). In parallel, rural areas face a dual constraint. On one hand, there is an urgent need for affordable infrastructure and housing to support sustainable development. On the other hand, access to low-carbon construction technologies remains limited. Rural communities also generate large quantities of biomass residues through agricultural, forestry, and related activities (Chen et al. 2023). These residues are often inadequately managed or disposed of through environmentally harmful practices such as open burning, which contributes to air pollution and greenhouse gas emissions. The conversion of such biomass into value-added materials has been identified as an effective approach to simultaneously address waste management challenges and promote sustainable construction practices in rural areas.

Waste-derived Biochar for Sustainable Rural Revitalization, First Edition. Edited by Yuqing Sun and Daniel C.W. Tsang.
Published 2026 by John Wiley & Sons, Inc.

Waste-derived biochar has received growing attention in this context. Biochar is a carbon-rich material produced through the thermal decomposition of biomass in oxygen-limited environments, a process commonly referred to as pyrolysis (Xu et al. 2024a). During this process, stable carbon is retained in the solid char fraction, while the emission of carbon dioxide is significantly lower than that from direct combustion. Since the carbon originates from atmospheric carbon dioxide absorbed by plants through photosynthesis, the final product is regarded as carbon-neutral or potentially carbon-negative. When used in long-lasting applications such as soil amendment or construction materials, biochar can contribute to long-term carbon sequestration. In construction applications, biochar produced from waste biomass can be used to partially replace conventional mineral components, thereby reducing the net carbon footprint of materials such as cement and concrete. In addition to its environmental benefits, biochar possesses favorable physicochemical properties. Its high porosity, large specific surface area, and the presence of mineral phases contribute to improved hydration reactions and enhanced mechanical performance in cementitious systems.

Recent research has demonstrated that biochar can function effectively as a supplementary cementitious material (SCM), enhancing properties such as compressive strength, durability, and microstructural stability while decreasing the need for clinker (Li et al. 2024). Biochar has also been investigated as a lightweight aggregate, allowing for the incorporation of larger volumes of biomass waste into structural materials. Moreover, the inclusion of biochar enables the development of specialized construction materials with additional functions, including improved thermal insulation, sound absorption, and electromagnetic shielding. These functional benefits are particularly valuable in rural regions, where locally available biomass can be processed into biochar and utilized to produce construction materials that are both environmentally sustainable and suited to local needs. The use of regional biomass resources in this manner supports the development of a circular rural economy by transforming waste into valuable inputs and reducing dependency on imported construction materials such as cement and gravel.

The chemical and physical characteristics of biochar produced from waste materials have been widely studied, and the influence of production parameters on its suitability for construction applications has been evaluated. Techniques for modifying and engineering biochar, including the adjustment of pyrolysis conditions, particle size control, and surface treatment, have been investigated to optimize performance in cement-based systems (Zhang et al. 2024). Applications in cement replacement, aggregate substitution, and advanced multifunctional composites have been explored. Environmental benefits such as carbon sequestration, pollution reduction, and waste minimization have been examined alongside economic factors, including cost-benefit analysis and the potential to stimulate rural economic growth. Nevertheless, challenges remain, particularly regarding the variability of biochar properties, uncertainty about long-term durability, and the absence of regulatory standards. Continued research is required to address these issues, explore practical implementation at scale, and develop supportive policy frameworks. By integrating waste-derived biochar into construction practices, it is possible to promote climate-resilient, resource-efficient, and economically viable infrastructure development in rural areas.

14.2 Properties of Waste-derived Biochar

Biochar derived from waste biomass possesses distinctive chemical and physical characteristics that underpin its effectiveness as a construction material additive. Understanding these properties is crucial, as they determine how biochar interacts within cementitious matrices and what benefits or challenges it may introduce.

14.2.1 Chemical Properties

Biochar primarily consists of carbon in stable aromatic forms and contains residual inorganic minerals originating from the feedstock. When produced at elevated temperatures, biochar typically contains over 60–80% carbon, making it chemically stable and resistant to biodegradation. Unlike untreated organic matter, its carbon structure remains intact in the alkaline environment of cement. Studies have shown that raw biomass, such as straw or wood fibers, can impair cement hydration and durability due to the release of organic acids and sugars. In contrast, pyrolysis eliminates these reactive compounds, enhancing compatibility with cement. The shared alkalinity between biochar and hydrated cement further inhibits decomposition, allowing the carbon to remain stably sequestered within the concrete matrix (Wen et al. 2023). A comparison of chemical compositions of biochar from different feedstocks is summarized in Table 14.1.

The ash fraction in biochar includes elements such as calcium, potassium, silica, and phosphates, with variations depending on the feedstock. Agricultural residues often yield biochar rich in silica, while biochar from poultry litter or sewage sludge contain more calcium and phosphates due to bone and manure content. These minerals can affect cement hydration. Ash containing alkali and alkaline earth metals may increase pH and stimulate pozzolanic reactions, whereas silica-rich biochar may function as micro-scale SCMs. However, some biochar may introduce undesirable effects. For instance, those high in soluble salts, such as peanut shell biochar, have been associated with increased mortar shrinkage due to impaired moisture retention. Such outcomes highlight the importance of feedstock selection and chemical composition (Wang et al. 2023). Generally, waste-derived biochar exhibit alkaline pH values between 9 and 12, which align with cementitious systems and may help buffer acid ingress, potentially improving resistance to carbonation and leaching.

Table 14.1 Comparison of chemical compositions of biochar from different feedstocks.

Feedstocks	SiO_2	Al_2O_3	CO_2	CaO	Fe_2O_3	MgO	K_2O	N_2O	LOI	References
Rice husk	15.77	0.03	85.57	0.19	0.05	0.04	0.88	N/A	0.47	Gupta et al. (2021)
Corn cob	34.91	5.31	N/A	4.06	2.3	3.1	4.1	12	45.02	Gómez-Vásquez et al. (2022)
Sewage sludge	61.00	10.5	N/A	6.1	19.6	1.77	N/A	N/A	1.03	Wang et al. (2022)
Waste plywood	20.03	6.26	N/A	63.57	2.59	2.68	0.62	0.33	3.92	Qin et al. (2021)
Forest wood	21.38	4.7	N/A	58.67	2.86	0.73	N/A	2.61	22.43	Maljaee et al. (2021)

14.2.2 Physical and Structural Properties

A defining physical characteristic of biochar is its porous structure. During pyrolysis, volatile compounds are released, leaving a carbon matrix rich in micropores and mesopores. A comparison of physical properties of biochar from different feedstocks is summarized in Table 14.2. This porosity often results in specific surface areas ranging from 100 to 300 m^2/g for biochar produced at 500–700 °C, though values depend on production parameters. The porous architecture benefits cement composites in several ways (Chen et al. 2022). First, biochar can act as an internal curing agent by absorbing mix water and gradually releasing it during hydration, maintaining internal humidity and reducing early-age cracking. This mechanism has been linked to increased hydration (up to 10–12%) and decreased permeability in hardened concrete. Second, the large surface area offers numerous nucleation sites for hydration products, accelerating cement reactions. For example, the addition of 5% fine biochar has been shown to enhance early heat evolution and promote calcium-silicate-hydrate formation, improving early strength development (Liu et al. 2022). However, highly porous or fine biochar may increase water demand, requiring mix adjustments such as superplasticizer additions to maintain workability. At higher dosages (e.g., 8%), delayed hydration peaks have been observed due to changes in mix rheology, emphasizing the need to balance particle characteristics with mix design.

Biochar's low density (typically 0.3–0.9 g/cm^3) is significantly lower than that of cement or sand, making it suitable for producing lightweight composites. Its inclusion can reduce structural weight and is especially useful in non-load-bearing applications. Depending on the production process, biochar can range from fine powders (<0.5 mm) to larger porous fragments. Fine particles (passing a 75 μm sieve) integrate well with cement paste, while coarser particles (1–5 mm) can substitute lightweight aggregates. Electron microscopy reveals biochar's preserved cellular structure, which promotes mechanical interlocking with cement and contributes to reduced permeability by creating complex fluid pathways. Incorporating small amounts of biochar has been reported to decrease water absorption by 50–60%, attributed to void filling and improved internal curing. Biochar also exhibits thermal stability and low thermal conductivity (Akinyemi and Adesina 2020). When used in construction materials, it enhances insulation performance; for instance, a 67% reduction in thermal conductivity was observed in biochar–clay panels, while concrete's conductivity dropped to 0.19 W/m·K with 1–2% biochar addition. Electrically, biochar possesses moderate conductivity due to its carbon structure and has been investigated for enabling sensing and electromagnetic shielding functionalities in concrete at low cost (Senadheera et al. 2023). Although its percolation threshold is higher than that of graphite or carbon fibers, effective electrical performance can still be achieved. Overall, the unique

Table 14.2 Comparison of physical properties of biochar from different feedstocks.

	Pyrolysis conditions		Physical properties			
Feedstock	Pyrolysis temperature (°C)	Residence time (min)	BET surface area (m^2/g)	Average pore diameter (nm)	Total pore volume of pores ($\times 10^{-3}$ cm^3/g)	References
Sewage sludge	400	120	33.4	9.4	783	Méndez et al. (2013)
	600	120	37.1	8.3	935	
Maize straw	300	180	3.7	13.2	12.5	
	600	180	353.5	1.8	165.9	
Pine saw dust	300	180	1.3	9.9	3.4	Li et al. (2018)
	600	180	371.2	1.6	121.8	
Rice straw	300	20	2.1	18.1	9.5	
	400	20	4.9	11.6	14.6	Tan et al. (2018)
	600	20	13.5	4.6	15.8	
	800	20	303.7	2.1	161	

combination of low density, high porosity, thermal resistance, and functional conductivity makes waste-derived biochar a versatile additive for enhancing the performance and sustainability of construction materials in rural applications.

14.3 Treatment and Engineering of Biochar

Biochar properties can be effectively tailored through controlled production and post-treatment processes. Optimizing pyrolysis conditions and feedstock selection, along with appropriate processing before use, is essential for enhancing its performance in cementitious materials. This section explores how production parameters, material origin, and functional modifications influence biochar characteristics, and discusses engineering strategies for maximizing their effectiveness in construction applications.

14.3.1 Feedstock Selection and Pyrolysis Conditions

Pyrolysis parameters, particularly temperature, heating rate, and residence time, critically influence biochar's suitability for cementitious applications. Among them, temperature is the most decisive factor. At low temperatures (around 300 °C), incomplete biomass decomposition leads to biochar with high volatile content and low carbon stability. As temperature rises to 500–800 °C, biochar becomes more carbon-rich, aromatic, and porous, while fixed carbon increases and volatile elements decrease. This enhances surface area and chemical stability but lowers yield. Temperatures above 800 °C produce highly stable carbon with minimal yield, making them less economical. Lower-temperature chars retain more functional groups but may contain tar residues that interfere with cement hydration. Optimal cement-compatible biochar is commonly produced around 500 °C, balancing yield, porosity, and stability by eliminating unstable organics such as hemicellulose and cellulose while preserving carbon structure (Wen et al. 2023).

Heating rate also affects biochar properties. Rapid pyrolysis, designed for bio-oil production, produces biochar with low surface area and collapsed pore structures, limiting its use in cement. In contrast, slow pyrolysis at 5–10 °C/min allows gradual devolatilization, forming more ordered carbon structures, higher porosity, and improved stability (Liu et al. 2022). It is generally preferred for construction use. Feedstock type further impacts ash content and elemental composition. Wood yields low-ash, high-carbon char; agricultural residues often produce silica- or potassium-rich ash; and manure-derived biochar contain more phosphorus and calcium. Pre-treatments like washing can reduce harmful salts, as demonstrated in studies on sodium-rich peanut shell biochar. Particle size also matters; fine feedstocks yield uniform, reactive char, while post-pyrolysis milling can further enhance surface area. However, ultra-fine chars may pose handling risks and often requires palletizations for safe application.

14.3.2 Functionalization or Modification Techniques

Methods such as steam or CO_2 activation at elevated temperatures are commonly used to increase porosity, especially in activated carbon production. Although this significantly enhances surface area, it also results in material loss and may be unsuitable for cement applications, where excessive surface area could increase water demand. For construction purposes, mild oxidation is more relevant, as it improves hydrophilicity. Biochar produced at high temperatures often exhibit hydrophobic behavior, which can hinder dispersion in cement paste. Surface oxidation, achieved through moderate heating in air or chemical treatments, introduces polar oxygen-containing groups that improve water wettability and interfacial bonding with cement (Senadheera et al. 2023).

Another approach involves preloading biochar with CO_2 prior to mixing. Biochar's porous structure allows CO_2 adsorption, and preliminary studies have reported changes in cement setting behavior when CO_2-loaded biochar is used. This suggests potential for biochar to act as a CO_2 carrier, releasing it during mixing or curing to react with calcium ions, potentially forming calcium carbonate in situ. Such a mechanism could enhance carbonation curing and contribute to additional CO_2 sequestration, although further investigation is required to verify and optimize this effect.

14.3.3 Stability and Safety Considerations

The handling and incorporation of biochar significantly influence its effectiveness in cementitious materials. Due to its high porosity, dry biochar readily absorbs mix water, which can reduce workability (Xu et al. 2024b). Pre-soaking the biochar in water or liquid admixtures before mixing is a common solution. This process saturates the internal pores, preventing water loss during mixing and providing internal curing. Studies consistently show improved hydration and strength when pre-wetted biochar is used. For instance, mortar incorporating pre-soaked biochar exhibited higher hydration and compressive strength compared to mixes with dry biochar. Another study found that adding only 1% moisture-conditioned biochar by cement weight reduced water sorptivity and permeability by 58–66% without affecting workability. These findings demonstrate that appropriate moisture conditioning enhances the densification of the concrete microstructure.

Biochar can also be combined with other SCMs for synergistic effects. Blending with silica fume or metakaolin has shown promising results. One study reported that a binary blend of 5% biochar and 10% silica fume accelerated hydration and increased 7-day cumulative heat evolution by ~20% compared to plain cement. Biochar provides nucleation sites that promote early hydration, while the pozzolanic activity of silica fume contributes to later strength development. Similarly, combining biochar with metakaolin helped produce a high-performance, carbon-negative concrete in another investigation. These outcomes highlight the potential of engineered mix designs, where biochar is integrated as part of a tailored blend to enhance both performance and sustainability (Chen et al. 2023). In rural settings, localized biochar production can be optimized based on available feedstocks and desired material properties. For example, a village-scale pyrolysis unit operating at 500 °C could produce stable, porous biochar from rice husks, which is then ground and pre-moistened for use in low-carbon bricks. Through careful control of production and handling, biochar can be engineered into a valuable component of sustainable construction materials.

14.4 Applications in Low-carbon Construction

Biochar can be incorporated into construction materials through three main approaches, as shown in Figure 14.1: (1) as a cement additive or partial binder replacement, (2) as a substitute for aggregate or filler, and (3) as a component in specialized materials such as insulating composites or pervious concrete. Each method leverages specific properties of biochar while contributing to carbon footprint reduction. In rural settings, these applications offer a pathway to convert local biomass waste into functional building materials, such as using agricultural residues to partially replace conventional components in village-scale concrete production (Wang et al. 2024). The following sections outline each application type, supported by recent performance data.

14.4.1 Biochar as a Cement Additive or Alternative

Biochar has been applied as a partial replacement for Portland cement to reduce clinker consumption and associated CO_2 emissions. The effect of biochar dosage on the compressive strength of cement mortar at 28 days is shown in Figure 14.2. At low replacement levels (typically $\leq$3% by cement weight), it has been shown to maintain or even improve compressive strength. Enhancements are attributed to the filler effect and the promotion of calcium-silicate-hydrate formation through nucleation (Legan et al. 2022). For example, a 1% biochar addition has increased compressive strength by up to 9%, while slight improvements in modulus and reductions in water absorption have also been reported. These results suggest that low-dosage biochar can densify the microstructure and improve overall concrete quality.

Figure 14.1 Application of biochar in low-carbon construction.

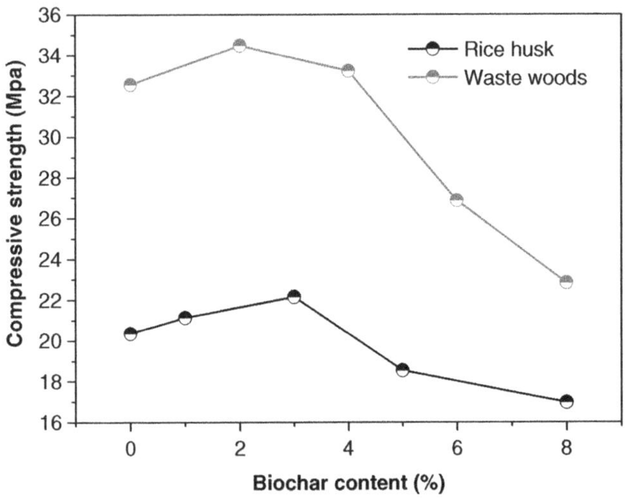

Figure 14.2 Effect of biochar dosage on compressive strength of cement mortar at 28 days. (Note: Rice husk (Aneja et al. 2022) and waste woods (Tan et al. 2022).)

Strength performance is generally optimal within the 3–8% replacement range. Studies have demonstrated that up to 10% biochar can be used without compromising strength, particularly when using fine, porous, and preconditioned materials. For instance, blends containing 5–10% rice husk or bagasse biochar have achieved higher compressive and tensile strengths compared to control mixes. In some cases, surface-treated biochar have produced strength increases of over 30%. However, results beyond 10% are inconsistent. Excessive replacement may reduce binder content and C-S-H formation, leading to strength loss (Wang et al. 2023). Still, within optimal dosages, biochar has consistently provided performance equal to or better than plain mixes.

Beyond mechanical strength, biochar enhances durability and thermal resistance. Its internal curing capacity reduces permeability and shrinkage. A 1% addition has lowered water penetration depth by up to 66%, while 2–5% additions have reduced drying shrinkage by 14–20%. Carbonation-related shrinkage and cracking in water-limited curing conditions have also been mitigated. Furthermore, biochar improves fire performance by releasing internal pressure and reducing cracking under heat exposure. Mortars with 1–2% biochar retained up to 20% more compressive strength after heating to 550 °C (Akinyemi and Adesina 2020). To maintain workability, particularly at higher dosages, pre-soaking and use of plasticizers are recommended. With appropriate mix design, biochar offers a practical, low-carbon alternative for rural construction without compromising concrete performance.

14.4.2 Biochar as an Aggregate or Filler Replacement

Biochar can serve as a partial or full replacement for traditional aggregates in concrete and mortar, allowing for greater incorporation of waste biomass. In this application, biochar acts as a lightweight aggregate, reducing composite density and adding functional benefits (Jiang et al. 2025). A study by Chen et al. (2022) demonstrated the potential of using wood-derived biochar to replace up to 30% of fine aggregate by mass, combined with metakaolin as a supplementary binder. This mix retained structural strength due to improved hydration and a denser matrix, while also achieving net-negative CO_2 emissions, sequestering up to 59 kg CO_2 per ton of concrete and generating an estimated US$35 profit per cubic meter through carbon credits and cost savings. These results confirm biochar's viability as a substantial aggregate replacement in environmentally optimized concrete.

Mechanical performance varies with biochar content and type. While biochar has lower strength than mineral aggregates, low to moderate replacement levels can maintain or improve strength due to internal curing and pore refinement (Chen et al. 2024). For example, concrete with 10–20% poultry litter biochar exhibited increased compressive strength, despite reduced density. Praneeth et al. (2021)

similarly reported strength gains with up to 40% biochar replacement. Biochar aggregates also improve thermal insulation and ease of handling due to reduced unit weight, with concrete becoming 15–30% lighter depending on the volume fraction. At higher contents, biochar creates a continuous, low-conductivity network, enhancing insulation and making it suitable for energy-efficient buildings in extreme climates. In addition, carbonation curing of biochar-amended concrete enhances both strength and CO_2 sequestration, with reported uptake reaching 65 kg CO_2 per ton of concrete and economic benefits up to US$41 per m^3.

Biochar has also been incorporated into pervious concrete, bricks, and insulation panels. In pervious concrete, biochar improves pollutant removal from runoff while preserving permeability. Some found that including 15–20 kg/m^3 of woody biochar significantly enhanced nitrogen and phosphorus removal. Masonry units and insulation panels with biochar content offer reduced thermal conductivity and weight, improving energy efficiency in rural housing. Nonetheless, high biochar contents may lower compressive strength and require adjusted mix designs. Workability challenges from angular, porous particles can be mitigated through water saturation and admixtures. Segregation can also be controlled by ensuring biochar is pre-wetted (Zhang et al. 2022). Overall, replacing mineral aggregates with biochar supports large-scale carbon sequestration, reduced raw material demand, and sustainable material sourcing, making it well-suited for rural, low-carbon construction.

14.4.3 Specialty Biochar-based Construction Materials

Biochar has driven the development of specialized construction materials beyond conventional concrete and mortar applications, utilizing its unique properties to enhance performance. These materials include thermal and acoustic insulation systems, electromagnetic shielding materials, high-performance concretes, and emerging 3D-printed construction technologies (De Schutter et al. 2018; Zhang et al. 2025b). The primary objectives are twofold: reducing carbon emissions and imparting new functionalities to building components. This dual purpose proves advantageous, particularly in rural areas, where multi-functionality can enhance community resilience by, for example, simultaneously insulating homes and sequestering carbon (Su et al. 2024; Zhou et al. 2025).

14.4.3.1 Thermal Insulating Composites

Biochar-based composites have been developed for construction applications, leveraging its porous, carbonaceous structure to enhance insulating properties (Lee et al. 2019). These materials have been combined with binders such as clay, lime, or gypsum to create insulating panels and plasters (Akinyemi and Adesina 2020). One study incorporated biochar from rice and coconut husks with natural clay, resulting in an exterior insulation layer with significantly reduced thermal conductivity. The biochar addition decreased the composite's thermal conductivity by up to 67%, lowering its heat transfer capability to one-third of its original value. Additionally, the composite's water vapor resistance increased by ~22%, preventing moisture ingress while maintaining breathability. In another study, adding 1–2% biochar to concrete reduced its thermal conductivity from the typical 0.6–1 W/m·K range to ~0.19 W/m·K, comparable to lightweight insulating concrete (Senadheera et al. 2023). This suggests that even structural elements can be made more insulating with minimal biochar content. In cold rural regions, biochar-amended concrete blocks could enhance home energy efficiency without synthetic insulation materials. Furthermore, biochar-based insulations exhibit fire resilience due to their non-flammable properties at typical building fire temperatures.

14.4.3.2 Acoustic Insulation

Research indicates that biochar's low density and porosity also enhance sound absorption, despite limited data on its acoustic properties. These characteristics enable porous carbon materials to dissipate acoustic energy within their pore networks. Biochar-containing concretes and panels are thus anticipated to exhibit superior sound absorption and noise reduction coefficients compared to conventional materials. Wen et al. (2023) highlighted the exploration of biochar-cement composites for acoustic insulation applications. For instance, lightweight biochar–gypsum panels could serve as interior walls that provide both thermal insulation and noise reduction between rooms. This dual functionality may benefit communal rural housing and livestock barns, where noise reduction can improve animal welfare. Although specific quantitative values were not provided, it is reasonable to assume that well-designed biochar composites can achieve acoustic performance comparable to commercial panels, given biochar's similarity to activated carbon, which is commonly used in sound absorption applications.

14.4.3.3 Electromagnetic Shielding Materials

Research has explored the use of biochar in developing electromagnetic interference (EMI) shielding construction materials, driven by increasing wireless communication and the need for cost-effective solutions. While high-tech materials like graphene or carbon nanotubes are used for EMI shielding, their high cost limits their application in rural areas. Biochar serves as a viable low-cost alternative due to its conductivity. Studies have shown that incorporating a small amount of biochar can significantly enhance the EMI shielding effectiveness of cement composites. For example, Khushnood et al. (2015) achieved a 353% increase in shielding effectiveness at 1.56 GHz with 0.5 wt% biochar in cement paste. Natalio et al. (2020) demonstrated that increasing biochar content in a lightweight biochar–cement–gypsum composite improved EMI shielding, particularly at higher frequencies (>4 GHz). Although the mechanisms (reflection by conductive networks or absorption by polarization losses) are still being studied, the practical implications are clear: biochar-infused walls or panels could act as EMI shields, protecting sensitive equipment from interference or preventing external signals from penetrating. This could be valuable for rural applications like shielding radio transmitter stations or reducing exposure to stray electromagnetic radiation in homes. Importantly, biochar achieves this at a fraction of the cost of nano-carbons, aligning with the goal of affordable rural technology.

14.4.3.4 Ultra-high-performance and Specialty Concretes

Biochar has been investigated for its potential to enhance high-performance concrete, including ultra-high-performance concrete (UHPC) and advanced formulations. UHPC, characterized by a low water-to-cement ratio, high cement content, and fine powders such as silica fume, is prone to self-desiccation and shrinkage. Biochar's internal curing capacity can mitigate autogenous shrinkage and potentially boost strength by promoting more complete hydration of ultra-fine cement particles. Additionally, biochar has been explored as an additive to improve the thixotropy and buildability of 3D-printed lightweight concrete, as reported by Zhang et al. (2022). A small dosage of biochar can enhance the fresh concrete mix's thixotropy, ensuring it flows when sheared but quickly stiffens at rest, which is ideal for maintaining shape in 3D-printed layers. The study also demonstrated that carbon-negative biochar can be incorporated into 3D printing mixes to reduce density and improve extrudability without compromising structural integrity (De Schutter et al. 2018). This innovative application shows promises for rapid, low-waste construction in remote areas, such as using mobile 3D printers to construct affordable rural housing with biochar-enhanced concrete, which is lighter and more stable for printing while also sequestering carbon within the structure.

14.4.3.5 Hybrid Biochar-PCMs

Research has explored the integration of biochar with phase change materials (PCMs) for energy storage in buildings (Zhang et al. 2025a). PCMs regulate indoor temperature by absorbing and releasing heat during phase transitions between solid and liquid states. However, they often face issues such as leakage and low thermal conductivity. Porous carbon materials like biochar can serve as supporting matrices for PCMs, enhancing their stability and heat transfer. The concept involves impregnating biochar pores with a type of PCM, such as wax or salt hydrate. Biochar stabilizes the PCM, preventing leakage when it melts, and slightly enhances thermal conductivity. Additionally, the surrounding biochar–concrete provides structural support (Liang et al. 2023). This material could be cast into walls or floor panels for passive thermal energy storage, smoothing out daily temperature fluctuations. This is particularly useful in climates with significant day–night temperature differences. Such hybrid innovations highlight biochar's versatility as a platform for multi-functional building materials.

14.5 Environmental and Economic Benefits

14.5.1 Carbon Sequestration and Emissions Reduction

Biochar offers significant advantages in building materials through carbon sequestration. Derived from pyrolyzed biomass, biochar converts CO_2-capturing biomass into a stable solid, locking away carbon for decades when embedded in materials like concrete or bricks. Each ton of biochar contains 70–80% carbon by weight, equivalent to storing 2.5–3 tons of atmospheric CO_2 (Liu et al. 2022). Additionally, biochar reduces cement usage, avoiding CO_2 emissions from cement production. These dual effects result in a low or negative carbon footprint for biochar concrete.

Life-cycle assessment studies have demonstrated impressive results. Chen et al. (2022) reported that a biochar-augmented concrete mix achieved net-negative CO_2 emissions, sequestering ~59 kg CO_2 per ton of concrete produced. This mix combined 30% biochar aggregate and supplementary metakaolin to achieve carbon-negative concrete production. Other analyses confirmed that biochar concrete can sequester 59–65 kg CO_2 per ton, highlighting its potential as a carbon sink. Unlike some carbon capture methods, CO_2 stored in buildings remains locked away for the structure's lifespan and can be preserved even after demolition (Song et al. 2023). Furthermore, biochar enhances CO_2 uptake through carbonation. Concrete carbonates slowly, but biochar increases internal surface area and CO_2 diffusion pathways. Accelerated carbonation curing can further enhance this process. Future research aims to optimize this mechanism, potentially tailoring biochar's pore structure for faster CO_2 absorption. For rural communities aiming to offset their carbon footprint, biochar-enhanced materials offer an accessible solution, aligning with global climate goals and potentially attracting funding or carbon credits for sustainable construction projects.

Biochar construction also provides pollution mitigation and waste reduction benefits. Utilizing waste biomass prevents air and water pollution. Agricultural residues, often open-burned, can be pyrolyzed into biochar for local construction, reducing air pollution. Similarly, diverting organic waste from decomposition avoids methane emissions. Biochar–concrete can even capture nutrients from runoff and sorb heavy metals from water. At end-of-life, biochar concrete can be recycled as a filtration medium for environmental remediation. During its service life, biochar may immobilize heavy metals in concrete, preventing leaching and enhancing environmental stability. Overall, biochar in construction encapsulates potential contaminants safely.

14.5.2 Circular Economy and Rural Waste Valorization

The utilization of waste-derived biochar in building materials exemplifies circular economic principles by converting local waste streams into valuable resources for industry. Agricultural and forestry waste, instead of being disposal problems, can be transformed into materials such as bricks, blocks, or concrete, thereby reducing the need for extracting and transporting conventional raw materials (Wang et al. 2023). For rural communities, this offers tangible benefits: farmers can generate additional income by selling crop residues for biochar production, while communities can save costs by producing their own construction aggregate locally, reducing the need for purchasing transported sand or gravel. Economically, this retains more value within local areas and fosters local enterprises. Small-scale pyrolysis units operated in rural towns can supply biochar to local construction projects, creating jobs in both biochar production and construction processes, including skilled roles in mixing design and quality control.

Environmentally, using biochar as a partial replacement for sand and gravel alleviates pressure on over-mined riverbeds and quarries, preserving ecosystems. Given that globally ~17.5 billion tons of aggregates are consumed annually, causing significant environmental impacts from mining and transport, biochar offers a lighter alternative that can replace several tons of sand or gravel per ton used. This helps protect natural landscapes and biodiversity in rural regions that might otherwise resort to local sand mining.

14.5.3 Economic Analysis and Cost Considerations

Cost is a critical factor for new construction materials. Many studies suggest that biochar-enhanced construction can be cost-competitive and economically beneficial under certain conditions. Chen et al. (2022) found that biochar concrete could yield satisfactory economic profits, as shown in Figure 14.3. After accounting for biochar production and metakaolin costs, the net cost per cubic meter was offset by reduced cement usage and potential carbon credit monetization. They estimated a profit of about US$35–36 per m^3 for the optimal mix, with another source indicating higher benefits (up to US$41 per m^3) for significant CO_2 uptake (Lin et al. 2023). These figures highlight biochar concrete's economic viability. In rural areas, using local biochar can reduce material costs, especially when replacing expensive cement with waste-derived biochar. Low-cost pyrolysis methods, using farm waste and renewable heat sources, can further reduce expenses (Labianca

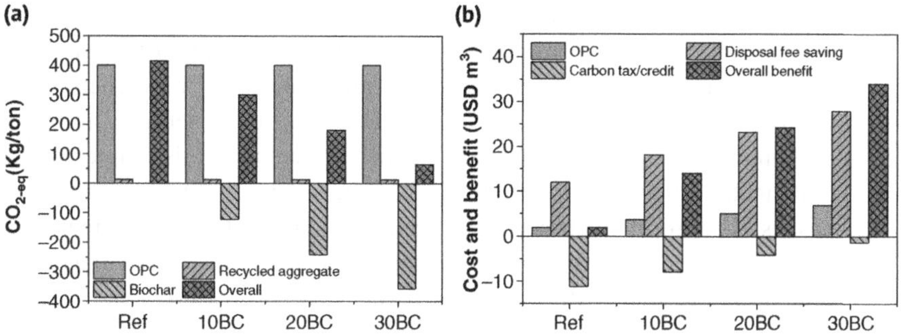

Figure 14.3 Life cycle assessment and cost-benefit analysis of biochar blocks.

et al. 2024). Additionally, biochar's multi-functionality can save costs in insulation, transportation, and handling. It may also improve structure durability, leading to lower maintenance and longer lifespan.

Initial implementation of biochar concrete may face cost challenges, such as setting up pyrolysis units or modifying mix designs. However, these can be addressed with planning and support from sustainability grants or carbon finance mechanisms. The Paris Agreement and carbon markets increasingly value carbon-storing materials. Rural communities could earn carbon credits by using biochar in construction. For example, a typical single-story house using 20 m^3 of biochar concrete could sequester ~1.2 tons of CO_2, potentially earning US$60 in carbon credits at US$50 per ton (Osman et al. 2024). In conclusion, biochar offers significant environmental benefits, such as reduced greenhouse emissions and improved waste management, alongside economic benefits like cost savings and potential profits from carbon credits. This aligns with rural revitalization goals, enabling climate-positive and economically sensible development. The synergy of these advantages makes biochar sustainable building material, moving construction practices toward regeneration rather than destruction.

14.6 Conclusion

The preceding discussion highlights biochar's multifaceted potential in rural development and climate action. Waste-derived biochar presents a promising approach to align rural progress with climate goals. By converting local biomass waste into a high-value construction material, rural communities can construct essential infrastructure in an environmentally friendly manner. This not only reduces carbon emissions and recycles waste but also potentially enhances building performance. Some conclusions are drawn as follows:

1. Waste-derived biochar offers a promising pathway to align rural development with climate action. By turning locally available biomass waste into a high-value construction material, rural communities can build needed infrastructure in a way that reduces carbon emissions, recycles wastes, and potentially improves building performance. The technical review presented in this chapter shows that biochar can enhance hydration, strength, and durability of cement composites at appropriate dosages, and even enable novel low-carbon materials like carbon-negative concrete and insulating composites. Environmentally, each structure built with biochar is partly a carbon reservoir and a testament to circular economic principles in action. Economically, with the right approach, biochar construction can be cost-effective and support local livelihoods, especially if carbon credit schemes or government incentives for green building are leveraged.
2. However, realizing this potential on a broad scale requires overcoming current challenges. Establishing standards, sharing practical knowledge, and instilling confidence through successful projects are all necessary steps. It calls for an interdisciplinary effort—materials scientists to refine biochar properties, engineers to optimize mix designs, economists to evaluate cost-benefit in various contexts, and policymakers to incorporate these innovations into building codes and rural development programs. Fortunately, the momentum is growing: the research community has rapidly expanded knowledge on biochar in construction over the past few years, as evidenced by the diverse studies cited (spanning cement chemistry, structural tests, life-cycle assessment, etc.). This growing knowledge base will feed into better guidelines and commercial products (for example, one could envision pre-packaged "biochar cement" blends for sale in rural markets in the future).
3. From a rural revitalization standpoint, the adoption of waste-derived biochar in construction could be transformative. It addresses three key pillars of sustainability: environmental quality (lower emissions and cleaner waste handling), economic vitality (local production and cost savings), and social well-being (better housing and infrastructure). By investing in training and small-scale technology, rural areas can leapfrog to more sustainable building practices without waiting for expensive industrial solutions to trickle down. In doing so, they also contribute to global climate mitigation efforts.
4. In closing, waste-derived biochar for construction exemplifies the idea that solutions to climate change can also be solutions to rural poverty and development. It turns liabilities into assets—carbon from a problem to a product, waste to wealth. While challenges remain, the evidence compiled in this chapter strongly supports the vision that with continued research, supportive policies, and community engagement, biochar-enhanced low-carbon

construction materials can become a mainstream reality. Such a development would not only help decarbonize one of the most carbon-intensive sectors (construction) but also empower rural communities to build a more sustainable and resilient future with their own hands and resources. The path forward will require dedication and innovation, but the foundation has been laid for waste-derived biochar to play a pivotal role in sustainable rural revitalization.

References

Akinyemi, B.A. and Adesina, A. (2020). Recent advancements in the use of biochar for cementitious applications: a review. *Journal of Building Engineering* 32: 101705.

Aneja, A., Sharma, R.L., and Singh, H. (2022). Mechanical and durability properties of biochar concrete. *Materials Today: Proceedings* 65: 3724–3730.

Chen, L., Huang, L., Hua, J. et al. (2023). Green construction for low-carbon cities: a review. *Environmental chemistry letters* 21 (3): 1627–1657.

Chen, L., Zhang, Y., Wang, L. et al. (2022). Biochar-augmented carbon-negative concrete. *Chemical Engineering Journal* 431: 133946.

Chen, L., Zhu, X., Zheng, Y. et al. (2024). Development of high-strength lightweight concrete by utilizing food waste digestate based biochar aggregate. *Construction and Building Materials* 411: 134142.

De Schutter, G., Lesage, K., Mechtcherine, V. et al. (2018). Vision of 3D printing with concrete—technical, economic and environmental potentials. *Cement and Concrete Research* 112: 25–36.

Du, T., Chen, J., Qu, F. et al. (2022a). Degradation prediction of recycled aggregate concrete under sulphate wetting–drying cycles using BP neural network. *Structures* 46: 1837–1850.

Du, T., Li, C., Wang, X. et al. (2022b). Effects of pipe diameter, curing age and exposure temperature on chloride diffusion of concrete with embedded PVC pipe. *Journal of Building Engineering* 57: 104957.

Gómez-Vásquez, R., Fernández-Ballesteros, E., and Camargo-Trillos, D. (2022). Biogenic nanoporous oxides recovery from by-products of bioenergy production: rice husks and corncob biochars. *Biomass and Bioenergy* 161: 106455.

Gupta, S., Muthukrishnan, S., and Kua, H.W. (2021). Comparing influence of inert biochar and silica rich biochar on cement mortar–hydration kinetics and durability under chloride and sulfate environment. *Construction and Building Materials* 268: 121142.

Jiang, X., Qu, F., Zhang, Y. et al. (2025). Cold-bonded biochar-cement lightweight aggregates for evaporation-enhanced permeable bricks. *Journal of Cleaner Production* 493: 144886.

Khushnood, R.A., Ahmad, S., Savi, P. et al. (2015). Improvement in electromagnetic interference shielding effectiveness of cement composites using carbonaceous nano/micro inerts. *Construction and building materials* 85: 208–216.

Labianca, C., Zhu, X., Ferrara, C. et al. (2024). A holistic framework of biochar-augmented cementitious products and general applications: technical, environmental, and economic evaluation. *Environmental Research* 245: 118026.

Lee, H., Yang, S., Wi, S. et al. (2019). Thermal transfer behavior of biochar-natural inorganic clay composite for building envelope insulation. *Construction and Building Materials* 223: 668–678.

Legan, M., Gotvajn, A.Ž., and Zupan, K. (2022). Potential of biochar use in building materials. *Journal of Environmental Management* 309: 114704.

Li, J., Cao, L., Yuan, Y. et al. (2018). Comparative study for microcystin-LR sorption onto biochars produced from various plant-and animal-wastes at different pyrolysis temperatures: influencing mechanisms of biochar properties. *Bioresource Technology* 247: 794–803.

Li, M., Zhu, X., Zhang, Y. et al. (2024). A multi-phase mechanical model of biochar–cement composites at the mesoscale. *Computer-Aided Civil and Infrastructure Engineering* 39 (23): 3552–3572.

Liang, Q., Pan, D., and Zhang, X. (2023). Construction and application of biochar-based composite phase change materials. *Chemical Engineering Journal* 453: 139441.

Lin, X., Li, W., Guo, Y. et al. (2023). Biochar-cement concrete toward decarbonisation and sustainability for construction: characteristic, performance and perspective. *Journal of Cleaner Production* 419: 138219.

Liu, J., Liu, G., Zhang, W. et al. (2022). Application potential analysis of biochar as a carbon capture material in cementitious composites: a review. *Construction and Building Materials* 350: 128715.

Maljaee, H., Paiva, H., Madadi, R. et al. (2021). Effect of cement partial substitution by waste-based biochar in mortars properties. *Construction and Building Materials* 301: 124074.

Méndez, A., Terradillos, M., and Gascó, G. (2013). Physicochemical and agronomic properties of biochar from sewage sludge pyrolysed at different temperatures. *Journal of Analytical and Applied Pyrolysis* 102: 124–130.

Natalio, F., Corrales, T.P., Feldman, Y. et al. (2020). Sustainable lightweight biochar-based composites with electromagnetic shielding properties. *ACS Omega* 5 (50): 32490–32497.

Osman, A.I., Farghali, M., Dong, Y. et al. (2024). Reducing the carbon footprint of buildings using biochar-based bricks and insulating materials: a review. *Environmental Chemistry Letters* 22 (1): 71–104.

Praneeth, S., Saavedra, L., Zeng, M. et al. (2021). Biochar admixtured lightweight, porous and tougher cement mortars: mechanical, durability and micro computed tomography analysis. *Science of The Total Environment* 750: 142327.

Qin, Y., Pang, X., Tan, K. et al. (2021). Evaluation of pervious concrete performance with pulverized biochar as cement replacement. *Cement and Concrete Composites* 119: 104022.

Qu, F., Zhang, Y., Li, M. et al. (2025). Resource recycling of industrial waste phosphogypsum in cementitious materials: pretreatment, properties, and applications. *Journal of Environmental Management* 376: 124291.

Senadheera, S.S., Gupta, S., Kua, H.W. et al. (2023). Application of biochar in concrete–a review. *Cement and Concrete Composites* 143: 105204.

Song, S., Liu, Z., Liu, G. et al. (2023). Application of biochar cement-based materials for carbon sequestration. *Construction and Building Materials* 405: 133373.

Su, Y., Qu, F., Meng, Y. et al. (2024). Microbial-induced carbonate precipitation (MICP) modified biochar for low-carbon cementitious materials. *Construction and Building Materials* 451: 138644.

Tan, K., Qin, Y., and Wang, J. (2022). Evaluation of the properties and carbon sequestration potential of biochar-modified pervious concrete. *Construction and Building Materials* 314: 125648.

Tan, Z., Zou, J., Zhang, L. et al. (2018). Morphology, pore size distribution, and nutrient characteristics in biochars under different pyrolysis temperatures and atmospheres. *Journal of Material Cycles and Waste Management* 20: 1036–1049.

Wang, L., Deng, J., Yang, X. et al. (2023). Role of biochar toward carbon neutrality. *Carbon Research* 2 (1): 2.

Wang, L., Nerella, V.N., Li, D. et al. (2024). Biochar-augmented climate-positive 3D printable concrete. *Communications Materials* 5 (1): 257.

Wang, Q., Li, J.-S., and Poon, C.S. (2022). An iron-biochar composite from co-pyrolysis of incinerated sewage sludge ash and peanut shell for arsenic removal: role of silica. *Environmental Pollution* 313: 120115.

Wen, J., Wang, B., Dai, Z. et al. (2023). New insights into the green cement composites with low carbon footprint: the role of biochar as cement additive/alternative. *Resources, Conservation and Recycling* 197: 107081.

Xu, W., Zhang, Y., Li, M. et al. (2024a). Durability and micromechanical properties of biochar in biochar-cement composites under marine environment. *Journal of Cleaner Production* 450: 141842.

Xu, W., Zhang, Y., Su, Y. et al. (2024b). Using waste to improve the weak: recycled seashell as an ideal way to regulate the interfacial transition zone in biochar-cement composites. *Construction and Building Materials* 444: 137765.

Zhang, Y., He, M., Wang, L. et al. (2022). Biochar as construction materials for achieving carbon neutrality. *Biochar* 4 (1): 59.

Zhang, Y., Li, M., Zhu, X. et al. (2025a). Enhanced thermal insulation of biochar-gypsum composites. *Cement and Concrete Composites* 159: 106013.

Zhang, Y., Xu, W., Zhu, X. et al. (2024). Scaling biochar solutions for urban carbon dioxide removal. *One Earth* 7 (9): 1481–1486.

Zhang, Y., Zhu, X., Li, M. et al. (2025b). 3D printing technology in concrete construction. *Nature Reviews Clean Technology* 1: 288–303.

Zhou, S., Yang, D., Zhang, Z. et al. (2025). Enhancing autonomous pavement crack detection: optimizing YOLOv5s algorithm with advanced deep learning techniques. *Measurement* 240: 115603.

15

Low-carbon Soil Remediation with Biochar and GGBS

*Weijian Xu, Jingyi Liang, Yuying Zhang, and Daniel C.W. Tsang**

Department of Civil and Environmental Engineering, The Hong Kong University of Science and Technology, Clear Water Bay, Hong Kong, China

*Corresponding author: cedan@ust.hk

15.1 Introduction

Soil contamination, primarily caused by industrial and agricultural activities, often involves toxic elements such as heavy metals, chlorides, and sulfates at concentrations exceeding environmental safety thresholds (Beiyuan et al. 2017). To mitigate these risks, soil remediation techniques—including in situ and ex situ methods like chemical-enhanced washing, electrokinetic extraction, and bioremediation—have been employed with varying success (Akcil et al. 2015). However, large-scale implementation of these methods often faces challenges related to cost and time efficiency.

Among remediation strategies, cement-based stabilization/solidification (S/S) has gained prominence due to its cost-effectiveness, rapid processing, and compatibility with diverse contaminants (Wang et al. 2020). Nevertheless, the presence of natural soil constituents and exogenous pollutants can disrupt hydration reactions, yielding low-quality stabilized materials. To achieve adequate mechanical strength and contaminant immobilization, high doses of ordinary Portland cement (OPC) are often required, leading to the formation of key hydration products such as calcium silicate hydrate (C-S-H) gel, ettringite, and monosulfate. However, OPC production is a major source of global CO_2 emissions, accounting for ~8% of the total (Andrew 2018). Consequently, developing low-carbon alternative binders is critical for sustainable soil remediation.

Waste-derived Biochar for Sustainable Rural Revitalization, First Edition. Edited by Yuqing Sun and Daniel C.W. Tsang.
Published 2026 by John Wiley & Sons, Inc.

Biochar has emerged as a promising stabilizing agent due to its cost efficiency, high metal adsorption capacity, and environmental benefits, including carbon sequestration and waste recycling (Zhang et al. 2024). The properties of biochar vary significantly depending on feedstock (e.g., wood waste, crop residues, manure) and production conditions (e.g., pyrolysis temperature, modification techniques), influencing its S/S performance (Oliveira et al. 2017; Wang et al. 2025). For instance, biochar derived from wheat straw may exhibit different stabilization efficiencies depending on soil origin, highlighting the need for systematic evaluation of feedstock-specific mechanisms.

To further reduce the carbon footprint of OPC, supplementary cementitious materials (SCMs) such as ground granulated blast furnace slag (GGBS), pulverized fly ash (PFA), and silica fume (SF) have been utilized as partial substitutes. These materials not only lower CO_2 emissions by avoiding additional clinkering but also repurpose industrial by-products. GGBS, in particular, enhances strength and reduces water absorption through pozzolanic reactions that form secondary C-S-H gels. Recent studies suggest that combining biochar, OPC, and GGBS offers an effective and sustainable approach for S/S treatment.

This review examines the latest advancements in biochar and GGBS applications for contaminated soil remediation. It also evaluates their feasibility in local contexts, considering site-specific constraints, material availability, and cost implications. Additionally, quality control measures, regulatory guidelines, and acceptance criteria for biochar/GGBS utilization in soil remediation are discussed.

15.2 Latest Developments and Applications of Biochar in Soil Remediation

15.2.1 Biochar Production

Biochar production is a versatile process that can be adapted to different operational scales, ranging from industrial facilities to small household systems, ensuring its applicability across diverse socioeconomic contexts. Various thermochemical conversion techniques, such as traditional pyrolysis (slow and fast pyrolysis), gasification, and microwave-assisted pyrolysis, are employed to generate biochar. Notably, slow pyrolysis stands out due to its economic viability, significant potential for climate change mitigation, and well-established technological readiness, positioning it as the leading approach in biochar manufacturing.

The characteristics of the feedstock, particularly when derived from waste biomass, are pivotal in determining the elemental makeup of the resulting biochar (Figure 15.1). Utilizing waste biomass rather than dedicated crops avoids issues related to deforestation and the displacement of agricultural land. Feedstocks can vary widely, including lignin- and cellulose-rich materials like crop residues

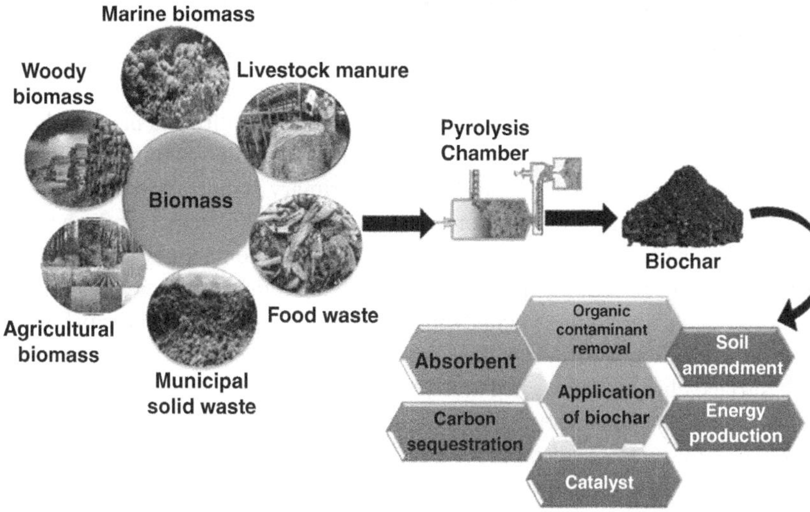

Figure 15.1 Functional biochar production from diverse biomass waste sources. *Source:* Ayaz et al. 2025/with permission of Springer Nature.

and wood waste, which yield carbon-dense biochar with aliphatic and aromatic structures, as well as nutrient-rich sources such as animal manure, sewage sludge, and anaerobic digestate, which contain higher levels of nitrogen, phosphorus, potassium, and mineral ash (e.g., Ca, Mg, Fe, S). The organic fraction of the feedstock dictates the carbon, hydrogen, and oxygen content in the biochar, while inorganic constituents influence nutrient availability and mineral speciation (He et al. 2022). By carefully selecting feedstocks or introducing supplementary additives, biochar with tailored properties can be engineered. This approach not only supports global biochar commercialization but also promotes sustainable waste valorization, aligning with circular bioeconomy principles.

Pyrolysis temperature is another key factor governing biochar properties. As temperature rises, organic matter undergoes dehydration, decomposition, and subsequent condensation and aromatization, leading to an enrichment of carbon alongside reductions in hydrogen and oxygen content. Concurrently, inorganic elements become more concentrated due to the loss of volatile organic components. The structural evolution of both organic and inorganic phases at elevated temperatures influences biochar functionality. Within the typical pyrolysis range (300–900 °C), higher temperatures enhance surface area and porosity due to thermal restructuring (He et al. 2022). Although greater energy input is required at higher temperatures, the resulting biochar exhibits improved carbon stability and net energy efficiency, contributing to overall negative greenhouse gas emissions.

15.2.2 Stabilization/Solidification of Potentially Toxic Elements (PTEs)

15.2.2.1 Remediation of Heavy Metals

Biochar stabilizes heavy metals resulting in reduced solubility and bioavailability of metals, which differs from remediation methods like soil washing, leaching, or excavation of heavy metals (Table 15.1). The primary stabilization mechanisms biochar implements are precipitation and the sorption of heavy metals. Such metal stabilization is possible due to biochar's structure, functional groups, and soluble metal salts (Ye et al. 2025). Depending on the pH, functional groups are protonated or deprotonated, providing biochar with an electrical charge. When the pH of the solution was raised from 3 to 8, the submerged biochar exhibited enhanced negative surface charges. These negatively charged functional groups adsorb heavy metal cations through electrostatic interactions. The aromatic structure of biochar is rich in electrons allowing for electrostatic attraction on electron-deficient metal cations through interactions. Functional groups (carboxyl, carbonyl, and hydroxyl) on the biochar surface increase the cation exchange capacity (CEC) of the biochar, and as these increases, so does the ability of the biochar to adsorb metal cations through the exchange of Ca, Mg, K, and Na. Precipitation reactions are also an essential part of heavy metal detoxification by biochar. Most biochar is alkaline and increases the soil solution pH, thus increasing the hydroxyl density on the biochar surface adsorbing Al or Mn. This allows for more hydroxyl-metal precipitates, thereby reducing the availability of the heavy metals in soil solution. Precipitates can also form from the mineral or ash content of biochar. Precipitation can occur with anions and carbonates in biochar with metals, forming more stable bonds than surface complexation (Figure 15.2).

15.2.2.2 Remediation of Organic Pollutants

Organic pollutants can react with the aromatic structure of biochar through the electrostatic attraction of nonpolar or polar forces, increasing potential with more surface area. Hydrogen bonding is possible through the biochar structure's C-, N-, H-, and O-containing functional groups. pH_{ZNC} can be changed to more neutral, negative, or positive through post-biochar treatments if the pollutant in question is polar, or ionized. If the biochar material chosen was high in metal compounds or was pre- or post-treated, then the presence of metal compounds on the biochar surface can also play a role in polar or ionized pollutant complexation in biochar surface or precipitate in soil solution. In the case of ionized impurities, some inner sphere adsorption can occur, reacting with functional groups and reducing their availability in solution. Finally, electrons on the aromatic structure of biochar can create electron donor interactions, the stronger bonds occurring between oppositely polarized molecules.

Table 15.1 Comparison of biochar types for heavy metal S/S.

Biochar		Contaminants	Dosage	Key findings	Mechanisms	References
General biochar	—	Cu^{2+}, Pb^{2+}	5% (w/w)	• Cu^{2+}: 51.05% reduction • Pb^{2+}: 25.36% reduction • Decreased mobility in soil washing residue post 14 days	Reduced mobility in soil	Shen et al. (2018b)
Wood-derived biochar	450 °C	PTEs (general)	0.5, 1.0, 2.0%	• Decreased mobility of PTEs in contaminated soil • Enhanced antioxidant activities • Improved immobilization of PTEs • Adjusted soil pH and conductivity	• High lignin and cellulose content increases biochar yield • Cost-effective and accessible • Carbon-rich structure enhances sorption capacity	Ali et al. (2019), Igalavithana et al. (2019), Liu et al. (2020)
Sugarcane bagasse biochar	500 °C, 2 h, 20 °C/min	Cd, Cr	1.5% (w/w)	Significant immobilization of Cd and Cr	• Effective at stabilizing contaminants	Bashir et al. (2018)
Sewage sludge biochar	—	Cu	Not specified	• Substantially reduced Cu leachability in TCLP tests • Adsorbed Cu transformed into $CuAl_2O_4$ spinel crystals via ceramic sintering	• Mineral composition facilitates Cu stabilization via exothermic reactions • Thermal sintering improves immobilization	Shen et al. (2018a)

(Continued)

Table 15.1 (Continued)

Biochar		Contaminants	Dosage	Key findings	Mechanisms	References
Lychee biochar	—	As, Cd, Pb, Zn	2.5, 5, 10%	• 5% dosage maximized plant biomass; 10% dosage inhibited growth • Reduced concentrations of Cd, Pb, As, and Zn in rhizosphere soil • Enhanced total accumulation of Cd, Pb, and As in plants (Zn decreased)	• Improved immobilization of PTEs in soil • Altered metal bioavailability and plant uptake	Jun et al. (2020)
Peanut hull/corn straw/kitchen waste biochar	—	Pb, Cd	Not specified	• Elevated soil pH • Reduced extractable Pb and Cd by up to 99.28% • Lowered Pb/Cd accumulation in plant tissues • Converted Pb/Cd into stable fractions	• Biochar's adsorption capacity and pH adjustments decreased metal mobility • Shifted metal fractions to less bioavailable forms	Xu et al. (2020)

Rice straw biochar	—	Ni	Not specified	• Increased soil pH • Decreased unstable Ni fractions • Highest Ni absorption capacity among tested amendments	• Abundant functional groups and large surface area enhanced Ni binding	Ali et al. (2020)
Sulfur-modified rice husk biochar	550 °C	Hg	—	• ~73% reduction in Hg • Lower Hg concentrations in leachates	• Sulfur groups enhance Hg binding	O'Connor et al. (2018)
High-temperature biochar	>500 °C	Pb, Zn, Cu, Cd, Cr, Ni, As	—	• Increased HM content in biochar • Low ecological risk at >800 °C	• Dehydration of cellulosic components at elevated temperature	Xing et al. (2019)

Source: Adapted from Tajudin et al. 2016.

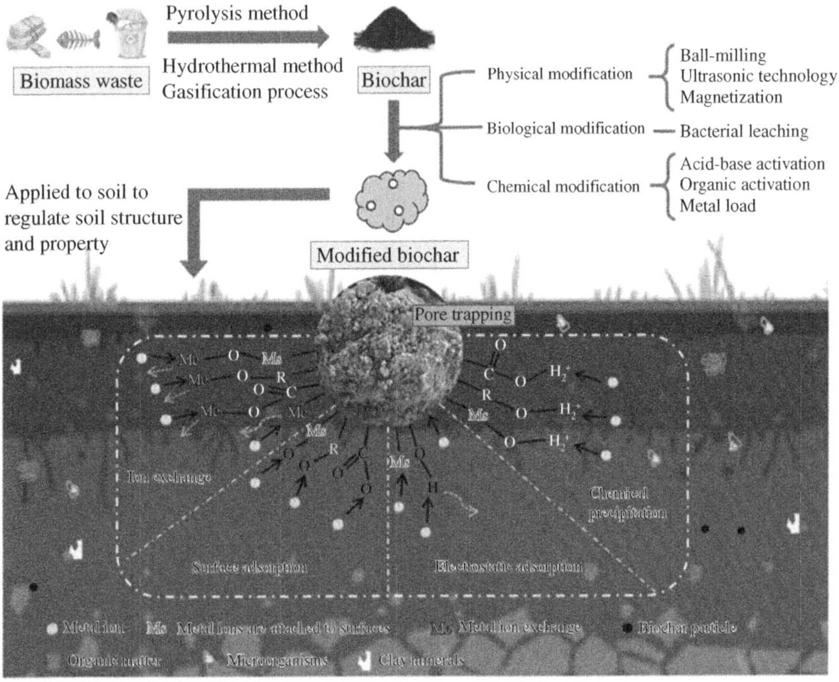

Figure 15.2 Biochar-based materials for the safe utilization and remediation of heavy metal-contaminated soil. *Source:* Ye et al. 2025/with permission of Elsevier.

Physical adsorption plays a vital role in non-ionic pollutant clean-up, and surface area and micropore size play a huge role in the capacity of biochar to adsorb contaminants. As biochar surface area increases, the porosity, aromaticity, and hydrophobicity increase. This means that higher-temperature treated biochar is favored for pollutant clean-up. This was shown as biochar derived from the same feedstock but at different temperatures had different sorption capacities of Deisopropylatrazine, as the higher temperature (700 °C) biochar had more adsorption capacity than low temperature (300 °C) produced biochar.

15.2.3 Field-scale Application of Biochar in Soil Remediation

15.2.3.1 Biochar-assisted Phytoremediation in Shuikoushan Mining Area, China

A field-scale soil remediation project was implemented in the Shuikoushan mining area (Hengyang City, Hunan Province, China), a region heavily contaminated with Pb, Cd, As, and Zn due to prolonged mining activities. The site experiences a subtropical monsoon climate with annual rainfall of 1400–1700 mm. To address the contamination, sunflowers (*Helianthus annuus*) were employed for phytoremediation, supplemented with biochar derived from lychee branches. Biochar was applied at rates of 0, 2.5, 5, and 10% (w/w) to evaluate its effects on heavy metal immobilization and plant growth.

The lychee biochar exhibited favorable characteristics for metal adsorption, including a porous structure, high carbon content (68.3%), and a surface area of 7.82 m^2/g. Functional groups such as —OH, C—H, and C=O further enhanced its metal-binding capacity. Application of biochar significantly reduced the bioavailability of heavy metals in the rhizosphere soil. When applied at 10%, the treatment decreased the concentrations of Pb, Cd, As, and Zn by 40.6, 31.6, 35.4, and 30.8%, respectively, compared to the untreated soil. Additionally, biochar improved soil fertility by increasing CEC, water retention, and nutrient availability. A 5% biochar dosage optimally boosted sunflower biomass (by 20–230% across roots, stems, leaves, and seeds), while excessive biochar (10%) inhibited growth due to reduced soil aeration.

Biochar influenced metal accumulation in sunflowers, promoting the uptake of Pb, Cd, and As while reducing Zn accumulation, suggesting antagonistic interactions. Metals were primarily sequestered in leaves (e.g., Pb: 626 mg/kg; Cd: 127 mg/kg under 2.5% biochar) and receptacles, whereas seed concentrations remained low (e.g., Cd: 2.62 mg/kg at 10% biochar), minimizing toxicity risks for potential agricultural reuse. Notably, the bioconcentration factor (BCF) for Cd in leaves exceeded 100, meeting the hyperaccumulator threshold and underscoring the sunflower's capability for Cd phytoextraction. Scaling these results to a hectare-level estimation, sunflowers could extract ~625 kg Cd, 164 kg Pb, 124 kg As, and 1724 kg Zn per growing season. Post-treatment assessments using the single-factor pollution index (P_i) confirmed a significant reduction in soil contamination levels, demonstrating the efficacy of this combined biochar-phytoremediation strategy.

This case study highlights the potential of biochar-assisted phytoremediation as a sustainable approach for heavy metal-contaminated soils, particularly in mining-affected regions. The synergistic use of sunflowers and lychee-derived biochar not only reduced metal bioavailability but also enhanced soil health, offering a scalable solution for environmental restoration.

15.2.3.2 Remediation of As-contaminated Technosol Using Biochar and Iron Sulfate, France

The study investigated the remediation of a technosol derived from a former tin mine in Abbaretz, France, abandoned since the 1950s (Simiele et al. 2020). The soil exhibited severe arsenic contamination (total As: 297 ± 30 mg/kg), acidic pH (5.15 ± 0.04), and low electrical conductivity (16 ± 0.3 μS/cm). Phytoavailable arsenic (3.3 ± 0.7 mg/kg) posed significant environmental risks, necessitating stabilization.

The remediation strategy combined hardwood-derived biochar (pyrolyzed at 500 °C, applied at 2% w/w) and iron sulfate (START®) to immobilize arsenic. Two application methods were tested: (1) partial biochar incorporation (upper 33 cm, AB + 1/3 BC) and (2) full homogenization (AB + BC). Iron sulfate (25 g/tube) was later added to promote arsenic adsorption via iron-oxide formation.

Biochar initially increased soil pore water (SPW) pH (~1 unit) and conductivity due to its alkaline nature (pH 8.46), but also temporarily elevated soluble arsenic via pH-induced desorption. Subsequent iron sulfate addition (2018) lowered SPW pH, enhancing arsenic co-precipitation with Fe oxides and reducing soluble arsenic by 55–75%. The AB + BC treatment demonstrated the lowest arsenic leaching, underscoring the importance of homogeneous biochar distribution.

Phytostabilization using Salicaceae species (*Populus euramericana*, *Salix purpurea*, *Salix viminalis*) revealed arsenic accumulation primarily in roots (up to 892 mg/kg in *S. viminalis*), with limited translocation to shoots. *S. purpurea* exhibited the highest tolerance, while *S. viminalis* showed growth inhibition under arsenic stress. Biochar improved root development in AB + BC, likely by mitigating arsenic toxicity in deeper soil layers.

15.3 Biochar-enhanced Cement for Stabilization/Solidification

15.3.1 The Role of Biochar in the Cement System

OPC is one of the most commonly used binders for S/S of contaminants, showing high efficiency, low permeability, and good durability (Chen et al. 2019). The cement hydrates include C-S-H, CH, and $CaCO_3$ (CC). The C-S-H and CH are the main components that are responsible for the S/S of toxic elements. The S/S mechanisms for toxic elements are (1) micro- or macro-encapsulation, (2) adsorption, and (3) chemical bonding (coprecipitation, divalent, surface complexation, precipitation) (Table 15.2). The degree of cement hydration plays a crucial role in determining the formation of C-S-H gel. Even in well-hydrated cement systems, residual unreacted clinker particles are commonly observed. These remaining anhydrous phases retain potential reactivity and can undergo further hydration when exposed to additional water or through the introduction of fine particulate additives that promote continued pozzolanic reactions.

Internal relative humidity (IRH) critically influences autogenous shrinkage and cracking, particularly in the absence of external curing. Cementitious materials dry via two primary mechanisms: water vapor diffusion (the dominant driver of internal water transfer) and capillary water consumption, which lowers IRH. Biochar incorporation helps maintain IRH for longer, slowing its decline

Table 15.2 The treatment performance by cement-based S/S approach.

Binder composition	Heavy metals	Key findings	Location	References
OPC, lime, and fly ash	Pb	• TCLP results showed Pb concentrations varied across acidic, neutral, and alkaline conditions • GANC of S/S materials leached all concentrations • In the MEP test, the average Pb leached was 4.8 mg/L	USA	Jing et al. (2004)
OPC	Pb, Zn, Cu, Fe, Mn	• The diffusion coefficients for all metals exceeded 11.5, indicating low mobility in the cement matrix • OPC samples achieved required strength after 28 days	India	Malviya and Chaudhary (2006)
Quicklime and fly ash	Pb	• Leaching tests indicated S/S treatment effectively immobilized Pb, complying with leachability index standards	USA	Moon and Dermatas (2006)
OPC with rice husk ash	Pb	• Combining RHA with OPC improved Pb remediation, reducing leachability compared to OPC alone	Malaysia	Yin et al. (2006)
OPC	Cd, Pb, Cu, Ni, As	• TCLP testing showed reductions in Cd, Pb, Zn, and Ni concentrations, while As levels remained unchanged and Cu increased	Slovenia	Voglar and Leštan (2010)
OPC, calcium aluminate cement, and pozzolanic cement	Cd, Pb, Zn, Cu, Ni, As	• Achieved soil leachability reductions for Cd, Pb, Cu, Ni, and As below acceptable limits, with mechanical strength up to 12 N/mm^2	Slovenia	

(Continued)

Table 15.2 (Continued)

Binder composition	Heavy metals	Key findings	Location	References
OPC with rice husk ash	Cr	• Compressive strength increased over time for both control and RHA-blended samples, peaking with 20% RHA blends	India	Jain (2011)
Fly ash (FA), bottom ash (BA), and OPC	Cr, Fe, Ni, Cu, Cd, Ba	• Cement-based S/S exhibited UCS values between 0.55 and 16.12 MPa • UCS decreased with lower cement content and partial addition of FA and BA	Greece	Anastasiadou et al. (2012)
Ordinary Portland Cement (OPC)	Pb	• Semi-dynamic leaching tests showed Pb concentrations dropped from 109, 83, and 71 mg to 37, 30, and 25 mg, respectively	China	Li et al. (2014)
OPC	Cu, Ni, Zn, Pb, Cd	• S/S treatment over 17 years met TCLP standards, with Ni remaining most stable at around 40% in soil	UK	Wang et al. (2014)
OPC, lime, montmorillonite, diatomite, sepiolite, and fly ash	Pb	• Results indicated that the optimal ratio of cement to fly ash and quicklime effectively remediated Pb levels of 10,000 mg/kg	China	Xi et al. (2014)

Source: Adapted from Tajudin et al. 2016.

and thereby improving hydration compared to biochar-free systems (Gupta et al. 2018). Researchers investigated biochar derived from mixed rice waste, wood waste, and food waste as a carbon-sequestering agent in cement mortar. Results showed that adding 1–2 wt% biochar from rice and food waste achieved comparable mechanical strength to samples without biochar. Notably, 1 wt% food waste biochar-reduced water penetration and absorptivity by 35 and 40%, respectively. Additionally, cement composites with wood biochar exhibited up to 20% higher compressive and tensile strength compared to control samples. Both food waste and wood waste biochar acted as reinforcements, improving ductility in cement mortar under flexural stress at failure.

Biochar particles with fine particle sizes can act as micro-scale fillers within composite materials, effectively occupying pore spaces to some extent. Nevertheless, an overabundance of biochar incorporation might adversely affect the material's structural integrity, primarily owing to the inherently porous and brittle characteristics of biochar itself. The hydration-enhancing effect of biochar appears independent of biomass type. For instance, 5% wood sawdust biochar increased hydration heat by 30% within 96 hours, whereas 2% dairy manure biochar (with 84% ash content, rich in phosphorus, calcium, and magnesium) reduced cumulative hydration heat by 15%, likely due to its interference with cement hydration (Gupta et al. 2020).

Biochar, particularly those with fine particle sizes, has been shown to promote the nucleation and growth of C-S-H phases in cementitious systems due to its high specific surface area. Research indicates that biochar derived from poultry litter can accelerate the development of a more compact microstructure during the early stages of hydration (Dixit et al. 2021). Therefore, biochar as an additive in concrete constructions made a contribution to carbon sequestering and may provide a sustainable way to enhance the hydration of cement composites.

15.3.2 Synergistic Effect of Biochar-cement on Stabilization/Solidification

The incorporation of biochar can enhance the S/S performance of cement-based composites. For instance, adding 1 wt% wood-derived biochar mitigates the adverse effects of contaminated dredged sediment while improving compatibility between OPC and the sediment. Researchers have evaluated the role of wood-derived biochar as a green additive in cement–biochar–sediment composites, analyzing their physicochemical properties (Wang et al. 2019; Wu et al. 2019). Results indicated that contaminated sediment and waste wood biochar could be recycled into green and sustainable construction materials with the incorporation of recycled aggregate and incineration sewage sludge ash.

Compressive strength results showed that incorporating biochar into the cement system could enhance the strength by prolonging the cement hydration process because of its internal curing effect and water retention capacity. The different chemical compositions of three different sources of sediments (medium contamination level) had a limited impact, whereas the particle size distribution of the soil affected the mechanical strength and pore structure of the S/S products. The addition of biochar facilitated the cement hydration process, but the porous structure may cause microcrack and degradation of mechanical strength.

15.3.3 Incorporation of Biochar with SCMs in Cement-based Stabilization/Solidification

The utilization of SCMs in biochar-modified cementitious systems offers a promising approach to improve S/S effectiveness while reducing carbon footprint. Based on their chemical characteristics, SCMs can be categorized into two distinct types. The first group comprises hydraulic materials such as GGBS. These materials are rich in calcium and exhibit self-cementing properties when combined with water, forming binding phases similar to OPC. The second category includes pozzolanic materials like SF and metakaolin (MK), which are characterized by high silica or alumina content. These materials do not possess inherent binding properties but react with CH in cementitious systems to produce secondary hydration products such as C-S-H or calcium aluminosilicate hydrate (C-A-S-H) gels.

GGBS is economical in large quantities, so it is suitable for premixed concrete, bulk on-site concrete production, and precast product manufacturing. GGBS produces lower heat during the hydration process, enhances resistance from chloride and sulfate attack compared with OPC. In addition, GGBS needs nowhere near as much energy to produce as OPC—less than one-fifth—and creates less than one-fifteenth of the carbon dioxide emissions. Using a ton of GGBS instead of Portland cement in concrete cuts CO_2 by about 650 kg. These carbon savings make GGBS an environmentally responsible choice for stabilizing soil (Samad and Shah 2017).

GGBS, characterized by its elevated calcium content, exhibits self-cementing properties similar to OPC upon hydration. Studies have demonstrated that the addition of GGBS in S/S systems substantially improves the encapsulation efficiency of heavy metals while concurrently increasing the structural integrity of the treated matrices (Wang et al. 2018a). Furthermore, GGBS modification offers multiple technological advantages, including reduced exothermic reaction during cement curing, as well as enhanced durability against corrosive agents such as chlorides and sulfates (Cheah et al. 2016). Besides, GGBS of small particle size can also function as fillers for densifying pore structure, facilitating the low permeability as well as physical encapsulation. Besides, GGBS is an Al-rich material (10–20% Al_2O_3), as Al_2O_3 is common in soil. The Al_2O_3-rich material usually presents high compatibility with soil, while the nature of amphoteric oxide could bond with toxic cation and anion in contaminated soil (Li et al. 2016; Li et al. 2021). Wang et al. (2018b) developed a novel and low-carbon approach that combines GGBS and OPC with phosphate-/sulfate-rich byproducts to produce low-alkalinity, high-compatibility, high-strength binary cement for soil S/S process. This binary cement presented superior environmental performance in terms of toxicity characteristic leaching procedure and semi-dynamic leaching tests, which reduced toxicity characteristic leaching procedure (TCLP) leachability of Pb by 86.9% and Pb diffusion coefficients by 69.4%. Therefore, GGBS demonstrates significant potential as a SCM when combined with cement for contaminant S/S applications.

The combination of biochar and GGBS would help form hydration products and improve strength development. Biochar–soil particle interface C-S-H gel forms between surfaces and creates strong networks that improve unconfined compressive strength. C-S-H forms on biochar surfaces and creates stronger bonds with nearby particles. Moreover, the Al atom introduced by GGBS participated in the recombination of dehydrated C-S-H and generated C-A-S-H. Meanwhile, the pozzolanic reaction products of GGBS can fill the pores of cured soil (Mu et al. 2024). Chen et al. (2022b) developed biochar-enhanced carbon-negative concrete by incorporating biochar with GGBS. The results showed that the replacement of cement by GGBS successfully compensated the 30 wt% lower cement content and generated additional C-S-H gel through the spontaneous hydration reaction. The 28-d strength achieved comparable values owing to pozzolanic reaction in the later stage. Li et al. (2021) investigated the geo environmental characteristics of a geogenic As-rich soil, and GGBS were employed for the S/S of the soil under field-relevant conditions. Results indicate that the use of 10% binder could effectively immobilize As and chemical stabilization/physical encapsulation jointly determined the leaching characteristics of the S/S soils. The increasing use of GGBS enhanced the strength of the 28-d cured S/S soils because of long-term pozzolanic reaction. The incorporation of biochar and GGBS demonstrated remarkable effectiveness in mitigating CO_2 emissions from cementitious materials. Substituting 30% of cement content with GGBS, in combination with biochar addition, enabled the composite blocks to achieve carbon-negative status (Chen et al. 2022b).

15.4 Key Parameters in Biochar-enhanced Soil S/S

The application of biochar in local contexts for soil remediation is promising due to its ability to improve soil quality, enhance crop yields, and sequester carbon. However, the effectiveness of biochar, along with GGBS, in soil remediation is influenced by several factors, including site limitations, supply availability, and cost implications.

15.4.1 Soil Properties

The effectiveness of biochar in soil remediation is highly dependent on the physicochemical properties of soil. Kwon et al. (2010) investigated the influence of particle size, density, and water adsorption of soil on the compressive strength and the results showed that the crushed soil particles were better than those from natural soil. Wang et al. (2019) explored the efficacy of wood waste biochar as a green admixture and assessed the influence of different physicochemical properties of dredged sediments on the mechanical performance. The results show that the particle size distribution of these dredged sediments influenced the pore structure and compressive strength of sediment products. Silt and clay in soil lead to large porosity and uneven pore size distribution. Additionally, fine soil may absorb relatively large volume of water and decrease the effective water-to-binder ratio. The physical improvements including altering the compactness and particle gradation of the soil using mechanical rolling or adding coarse aggregates like sand and gravel can be applied to enhance the soil.

The pH of soil is the key factor controlling the forms of the heavy metals in the soil. Biochar is usually alkaline, so after the application of biochar, pH of the soil will significantly increase, which is also one of the key factors for the biochar immobilization of heavy metals in the soil (He et al. 2022). Rees et al. (2015) demonstrated that the heavy metals in the soil were immobilized by biochar within a short time mainly through the internal diffusion of the biochar particles and the increase of soil pH. Lucchini et al. (2014) found that both the wood biochar and its ash content led to significant increase of the pH value of soil at the early stage of application. The field test results showed that after biochar treatment, the soil pH began to increase but is followed by a decline, but the change of bioavailable Cu and Cd concentrations was opposite, indicating the decrease of alkalinity of biochar caused by leaching reduced the biochar availability (Cui et al. 2016). Li et al. (2021) systematically evaluated the influence of pH on the S/S efficiency of geogenic As in contaminated soils using biochar and GGBS as partial substitutes for OPC. The result suggested that geogenic As could be effectively immobilized at the pH range of 5.5–6.5.

The moisture content of soil also affects the S/S performance by adjusting the hydration degree. The effect of moisture content of the soil on TCLP-As was also investigated by Li et al. (2021). The S/S soils with a moisture content of 20% showed a slightly lower TCLP-As compared with other samples at both 7-d and 28-d. Therefore, it can be concluded that the physical encapsulation of S/S mix played a major role for the immobilization of As in soils due to the more hydration products and denser structure of samples with 20% moisture content. However, the high moisture content in soil may induce more capillary pores in the matrix of S/S products that reduce the strength development.

15.4.2 Biochar Substitution Amount

The impact of biochar substitution on cement-based S/S of contaminated soils is a multifaceted topic that involves understanding the interactions between biochar, cement, and soil contaminants. The substitution amount of biochar plays a critical role in determining the effectiveness of the S/S process, influencing factors such as unconfined compressive strength (UCS), leachability, and microstructural characteristics of the treated soils.

Biochar addition to cement-stabilized soil has been shown to promote cement hydration reactions and increase particle agglomeration, which enhances the mechanical strength of the treated soil. An optimal biochar dosage of 5% was found to significantly improve the early growth speed of UCS in Cd-contaminated soils (Meng et al. 2024). In Pb-contaminated soils, the addition of 1–5 wt% biochar significantly improved the UCS, although excessive biochar content could have a detrimental effect on strength (Zou et al. 2023).

Biochar's fine-grained nature and high sorption capacity contribute to its ability to improve soil structure and mechanical properties, making it a viable alternative to traditional fillers like sand. The incorporation of biochar in cement-based S/S systems has been shown to reduce the leachability of heavy metals. For instance, a 5% biochar addition in Cd-contaminated soils significantly reduced Cd-leaching concentration and improved the stabilization rate, achieving a stabilization rate of over 99.8% with a 10% cement dosage (Meng et al. 2024). Similarly, in Pb-contaminated soils, increasing biochar content led to a significant decrease in leaching concentration, falling below hazardous waste identification standards (Zou et al. 2023).

The stabilization effects of biochar–cement composites are attributed to the formation of calcium silicate hydrate, calcium aluminate hydrate, and ettringite, which fill soil and biochar pores and encapsulate or adsorb contaminants like metal cations (Meng et al. 2024). Biochar's porous structure and high surface area enhance its ability to adsorb heavy metals and organic pollutants, while its interaction with cementitious materials contributes to the formation of stable chemical bonds that immobilize contaminants (Chen et al. 2022a).

The use of biochar in S/S processes aligns with green remediation strategies by reducing the carbon footprint associated with traditional cement use. Biochar is a carbon-negative material that can sequester CO_2, offering a sustainable alternative to conventional binders (Chen et al. 2024; Wang et al. 2020). The economic feasibility of biochar as a partial replacement for cement depends on the source of the feedstock and the local availability of biochar, which can potentially make it a cheaper option than cement or sand. However, excessive biochar content can negatively impact the mechanical strength of treated soils, and the variability in biochar properties depending on the feedstock and production process can lead to inconsistent results. Additionally, while biochar can enhance the stabilization of certain contaminants, its effectiveness may vary with different types of pollutants and soil conditions. Therefore, optimizing the biochar dosage and understanding its interactions with specific contaminants are crucial for maximizing the benefits of biochar in S/S applications.

15.4.3 Ratio of Binder to Soil

The ratio of binder to soil is a critical factor in the effectiveness of cement-based S/S of contaminated soils. This ratio influences the mechanical strength, leachability of contaminants, and overall stability of the treated soil. The optimal binder-to-soil ratio varies depending on the type of contaminants, the specific binders used, and the desired outcomes of the S/S process. The following sections explore these aspects in detail.

For chromium-contaminated soils, a cement-to-soil ratio of 35% was found to be optimal, achieving a high curing rate of 98.27% and reducing Cr (VI) leaching to 0.89 mg/L. This ratio ensures effective immobilization of chromium while maintaining economic feasibility (Yang et al. 2011). In a study comparing different binders, a 10% binder content was effective in immobilizing pollutants and achieving satisfactory compressive strength. Calcium aluminate cement was particularly effective, providing higher strength and reduced leachability compared to other binders (Bougharraf et al. 2018).

The UCS of treated soils generally increases with higher binder content. For instance, increasing the soda residue content in a cement/soda residue mix improved soil strength significantly, with UCS values reaching up to 6.48 MPa (Zha et al. 2020). The combination of binders, such as Portland cement with GGBS and bentonite, can enhance mechanical properties. The use of a 10% OPC–GGBS mixture supplemented with 10% bentonite significantly increased the retention of toxic metals and hydrocarbon compounds (Bougharraf et al. 2018).

The binder-to-soil ratio significantly affects the leachability of heavy metals. Lower water/cement ratios (e.g., 0.5 : 1) have been shown to better immobilize metals like Cu, Ni, and Pb compared to higher ratios (1 : 1) (Wang et al. 2018a). The long-term effectiveness of S/S treatments is influenced by the binder ratio. For example, the use of soda residue in combination with cement showed improved long-term stability under wetting-drying cycles, reducing Cr^{3+} leachability over time (Zha et al. 2020).

The choice of binder and its ratio to soil must balance cost and performance. Cement-based S/S is generally cost-effective, but the inclusion of alternative binders like fly ash or bone meal can further reduce costs while maintaining performance (Liu et al. 2024). The binder-to-soil ratio must also ensure compliance with environmental regulations regarding leachate concentrations. For instance, a cement-to-soil ratio of less than 35% was found insufficient for Polychlorinated biphenyl (PCB) oil-contaminated soils, highlighting the need for adequate binder content to meet regulatory standards (Yilmaz et al. 2003). While the binder-to-soil ratio is crucial for the success of S/S treatments, it is not the only factor. The type of binder, the nature of the contaminants, and environmental conditions also play significant roles. Additionally, the long-term performance of

S/S-treated soils can be influenced by interactions between the soil matrix and environmental factors, which may affect leach rates and mechanical stability over time (Cook and Batchelor, 1996).

15.4.4 Ratio of GGBS in Binder

The impact of the ratio of GGBS in the binder on the cement-based S/S of contaminated soils is significant, as it influences the mechanical properties and leachability of contaminants. GGBS, when used in combination with other materials, enhances the immobilization of heavy metals and improves the structural integrity of the treated soils. The effectiveness of GGBS in S/S processes is attributed to its pozzolanic activity and the formation of stable hydration products that encapsulate contaminants.

The inclusion of GGBS in the binder enhances the UCS of the solidified soil. For instance, a study using a geopolymer binder composed of fly ash and GGBS showed a steady increase in UCS with higher GGBS content and longer curing times (Wang et al. 2021). In another study, the optimal mixture ratio of thermally modified waste concrete powder and GGBS was found to be 9:1, achieving a compressive strength of 4.2 MPa after 28 days of curing (Mu et al. 2024). The use of GGBS in combination with cement kiln dust also significantly improved the mechanical properties of soft soils, with a 6% GGBS content leading to a 5.5-fold increase in strength after 28 days (Al-Khafaji et al. 2023).

GGBS is effective in reducing the leachability of heavy metals such as Pb and Cd. The leached concentrations of these metals decreased significantly with increasing GGBS content in the binder (Wang et al. 2021). A 20% replacement of OPC with GGBS was found to be most feasible for reducing the leachability of As in geogenic As-containing soils, due to the encapsulation effect and interlocking network of hydration products (Li et al. 2017). The formation of stable mineral phases such as C-(N)-A-S–H gels and the encapsulation of contaminants within these structures are key mechanisms by which GGBS contributes to the immobilization of heavy metals (Wang et al. 2021; Li et al. 2017).

Microscopic analyses, including scanning electron microscopy (SEM) and X-ray diffraction (XRD), have shown that the addition of GGBS leads to the formation of dense microstructures that enhance the encapsulation of contaminants (Wang et al. 2021; Li et al. 2017). The presence of GGBS in the binder results in the formation of hydration products such as C-S-H and ettringite, which are crucial for the stabilization of heavy metals like Pb (Qi et al. 2022).

The use of GGBS as a partial replacement for OPC not only improves the S/S performance but also reduces carbon emissions, making it an environmentally friendly option (Mu et al. 2024). Economically, the incorporation of GGBS can be cost-effective, as it utilizes industrial by-products and reduces the reliance on traditional cement, which is more expensive (Yang et al. 2011). While the use of GGBS in S/S processes offers numerous benefits, it is important to consider the specific soil and contaminant characteristics when determining the optimal GGBS ratio. The effectiveness of GGBS can vary depending on the type of contaminants and the environmental conditions of the site. Additionally, the long-term stability and durability of the solidified matrix should be evaluated to ensure the sustained immobilization of contaminants.

15.5 Supply Availability of Biochar

Biochar is derived from a variety of biomass sources, including agricultural byproducts and organic waste. The availability of these feedstocks is essential for biochar production, with regions rich in agricultural waste or forestry residues being more likely to maintain a consistent supply.

Generally, producing biochar is more cost-effective than traditional soil remediation methods. Several factors influence the costs associated with biochar production, such as the source of feedstock, the conditions of the pyrolysis process (including peak temperature and duration of heating), local energy prices, and labor costs. Additionally, rapidly growing plants with low nutrient needs, as well as waste materials like sludge from wastewater treatment, can be utilized for biochar production with minimal preparation costs.

However, when biochar is used for specific applications like pollutant removal, additional treatments, such as surface modifications with other compounds to enhance adsorption properties, may be necessary. The proportion of biochar used for various soil applications is crucial, as it can significantly impact overall costs. The local availability of feedstock can lower the final price of biochar compared to most additive-based treatment technologies. While the initial costs of soil remediation using biochar may be high, the long-term advantages—such as enhanced soil fertility and increased crop yields—can outweigh these expenses, making it a financially viable choice for soil rehabilitation. Therefore, biochar presents an economical solution for managing soil pollutants.

15.5.1 Feedstock Procurement

The expenses associated with acquiring feedstock can be determined in two ways: (1) the direct production cost if the material is cultivated specifically for pyrolysis, or (2) the opportunity cost if existing feedstock is redirected from other uses (Kung et al. 2013). For instance, corn stover procurement costs align with market rates, typically ranging from $82 to $100 per ton (Thompson and Tyner 2014). Brewery byproducts, such as spent grains, may be allocated to either bioenergy or animal feed, resulting in a higher opportunity cost of $91 (Field et al. 2013;

Shackley et al. 2015). Conversely, some feedstocks, like municipal solid waste, may have a negative acquisition cost if they require costly disposal. Research indicates that feedstock procurement is a critical variable in biochar's financial viability (Brown et al. 2011; Shackley et al. 2015). Shifting from high-value feedstocks (e.g., corn stover, brewery residues) to waste-based materials (e.g., forestry residues, yard waste) could enhance profitability, particularly in forestry applications where feedstock costs may be offset by disposal savings.

15.5.2 Feedstock Transportation

Transportation costs depend linearly on the distance between feedstock sources and pyrolysis facilities. While not a dominant factor in cost-benefit assessments (Campbell et al. 2018; Shackley et al. 2015), biomass transport remains expensive, though economies of scale can mitigate expenses. For an average hauling distance of 9.22 miles, transport may contribute around 11% to total production costs (Kung et al. 2013). Consequently, early-stage biochar markets are likely to be localized, with feedstock proximity playing a pivotal role (Dumortier et al. 2020). Mobile pyrolysis units could reduce reliance on nearby feedstocks, while co-location with industries like wood processing may eliminate transport costs entirely (Kim et al. 2015). Until such solutions are widely adopted, pyrolysis projects must prioritize locally available feedstocks.

15.5.3 Feedstock Preprocessing

Research on preprocessing methods for biochar production remains limited. Requirements vary by technology: slow pyrolysis necessitates only drying, gasification with power generation demands drying and coarse chipping, and fast pyrolysis for liquid fuels requires drying, chipping, and fine grinding. Phosphoric acid treatment can enhance biochar's value as a phosphorus-rich fertilizer, offsetting acid costs. While most studies acknowledge preprocessing (e.g., drying, chipping), specific cost data are scarce (Field et al. 2013). Campbell et al. (2018) estimated feedstock costs ranging from $0 (unchipped waste) to $80 (preprocessed material). Kung et al. (2013) suggest preprocessing accounts for ~6% of total costs, though sensitivity analyses often hold these costs constant (Shackley et al. 2015). A comparative analysis of drying versus chipping expenses would be valuable, as their energy demands are inversely related—woody biomass requires less drying but more chipping, while moist waste (e.g., sludge) consumes more energy in drying but minimal chipping.

15.5.4 Pyrolysis Costs: Infrastructure and Operations

Pyrolysis expenses are highly uncertain but offer potential for cost reductions. Basic methods like pit or mound kilns require minimal capital, relying on simple earthworks to control oxygen and heat (Lehmann and Joseph 2024). More advanced systems (e.g., brick/metal kilns with continuous feed) improve efficiency at higher costs. Shackley et al. (2015) report pyrolysis costs between $40 and $450 per ton across 10 facilities. Slow pyrolysis recovers heat but yields negligible bio-oil, whereas fast pyrolysis' profitability hinges on bio-oil output, feedstock price, and capital costs (Brown et al. 2011; Shackley et al. 2015). Biochar's secondary role in fast pyrolysis makes its value less influential. Fast pyrolysis requires ~45% higher capital than slow pyrolysis but generates higher-value bio-oil (53% yield vs. <1%) (Brown et al. 2011). Notably, this study incorporated escalating carbon offset values ($20–$60 over 30 years).

Supply chain costs are substantial. McCarl et al. (2012) estimate that for syngas/bio-oil energy systems, feedstock procurement ($59.44 per ton, 39.7%) and pyrolysis/plant operations ($90.08 per ton, 60.1%) comprise the total $149.91 per ton cost. Similarly, Dickinson et al. (2015) attribute ~50% of costs to feedstock handling and 50% to pyrolysis.

15.5.5 Biochar Industry Overview

Biochar enhances crop yields by improving input efficiency. Early adoption is expected in regions with poor soil quality, high-value crops, or proximity to biochar sources. Carbon sequestration incentives and wildfire risk reduction further boost adoption. Optimal conditions combine low feedstock costs, high crop value, and carbon credits (Table 15.3).

Commercial adoption depends on integrating biochar with bioenergy systems and fostering supply chain collaboration to balance farmer profits and societal benefits. Barriers include limited financial incentives (e.g., carbon credits) and uncertainty about returns. Additionally, low awareness has confined adoption to developed nations, despite biochar's potential in developing regions. Recent advancements have reduced production costs. In California, biochar costs $200–$1000 per ton (avg. $400) (Thengane et al. 2021). Market prices have dropped from $2850 per ton (2013) to $600–$1300 per ton (2021), with further declines possible as carbon markets mature (Thengane et al. 2021).

15.6 GGBS Supply Availability

15.6.1 Availability of GGBS

GGBS has gained prominence as a SCM due to its ability to reduce the carbon footprint of concrete when substituted for Portland cement clinker (Figure 15.3). However, its supply is inherently constrained by the limited availability of blast

Table 15.3 Cost of biochar application in soil.

Biomass	Pyrolysis conditions	Soil applications	Remarks	References
Straw	500 °C, 8 h	Soil amendment	• Evaluated crop yield and carbon sequestration potential • Enhanced crop growth and significantly reduced N_2O and CH_4 emissions over five cycles • Net + decreased by 0.6–19 times post-application	Bi et al. (2019)
Maize stalk	500 °C (biochar kiln)	Soil amendment	• Demonstrated high adsorption efficiency for flubendiamide in Indian soils • Cost-effective application at 5% effectively adsorbed the pesticide	Das and Mukherjee (2020)
Rice straw, seaweeds, etc.	—	Nutrient cycling (N, P, K)	• Investigated biochar's role in nutrient cycling • Improved soil N transformation, reducing NH_3 and N_2O emissions, and enhanced K and P availability	Jindo et al. (2020)
Wood	600–700 °C, 12 h	Orchard soil amendment	• Increased avocado growth and fruit yield over three years. • Cost-benefit analysis indicated a net benefit of US$8581 per hectare	Joseph et al. (2020)
Wheat straw, corn, etc.	—	Soil amendment	• Assessed financial viability in low-income nations • Biochar can mitigate land degradation in tropical regions • Average cost-effective biochar in low-income countries: US$58/Mg CO_2e	Robb et al. (2020)

Source: Adapted from Kamali et al. 2022.

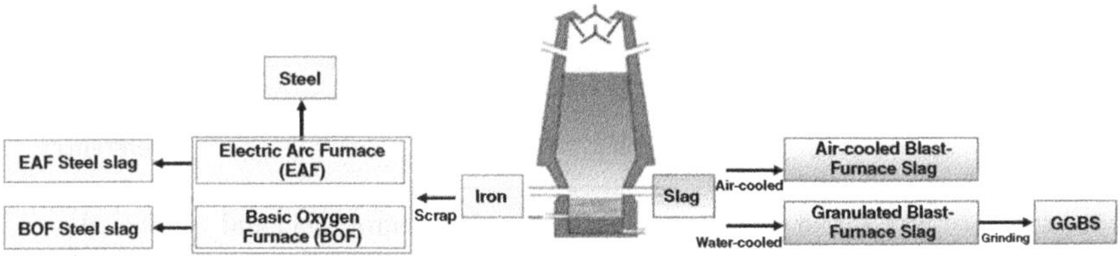

Figure 15.3 Flow chart of GGBS production. *Source:* Adapted from Miyamoto et al. 2015.

furnace slag and existing utilization rates. Current estimates place global GGBS production between 300 and 407 Mt annually (Arnold et al. 2023), dwarfed by clinker production (3280–3840 Mt/year), resulting in an 8 : 1 to 12 : 1 clinker-to-GGBS ratio. This imbalance persists because GGBS supply is tied to steel industry outputs, with 86–99% of slag already granulated for use. While GGBS production may rise by 25% by 2030 due to improved granulation rates (Arnold et al. 2023), it will remain a finite resource, as stockpiles of usable slag are negligible (<1 Mt in the UK) and often unsuitable due to aging or air-cooling (UK BEIS 2020).

Critically, GGBS is already near-maximum global utilization, meaning regional demand surges (e.g., for low-carbon concrete) may simply divert supply rather than reduce net emissions. For instance, substituting imported GGBS for local clinker in one region could displace its use elsewhere, negating global GHG benefits. This scarcity underscores the need for strategic allocation: GGBS should prioritize applications where its technical benefits (e.g., durability in marine environments) are indispensable, rather than blanket use for carbon reduction. Alternatives like abundant SCMs (e.g., fly ash) or clinker efficiency measures (e.g., optimized mix designs) may offer more scalable decarbonization pathways. The disconnect between local life-cycle assessment (LCA) metrics (favoring GGBS) and global resource limits calls for revised standards to align GGBS use with system-level carbon reduction (Kelly 2023).

15.6.2 Cost Reduction and Economic Benefits of GGBS

GGBS is generally less expensive than Portland cement, which can lead to a reduction in the overall cost of concrete production. Studies have shown that replacing a portion of cement with GGBS can result in significant cost savings. For instance, a study on residential buildings demonstrated a cost reduction of 22.61% when GGBS was used as a partial replacement for cement (Dubey et al. 2019). The production of GGBS requires less energy compared to cement, which translates into lower production costs. This energy efficiency, combined with the reduced need for raw materials, contributes to the economic viability of using GGBS in concrete. Moreover. GGBS enhances the durability of concrete, which can lead to lower maintenance costs over the lifespan of a structure. Its use in high-strength concrete mixes has been shown to improve mechanical properties and resistance to chemical attacks, thereby extending the service life of concrete structures.

While GGBS offers long-term cost benefits, the initial investment in sourcing and processing the material can be higher, especially if superfine slag is used, which is more expensive than regular cement. The cost-effectiveness of GGBS can vary depending on regional availability and market conditions. In some regions, the cost savings may not be fully realized due to transportation and processing costs (Tang et al. 2015).

15.7 Environmental Benefits

Biochar-GGBS applications do more than just trap contaminants—they bring real environmental advantages compared to old-school remediation methods. These new technologies are a big step forward in treating soil sustainably and show clear ecological benefits. The cement industry pumps out about 7% of global CO_2 emissions. GGBS concrete tells a different story with just 287 kg CO_2 eq per m^3, while traditional Portland cement concrete hits 386 kg/m^3. This means carbon emissions drop by 55–75%. Adding alum sludge biochar to cement mixtures cuts global warming potential by another 32.2% compared to regular mixes. The use of GGBS in concrete significantly reduces the carbon footprint of construction projects. By substituting cement with GGBS, CO_2 emissions associated with cement production are decreased, aligning with global sustainability goals. GGBS is a recycled material, and its use in concrete helps in the sustainable disposal of industrial waste. This not only conserves natural resources but also reduces the environmental impact of waste disposal (Nicula et al. 2023). Studies reveal that concrete incorporating GGBS emits 37% fewer carbon emissions than fly ash concrete and 50% less than glass powder concrete (Nicula et al. 2023).

Recent studies have suggested positive effects of biochar for soil application on global warming potential (GWP) reduction. For instance, Bi et al. demonstrated that biochar derived from straw had a great potential on Net GWP (a 0.6–19-fold decrease after the application of biochar). The authors have stated that due to the low trade price of carbon dioxide (0.21 × 103 CNY/Mg), the benefits of such effects on GWP may not be considerable under the current situation (Bi et al. 2019). However, when biochar is utilized for soil decontamination purposes, indirect benefits (such as environmental clean-up and ecosystem protection) rather than direct advantages (such as promoting the crop production) need to highlighted.

15.8 Conclusion

The integration of biochar and GGBS presents a transformative approach to soil remediation, combining contaminant immobilization with carbon sequestration and waste valorization. Biochar's adsorption capacity and GGBS's pozzolanic reactivity synergistically enhance S/S performance, reducing heavy metal mobility and enhancing organic pollutant sequestration. GGBS-biochar systems cut CO_2 emissions by 30–75% compared to OPC, while biochar production repurposes agricultural/industrial waste.

Local feedstock use (e.g., rice straw, sewage sludge) lowers costs, though pre-treatment and pyrolysis scalability require optimization. In future studies, the guidelines for biochar feedstock selection, pyrolysis conditions, and application ratio should be developed. The long-term field studies are needed to assess contaminant leaching and soil health impacts. Carbon pricing and subsidies could accelerate adoption, particularly in developing economies. By addressing these gaps, biochar-GGBS systems can transition from laboratory success to scalable, eco-engineered solutions for global soil remediation.

References

Akcil, A., Erust, C., Ozdemiroglu, S. et al. (2015). A review of approaches and techniques used in aquatic contaminated sediments: metal removal and stabilization by chemical and biotechnological processes. *Journal of Cleaner Production* 86: 24–36.

Al-Khafaji, R., Dulaimi, A., Jafer, H. et al. (2023). Stabilization of soft soil by a sustainable binder comprises ground granulated blast slag (GGBS) and cement kiln dust (CKD). *Recycling* 8 (1): 10.

Ali, A., Guo, D., Jeyasundar, P.G.S.A. et al. (2019). Application of wood biochar in polluted soils stabilized the toxic metals and enhanced wheat (*Triticum aestivum*) growth and soil enzymatic activity. *Ecotoxicology and Environmental Safety* 184: 109635.

Ali, U., Shaaban, M., Bashir, S. et al. (2020). Rice straw, biochar and calcite incorporation enhance nickel (Ni) immobilization in contaminated soil and Ni removal capacity. *Chemosphere* 244: 125418.

Anastasiadou, K., Christopoulos, K., Mousios, E. et al. (2012). Solidification/stabilization of fly and bottom ash from medical waste incineration facility. *Journal of Hazardous Materials* 207: 165–170.

Andrew, R.M. (2018). Global CO_2 emissions from cement production. *Earth System Science Data* 10 (1): 195–217.

Arnold, W., Astel, P., Forman, T. et al. (2023). The efficient use of GGBS in reducing global emissions. *Institution of Structural Engineers, United Kingdom*.

Ayaz, M., Muntaha, S.T., Baltrėnaitė-Gedienė, E. et al. (2025). Biochar and carbon-negative technologies: exploring opportunities for climate change mitigation. *Biochar* 7 (1): 17.

Bashir, S., Hussain, Q., Akmal, M. et al. (2018). Sugarcane bagasse-derived biochar reduces the cadmium and chromium bioavailability to mash bean and enhances the microbial activity in contaminated soil. *Journal of Soils and Sediments* 18: 874–886.

Beiyuan, J., Awad, Y.M., Beckers, F. et al. (2017). Mobility and phytoavailability of As and Pb in a contaminated soil using pine sawdust biochar under systematic change of redox conditions. *Chemosphere* 178: 110–118.

Bi, Y., Cai, S., Wang, Y. et al. (2019). Assessing the viability of soil successive straw biochar amendment based on a five-year column trial with six different soils: views from crop production, carbon sequestration and net ecosystem economic benefits. *Journal of Environmental Management* 245: 173–186.

Bougharraf, N., Louati, D., Mosbahi, M. et al. (2018). Comparison of the effectiveness of different binders in solidification/stabilization of a contaminated soil. *Arabian Journal of Geosciences* 11 (13): 1–11.

Brown, T.R., Wright, M.M., and Brown, R.C. (2011). Estimating profitability of two biochar production scenarios: slow pyrolysis vs fast pyrolysis. *Biofuels, Bioproducts and Biorefining* 5 (1): 54–68.

Campbell, R.M., Anderson, N.M., Daugaard, D.E. et al. (2018). Financial viability of biofuel and biochar production from forest biomass in the face of market price volatility and uncertainty. *Applied Energy* 230: 330–343.

Cheah, C.B., Chung, K.Y., Ramli, M. et al. (2016). The engineering properties and microstructure development of cement mortar containing high volume of inter-grinded GGBS and PFA cured at ambient temperature. *Construction and Building Materials* 122: 683–693.

Chen, L., Wang, L., Cho, D.-W. et al. (2019). Sustainable stabilization/solidification of municipal solid waste incinerator fly ash by incorporation of green materials. *Journal of Cleaner Production* 222: 335–343.

Chen, L., Wang, L., Xiong, X. et al. (2022a). Biochar for green and sustainable stabilization/solidification. *Low Carbon Stabilization and Solidification of Hazardous Wastes* 65–73.

Chen, L., Zhang, Y., Wang, L. et al. (2022b). Biochar-augmented carbon-negative concrete. *Chemical Engineering Journal* 431: 133946.

Chen, L., Wang, L., and Tsang, D.C. (2024). Biochar-augmented binders for sustainable stabilization/solidification of wastes. *Biochar Application in Soil to Immobilize Heavy Metals* 163–190.

Cook, E.R. and Batchelor, B. (1996). Stabilization/solidification of hazardous wastes in soil matrices. In: *Advances in Porous Media* (ed. M. Yavuz Corapcioglu), 3: 307–359. Elsevier.

Cui, L., Pan, G., Li, L. et al. (2016). Continuous immobilization of cadmium and lead in biochar amended contaminated paddy soil: a five-year field experiment. *Ecological Engineering* 93: 1–8.

Das, S.K. and Mukherjee, I. (2020). Low cost biomass derived biochar amendment on persistence and sorption behaviour of flubendiamide in soil. *Bulletin of Environmental Contamination and Toxicology* 105 (2): 261–269.

Dickinson, D., Balduccio, L., Buysse, J. et al. (2015). Cost-benefit analysis of using biochar to improve cereals agriculture. *GCB Bioenergy* 7 (4): 850–864.

Dixit, A., Verma, A., and Dai Pang, S. (2021). Dual waste utilization in ultra-high performance concrete using biochar and marine clay. *Cement and Concrete Composites* 120: 104049.

Dubey, S., Singh, A., and Kushwah, S.S. (2019). Utilization of iron and steel slag in building construction. *AIP Conference Proceedings*, (Vol. 2158, No. 1). AIP Publishing.

Dumortier, J., Dokoohaki, H., Elobeid, A. et al. (2020). Global land-use and carbon emission implications from biochar application to cropland in the United States. *Journal of Cleaner Production* 258: 120684.

Field, J.L., Keske, C.M., Birch, G.L. et al. (2013). Distributed biochar and bioenergy coproduction: a regionally specific case study of environmental benefits and economic impacts. *GCB Bioenergy* 5 (2): 177–191.

Gupta, S., Kua, H.W., and Low, C.Y. (2018). Use of biochar as carbon sequestering additive in cement mortar. *Cement and Concrete Composites* 87: 110–129.

Gupta, S., Palansooriya, K.N., Dissanayake, P.D. et al. (2020). Carbonaceous inserts from lignocellulosic and non-lignocellulosic sources in cement mortar: preparation conditions and its effect on hydration kinetics and physical properties. *Construction and Building Materials* 264: 120214.

He, M., Xu, Z., Hou, D. et al. (2022). Waste-derived biochar for water pollution control and sustainable development. *Nature Reviews Earth & Environment* 3 (7): 444–460.

Igalavithana, A.D., Kwon, E.E., Vithanage, M. et al. (2019). Soil lead immobilization by biochars in short-term laboratory incubation studies. *Environment International* 127: 190–198.

Jain, N. (2011). Solidification and leachability of Cr (VI) in rice husk ash-blended cement. *International Scholarly Research Notices* 2011 (1): 183158.

Jindo, K., Audette, Y., Higashikawa, F.S. et al. (2020). Role of biochar in promoting circular economy in the agriculture sector. Part 1: A review of the biochar roles in soil N, P and K cycles. *Chemical and Biological Technologies in Agriculture* 7: 1–12.

Jing, C., Meng, X., and Korfiatis, G.P. (2004). Lead leachability in stabilized/solidified soil samples evaluated with different leaching tests. *Journal of Hazardous Materials* 114 (1–3): 101–110.

Joseph, S., Pow, D., Dawson, K. et al. (2020). Biochar increases soil organic carbon, avocado yields and economic return over 4 years of cultivation. *Science of the Total Environment* 724: 138153.

Jun, L., Wei, H., Aili, M. et al. (2020). Effect of lychee biochar on the remediation of heavy metal-contaminated soil using sunflower: a field experiment. *Environmental Research* 188: 109886.

Kamali, M., Sweygers, N., Al-Salem, S. et al. (2022). Biochar for soil applications-sustainability aspects, challenges and future prospects. *Chemical Engineering Journal* 428: 131189.

Kelly, F. (2023). A review of GGBS use in the UK and its role in reducing embodied carbon. *The Structural Engineer* 101 (7): 24.

Kim, D., Anderson, N.M., and Chung, W. (2015). Financial performance of a mobile pyrolysis system used to produce biochar from sawmill residues. *Forest Products Journal* 65 (5–6): 189–197.

Kung, C.-C., McCarl, B.A., and Cao, X. (2013). Economics of pyrolysis-based energy production and biochar utilization: a case study in Taiwan. *Energy Policy* 60: 317–323.

Kwon, H.-M., Le, A.T., and Nguyen, N.T. (2010). Influence of soil grading on properties of compressed cement-soil. *KSCE Journal of Civil Engineering* 14 (6): 845–853.

Lehmann, J. and Joseph, S. (eds.) (2024). *Biochar for Environmental Management: Science, Technology and Implementation*. Taylor & Francis.

Li, J.S., Beiyuan, J., Tsang, D.C. et al. (2017). Arsenic-containing soil from geogenic source in Hong Kong: leaching characteristics and stabilization/solidification. *Chemosphere* 182: 31–39.

Li, J. S., Chen, L., Zhan, B. et al. (2021). Sustainable stabilization/solidification of arsenic-containing soil by blast slag and cement blends. *Chemosphere* 271: 129868.

Li, J.-S., Xue, Q., Wang, P. et al. (2014). Effect of drying-wetting cycles on leaching behavior of cement solidified lead-contaminated soil. *Chemosphere* 117: 10–13.

Li, P., Pan, S.-Y., Pei, S. et al. (2016). Challenges and perspectives on carbon fixation and utilization technologies: an overview. *Aerosol and Air Quality Research* 16 (6): 1327–1344.

Li, Y., Yu, H., Liu, L. et al. (2021). Application of co-pyrolysis biochar for the adsorption and immobilization of heavy metals in contaminated environmental substrates. *Journal of Hazardous Materials* 420: 126655.

Liu, N., Zhang, Y., Xu, C. et al. (2020). Removal mechanisms of aqueous Cr (VI) using apple wood biochar: a spectroscopic study. *Journal of Hazardous Materials* 384: 121371.

Liu, Y.-Z., Zhang, C., Wu, K. et al. (2024). Sustainable stabilization/solidification of Cd and Pb in industrially heavy metal-contaminated site soils using a novel binder incorporating bone meal and fly ash. *Construction and Building Materials* 416: 135162.

Lucchini, P., Quilliam, R.S., DeLuca, T.H. et al. (2014). Increased bioavailability of metals in two contrasting agricultural soils treated with waste wood-derived biochar and ash. *Environmental Science and Pollution Research* 21: 3230–3240.

Malviya, R. and Chaudhary, R. (2006). Factors affecting hazardous waste solidification/stabilization: a review. *Journal of Hazardous Materials* 137 (1): 267–276.

McCarl, B.A., Peacocke, C., Chrisman, R. et al. (2012). Economics of biochar production, utilization and greenhouse gas offsets. *Biochar for Environmental Management* (pp. 373–390). Routledge.

Meng, S., Ge, C., Zhao, G. et al. (2024). Sustainable stabilization of cadmium (Cd)-contaminated soil with biochar-cement: evaluation of strength, leachability, microstructural characteristics, and stabilization mechanism. *Soil and Sediment Contamination: An International Journal* 34 (5): 1–18.

Miyamoto, T., Torii, K., Akahane, K. et al. (2015). Production and use of blast furnace slag aggregate for concrete. *Nippon Steel & Sumitomo Metal Technical Report* 109: 102.

Moon, D.H. and Dermatas, D. (2006). An evaluation of lead leachability from stabilized/solidified soils under modified semi-dynamic leaching conditions. *Engineering Geology* 85 (1–2): 67–74.

Mu, F., Ma, Z., Lang, L. et al. (2024). Solidification/stabilization of Pb-contaminated soil by using low-carbon binder derived from dehydrated waste concrete powders and GGBS. *Developments in the Built Environment* 17: 100357.

Nicula, L.M., Manea, D.L., Simedru, D. et al. (2023). Potential role of GGBS and ACBFS blast furnace slag at 90 days for application in rigid concrete pavements. *Materials* 16 (17): 5902.

Oliveira, F.R., Patel, A.K., Jaisi, D.P. et al. (2017). Environmental application of biochar: Current status and perspectives. *Bioresource Technology* 246: 110–122.

O'Connor, D., Peng, T., Li, G. et al. (2018). Sulfur-modified rice husk biochar: a green method for the remediation of mercury contaminated soil. *Science of The Total Environment* 621: 819–826.

Rees, F., Germain, C., Sterckeman, T. et al. (2015). Plant growth and metal uptake by a non-hyperaccumulating species (*Lolium perenne*) and a Cd-Zn hyperaccumulator (*Noccaea caerulescens*) in contaminated soils amended with biochar. *Plant and Soil* 395: 57–73.

Qi, W., Ren, Q., Zhao, Q. et al. (2022). Effectiveness of soda residue-activated GGBS as alternative binder on compressive strength and workability of cemented paste backfills: reuse of multi-source solid wastes. *Construction and Building Materials* 348: 128594.

Robb, S., Joseph, S., Abdul Aziz, A. et al. (2020). Biochar's cost constraints are overcome in small-scale farming on tropical soils in lower-income countries. *Land Degradation & Development* 31 (13): 1713–1726.

Samad, S. and Shah, A. (2017). Role of binary cement including Supplementary Cementitious Material (SCM), in production of environmentally sustainable concrete: a critical review. *International Journal of Sustainable Built Environment* 6 (2): 663–674.

Shackley, S., Clare, A., Joseph, S. et al. (2015). Economic evaluation of biochar systems: current evidence and challenges. *Biochar for Environmental Management* (pp. 813–851). Routledge.

Shen, T., Tang, Y., Lu, X.-Y. et al. (2018a). Mechanisms of copper stabilization by mineral constituents in sewage sludge biochar. *Journal of Cleaner Production* 193: 185–193.

Shen, Z., Hou, D., Zhao, B. et al. (2018b). Stability of heavy metals in soil washing residue with and without biochar addition under accelerated ageing. *Science of The Total Environment* 619: 185–193.

Simiele, M., Lebrun, M., Miard, F. et al. (2020). Assisted phytoremediation of a former mine soil using biochar and iron sulphate: effects on as soil immobilization and accumulation in three Salicaceae species. *Science of The Total Environment* 710: 136203.

Tajudin, S.A., Azmi, M.M., and Nabila, A.T.A. (2016). Stabilization/solidification remediation method for contaminated soil: a review. *IOP Conference Series: Materials Science and Engineering* (Vol. 136, No. 1), 012043. IOP Publishing.

Tang, K., Millard, S., and Beattie, G. (2015). Technical and economical feasibility of using GGBS in long-span concrete structures. *Advances in Concrete Construction* 3 (1): 001.

Thengane, S.K., Kung, K., Hunt, J. et al. (2021). Market prospects for biochar production and application in California. *Biofuels, Bioproducts and Biorefining* 15 (6): 1802–1819.

Thompson, J.L. and Tyner, W.E. (2014). Corn stover for bioenergy production: cost estimates and farmer supply response. *Biomass and Bioenergy* 62: 166–173.

Voglar, G.E. and Leštan, D. (2010). Solidification/stabilisation of metals contaminated industrial soil from former Zn smelter in Celje, Slovenia, using cement as a hydraulic binder. *Journal of Hazardous Materials* 178 (1–3): 926–933.

Wang, X., Zou, T., Lian, J. et al. (2025). Simultaneous mitigation of cadmium contamination and greenhouse gas emissions in paddy soil by iron-modified biochar. *Journal of Hazardous Materials* 488: 137430

Wang, F., Xu, J., Zhang, Y. et al. (2021). MgO-GGBS binder–stabilized/solidified PAE-contaminated soil: strength and leachability in early stage. *Journal of Geotechnical and Geoenvironmental Engineering* 147 (8): 04021059.

Wang, F., Wang, H., and Al-Tabbaa, A. (2014). Leachability and heavy metal speciation of 17-year old stabilised/solidified contaminated site soils. *Journal of Hazardous Materials* 278: 144–151.

Wang, F., Wang, F., Shen, Z. et al. (2018a). An evaluation of stabilised/solidified contaminated model soil using PC-based and MgO-based binders under semi-dynamic leaching conditions. *Environmental Science and Pollution Research* 25 (16): 16050–16060.

Wang, L., Chen, L., Guo, B. et al. (2020). Red mud-enhanced magnesium phosphate cement for remediation of Pb and as contaminated soil. *Journal of Hazardous Materials* 400: 123317.

Wang, L., Chen, L., Tsang, D.C.W. et al. (2019). The roles of biochar as green admixture for sediment-based construction products. *Cement and Concrete Composites* 104: 103348.

Wang, L., Yu, K., Li, J.-S. et al. (2018b). Low-carbon and low-alkalinity stabilization/solidification of high-Pb contaminated soil. *Chemical Engineering Journal* 351: 418–427.

Wu, C., Shi, L., Xue, S. et al. (2019). Effect of sulfur-iron modified biochar on the available cadmium and bacterial community structure in contaminated soils. *Science of The Total Environment* 647: 1158–1168.

Xi, Y., Wu, X., and Xiong, H. (2014). Solidification/stabilization of Pb-contaminated soils with cement and other additives. *Soil and Sediment Contamination: An International Journal* 23 (8): 887–898.

Xing, J., Li, L., Li, G. et al. (2019). Feasibility of sludge-based biochar for soil remediation: characteristics and safety performance of heavy metals influenced by pyrolysis temperatures. *Ecotoxicology and Environmental Safety* 180: 457–465.

Xu, C., Zhao, J., Yang, W. et al. (2020). Evaluation of biochar pyrolyzed from kitchen waste, corn straw, and peanut hulls on immobilization of Pb and Cd in contaminated soil. *Environmental Pollution* 261: 114133.

Yang, W., Wang, L., Tan, W.H. et al. (2011). Cement-based solidification/stabilization of high concentration chromium-contaminated soil. *Advanced Materials Research* 414: 232–237.

Ye, L., Gao, G., Li, F. et al. (2025). A comprehensive review on biochar-based materials for the safe utilization and remediation of heavy metal-contaminated agricultural soil and associated mechanisms. *Journal of Environmental Chemical Engineering* 13 (3): 116179.

Yin, C.-Y., Mahmud, H.B., and Shaaban, M.G. (2006). Stabilization/solidification of lead-contaminated soil using cement and rice husk ash. *Journal of Hazardous Materials* 137 (3): 1758–1764.

Yilmaz, O., Ünlü, K., and Cokca, E. (2003). Solidification/stabilization of hazardous wastes containing metals and organic contaminants. *Journal of Environmental Engineering* 129 (4): 366–376.

Zha, F., Fanghua, Z., Kang, B. et al. (2020). Experimental investigation of cement/soda residue for solidification/stabilization of Cr-contaminated soils. *Advances in Civil Engineering* 2020: 1–13.
Zhang, Y.Y., Xu, W.J., Zhu, X.H. et al. (2024). Scaling biochar solutions for urban carbon dioxide removal. *One Earth* 7 (9): 1481–1486.
Zou, Z., Qin, Y., Zhang, T. et al. (2023). Enhancing road performance of lead-contaminated soil through biochar-cement solidification: an experimental study. *Journal of Environmental Management* 348: 119315.

16

Technical and Economic Analysis of Biochar Technologies

Le Fang[1,2,3], Yifan Xing[1,2,3], and Yingying Han[1,2,3]

[1] *Research and Development Center for Watershed Environmental Eco-Engineering, Beijing Normal University, Zhuhai 519087, P. R. China*
[2] *State Key Laboratory of Wetland Conservation and Restoration, School of Environment, Beijing Normal University, Beijing 100875, P. R. China*
[3] *Key Laboratory of Coastal Water Environmental Management and Water Ecological Restoration of Guangdong Higher Education Institutes, Beijing Normal University, Zhuhai, P. R. China*

16.1 Introduction

In recent years, sustainable biochar production has gained significant interest due to its extensive physicochemical features and potential wide application. Raw materials utilized include organic wastes such as livestock and poultry manure, crop residues, sludge, and municipal wastes. Carbon fixation in biomass is possible due to varying physicochemical properties, organic and inorganic compositions, and their capacity for carbon conversion under specified conditions (Xie et al. 2022).

Biochar serves as a significant byproduct rich in solid carbon and can be utilized in diverse sectors including businesses and agriculture. Produced biochar possesses high energy and is used as a coal substitute in industries for fuel production. The microporous structure and nutrition content of biochar can enhance soil quality and accelerates carbon sequestration, while also delay nutrient degradation (Matuštík et al. 2022; Shi et al. 2021; Yang et al. 2021). It is employed as an immobilization agent in purification process to remove hazardous and heavy metals, and as an economical chemical removal agent (Awogbemi and Kallon 2023). Therefore, biochar presents a multifaceted opportunity for sustainable development across agriculture, environmental management, waste reduction, and business innovation. Its diverse applications underline the importance

Waste-derived Biochar for Sustainable Rural Revitalization, First Edition. Edited by Yuqing Sun and Daniel C.W. Tsang.
Published 2026 by John Wiley & Sons, Inc.

of integrating biochar into various sectors to promote ecological balance and economic viability.

To ensure consistent biochar market supply, the technical and economic cost measurement of the biochar production system must be evaluated. Linking carbon markets to agricultural production can be a mutually beneficial strategy. Biochar application reduces traditional fertilizers usage and recycles agricultural waste. Globally, agricultural waste can generate 373 million tons of biochar annually, sequestering 0.55 Pg of CO_2 (Ganguly et al. 2022). Economic benefit analysis of straw biochar revealed that, considering the carbon sequestration potential, preparing straw-hydrated charcoal with a reaction condition of a solid–liquid ratio of 2 : 1 could increase single-process income by 1356 yen (Si et al. 2023).

Consequently, various technologies, policies, and market mechanisms are being designed to facilitate carbon sequestration and carbon market trading for sustainable agriculture. This chapter aims to present the following insights: (1) techno-economic analysis of biochar technologies in production process; (2) techno-economic analysis of biochar application scenarios; (3) element circulation and sustainable development; and (4) current limitations and future perspectives. This study provides a theoretical framework for further development.

16.2 Techno-economic Analysis of Biochar Technologies in Production Process

16.2.1 Thermochemical Conversion Process Analysis Instruments and Techniques

Thermochemical processing for biochar production offers a sustainable biomass transformation and waste management option. This include torrefaction, hydrothermal carbonization (HTC), pyrolysis (including slow [300–600 °C, <0.5 °C/s], fast [500–1000 °C, >100 °C/s] and flash [900–1200 °C, 10–1000 °C/s], microwave pyrolysis [MWP]), and gasification. Torrefaction enhances biomass fuel characteristics for independent burning (Jha et al. 2022). HTC transforms moist waste into char (Yoganandham et al. 2020). Pyrolysis produces three primary pyrolysates: pyrolytic gas, pyrolytic oil, and char (Choi et al. 2023). Gasification converts carbonaceous waste into H_2, CO, and CO_2 (Valizadeh et al. 2022). Various feedstocks, including lignocellulosic biomass (LCB), algal, food residues, and solid municipal waste, can be converted into biochar using different thermochemical processes. Supraja et al. (2023) concluded various thermochemical conversion processes for biomass valorization (Figure 16.1).

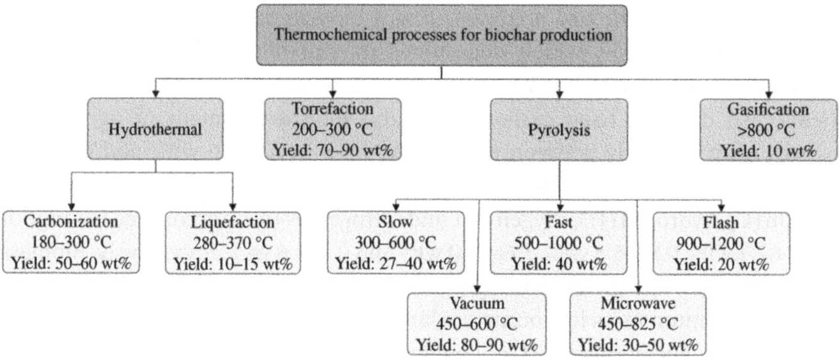

Figure 16.1 Various thermochemical conversion processes for biomass valorization.

Table 16.1 Analysis methods employed on the thermochemical conversion of biochar.

Analysis methods	Instruments	Properties	References
Proximate analysis	ASTM D1762-84	Moisture, ash, volatile matter, and fixed carbon contents	Alfattani et al. (2022); Naqvi et al. (2021)
Ultimate analysis	ASTM standard D5373-08, CHNS elemental analyzer	C, H, N, O	Alfattani et al. (2022)
Thermogravimetric analysis (TGA/DSC)	Thermogravimetric analyzer	Mass/heat change during heating in different atmospheres	Alfattani et al. (2022); Chen et al. (2024); Priyadarsini et al. (2023)

Different feedstock-derived biochars exhibit distinct physicochemical characteristics. Statistically analyzed, these characteristics, along with the energy and resource recovery potential of biomass resources pyrolysis, were elucidated. A detailed description of analytical instrument and methodologies for biochar thermochemical conversion is summarized in Table 16.1. This encompass proximate analysis, ultimate analysis, thermogravimetric analysis (TGA), and response analysis. Proximate and ultimate analyses were carried out according to American Society for Testing and Materials (ASTM) standards, which were acting as an evaluation tool for composites. TGA, as one of the highly respected thermal analysis tools in solid thermal degradation research, has gained wide recognition for its accuracy and reliability in various biomass pyrolysis studies (Naron et al. 2017). Data acquired from the TGA experiment underwent pyrolysis kinetic analysis.

To be specific, primary analysis of any biomaterial consists of proximate and ultimate analysis, which determines its suitability for its conversion to value-added products. The proximate analysis revealed that the volatile matter (VM), ash content, fixed carbon (FC), and moisture content of agricultural wastes were generally between 65–90 wt%, 1–15 wt%, 3–26 wt%, and less than 10 wt%, respectively (Li et al. 2023b). Ultimate analysis estimated that the carbon (C), hydrogen (H), oxygen (O), and nitrogen (N) were accounted in range of 55.3–82.7 wt%, 0.89–5.5 wt%, 1.6–23.0 wt% and 0.1–1.6 wt%, respectively. Biomass pyrolysis in TGA progresses through three stages: drying (150 °C), decomposition (where maximum mass loss occurred), and finally carbonization (600 °C) (Gupta et al. 2017). Furthermore, TGA coupled with FT-IR spectroscopy (TG-FTIR) enables qualitative evaluation of gaseous functional groups, further integrated with kinetic models offers comprehensive understanding of degradation behaviors, kinetics, and properties of volatile compounds (Kumar et al. 2024).

16.2.2 Techno-economic Analysis Models to Simulate Biochar Production

Thermochemical biochar production is a sustainable biomass conversion strategy and waste management technique. Its commercial application merits further investigation. Machine learning can optimize the efficiency and reduce production costs of biochar production by designing, predicting, and optimizing the process. This process includes process optimization, predictive analytics, cost reduction, data integration and analysis, marked analysis and demand forecasting, feedstock selection and preprocessing, and enhanced research and development.

Table 16.2 summarizes the biochar production simulation models. Aspen Plus can simulate the production process, offering numerous benefits for new technology and production condition optimization (Saeed et al. 2020; Treeyawetchakul 2020). Extreme Gradient Boosting (XGBoost) and Dense Neural Network (DNN) exhibited superior predictive capabilities for pyrolysis yields. Artificial neural network (ANN) demonstrated exceptional estimation accuracy in estimating the high heating value (HHV) of the biochar production process. These models consider not only pyrolysis temperature, feedstock composition, heating rate, and residence time, which directly influence the physicochemical properties of biochar, but also economic factors like energy consumption, scalability, and the market value of biochar. Consequently, they optimize the production process to ensure technical and economical feasibility. When optimizing biochar production, the primary objectives are maximizing biochar yield and quality while minimizing production costs and environmental impact. The aforementioned models have been utilized to optimize biochar production, including designing experiments to determine optimal pyrolysis parameters and processes, considering the interconnected elements (Chettri et al. 2024).

Table 16.2 Models employed on the production of biochar.

Model	Biochar type	Function	Regression coefficient (R^2)	References
Aspen Plus	Rice straw and sugarcane bagasse	Construct a biomass pyrolysis flowchart	—	Liu et al. (2022)
Extreme Gradient Boosting (XGBoost)	Algal biochar	Predict biochar yield	0.84	Leng et al. (2022); Pathy et al. (2020)
Dense Neural Network (DNN)	Plastics and biomass	Predict biochar and bio-oil yield	0.96	Alabdrabalnabi et al. (2022)
Artificial neural network (ANN)	—	Predicting biochar's HHV	0.9651	Çakman et al. (2024)

By leveraging machine learning, biochar production processes can become more efficient, cost-effective, and environmentally friendly. The integration of machine learning not only improves the production quality and yield but also supports sustainable practices in agriculture and waste management. As the field of machine learning continues to evolve, its applications in biochar production will likely expand, offering even more innovative solutions.

16.2.3 Cost and Energy Consumption Evaluation of Biochar Production

The fiscal feasibility and energy input associated with biochar production are key barriers to its widespread adoption. To enhance the market acceptance of biochar, it is essential to conduct a comprehensive analysis of its financial viability throughout various stages of the production and distribution process, including processing, pyrolysis, storage, and transportation.

Addressing the fiscal feasibility of biochar production requires a multidisciplinary approach, incorporating economic, environmental, and technological analyses. By thoroughly investigating the financial viability at each stage of the production and distribution process, stakeholders can identify opportunities for cost reduction, enhance profit margins, and promote broader adoption of biochar as a sustainable solution in various industries.

Expenses can be bifurcated into two main segments: fixed and variable costs. Fixed biochar production costs comprise mobile pyrolysis unit, preprocessing machinery, pyrolysis setup, transportation, water tank, and storage facility among others. Variable costs include fuel, oil and lubricants, labor, and miscellaneous costs. Fixed and Variable costs are valued at US$754.68 and US$1542.16/Mg, respectively (Nematian et al. 2021). Monte Carlo (MC) provides a realistic assessment of biochar production costs and evaluates uncertainty. The maximum, average, and minimum biochar production costs can be calculated via a stochastic analysis using @Risk software from Palisade Corporation. A stochastic MC simulation indicates a likely range of biochar production costs between US$448.78 and US$1846.96/Mg, with 90% probability that costs will fall between US$571 and US$1455/Mg (Nematian et al. 2021).

Pyrolysis operates as a thermal energy system, necessitating a total energy input. Of this, 51.6% is allocated to thermal energy requirements, with 48.4% utilized for electrical needs, predominantly heating the pyrolysis reactor (Shahbaz et al. 2024). The thermal loss analysis results of the whole process operation system can be calculated according to operation process parameters of biochar production process parameters. Previous research indicates the energy consumption of sludge (Meng et al. 2024), wheat and corn (Zhao et al. 2011), sycamore and pine (Adam et al. 2017), lignocellulosic biomass (Cao et al. 2024) in biochar production process is estimated as 23,550 MJ/h, 0.462–0.523 kW·h/(kg straw), 3.5–4.2 kJ/g, 0.387 $GJ/t_{aromatics}$, respectively. Therefore, wheat and corn, globally significant food crops, can provide ample crop residue resources for low-cost, energy-efficient biochar production, promoting Rural Revitalization Sustainability.

16.3 Techno-economic Analysis of Biochar Application Scenarios

16.3.1 Soil Application

Biochar is predominantly employed for soil modification due to its high-water absorption and surface area with an activated carbon role (Chen et al. 2020b). In China, the annual carbon sequestration potential of biochar production was estimated at ~4.98×10^8 t CO_2e (Yang et al. 2021). Biochar amendments significantly improved chemical properties of soils (pH, cation exchange capacity, and organic carbon), physical properties (bulk density and porosity), and microbial diversity (bacterial diversity and fungal diversity) (Singh et al. 2022). More importantly, the field application of biochar has an influence on soil fertility and crop productivity. Biochar can increase the cation exchange capacity of soils, allowing them to retain more nutrients (such as calcium, magnesium, potassium, and ammonium) that are essential for plant growth.

Significantly, biochar application boosts crop productivity via expanding soil organic carbon (SOC) and soil inorganic carbon (SIC) contents, affecting plant growth. A 10-year field study observed consecutive biochar application significantly increases the contents and stocks of the SIC, organic carbon content (POC), and SOC content, respectively, by 3.2–24.3%, 38.2–166.2%, and 15.8–82.2% (Shi et al. 2021). Meta-analysis studies report a yield increase of 17–53% across various crops (wheat, maize, canola, barley, rice, sorghum, tomato, groundnut, faba bean, turnip, and peanuts) post biochar application, indicating a positive priming effect on native SOC. More importantly, the specific benefits can vary based on soil type, biochar characteristics, and crop species.

Biochar can serve as a slow-release nutrient source, gradually supplying essential nutrients to crops over time, which can reduce the need for synthetic fertilizers. According to the calculations of Yang et al. (2021), application of 337.83 kg biochar (from 1 t biomass) could reduce 0.10 kg of N-fertilizer, 0.02 kg of P-fertilizer, and 0.01 kg of K-fertilizer demand per year. And Matuštík et al. (2022) set that N and P fertilizers can be reduced by 25% after the addition of biochar to the soil, which would result in a reduction of nitrogen fertilizer by 25 kg and phosphorus fertilizer by 7.75 kg/ha. Hence, biochar presents the potential to curtail excessive nitrogen and phosphorus fertilization, but the effectiveness of biochar can vary based on its source, production methods, application rates, and local soil conditions. Therefore, it is essential for farmers and agronomists to consider these factors when implementing biochar in their agricultural practices.

16.3.2 Use in Wastewater Treatment

Biochar's potential as an economical, eco-friendly adsorbent for wastewater treatment and purification is gaining attention. Using biochar in wastewater treatment can help mitigate water pollution, improve soil health when applied to land, and reduce the need for chemical coagulants or flocculants, making the process more environmentally friendly. It efficiently eradicates various pollutants from wastewater including domestic, industrial, and agricultural sources such as pathogens, pesticides, organic contaminants, inorganic ions, dyes, and others (Awogbemi and Kallon 2023).

The removal efficiency of phosphorus (Xu et al. 2023), heavy metals (Das et al. 2023), antibiotics (Fan et al. 2018), phenol (Mishra et al. 2019; Mohammed et al. 2018), dyes (Saniya et al. 2020; Sawalha et al. 2022), and pesticides (Mojiri et al. 2022; Samy et al. 2022) in wastewater by adding biochar can reach 93.9–99%, 52.6–95.5%, 56.9%, 81–96%, 70–99.6%, 46.9–93.2%, respectively. Its capacity to adsorb contaminants can vary based on the feedstock used and the pyrolysis conditions. The removal efficiency of contaminants mainly depends on the source, type, physical and chemical properties, and dosage of the adsorbents. Biochar can be modified to enhance its adsorption properties, allowing for tailored applications depending on the specific contaminants present in wastewater. This adaptability makes it suitable for various treatment scenarios. Most importantly, after use, biochar can be regenerated through methods such as thermal treatment or washing, making it reusable and extending its lifespan as an adsorbent.

Using biochar in wastewater treatment can not only help mitigate water pollution, improve soil health when applied to land, but also reduce the need for chemical coagulants or flocculants, making the process more environmentally friendly. As the use of biochar in wastewater treatment grows, regulatory frameworks are evolving to ensure its safe and effective use, addressing concerns about leachate and potential contaminants. Therefore, enhancing biochar's performance can be achieved through techniques like calcination, chemical oxidation, reduction, metal impregnation, ozone oxidation, low-temperature plasma, and other modifications (Awogbemi and Kallon 2022).

Overall, biochar presents a promising solution for wastewater treatment, combining economic benefits with environmental sustainability. Continued research and development efforts will be vital to fully realize its potential and integrate it into mainstream wastewater management practices.

16.3.3 Other Application Scenario

In addition to soil amelioration and water purification, biochar can also be used in the realms of energy, chemistry, etc. The operational conditions and reactor design can be tailored to accommodate biochar's diverse applications. Figure 16.2 illustrates the conversion pathway from waste into various biochar applications (Anand et al. 2022). Specifically, biochar can also be used in the following ways:

1. as substitutes for electrode materials, cathode catalysts, and proton exchange membranes (PEM) in MFC (Chakraborty et al. 2020). MFC harnesses energy from organic compounds by converting wastewater into electrical energy. Biochar could be utilized as an electrode material for MFC and a cathode catalyst. The price of a biochar-based electrode is lower than commercial ones. The material cost of N/Fe-C was about $0.03–0.08/g, which is a thousand times lower than a commercial platinum electrode (Cao et al. 2016).
2. as catalysts for the transesterification of oils for biodiesel production (Kant Bhatia et al. 2021). Biodiesel is considered as a favorable fuel because of its high energy density and presence of C14–C20 long carbon chain fatty acids. The porous structure of biochar allows easy access of reactants to the active site to facilitate the transesterification process, and biochar's hydrophobic surface helps remove unwanted products generated during catalytic reactions.
3. as additives in cementitious products (Labianca et al. 2024). Biochar can boost the compressive strength of the final product. Incorporating biochar with recycled materials can significantly mitigate environmental impacts, particularly global warming, making biochar-augmented cementitious products and their application as carbon-negative resources feasible.

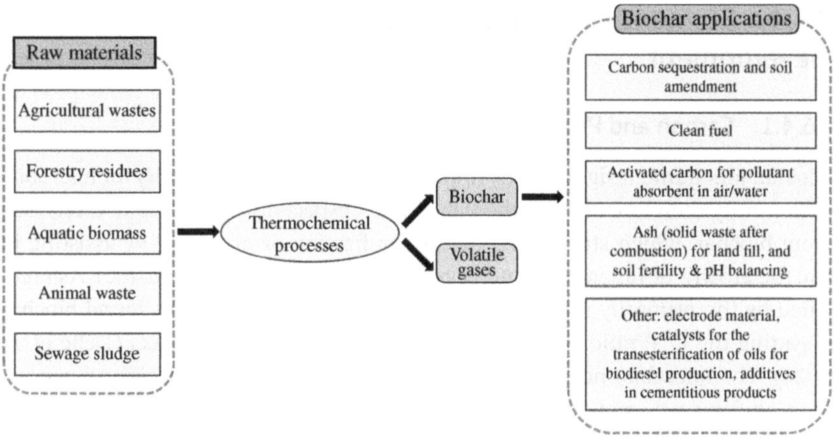

Figure 16.2 Conversion pathway from crop residue to various applications of biochar.

16.3.4 Economics of Biochar Application

Despite broad application, current market conditions indicate biochar applications are economically inviable. This is associated with the high capital costs of pyrolysis plants and low government incentives (Rajabi Hamedani et al. 2019). Techno-economic analysis (CBA) has frequently explored hypothetical scenarios for enhanced economics. For instance, Haeldermans et al. (2020) compared biochar production from conventional pyrolysis (CP) and MWP through techno-economic evaluation. Minimum prices varied from US$454 to US$871/ton for CP-biochar and US$588 to US$1020/ton for MWP-biochar (assuming a EUR/US$ currency exchange rate of 1.04). While CP is a streamlined technology, MWP-biochar demonstrated superior quality and technical feasibility. Additionally, biochar price per ton was a crucial factor for biochar production plants and was highly influenced by government carbon taxes.

The economics of biochar production systems has been assessed with respect to raw material or feedstock used, the conversion technology employed, carbon sequestration subsidies, and carbon credits reflecting the social value of GHG emission reductions. Adopting smart farming practices can enhance crop yields and farmer economic status while mitigating climate change impacts (Haeldermans et al. 2020). Compared to inorganic fertilizers, biochar offers long-term agricultural enhancement. When combined with plant growth-promoting rhizobacteria and N-P-K fertilizer, wheat crops' grain yield and economic outcomes improved significantly. A shift in grain yield from 4.54 to 4.70 ton/ha resulted to a rise of net benefit from US$293 to US$438/ha (i.e., 50% relative increment) and a relative increase of 50% (Ijaz et al. 2019). This underscores the potential for biochar to contribute significantly to biochar production profitability.

16.4 Element Circulation and Sustainable Development

16.4.1 Carbon and Phosphorus Cycle

Biochar amendment significantly improves soil's P and C availability for microbial community therefore altering the P dynamics, especially for the labile C released from biochar, which stimulating the mineralization of organic P by assisting in the the growth of fungal taxa and keystone bacterial (Chen et al. 2020c). As identified by the primarily microbial predictor of soil labile P, studies found biochar elevating the C : P ratio of soil, therefore shaping the keystone taxa (Deng et al. 2024). The microbial network complexity was increased; specifically, a keystone taxa shifting was noticed from oligotrophs to copiotrophs (like Rhizobiales, Sphingomonadales, and Rhodospirillales) (Deng et al. 2024). With increasing pyrolysis temperature, the dehydration, decarboxylation, and polymerization of OP intensify, releasing organic functional groups attached to phosphate groups and resulting in the formation of orthophosphate and pyrophosphate (Huang and Tang 2015; Qian and Jiang 2014; Sun et al. 2018). In addition, the positions formerly occupied by these organic functional groups are reoccupied by metal cations (Ca, Fe, and Al); therefore, the orthophosphate and metal cations formed with crystalline are adsorbed onto the surface of biochars.

Biochar is enriched with phosphorus, which would afford phosphorus for soil and assist soil bacterials and plants growing. There are around 4–9% of P in sewage, majorly in form of organic P (OP) and inorganic P (IP). And the OP contains various forms such as nucleic acid, phospholipid, and phosphoprotein, while IP could be further classified into apatite inorganic P (AP) and non-apatite inorganic P (NAIP). In sewage sludge singular pyrolysis process, Fe and Al readily adsorb P, which produce iron phosphate and aluminum phosphate that are hard to release. With the energy input, the bonding of P in organic contained substances is broken and the released phosphorus would be resource-oriented approached by adding of CaO or MgO (Li et al. 2023a). Adding the alkalines such as CaO would stimulate the NAIP transforming into AP; the reaction sequence of metal ions and P is as follows: $Ca^{2+} > Mg^{2+} > Al^{3+} > Fe^{3+}$ (Chen et al. 2022; Chen et al. 2020a; Li et al. 2023a).

The available P of soil was increased in the short term, which was released from the biochar. In addition, the increasing of P was persistent over time, which could partially attribute to the enhanced microbial activities, nourished by released carbon from biochar, promotes the activation of soil phosphorus by phosphorus-solubilizing microorganisms. The phoD-harboring bacterial community dominates phosphorus cycling, transferring phosphorus from biochar to soil (Xue et al. 2024).

16.4.2 Environmental Impact Assessment of Biochar Production

Life-cycle assessment (LCA) is a tool routinely used to assess the environmental impacts of biochar production via pyrolysis processes, which adopts a whole lifecycle perspective. In manufactured products, the environmental impacts are evaluated from the extraction and processing of raw material via the product's manufacture, distribution, and utilization, to the reprocessing or final disposal of the materials composing it (Ilgin and Gupta 2010). For example, raw materials' extraction, processing, transport, and product manufacturing were included in the system boundaries (Figure 16.3) for the biochar production (Labianca et al. 2024).

GaBi 8.70, a LCA tool offered by the Thinkstep, served as our background data source and computational executor for the impact categories. Comprehensive details about GaBi are accessible at LCA For Experts I Sphera. CML 2001, a more transparent and less uncertain GaBi method, assessed the 10 environmental impacts of pyrolysis application, including resource consumption, environmental damage, and toxicity potential (Yang et al. 2021). The LCA results indicated that elevating the ratio of biochar in the mixture resulted in reduced global warming potential (Labianca et al. 2024). As per Sun et al. LCA of wastewater microalgae biocarbon with the L (leaching) A (physical activation) process revealed a substantial global warming mitigation potential of 1.34–2.94 tonnes of CO_2-eq/yr for every tonne of biochar produced (Sun et al. 2024).

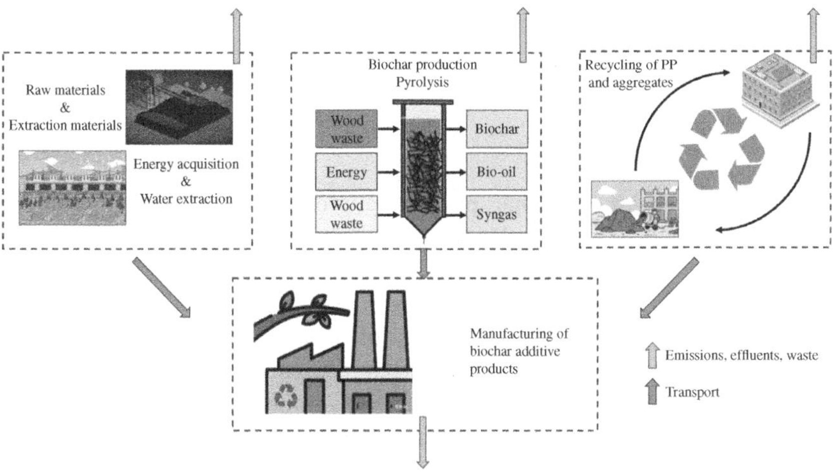

Figure 16.3 System boundaries.

16.5 Current Limitations and Future Perspectives

Due to the environmental impact of fossil energy, escalating sustainable development demands, and the necessity of transitioning to renewables, the implementation of biochar technology is crucial. As per the National Energy Administration, China aims to elevate its energy from renewable sources during the 14th Five-Year Plan (2021–2025).

The energy consumption of pyrolysis process determines the success of biochar technology. Innovative thermo-chemical techniques, including solar pyrolysis, synergetic pyrolysis with cement kiln, and MWP, are being explored for biochar production. These techniques hold promise to significantly decrease pyrolysis energy consumption, thereby minimizing emissions and environmental impact.

Despite biochar's agricultural, environmental, and energy advantages, ecological risks persist due to pollutants like polycyclic aromatic hydrocarbons (PAHs), volatile organic compounds (VOCs), polychlorinated dibenzo-p-dioxins (PCDDs), persistent free radicals (PFRs), toxic metals, and water-soluble organic compounds (WSOCs). Future research should concentrate on safer biochar production methods, such as slow pyrolysis, to reduce PAHs and optimized temperature conditions to manage PCDDs and VOCs. Additionally, long-term ecological studies are essential to evaluate biochar's impact on soil microorganisms, plant growth, and soil health. Establishing guidelines for biochar's safe application is crucial to ensure its benefits outweigh any adverse effects. Interdisciplinary research combining environmental, agricultural, and material sciences can aid in addressing biochar's complex challenges.

16.6 Summary

This chapter explored biochar production process technology-and-economy analysis coupled with its application. Furthermore, the dialogue focused on element circulation and sustainable progression. Biochar, beyond acting as an eco-friendly adsorbent, could potentially boost rural economic development. This analysis aimed at encouraging the use of biochar in the Rural Revitalization Sustainability and bridging the gap between its production and utilization. The production variables and feedstock considerably influence the biochar yield and properties. Consequently, biochar features can be tailored efficiently and flexibly, facilitating its implementation in innumerable scenarios. Additionally, the agricultural economic feasibility of biochar usage hinges on factors, including the biochar application rate, the expense of transport and field application, the crop yield, and the anticipated benefit of carbon sequestration. Initiatives such as subsidizing biochar's initial capital expenditure and assigning a non-negligible carbon abatement price in future pricing models will bolster the economical rewards for rice farmers.

References

Adam, M., Beneroso, D., Katrib, J. et al. (2017). Microwave fluidized bed for biomass pyrolysis. Part II: effect of process parameters. *Biofuels, Bioproducts and Biorefining* 11 (4): 613–624.

Alabdrabalnabi, A., Gautam, R., and Mani Sarathy, S. (2022). Machine learning to predict biochar and bio-oil yields from co-pyrolysis of biomass and plastics. *Fuel* 328: 125303.

Alfattani, R., Shah, M.A., Siddiqui, M.I. et al. (2022). Bio-char characterization produced from walnut shell biomass through slow pyrolysis: sustainable for soil amendment and an alternate bio-fuel. *Energies* 15: 1.

Anand, A., Kumar, V., and Kaushal, P. (2022). Biochar and its twin benefits: crop residue management and climate change mitigation in India. *Renewable and Sustainable Energy Reviews* 156: 111959.

Awogbemi, O. and Kallon, D.V.V. (2022). Pretreatment techniques for agricultural waste. *Case Studies in Chemical and Environmental Engineering* 6: 100229.

Awogbemi, O. and Kallon, D.V.V. (2023). Progress in agricultural waste derived biochar as adsorbents for wastewater treatment. *Applied Surface Science Advances* 18: 100518.

Çakman, G., Gheni, S., and Ceylan, S. (2024). Prediction of higher heating value of biochars using proximate analysis by artificial neural network. *Biomass Conversion and Biorefinery* 14 (5): 5989–5997.

Cao, B., Lu, W., Shen, W. et al. (2024). Sustainability of biomass pyrolysis for bio-aromatics and bio-phenols production: life cycle assessment of stepwise catalytic approach. *Energy Conversion and Management* 319: 118912.

Cao, C., Wei, L., Su, M. et al. (2016). Low-cost adsorbent derived and in situ nitrogen/iron co-doped carbon as efficient oxygen reduction catalyst in microbial fuel cells. *Bioresource Technology* 214: 348–354.

Chakraborty, I., Sathe, S.M., Dubey, B.K. et al. (2020). Waste-derived biochar: applications and future perspective in microbial fuel cells. *Bioresource Technology* 312: 123587.

Chen, J., Aihemaiti, A., Xia, Y. et al. (2022). The effect of soil amendment derived from P-enhanced sludge pyrochar on ryegrass growth and soil microbial diversity. *Science of The Total Environment* 813: 152526.

Chen, J., Tang, S., Yan, F. et al. (2020a). Efficient recovery of phosphorus in sewage sludge through hydroxylapatite enhancement formation aided by calcium-based additives. *Water Research* 171: 115450.

Chen, L., Tu, Z., Chen, Y. et al. (2024). Bioenergy and value-added chemicals of banana peel waste (BPW): deeper insights from thermal kinetics, thermodynamics, in-situ volatile products analysis, and bio-chars application for Cd(II) highly-efficient removal. *Biomass and Bioenergy* 185: 107238.

Chen, W.-H., Lee, K.-T., Chih, Y.-K. et al. (2020b). Novel renewable double-energy system for activated biochar production and thermoelectric generation from waste heat. *Energy & Fuels* 34 (3): 3383–3393.

Chen, Y., Sun, R., Sun, T. et al. (2020c). Evidence for involvement of keystone fungal taxa in organic phosphorus mineralization in subtropical soil and the impact of labile carbon. *Soil Biology and Biochemistry* 148: 107900.

Chettri, D., Boro, D., Chirania, M. et al. (2024). Integrating biochar production in biorefineries: towards a sustainable future and circular economy. *Biofuels, Bioproducts and Biorefining* 18: 2156–2176.

Choi, H., Kim, Y.T., Tsang, Y.F. et al. (2023). Integration of thermochemical conversion processes for waste-to-energy: a review. *Korean Journal of Chemical Engineering* 40 (8): 1815–1821.

Das, S.K., Ghosh, G.K., and Avasthe, R. (2023). Conversion of crop, weed and tree biomass into biochar for heavy metal removal and wastewater treatment. *Biomass Conversion and Biorefinery* 13 (6): 4901–4914.

Deng, J., He, D., Zhu, X. et al. (2024). Biochar amendment shifts bacterial keystone taxa regulating soil phosphorus dynamics. *Applied Soil Ecology* 201: 105521.

Fan, S., Wang, Y., Li, Y. et al. (2018). Removal of tetracycline from aqueous solution by biochar derived from rice straw. *Environmental Science and Pollution Research* 25 (29): 29529–29540.

Ganguly, A., Brown, R.C., and Wright, M.M. (2022). Techno-economic and greenhouse gas emission assessment of carbon negative pyrolysis technology. *Green Chemistry* 24 (23): 9290–9302.

Gupta, G., Ram, M., Bala, R. et al. (2017). Pyrolysis of chemically treated corncob for biochar production and its application in Cr(VI) removal. *Environmental Progress & Sustainable Energy* 37: 1606–1617.

Haeldermans, T., Campion, L., Kuppens, T. et al. (2020). A comparative techno-economic assessment of biochar production from different residue streams using conventional and microwave pyrolysis. *Bioresource Technology* 318: 124083.

Huang, R. and Tang, Y. (2015). Speciation dynamics of phosphorus during (Hydro) thermal treatments of sewage sludge. *Environmental Science & Technology* 49 (24): 14466–14474.

Ijaz, M., Tahir, M., Shahid, M. et al. (2019). Combined application of biochar and PGPR consortia for sustainable production of wheat under semiarid conditions with a reduced dose of synthetic fertilizer. *Brazilian Journal of Microbiology* 50 (2): 449–458.

Ilgin, M.A. and Gupta, S.M. (2010). Environmentally conscious manufacturing and product recovery (ECMPRO): a review of the state of the art. *Journal of Environmental Management* 91 (3): 563–591.

Jha, S., Okolie, J.A., Nanda, S. et al. (2022). A review of biomass resources and thermochemical conversion technologies. *Chemical Engineering & Technology* 45 (5): 791–799.

Kant Bhatia, S., Palai, A.K., Kumar, A. et al. (2021). Trends in renewable energy production employing biomass-based biochar. *Bioresource Technology* 340: 125644.

Kumar, A., Jamro, I.A., Wang, J. et al. (2024). Co-pyrolysis of microalgae residue and sewage sludge: an in-depth characterization of kinetics, drivers, and gas-oil-char behaviors. *Journal of Analytical and Applied Pyrolysis* 179: 106438.

Labianca, C., Zhu, X., Ferrara, C. et al. (2024). A holistic framework of biochar-augmented cementitious products and general applications: technical, environmental, and economic evaluation. *Environmental Research* 245: 118026.

Leng, L., Yang, L., Lei, X. et al. (2022). Machine learning predicting and engineering the yield, N content, and specific surface area of biochar derived from pyrolysis of biomass. *Biochar* 4 (1): 63.

Li, J., Li, Y., Liu, F. et al. (2023a). Pyrolysis of sewage sludge to biochar: transformation mechanism of phosphorus. *Journal of Analytical and Applied Pyrolysis* 173: 106065.

Li, Y., Gupta, R., Zhang, Q. et al. (2023b). Review of biochar production via crop residue pyrolysis: development and perspectives. *Bioresource Technology* 369: 128423.

Liu, Y., Yang, X., Zhang, J. et al. (2022). Process simulation of preparing biochar by biomass pyrolysis via aspen plus and its economic evaluation. *Waste and Biomass Valorization* 13: 1–14.

Matuštík, J., Pohořelý, M., and Kočí, V. (2022). Is application of biochar to soil really carbon negative? The effect of methodological decisions in life cycle assessment. *Science of The Total Environment* 807: 151058.

Meng, Z., Xiong, H., Zhu, C. et al. (2024). Study on the process technology of energy saving and consumption reducing for municipal sludge-based biochar. *Renewable Energy* 220: 119606.

Mishra, S., Yadav, S.S., Rawat, S. et al. (2019). Corn husk derived magnetized activated carbon for the removal of phenol and para-nitrophenol from aqueous solution: interaction mechanism, insights on adsorbent characteristics, and isothermal, kinetic and thermodynamic properties. *Journal of Environmental Management* 246: 362–373.

Mohammed, N.A.S., Abu-Zurayk, R.A., Hamadneh, I. et al. (2018). Phenol adsorption on biochar prepared from the pine fruit shells: equilibrium, kinetic and thermodynamics studies. *Journal of Environmental Management* 226: 377–385.

Mojiri, A., Zhou, J.L., Nazari V, M. et al. (2022). Biochar enhanced the performance of microalgae/bacteria consortium for insecticides removal from synthetic wastewater. *Process Safety and Environmental Protection* 157: 284–296.

Naqvi, S.R., Tariq, R., Shahbaz, M. et al. (2021). Recent developments on sewage sludge pyrolysis and its kinetics: resources recovery, thermogravimetric platforms, and innovative prospects. *Computers & Chemical Engineering* 150: 107325.

Naron, D.R., Collard, F.X., Tyhoda, L. et al. (2017). Characterisation of lignins from different sources by appropriate analytical methods: introducing thermogravimetric analysis-thermal desorption-gas chromatography–mass spectroscopy. *Industrial Crops and Products* 101: 61–74.

Nematian, M., Keske, C., and Ng'ombe, J.N. (2021). A techno-economic analysis of biochar production and the bioeconomy for orchard biomass. *Waste Management* 135: 467–477.

Pathy, A., Meher, S., and Balasubramanian, P. (2020). Predicting algal biochar yield using eXtreme Gradient Boosting (XGB) algorithm of machine learning methods. *Algal Research* 50: 102006.

Priyadarsini, A., Swain, B., Mishra, A. et al. (2023). Study on biofuel efficiency of tropical banana leaf biomass using spectroscopy, kinetic and thermodynamic parameters. *Bioresource Technology Reports* 23: 101522.

Qian, T.-T. and Jiang, H. (2014). Migration of phosphorus in sewage sludge during different thermal treatment processes. *ACS Sustainable Chemistry & Engineering* 2 (6): 1411–1419.

Rajabi Hamedani, S., Kuppens, T., Malina, R. et al. (2019). Life cycle assessment and environmental valuation of biochar production: two case studies in Belgium. *Energies* 12: 2166.

Saeed, S., Shafeeq, A., Raza, W. et al. (2020). Thermal performance analysis of ionic liquid-pretreated spent coffee ground using aspen plus®. *Chemical Engineering & Technology* 43 (12): 2447–2456.

Samy, M., Gar Alalm, M., Ezeldean, E. et al. (2022). Solar-light-driven ZnO/biochar treatment of pesticides contaminated wastewater: a practical and computational study. *Energy Science & Engineering* 10 (12): 4708–4725.

Saniya, A., Sathya, K., Nagarajan, K. et al. (2020). Modelling of the removal of crystal violet dye from textile effluent using *Murraya koenigii* stem biochar. *Desalination and Water Treatment* 203: 356–365.

Sawalha, H., Bader, A., Sarsour, J. et al. (2022). Removal of dye (Methylene blue) from wastewater using bio-char derived from agricultural residues in Palestine: performance and isotherm analysis. *Processes* 10: 2039.

Shahbaz, M., Alherbawi, M., Okonkwo, E.C. et al. (2024). Evaluating negative emission technologies in a circular carbon economy: a holistic evaluation of direct air capture, bioenergy carbon capture and storage and biochar. *Journal of Cleaner Production* 466: 142800.

Shi, S., Zhang, Q., Lou, Y. et al. (2021). Soil organic and inorganic carbon sequestration by consecutive biochar application: results from a decade field experiment. *Soil Use and Management* 37 (1): 95–103.

Si, H., Zhao, C., Wang, B. et al. (2023). Liquid-solid ratio during hydrothermal carbonization affects hydrochar application potential in soil: based on characteristics comparison and economic benefit analysis. *Journal of Environmental Management* 335: 117567.

Singh, H., Northup, B.K., Rice, C.W. et al. (2022). Biochar applications influence soil physical and chemical properties, microbial diversity, and crop productivity: a meta-analysis. *Biochar* 4 (1): 8.

Sun, D., Hale, L., Kar, G. et al. (2018). Phosphorus recovery and reuse by pyrolysis: applications for agriculture and environment. *Chemosphere* 194: 682–691.

Sun, J., Benavente, V., and Mašek, O. (2024). Synergistic valorisation of wastewater microalgae: sulphuric acid-assisted phosphorus recovery and enhanced electrochemical performance of biochar for supercapacitors. *Chemical Engineering Journal* 494: 152986.

Supraja, K.V., Kachroo, H., Viswanathan, G. et al. (2023). Biochar production and its environmental applications: recent developments and machine learning insights. *Bioresource Technology* 387: 129634.

Treeyawetchakul, C. (2020). Preliminary modified biodiesel production by coupling reactive distillation with a steam reformer via aspen plus®. *IOP Conference Series: Materials Science and Engineering* 778 (1): 012064.

Valizadeh, S., Hakimian, H., Farooq, A. et al. (2022). Valorization of biomass through gasification for green hydrogen generation: a comprehensive review. *Bioresource Technology* 365: 128143.

Xie, Y., Wang, L., Li, H. et al. (2022). A critical review on production, modification and utilization of biochar. *Journal of Analytical and Applied Pyrolysis* 161: 105405.

Xu, C., Liu, R., and Chen, L. (2023). Removal of phosphorus from domestic sewage in rural areas using oyster shell-modified agricultural waste–rice husk biochar. *Processes* 11: 2577.

Xue, P., Hou, R., Fu, Q. et al. (2024). Mechanisms of phosphorus activation in charosphere and non-charosphere: the priming effect of biochar. *Soil and Tillage Research* 244: 106195.

Yang, Q., Mašek, O., Zhao, L. et al. (2021). Country-level potential of carbon sequestration and environmental benefits by utilizing crop residues for biochar implementation. *Applied Energy* 282: 116275.

Yoganandham, S.T., Sathyamoorthy, G., and Renuka, R.R. (2020). Chapter 8–Emerging extraction techniques: hydrothermal processing. In: *Sustainable Seaweed Technologies* (ed. M.D. Torres, S. Kraan, and H. Dominguez), 191–205. Elsevier.

Zhao, X., Zhang, J., Song, Z. et al. (2011). Microwave pyrolysis of straw bale and energy balance analysis. *Journal of Analytical and Applied Pyrolysis* 92 (1): 43–49.

17

ESG Perspective and Biodiversity Impact of Waste-derived Biochar

*Maheshika Senanayake[1,3], Pavani Dulanja Dissanayake[2], Jay Hyuk Rhee[3,4], Meththika Vithanage[5], and Yong Sik Ok[1,3]**

[1] *Korea Biochar Research Center, Department of Environmental Science and Ecological Engineering, Korea University, Seoul, Republic of Korea*
[2] *Department of Soil Science, Faculty of Agriculture, University of Peradeniya, Peradeniya, Sri Lanka*
[3] *International ESG Association (IESGA), Seoul, Republic of Korea*
[4] *School of Business Administration, Korea University, Seoul, Republic of Korea*
[5] *Department of Geochemistry, Geological Survey of Denmark and Greenland, Copenhagen K, Denmark*

*Corresponding author: yongsikok@korea.ac.kr

Graphical Abstract

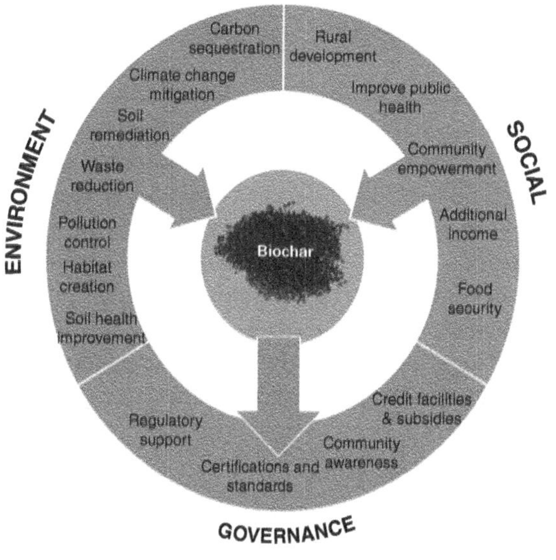

Waste-derived Biochar for Sustainable Rural Revitalization, First Edition. Edited by Yuqing Sun and Daniel C.W. Tsang.
Published 2026 by John Wiley & Sons, Inc.

17.1 Introduction

Biochar, a carbon-rich material derived from organic biomass under oxygen-limited conditions, is increasingly recognized for its broad environmental benefits, particularly within the framework of Sustainable Rural Revitalization (International Biochar Initiative 2025). Biochar is gaining attention as a sustainable solution in agriculture and waste management. Biochar functions as a bridge between traditional knowledge and modern sustainability. According to recent studies, the global biochar market was valued at $680.84 million in 2023 and is anticipated to grow from $763.48 million in 2024 to $2097.72 million by 2032, showing a compound annual growth rate (CAGR) of 13.47% over the forecast period (Insights 2025). Biochar application can increase crop yields by around 16% on average while reducing greenhouse gas (GHG) emissions by 2.1 Gt annually, highlighting its influence in climate change mitigation and sustainable farming practices (Jiang et al. 2024; World Economic Forum 2023).

This chapter will explore the multifaceted benefits and considerations of waste-derived biochar referring to ESG pillars and biodiversity, emphasizing its potential to minimize pollution, enhance waste recycling, support biodiversity, generate economic value, and align with evolving policy frameworks. By focusing on these pillars, the chapter aims to provide a comprehensive analysis of how waste-derived biochar can contribute to sustainability and achieve nature-positivity by supporting biodiversity conservation and ecosystem health.

17.2 Environmental (E) Perspective

Biochar is considered a promising product for many environmental issues. Biochar not only provides a solution for waste generated in rural areas but also contributes to carbon sequestration acting as a carbon sink, improving environmental quality while enhancing soil productivity and resilience. Biochar significantly supports the Environmental (E) pillar of Environmental, Social, and Governance (ESG) standards and aligns with numerous United Nations Sustainable Development Goals (SDGs). Thus, alignment of biochar with the Environmental pillar in ESG provides rural areas with a sustainable pathway to revitalize their agricultural ecosystems, adapting to climate challenges and, ultimately, driving to long-term ecological balance while improving the livelihood of rural community.

17.2.1 Promote Waste Reduction and Recycling

Biochar can be produced from a variety of organic wastes, including agricultural residues (corn stover, rice husks, and rice straw), forestry byproducts (sawdust and wood chips), animal manure (poultry manure and cow dung), garden waste, and food waste (Ighalo et al. 2025). At present, production of biochar through food waste has been recognized as an important method for minimizing environmental risks while improving the economic profitability (Pradhan et al. 2024). Previously, biochar production focused on selecting economical and locally abundant cheaper and natural feedstocks. Nevertheless, recent research explores production of biochar from challenging feedstocks such as plastics (Adeniyi et al. 2024), municipal solid waste (Gan et al. 2022) and compost residue (Ranaweera et al. 2025), sewage sludge (Wang et al. 2021), and hospital waste (Igalavithana et al. 2022) as an innovative approach to manage these persistent environmental burdens. Transforming such waste into biochar reduces waste volume in landfills and pollution loading as well. This approach aligns with the principles of sustainable production and consumption. Biochar production using waste materials supports rural revitalization by creating local opportunities for waste management and biochar production. In addition, it encourages the establishment of small-scale biochar production facilities, which can generate employment while ensuring soil quality in rural areas. Moreover, the increase in application of biochar to agricultural fields in rural areas enhances the productivity of lands. This dual benefit of waste management and economic empowerment directly contributes to poverty alleviation and sustainable rural development complying with the SDG 1 and 11, i.e., no poverty and sustainable cities and communities, respectively (Figure 17.1).

17.2.2 Mitigation of Climate Change

Biochar has emerged as a promising material for carbon sequestration and reduction of GHG emissions (Xia et al. 2023). Afforestation has already been identified as tradable "carbon offsets" under the Kyoto Protocol. However, heating plant biomass in the absence of oxygen transforms plant materials into biochar, which has twice the carbon content of regular biomass while locking up rapidly decomposing carbon in plant biomass in a much more durable form (Lehmann 2007). Application of pyrogenic organic matter reduces the mineralization of soil organic matter (SOM), thereby increasing native soil organic carbon stocks (Joseph et al. 2021). Wang et al. (2016) observed decrease in mineralization of SOM by 20% with application of biochar produced with crop residues (Wang et al. 2016). In addition, 19% decrease in SOM mineralization has been observed with the application of biochar produced through fast pyrolysis or at low temperatures (200–375 °C) (Wang et al. 2016). Biochar-induced positive priming could occur directly through enhanced microbial activity and enzyme production driven by easily degradable carbon in biochar or nutrient mining by microbes, and indirectly through reduction of soil acidity, alleviation of nutrient limitations, changing microbial habitats and soil faunal activity, and improved soil aeration

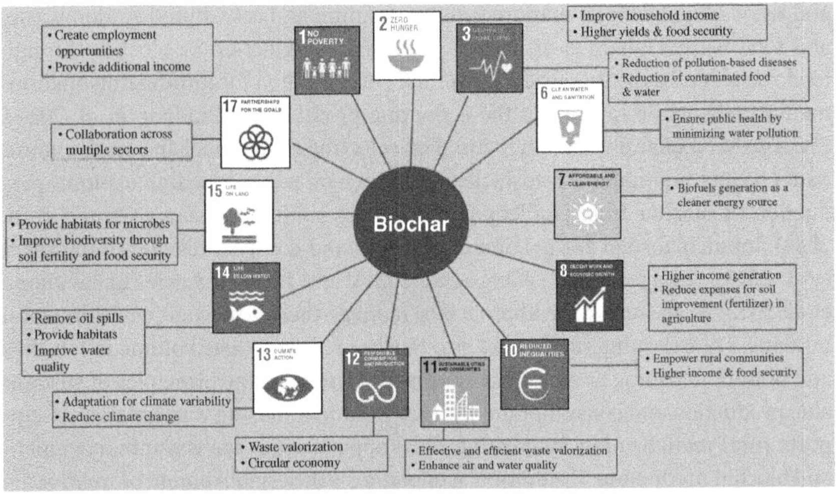

Figure 17.1 Alignment of Biochar with the Sustainable Development Goals (SDGs).

(Joseph et al. 2021). Furthermore, biochar application can reduce the emission of methane (CH_4) and nitrous oxide (N_2O) from agricultural fields, demonstrating its role as a climate change mitigation strategy (Shakoor et al. 2021). Full life cycle impacts of biochar include emissions associated with transportation, storage, and incorporation, which contribute minimally to overall GHG emissions. Use of feedstock such as waste result in GHG emissions of 0–1.6 tCO_2e per ton of biomass in biochar production, whereas the addition of biochar to flooded rice systems can reduce emissions by up to 3.9 tCO_2e per ton when compared to addition of raw feedstock (Lehmann et al. 2021). Hence, biochar supports rural farmers to adapt climate variability, which aligns with SDG 13, i.e., climate action, ensuring sustainable livelihoods.

17.2.3 Remediation of Contaminated Soil and Water

Biochar has gained a considerable attention due to its successful applications in environmental remediation throughout the world (Yining et al. 2022). Unique properties of biochar, including high surface area, porosity, and presence of functional groups, make it highly effective in adsorbing organic and inorganic contaminants present in soil, water, and air (El-Naggar et al. 2024; Zheng et al. 2022). Biochar reduces the mobility and bioavailability of certain contaminants mitigating their adverse environmental impacts. A study reveals that incorporating biochar into DDT-contaminated soil reduced the uptake of DDT by earthworms by around 50%, showing the biochar's ability to immobilize pollutants (Norrman 2025). However, adsorption capacities of biochar vary with the properties of biochar, which is governed by feedstock type and production conditions. For instance, biochar produced at temperatures higher than 500 °C tend to be less polar and more aromatic due to the loss of O- and H-containing functional groups facilitating adsorption of organic compounds (Ahmad et al. 2014). Application of biochar could immobilize cationic metal(lloid)s such has cadmium, lead through physisorption, outer layer (co)precipitations, and surface complex formation with active functional groups (El-Naggar et al. 2019). Further, biochar-induced increase of soil pH could decrease the bioavailability of cationic metals or metal(loid)s (Van Nguyen et al. 2022). On the contrary, high pH in biochar-applied soils could result in significant increase in the mobility of anionic metal(loid)s such as Arsenic (Van Nguyen et al. 2022).

Nevertheless, advances in biochar research led to modifying the structure and surface properties of biochar to improve its environmental benefits and effectiveness in remediation (Weijuan et al. 2024). These modifications include increasing surface area, altering pore structure, and changes in functional groups. Production of engineered/modified biochar through these modifications significantly increase biochar's capacity to adsorb contaminants (Murtaza et al. 2024). However, feedstock selection and customized production need to be ensured considering the type of contaminants and environmental conditions (He et al. 2022).

In rural communities, biochar can be used in remediation of contaminated environments as well as water filtration systems to provide access to clean water, addressing critical environmental issues and challenges in public health-preserving ecosystems. A review of 31 field and mesocosm studies further confirmed that biochar can be integrated into water remediation technologies for metals, nutrients, pesticides, and pharmaceuticals (Fang et al. 2025). Another study by Hassan et al. (2024) has found that engineered biochar modified with clay minerals, nanoparticles, polymers, surfactants, and oxidants/reductants can efficiently remove a wide range of contaminants, including heavy metals, pesticides, dyes, PAHs, PFAS, and agrochemicals from soil and water. These applications will ensure achieving several SDGs, i.e., good health and well-being (SDG 3), clean water and sanitization (SDG 6), sustainable cities and communities (SDG 11), life below water (SDG 14), and life on land (SDG 15) (Figure 17.1).

17.2.4 Improvement of Soil Fertility and Crop Productivity

Biochar significantly enhances soil fertility and productivity, making it a valuable amendment for sustainable agriculture supporting food security. A wide range of effects of biochar on physical, biological, and chemical soil properties and functions, and on plant growth have been reported in numerous studies (Melo et al. 2022). Biochar application to soil provides several benefits to soil, including increase of cation exchange capacity, improvement of nutrient retention, increase of buffering capacity, increase of organic matter content, improvement of water-holding capacity, improvement of aggregate stability, reduction of bulk density, alteration of pH, and alteration of soil microbial diversity and community composition (Joseph et al. 2021; Melo et al. 2022). This could result in an increase of crop productivity. A meta-analysis proved that application of only biochar increased crop yield on average by 25.3% $\pm$ 3.2, whereas co-application of biochar with inorganic and organic fertilizers led to an additional 12.1% increase in yield (Bai et al. 2022). However, the highest yield increase was found with the application of biochar produced at temperatures of 401–500 °C, with C:N between 30 and 100, and when biochar was applied to soils having pH $\leq$ 5 (Bai et al. 2022). Another study showed 12.04% increase in nitrogen use efficiency and 10.73% increase in rice yields with the application of biochar (Liu et al. 2022). These findings underline the role of biochar in boosting productivity making it a valuable amendment for sustainable agriculture to ensure achieving no poverty (SDG 1) and zero hunger (SDG 2).

17.2.5 Renewable Energy Production

Biochar production process generates renewable energy in the form of syngas and bio-oil and biochar itself can act as energy storage material. In addition to production of electricity and heat, biochar can be used in anaerobic digestion (You et al. 2022). However, the yield of the products or byproducts or energy produced through biochar production process varies depending on the type of feedstock, method of production, and particle size (Kung and Mu 2019). These applications minimize the use of non-renewable fossil fuels while reducing GHG emissions. Moreover, unique structural and chemical properties of biochar make it a promising material for energy storage applications, such as in supercapacitors and rechargeable battery systems (You et al. 2022). For rural areas, these renewable energy options provide affordable, accessible, and sustainable energy solutions, promoting economic growth and energy independence. Additionally, integrating biochar systems with renewable energy production creates job opportunities in biomass collection, pyrolysis operations, and energy distribution, contributing to rural development. This multifaceted approach supports affordable and clean energy (SDG 7), zero hunger (SDG 2), and climate action (SDG 13). Through these impacts, biochar production not only revitalizes rural communities but also fosters sustainable development on a broader scale (Figure 17.2).

17.3 Social (S) Perspective

Biochar produced from waste is a catalyst for sustainable agriculture (Sirohi et al. 2025). It has multifaceted social benefits, including economic value and rural development, empowering local communities, and improving public health

Figure 17.2 Coconut tree sources and other plant sources used for biochar production in auto-thermal pyrolizer. Photo credit: Prof. Meththika Vithanage.

(Table 17.1). By transforming food waste, municipal waste and agricultural waste into biochar, communities can overcome waste management challenges and convert this waste into a valuable economic source.

17.3.1 Economic Value and Rural Development

Waste-derived biochar has strong economic potential, including cost savings, revenue generation, and promoting sustainable business models (Li and Skelly 2023). Municipal waste usually has higher cost incurred due to waste collection, waste burial, and maintenance cost (Sughosh et al. 2021). According to a study by Kazemi et al. (2021), transportation accounted for 43% of management costs, followed by sidewalk cleaning and river dredging at 26%. Burials of waste contributed 10%, highlighting that waste collection, burial, and maintenance significantly drive costs in municipal solid waste management. If this waste is transformed into biochar, it can generate an income, since the biochar has economic value. According to a techno-economic assessment with Monte Carlo risk analysis conducted on 3-ton/hr biochar plants indicate selling prices of biochar ranges from US$468 to 926/ton (Kazemi et al. 2022). This shows the potential economic benefit of municipal waste-derived biochar. In addition, char produced from various solid wastes through thermochemical processes can be used as both a material and energy source (Santos et al. 2024). During the pyrolysis process bio-oil and syngas are generated, which can be used as renewable energy sources (Panwar

2024). These renewable energy sources provide additional revenue apart from waste-derived biochar. This approach promotes waste recycling and advances a circular economy by converting waste into valuable products (Emenike et al. 2024).

Rural farmers in developing countries, who primarily rely on agriculture, often have lower income levels than those in other professions (Run et al. 2024). These farmers spend a significant portion of their earnings on fertilizers to maintain soil nutrients (Małgorzata and Grzegorz 2022). By producing biochar from food waste, sewage sludge, and agricultural residues, which are already in their fields, they can enhance soil fertility (Li et al. 2019) while reducing the dependency on costly fertilizers. This practice not only saves money but also creates an additional income stream through the sale of excess biochar. Furthermore, it is an act of aligning with the regenerative agricultural principles supporting for a circular economy. Converting waste into biochar thus generates economic value from materials that would otherwise be discarded, contributing to a more sustainable, circular economy. The biochar production process can generate employment opportunities in waste management and agriculture (Ajien et al. 2023; Li and Skelly 2023). Unemployment is a significant challenge in many countries, particularly in developing nations (Siddiqa 2021). Rural communities are especially affected due to limited access to employment opportunities compared to urban areas. Establishing biochar production from waste in rural regions can help address this issue by creating various job opportunities, including roles in waste collection and processing (Potnuri et al. 2023).

A study conducted in Sri Lanka introduced a locally developed auto-thermal pyrolizer for producing biochar from municipal solid waste compost residue and other feedstocks at pilot scale (200–400 kg per batch) (Figures 17.2 and 17.3). The resulting biochars showed high carbon content (52–74%), stable aromatic structures (H/C < 0.7), and functional groups suitable for contaminant remediation, with yields of 27–49%. Coconut husk biochar offered the highest carbon credit potential, and since Sri Lanka has abundant coconut trees with husks available in almost every rural household, this represents a highly viable solution for carbon sequestration, pollution control (e.g., landfill leachate treatment), and integrating environmental benefits with financial incentives (Ranaweera et al. 2025).

Figure 17.3 Auto-thermal pyrolizer. Photo credit: Prof. Meththika Vithanage.

17.3.2 Community Engagement and Awareness

Educating rural communities about biochar produced from waste is essential for encouraging sustainable practices, meeting energy requirements, boosting agricultural output, and minimizing environmental pollution (Table 17.1). By involving communities in educational initiatives, we can inspire innovation and enable them to make effective use of local biomass resources (Matsagar and Wu 2022). Providing education and practical knowledge to the community on biochar can increase community engagement and awareness among the rural community about the opportunity to produce biochar by waste valorization. There are many global examples of success of community engagement for biochar preparation. One partnership program educated the Indonesian rural community on converting biomass waste into biochar using a double-tube reactor, giving hands-on experience with carbonizing wood chips and coconut shells. Participants showed strong interest in using biochar for household energy, agriculture, and income generation, recognizing its valuable benefits (Mangalla et al. 2023). Another workshop Training in Sub-Saharan Africa shows an example for empowering the community by enhancing their knowledge and skills in biochar technology (Mohammed et al. 2024).

When the community has knowledge on biochar production, byproducts, and market viability, they tend to use these technologies to valorize their food waste, crop residues, and sewage sludge to produce biochar focusing on the income as well as self-sufficient economic practices. Then this knowledge can transfer from generation to generation leading to future communities becoming more equipped with the technologies to valorize waste into biochar finally making it a circular economy.

17.3.3 Public Health and Food Security

Converting waste into biochar reduces the health risks associated with improper waste disposal, which can pose significant threats to human health. In addition, the application of biochar in crop production enhances soil fertility and agricultural productivity, subsequently contributing to food security (Fytili and Zabaniotou 2018). Numerous studies have examined occupational health concerns affecting humans, animals, and birds, particularly those residing near waste disposal sites. Beyond their carcinogenic effects, various elements found in decomposed food waste can impact the nervous system, liver, kidneys, lungs, heart, skin, and reproductive health (Solomon and Isaac Kow 2023). Health issues associated with municipal solid waste management also include increased risks of adverse birth and neonatal outcomes, mortality, respiratory diseases, and negative mental health effects, particularly for residents near landfills, incinerators, and

Table 17.1 Successfully implemented biochar projects for community and rural development.

Project/research article	Country	Year	Project progress	References
Stockholm Biochar Project	Sweden	2017	Produced more than 100 tons of biochar, capturing carbon equivalent to removing ~700 cars from the road annually and scaling up for higher goals.	Johns Hopkins University (2021)
Google and Varaha Company	India	2025	Google has planned to purchase 100,000 tons of carbon dioxide removal credits from Varaha company in India. Smallholder farmers will utilize biochar produced to improve soil fertility and water retention of sandy soils.	Singh (2025)
Centre for International Forestry Research and World Agroforestry (CIFOR-ICRAF) and German Development Cooperation (GIZ)	India	2023	A pilot project in eastern India empowered women through gender-inclusive training to produce and use biochar using smart cookstoves, improving soil fertility and crop yields. The initiative also reduced indoor air pollution and health risks by replacing traditional cooking methods.	Gaur (2024)
Sustaining agricultural economies: regional economic impacts of biochar production from waste orchard biomass in California's Central Valley	United States	2022	In California's central valley area biochar production from waste orchard biomass could generate 16.56–17.69 new jobs annually while contributing toward higher labor income and higher GDP.	Nematian et al. (2024)
Diffusion and adoption of biochar innovations in agriculture at Gunungsari Village, Bumiaji Subdistrict, Batu City, East Java, Indonesia	Indonesia	2022	According to the results of the study production of the biochar was a solution for two challenges the farmers were facing. They are addressing the issue of excess animal feed waste and overcoming the scarcity of fertilizer for agricultural fields.	Lestari et al. (2022)

Production, analysis, and field application of biochar in Nepal (a case study of Padali Village, Lamatar-7, Lalitpur)	Nepal	2011	Biochar production was done using rice husk, wheat husk, and pine needles. Field trials on leaf mustard revealed that the soil in plots amended with biochar had higher pH levels and greater concentrations of nitrogen (N), phosphorus (P), potassium (K), and organic matter (OM) compared to the control plots without any fertilizer application.	Karna et al. (2011)
Factors affecting biochar technology adoption by vegetable farmers in the Kwahu East District of Ghana	Ghana	2015	The study found that vegetable farming using biochar in Akwasiho was more profitable during both the 2011 and 2012 growing seasons than farming practices that did not incorporate biochar.	Ninson (2015)
A project by The African Christians Organization Network (ACON)	Kenya	2016	ACON, in partnership with international experts, conducted successful field trials demonstrating that biochar significantly improved soil moisture retention and crop yields, even during drought conditions. ACON has trained hundreds of local farmers on biochar application, with a goal of reaching 1000 farmers by 2010, empowering communities.	International Biochar Initiative (2016)
Improving Agricultural Resilience to Salinity through Development and Promotion of Pro-poor Technologies (RESADE) project, led by the International Center for Biosaline Agriculture (ICBA)	Botswana, Gambia, Liberia, Mozambique, Namibia, Sierra Leone, and Togo	2024	The project was initiated in partnership with the International Fund for Agricultural Development and the Arab Bank for Economic Development in Africa. Smallholder farmers are trained through farmer field schools to produce and apply biochar using locally available agricultural waste and biomass. Research conducted under the project has demonstrated increases in crop biomass yield, improvement in soil water retention, and enhanced germination rates when biochar is applied to sandy and salt-affected soils, supporting both improved agricultural productivity and the long-term sequestration of carbon in soils.	ICBA (2024)

(Continued)

Table 17.1 (Continued)

Project/research article	Country	Year	Project progress	References
A study by International Crops Research Institute for the semi-arid tropics	India	2024	ICRISAT conducted a project on decentralized biochar production technology aimed at helping smallholder farmers and Self-Help Groups manage crop residues. By the project farmers received benefits such as reduced dependence on residue burning, improved soil health, increased nutrient and water retention, and cost savings.	ICRISAT (2024)
Biochar production from municipal solid waste using an auto-thermal pyrolizer	Sri Lanka	2024	This study developed an auto-thermal pyrolizer to convert municipal solid waste compost residue and other feedstocks into biochar, yielding 27–49% with high carbon content (52–74%) and stability. The biochars exhibited aromatic structures, diverse functional groups, and high porosity, making them suitable for contaminant remediation, compost enhancement, and greenhouse gas reduction, with coconut husk showing the greatest carbon credit potential.	Ranaweera et al. (2025)

dumpsites/open burning sites (Vinti et al. 2021). Therefore, valorizing waste into biochar is a promising solution to reduce the waste and improve public health. Additionally, by reducing nutrient runoff (nitrogen and phosphorus) into waterways, biochar helps prevent algal blooms that can contaminate drinking water and produce toxins harmful to human health.

One most important factor for public health is water. Heavy metals in water pose significant risks to public and environmental health. Advancements in utilizing low-cost biochar-based adsorbents, synthesized from diverse feedstocks, have proven to be highly effective in removing heavy metals from water (Moureen et al. 2024). Furthermore, by binding toxic substances in soil, biochar can prevent those toxins from entering to the water sources ensuring food security and reducing water toxicity (Fasih Ullah et al. 2022). Rural communities can use simple biochar filters to treat water or rainwater well, which is an affordable solution to improve water quality.

Biochar enhances soil properties, increasing water-holding capacity and organic matter, which improves plant growth under drought stress. This contributes to food security by promoting sustainable agricultural practices and optimizing resource use in the face of climate change challenges (Abd El-Moaty et al. 2024; Pandian et al. 2024). Its ability to sequester carbon and reduce GHG emissions further supports sustainable agricultural practices essential for ensuring global food security (Upadhyay et al. 2024).

Through waste-derived biochar, we can support below SDGs with reference to social pillar (THE 17 GOALS | Sustainable Development 2024; Figure 17.1). SDG 1: No poverty: Create employment opportunities: Provide additional income for rural communities. SDG 2: Zero hunger: By enhancing the income, empowering the rural community can contribute to reducing hunger. Application of biochar in croplands can reduce the fertilizer use, which is an additional cost, boosting the crop yield finally contributing the food security. SDG 3: Good health and well-being: Municipal waste and sewage sludge often contribute to land pollution, air pollution and water pollution eventually leading to public epidemics like cholera, dengue, typhoid, and leptospirosis. Other than that food contamination and food poisoning can also be caused due to waste. Therefore, turning this waste which is a threat to public health into biochar can support this goal. SDG 6: Clean water and sanitation: As the water pollution from municipal waste and sewage sludge can cause threat to the public sanitation, transforming this waste into biochar can support this goal. SDG 8: Decent work and economic growth: As per the economic gain from biochar and the cost reduction from maintaining waste this goal also can be addressed through waste-derived biochar. SDG 10: Reduced inequalities: Encouraging the production of biochar from waste can help reduce social inequalities by empowering rural communities and enhancing their income. This approach provides economic opportunities, supporting underserved populations in achieving greater financial stability and inclusion. SDG 11: Sustainable cities and communities: Biochar can significantly contribute to the development of sustainable urban and rural environments. It serves as an effective waste management solution by transforming waste into a valuable product, thereby diverting it from landfills and minimizing the environmental impacts of waste disposal. Additionally, biochar can enhance air and water quality in both urban and rural areas while empowering rural communities through sustainable practices and improved agricultural productivity. SDG Goal 12: Responsible consumption and production: The generation and application of biochar from waste materials supports responsible consumption and production. By turning waste into a valuable resource and renewable energy sources as byproducts, waste-derived biochar promotes the principles of a circular economy.

17.4 Governance (G) Perspective

The governance of biochar production and application is critical in ensuring its effectiveness, safety, and alignment with sustainability objectives. Non-uniform physicochemical properties and inconsistent performances of biochar that occurs due to the inherent variability of feedstocks, and production processes is one of the key challenges in biochar governance. Therefore, establishing quality standards is important to promote usage and improve sustainability. When referring to the global situation, different countries have different standards to promote biochar production and usage. To address this, robust quality standards, certification protocols, and traceability systems are essential. Moreover, ensuring transparent pricing, supply chain traceability, and active stakeholder involvement from farmers and local communities to private companies and policymakers is essential for creating a sustainable and economically viable biochar market.

17.4.1 Country-wise Regulations and Voluntary Standards

In the absence of comprehensive regulations, voluntary standards have emerged in Europe and the United States, promoting best practices in biochar production and application (Meyer et al. 2017). One step is defining biochar in law to ensure safe and effective use (European Commission 2021). The European Union is actively developing frameworks to integrate carbon removals, including biochar, into its climate strategies. The upcoming 2040 Climate Targets and the Industrial Carbon Management Strategy aim to incorporate 75 million tons of CO_2 removals, with biochar being a significant component (European Union 2024). European Biochar Certificate (EBC) is a voluntary certification system that sets environmental and production standards for biochar use across Europe. It ensures that biochar products meet specific quality criteria, promoting sustainable practices in biochar production and application (The European Biochar Certificate [EBC] 2024). Regarding carbon removal, Biochar is now considered an effective way of removing carbon, by many carbon markets (Puro Earth 2025). In the United States, US Biochar Initiative (USBI) promotes the sustainable production and use of biochar through research, policy advocacy, and education. It serves as a hub for stakeholders interested in biochar applications and policies in the United States (NRDC 2010). Other than that, in the United States, various programs offer financial incentives and non-financial support for biochar production. A study by Pourhashem et al. (2019) have identified three types of programs to support biochar production: commercial financial incentives to promote investment, non-financial policy support to create a conducive regulatory environment, and research and development funding to drive innovation. Countries like Cameroon have developed national biochar policies to promote its use in agriculture and waste management. These policies aim to enhance soil fertility, manage agricultural residues, and contribute to carbon sequestration efforts. Nations such as Ghana have entered bilateral agreements under Article 6.4 of the Paris Agreement, focusing on carbon offset projects that include biochar production and application (USBI 2024).

When referring to Asia, China has implemented policies promoting biochar in rural biogas projects and agricultural waste management. The government supports research and development to explore biochar's role in carbon sequestration and soil enhancement (International Biochar Initiative 2025). The South Korean government announced a comprehensive plan in December 2021 to achieve carbon neutrality in the agri-food sector by 2050, outlining measures to reduce GHG emissions through food production, distribution, consumption, and energy conversion (Oh 2022).

17.4.2 Global Policy Frameworks and Standards

The World Bank has recognized the potential of biochar systems to contribute to sustainable agriculture and climate resilience. In its report "Biochar Systems for Smallholders in Developing Countries," the World Bank discusses how biochar can offer a "triple win" by enhancing energy production, mitigating climate change, and improving food security (Scholz et al. 2014).

United Nations Environment Program Finance Initiative (UNEP FI) has been leading efforts to develop the best practices for identifying, measuring, disclosing, and managing climate risks, which may indirectly support biochar projects aimed at climate mitigation (Climate Risk and TCFD Programme 2024). International Biochar Initiative (IBI) and European Biochar Certificate (EBC) support to regulate biochar quality by setting the production standards (The European Biochar Certificate [EBC] 2024; International Biochar Initiative 2025). Production and usage of biochar is also supported by UN SDG programs, Paris Agreement-aligned climate strategies, and FAO guidelines for sustainable agriculture (FAO 2021; United Nations 2022; United Nations Climate Change Convention 2023).

17.4.3 Corporate Support for Biochar

Corporate entities are also integrating biochar into their sustainability and ESG strategies, to reduce their own environmental footprint as well as to support global climate goals (Zilberman et al. 2023). Google has signed agreements to purchase carbon removal credits from India (Table 17.1) through a biochar project (Segal 2025). Microsoft purchases biochar carbon removal credits from Mexico, planning to achieve net zero carbon status by 2030 (Jennifer 2024a). Shopify, a largest e-commerce platform, has invested in biochar-related products to tackle climate change (Shopify 2024). Shell PLC, the worldwide gas company, is supporting biochar production projects with the aim of purchasing carbon removal credits (Jennifer 2024b).

When referring to the governance (G pillar) aspect of biochar, its production from municipal food waste contributes directly to SDG 12 by promoting responsible consumption and production through circular economy, waste valorization, and sustainable resource management. Achieving widespread adoption and implementation of waste-derived biochar requires collaboration across multiple sectors, aligning with SDG 17 by strategic partnerships between governments, research institutions, private enterprises, and local communities to drive innovation and scale sustainable solutions (THE 17 GOALS | Sustainable Development 2024).

17.5 Biodiversity

Waste-derived biochar can support the improvement of biodiversity in plants as well as soil organisms. Biochar's effects on biodiversity vary widely depending on feedstock, pyrolysis conditions, soil type, application rate, and environmental context (Ali and Al-Wahaibi 2025b). According to a study by Yuan et al. (2024), diverse plant communities benefit from multiple biochar types, leading to improved soil health and reduced contaminants (Yuan et al. 2024). Biochar significantly enhances native tree growth, survivorship, and species diversity in invaded ecosystems. In the study, biochar treatments increased native tree performance and suppressed invasive strawberry guava regeneration, demonstrating its potential to promote biodiversity in tropical island environments (Sujeeun and Thomas

2022). A study by Ali and Al-Wahaibi (2025) highlights biochar helps protect biodiversity by enhancing plant growth, soil fertility, and microbial activity, even under drought. By supporting both above- and belowground diversity, it strengthens ecosystem resilience and aids in restoring degraded arid environments (Prabha 2025). Beyond agriculture, biochar can suppress invasive species (by adsorbing allelochemicals) and help native species re-establish (Ali and Al-Wahaibi 2025b).

Biochar application enhances soil biodiversity, significantly correlating with soil multifunctionality (Xiao et al. 2024). Biochar boosts biodiversity by providing habitats for microorganisms within its porous structure and improving the formation of stable soil micro-aggregates. It enhances soil quality, supports a diverse microbial community, and aids in nutrient recycling, thereby promoting a healthier and more productive soil ecosystem (Bolan et al. 2024). A study by Zhao et al. (2024) shows that biochar application raised soil pH, total carbon, and nitrogen at all doses while enhancing microbial alpha-diversity in a dose-dependent manner. At >2% addition, it significantly altered bacterial beta-diversity, increasing Acidobacteria, Bacteroidetes, Gemmatimonadetes, and Proteobacteria. In contrast, Verrucomicrobia and Basidiomycota declined in relative abundance. According to a meta-analysis by Adirianto and Bachtiar (2023), adding biochar requires careful attention to the type of biochar and the amount applied to promote microbial diversity (Adirianto and Bachtiar 2023). The type of biochar, the rate of application, and the type of soil are key factors that affect how biochar influences soil microbial diversity. A study investigating the effects of biochar on soil fauna and microbial community responses found that biochar with a higher pH and biochar produced through conventional pyrolysis had a positive impact on soil fauna. Medium-sized animals benefited the most from biochar addition. Animals with exoskeletons generally exhibited favorable responses to biochar, while those without exoskeletons showed mixed or unfavorable responses (Li et al. 2024).

Biochar can bind oil and other pollutants in contaminated soils, reducing their bioavailability and toxicity and supporting environment for the recovery of soil microbes and plants (Abbas et al. 2024). Oil spills in marine environment cause threats to marine biodiversity including aquatic plants, fish, and other marines species (de Lima Silva et al. 2024). Adding biochar as an oil absorbent for oil spills in marine environment is in research and it has shown positive results. Magnetic biochar composite (MBCC) derived from Heglig tree bark proves to remove oil spills from seawater (Amar et al. 2024). A pilot-scale study during the X-Press Pearl maritime disaster in Sri Lanka demonstrated that biochar derived from bio-energy industries rapidly removed heavy fuel oil within 30 minutes, confirming their potential as cost-effective materials for oil spill remediation (Pallewatta et al. 2023). Another study conducted by Ariyanti et al. (2023) shows that biochar, particularly from rice husk and sugarcane bagasse, effectively absorbs oil spills in marine environments, achieving up to 84% removal efficiency (Ariyanti et al. 2023). Using biochar to absorb oil can be an economically viable and environmentally friendly solution for biodiversity protection (Chanaka et al. 2022). Therefore, it is evident that there are multiple ways that engineered biochar and advanced biochar derived from waste materials can be used for biodiversity protection and enhancement.

17.6 Challenges and Opportunities

The significant challenges in waste-derived biochar are variable characteristics and composition of the feed stock which provide a significant impact on the properties and performance of the biochar (Ippolito et al. 2020). Developing standardization protocols and employing advanced sorting technologies can effectively separate biomass from contaminants, ensuring consistent biochar properties. Pre-treatment processes, such as drying, crushing, and sieving, are essential to prepare feedstock under specific conditions required for each process, thereby enhancing the uniformity and quality of the resulting biochar.

Some sources like wetland plant residue biomass can contain a lot of moisture, which paves the way to bear higher costs for transportation and difficulties in pyrolysis (Cui et al. 2022). Using drying technologies can substantially enhance pyrolysis efficiency and biochar quality. Research suggests that reducing biomass moisture content leads to a more efficient pyrolysis process, resulting in higher quality biochar (Chun et al. 2021).

Another obstacle to the widespread adoption of the biochar system has been the absence of a supportive policy frameworks and globally harmonized standards, which has led to low demand for biochar as a soil amendment (Verde and Chiaramonti 2021). Developing standardized quality protocols, regulatory frameworks, and incentive schemes is crucial for ensuring safe, consistent, and effective biochar production.

Limited awareness and knowledge of biochar production and application among rural communities and smallholder farmers restrict its adoption. Capacity-building programs, training workshops, and community engagement initiatives can empower stakeholders, create employment opportunities, and enhance food security by improving soil fertility and crop productivity. Moreover, integrating waste-to-biochar systems with circular economy models offers pathways for sustainable income generation and rural development, advancing the social dimension of ESG.

Scaling up biochar applications from laboratory and pilot studies to real-world conditions remains a major challenge for both water and soil remediation. Establishing biochar-based systems at full scale requires technical validation, infrastructure, and policy support. Developing sustainable biochar markets in rural communities is also critical, as biochar could provide environmental and socio-economic benefits if demand mechanisms are strengthened. Furthermore, in complex contamination scenarios such as landfill leachate, biochar alone often shows limited removal capacity, indicating its role as a subprocess within multi-step treatment systems rather than a standalone solution. These challenges highlight the need for context-specific scale-up strategies, integrated treatment approaches, and supportive market and policy frameworks to fully realize biochar's remediation potential.

17.7 Conclusion

The production and application of waste-derived biochar offer a promising pathway to address pressing ESG challenges while simultaneously contributing to biodiversity conservation. From an environmental standpoint, waste-derived biochar promotes waste reduction, mitigates climate change, enhances soil fertility, and aids in the remediation of contaminated environments. It supports the circular economy by transforming waste into valuable resources, promoting renewable energy production, and reducing pollution. Socially, biochar technologies contribute to rural revitalization by generating employment, supporting community development, and improving public health through cleaner air and better agricultural productivity. These advancements directly align with food security goals by enhancing soil health and increasing crop yields. Additionally, biochar initiatives encourage community engagement and education, improving awareness about sustainable agricultural practices. From a governance perspective, the challenges surrounding the variability of biochar properties highlight the urgent need for standardized quality protocols and supportive policy frameworks. Establishing regulatory frameworks and voluntary standards can help ensure consistent biochar quality and promote its adoption at scale. Collaborative efforts among governments, private enterprises, research institutions, and non-profits are essential to create an enabling environment for biochar technologies.

To fully realize the potential of waste-derived biochar, stakeholders must take decisive action. Governments should provide policy incentives and financial support, while the private sector can invest in advanced production technologies and market development. Research institutions play a crucial role in innovation and knowledge dissemination, and non-profits can drive community engagement and education initiatives. These collaborative efforts can help overcome existing challenges, scale biochar production, and promote its widespread adoption. Investing in waste-derived biochar technologies represents a strategic and impactful approach to achieving sustainable development and supporting biodiversity conservation. By transforming waste into a valuable resource, biochar production not only addresses environmental degradation but also contributes to rural revitalization and climate resilience. Stakeholders across sectors must join forces to harness the full potential of this technology, ensuring a sustainable, biodiverse, and resilient future for generations to come.

Funding

This research was supported by a Basic Science Research Program through the National Research Foundation of Korea (NRF) funded by the Ministry of Education (NRF 2021R1A6A1A10045235). This work was supported by the Technology Innovation Program (Project No. 00432915, Development of biodegradable polymer and their applications using high performance enzyme activation technologies for acceleration of biodegradability) funded by the Ministry of Trade, Industry and Energy (MOTIE, Korea). This work was supported under the framework of the international cooperation program managed by the National Research Foundation of Korea (NR059458). This work was supported by OJEong Resilience Institute, Korea University.

Declaration of Competing Interest

The authors declare that they have no known competing financial interests or personal relationships that could have appeared to influence the work reported in this chapter.

Data Availability

The authors confirm that the data supporting the findings of this study are available within the chapter.

Declaration of Generative AI and AI-assisted Technologies in the Writing Process

During the preparation of this work the authors used ChatGPT 3.5 to correct English. After using this tool, the authors reviewed and edited the content as needed and take full responsibility for the content of the publication.

References

Abbas, H.M.M., Rais, U., Altaf, M.M. et al. (2024). Microbial-inoculated biochar for remediation of salt and heavy metal contaminated soils. *Science of the Total Environment* 954: 176104.

Abd El-Moaty, H.I., El-Dissouky, A., Elhusseiny, A.F. et al. (2024). Low-cost nano biochar: a sustainable approach for drought stress mitigation in faba bean (*Vicia faba* L.). *Frontiers in Plant Science* 15: 1438893.

Adeniyi, A.G., Iwuozor, K.O., Emenike, E.C. et al. (2024). Thermochemical co-conversion of biomass-plastic waste to biochar: a review. *Green Chemical Engineering* 5 (1): 31–49.

Adirianto, B. and Bachtiar, T. (2023). Effect of biochar in soil on microbial diversity: a meta-analysis. *IOP Conference Series: Earth and Environmental Science* 1263: 012047.

Ahmad, M., Rajapaksha, A.U., Lim, J.E. et al. (2014). Biochar as a sorbent for contaminant management in soil and water: a review. *Chemosphere* 99: 19–33.

Ajien, A., Idris, J., Md Sofwan, N. et al. (2023). Coconut shell and husk biochar: a review of production and activation technology, economic, financial aspect and application. *Waste Management & Research* 41 (1): 37–51.

Ali, H.E., & Al-Wahaibi, A.M. (2025). High species diversity and biochar can mitigate drought effects in arid environments. *Frontiers in Plant Science*, 16. https://doi.org/10.3389/fpls.2025.1563585.

Amar, I.A., Alzarouq, A., Mohammed, W. et al. (2024). Oil spills removal from seawater surface by magnetic biochar composite derived from Heglig tree bark and cobalt ferrite. *World Journal of Engineering* 21 (6): 1064–1073.

Ariyanti, D., Widiasa, I.N., Widiyanti, M. et al. (2023). Agricultural waste-based magnetic biochar produced via hydrothermal route for petroleum spills adsorption. *International Journal of Renewable Energy Development* 12 (3): 499–507.

Bai, S.H., Omidvar, N., Gallart, M. et al. (2022). Combined effects of biochar and fertilizer applications on yield: a review and meta-analysis. *Science of the Total Environment* 808: 152073.

Bolan, S., Sharma, S., Mukherjee, S. et al. (2024). Biochar modulating soil biological health: a review. *Science of the Total Environment* 914: 169585.

Chanaka, N., Prashan, M.R., Vishmi, S.T. et al. (2022). Biochar and biochar composites for oil sorption. In: *Sustainable Biochar for Water and Wastewater Treatment* (ed. M. Dinesh, U.P. Charles, and E.M. Todd), 527–554. Elsevier. https://doi.org/10.1016/B978-0-12-822225-6.00010-5.

Chun, Y., Lee, S.K., Yoo, H.Y. et al. (2021). Recent advancements in biochar production according to feedstock classification, pyrolysis conditions, and applications: a review. *BioResources* 16 (3): 6512.

Climate Risk and TCFD Programme. (2024). https://www.unepfi.org/climate-change/tcfd (retrieved 30 January 2025).
Cui, X., Wang, J., Wang, X. et al. (2022). Biochar from constructed wetland biomass waste: a review of its potential and challenges. *Chemosphere* 287: 132259.
de Lima Silva, F.J., Attademo, F.L.N., Gavilan, S.A. et al. (2024). Overview of oil spills worldwide and impacts on marine megafauna. *Brazilian Journal of Case Reports* 4 (1): 78–94.
El-Naggar, A., Jiang, W., Tang, R. et al. (2024). Biochar and soil properties affect remediation of Zn contamination by biochar: a global meta-analysis. *Chemosphere* 349: 140983.
El-Naggar, A., Lee, M., Hur, J. et al. (2019). Biochar-induced metal immobilization and soil biogeochemical process: An integrated mechanistic approach. *The Science of the Total Environment* 698: 134112. https://doi.org/10.1016/j.scitotenv.2019.134112.
Emenike, E.C., Iwuozor, K.O., Ighalo, J.O. et al. (2024). Advancing the circular economy through the thermochemical conversion of waste to biochar: a review on sawdust waste-derived fuel. *Biofuels* 15 (4): 433–447.
European Commission. (2021). https://ec.europa.eu/info/law/better-regulation/have-your-say/initiatives/12136-Fertilising-products-pyrolysis-and-gasification-materials_en (retrieved 14 February 2025).
European Union. (2024). https://climate.ec.europa.eu/carbon-removals-and-carbon-farming/eu-carbon-removal-certification-biochar-methodology-2024-06-18_en (retrieved 10 January 2025).
Fang, J., Wang, D., Wilkin, R. et al. (2025). Realistic and field scale applications of biochar for water remediation: A literature review. *Journal of Environmental Management* 385: 125524. https://doi.org/10.1016/j.jenvman.2025.125524.
FAO. (2021). https://www.fao.org/family-farming/detail/en/c/1612676/ (retrieved 15 August 2024).
Fasih Ullah, H., Xiukang, W., Usman, Z. et al. (2022). Biochar application for remediation of organic toxic pollutants in contaminated soils; an update. *Ecotoxicology and Environmental Safety* 248: 114322. https://doi.org/10.1016/j.ecoenv.2022.114322.
Fytili, D. and Zabaniotou, A. (2018). Circular economy synergistic opportunities of decentralized thermochemical systems for bioenergy and biochar production fueled with agro-industrial wastes with environmental sustainability and social acceptance: a review. *Current Sustainable/Renewable Energy Reports* 5: 150–155.
Gan, Z., Zhuang, X., Cen, K. et al. (2022). Co-pyrolysis of municipal solid waste and rice husk gasification tar to prepare biochar: an optimization study using response surface methodology. *Fuel* 330: 125574.

Gaur, S. (2024). Gender-based participation spurs pilot biochar industry in eastern India. *Forest News*. https://forestsnews.cifor.org/85785/gender-based-participation-spurs-pilot-biochar-industry-in-eastern-india?fnl=en

Hassan, M., Wang, B., Wu, P. et al. (2024). Engineered biochar for in-situ and ex-situ remediation of contaminants from soil and water. *The Science of the Total Environment, 957*, 177384. https://doi.org/10.1016/j.scitotenv.2024.177384.

He, M., Xu, Z., Hou, D. et al. (2022). Waste-derived biochar for water pollution control and sustainable development. *Nature Reviews Earth & Environment* 3 (7): 444–460.

ICBA. (2024). https://www.biosaline.org/news/2024-08-12-15999?utm (retrieved 04 April 2025).

ICRISAT. (2024). ICRISAT study unveils decentralised Biochar production system for farm-level sustainability. *Agro Spectrum*. https://agrospectrumindia.com/2024/03/29/icrisat-study-unveils-decentralised-biochar-production-system-for-farm-level-sustainability.html

Igalavithana, A.D., Yuan, X., Attanayake, C.P. et al. (2022). Sustainable management of plastic wastes in COVID-19 pandemic: the biochar solution. *Environmental Research* 212: 113495.

Ighalo, J.O., Ohoro, C.R., Ojukwu, V.E. et al. (2025). Biochar for ameliorating soil fertility and microbial diversity: from production to action of the black gold. *iScience* 28 (1): 111524.

Insights, F.B. (2025). *Biochar Market Size*. https://www.fortunebusinessinsights.com/industry-reports/biochar-market-100750

International Biochar Initiative. (2016). https://biochar-international.org/profile-using-improved-cookstoves-and-biochar-in-western-kenya/?utm_source=chatgpt.com (retrieved 05 April 2025).

International Biochar Initiative. (2025). https://biochar-international.org/ (retrieved 30 January 2025).

Ippolito, J.A., Cui, L., Kammann, C. et al. (2020). Feedstock choice, pyrolysis temperature and type influence biochar characteristics: a comprehensive meta-data analysis review. *Biochar* 2: 421–438.

Jennifer, L. (2024a). Microsoft to purchase 95,000 biochar carbon removal credits from the next 150. https://carboncredits.com/microsoft-to-purchase-95000-biochar-carbon-removal-credits-from-the-next-150/#:~:text=The%20Next%20150-,Microsoft%20to%20Purchase%2095%2C000%20Biochar%20Carbon%20Removal%20Credits%20from%20The,lending%20supported%20by%20creditworthy%20offtakes.%E2%80%9D

Jennifer, L. (2024b). Shell to buy 22,500 biochar removal credits from the next 150. https://carboncredits.com/shell-to-buy-22500-biochar-removal-credits-from-the-next-150/

Jiang, Y., Li, T., Xu, X. et al. (2024). A global assessment of the long-term effects of biochar application on crop yield. *Current Research in Environmental Sustainability* 7: 100247. https://doi.org/10.1016/j.crsust.2024.100247.

Johns Hopkins University. (2021). https://bloombergcities.jhu.edu/news/solution-spotlight-turning-garden-waste-carbon-sink-stockholm (retrieved 15 February 2025).

Joseph, S., Cowie, A.L., Van Zwieten, L. et al. (2021). How biochar works, and when it doesn't: a review of mechanisms controlling soil and plant responses to biochar. *GCB Bioenergy* 13 (11): 1731–1764.

Karna, R., Bajracharya, T., and Thakur, N. (2011). Production, analysis and field application of biochar in Nepal (A case study of Padali Village, Lamatar-7, Lalitpur). Academia Edu. https://www.academia.edu/15906152/Production_Analysis_and_Field_Application_of_Biochar_in_Nepal_A_Case_Study_of_Padali_Village_Lamatar_7_Lalitpur_

Kazemi, Z., Jafari, A.J., Leyli, E.K. et al. (2021). Analysis of management costs of waste organization and identification of factors affecting them with a factor analysis model in one of the northern cities of Iran. *Journal of Hazardous Toxic and Radioactive Waste* 26 (2). https://doi.org/10.1061/(asce)hz.2153-5515.0000678.

Kazemi, Z., Jafari, A.J., Leyli, E.K. et al. (2022). Analysis of management costs of waste organization and identification of factors affecting them with a factor analysis model in one of the northern cities of Iran. *Journal of Hazardous, Toxic, and Radioactive Waste* 26 (2): 04021062.

Kung, C.-C. and Mu, J.E. (2019). Prospect of China's renewable energy development from pyrolysis and biochar applications under climate change. *Renewable and Sustainable Energy Reviews* 114: 109343.

Lehmann, J. (2007). A handful of carbon. *Nature* 447 (7141): 143–144.

Lehmann, J., Cowie, A., Masiello, C.A. et al. (2021). Biochar in climate change mitigation. *Nature Geoscience* 14 (12): 883–892.

Lestari, A.M., Wisadirana, D., Kanto, S. et al. (2022). Diffusion and adoption of biochar innovations in agriculture at Gunungsari village, Bumiaji subdistrict, Batu city, East Java, Indonesia. *RA Journal of Applied Research* 8 (4): 303–313. https://doi.org/10.47191/rajar/v8i4.14.

Li, S., Harris, S., Anandhi, A. et al. (2019). Predicting biochar properties and functions based on feedstock and pyrolysis temperature: a review and data syntheses. *Journal of Cleaner Production* 215: 890–902.

Li, S. and Skelly, S. (2023). Physicochemical properties and applications of biochars derived from municipal solid waste: a review. *Environmental Advances* 13: 100395.

Li, T., Jiao, Y., Liu, T. et al. (2024). Effects of biochar addition on soil fauna communities—a meta-analysis. *Soil Use and Management* 40 (3): e13096.

Liu, Y., Li, H., Hu, T. et al. (2022). A quantitative review of the effects of biochar application on rice yield and nitrogen use efficiency in paddy fields: a meta-analysis. *Science of the Total Environment* 830: 154792.

Low, Y.W. and Yee, K.F. (2021). A review on lignocellulosic biomass waste into biochar-derived catalyst: current conversion techniques, sustainable applications and challenges. *Biomass and Bioenergy* 154: 106245.

Małgorzata, M. and Grzegorz, I. (2022). Toward increasing efficiency of fertilization. In: *Smart Agrochemicals for Sustainable Agriculture*, (ed. C. Katarzyna and S. Agnieszka), 139–162. Academic Press. https://doi.org/10.1016/B978-0-12-817036-6.00005-4.

Mangalla, L.K., Sisworo, R.R., Sudia, B. et al. (2023). Socialization and training on production of bio-charcoal made from biomass waste using a double-tube reactor. *Abdimas: Jurnal Pengabdian Masyarakat Universitas Merdeka Malang* 8 (1): 1–12.

Matsagar, B.M. and Wu, K.C.-W. (2022). Agricultural waste-derived biochar for environmental management. In: *Biochar in Agriculture for Achieving Sustainable Development Goals* 3–13. Elsevier.

Melo, L.C.A., Lehmann, J., Carneiro, J.S.S. et al. (2022). Biochar-based fertilizer effects on crop productivity: a meta-analysis. *Plant and Soil* 472 (1): 45–58.

Meyer, S., Genesio, L., Vogel, I. et al. (2017). Biochar standardization and legislation harmonization. *Journal of Environmental Engineering and Landscape Management* 25 (2): 175–191.

Mohammed, S., Fatumah, N., Abasi, K. et al. (2024). Co-designing sustainable biochar business models with sub-Saharan African communities for inclusive socio-economic transformation. *Scientific Reports* 14 (1): 15802.

Moureen, A., Waqas, M., Khan, N. et al. (2024). Untapped potential of food waste derived biochar for the removal of heavy metals from wastewater. *Chemosphere* 356: 141932.

Murtaza, G., Ahmed, Z., Valipour, M. et al. (2024). Recent trends and economic significance of modified/functionalized biochars for remediation of environmental pollutants. *Scientific Reports* 14 (1): 217. https://doi.org/10.1038/s41598-023-50623-1.

Nematian, M., Ng'ombe, J.N., and Keske, C. (2024). Sustaining agricultural economies: regional economic impacts of biochar production from waste orchard biomass in California's Central Valley. *Environment, Development and Sustainability* 26 (12): 30701–30721. https://doi.org/10.1007/s10668-023-03984-6.

Ninson, D. (2015). Factors affecting biochar technology adoption by vegetable farmers in the Kwahu East District of Ghana. University of Ghana. https://ugspace.ug.edu.gh/items/6b2fa624-8b40-4caa-b3bd-d577cd6e0bdb

Norrman, J. (2025). Chalmers University: biochar reduces the risks of DDT-contaminated soil. https://www.chalmers.se/en/current/news/ace-biochar-reduces-the-risks-of-ddt-contaminated-soil/

NRDC. (2010). Natural Resources Defense Council. https://www.nrdc.org/ (retrieved 23 November 2024).

Oh, S. (2022). Korea's low carbon green growth policy and transition to low protein feed. USDA Foreign Agricultural Service. https://apps.fas.usda.gov/newgainapi/api/Report/DownloadReportByFileName?fileName=Korea%27s%20Low%20Carbon%20Green%20Growth%20Policy%20and%20Transition%20to%20Low%20Protein%20Feed_Seoul_Korea%20-%20Republic%20of_KS2022-0014.pdf

Pandian, K., Vijayakumar, S., Mustaffa, M.R.A.F. et al. (2024). Biochar–a sustainable soil conditioner for improving soil health, crop production and environment under changing climate: a review. *Frontiers in Soil Science* 4: 1376159.

Panwar, N. (2024). Pyrolysis technologies for biochar production in waste management: a review. *Clean Energy* 8 (4): 61–78.

Pallewatta, S., Samarasekara, S.M., Rajapaksha, A.U. et al. (2023). Oil spill remediation by biochar derived from bio-energy industries with a pilot-scale approach during the X-Press Pearl maritime disaster. *Marine Pollution Bulletin* 189: 114813. https://doi.org/10.1016/j.marpolbul.2023.114813.

Potnuri, R., Surya, D.V., Rao, C.S. et al. (2023). A review on analysis of biochar produced from microwave-assisted pyrolysis of agricultural waste biomass. *Journal of Analytical and Applied Pyrolysis* 173: 106094.

Pourhashem, G., Hung, S.Y., Medlock, K.B. et al. (2019). Policy support for biochar: review and recommendations. *GCB Bioenergy* 11 (2): 364–380.

Prabha, S.V. (2025, May 23). Biochar as a catalyst for harmony with nature and sustainable development in biodiversity conservation. *Biochar Today*. https://biochartoday.com/blog/biochar-as-a-catalyst-for-harmony-with-nature-and-sustainable-development-in-biodiversity-conservation/

Pradhan, S., Parthasarathy, P., Mackey, H.R. et al. (2024). Food waste biochar: a sustainable solution for agriculture application and soil–water remediation. *Carbon Research* 3 (1): 41.

Puro Earth. (2025). https://puro.earth/biochar#:~:text=Biochar:%20A%20Solution%20for%20Combatting,by%20a%20reputable%20third%20party (retrieved 16 March 2025).

Ranaweera, M.S.K., Edirisinghe, E.A.N.K., Narayana, M. et al. (2025). Locally developed pyrolizer for carbonization of municipal solid waste compost residue: biochar characteristics and carbon credits. In *Selected Proceedings of the 2nd International Engineering Research Symposium* (pp. 203–216). https://doi.org/10.1007/978-981-96-1399-1_17.

Run, Y., Sun, Y., and Wang, D. (2024). Rural development policy, agricultural producer services and wage disparities in the context of developing economies. *South Asian Journal of Social Studies and Economics* 21: 57–65. https://doi.org/10.9734/sajsse/2024/v21i2773.

Santos, S.M., Gonçalves, M., Brito, P. et al. (2024). Waste-derived chars: a comprehensive review. *Waste* 2 (3): 218–239.

Scholz, S.B., Sembres, T., Roberts, K. et al. (2014). *Biochar Systems for Smallholders in Developing Countries: Leveraging Current Knowledge and Exploring Future Potential for Climate-Smart Agriculture*. https://doi.org/10.1596/978-0-8213-9525-7.

Segal, M. (2025). Google signs largest-ever biochar carbon removal purchase deals. https://www.esgtoday.com/google-signs-largest-ever-biochar-carbon-removal-purchase-deals/#:~:text=Google%20announced%20that%20is%20has,of%20200%2C000%20tonnes%20by%202030

Shakoor, A., Arif, M.S., Shahzad, S.M. et al. (2021). Does biochar accelerate the mitigation of greenhouse gaseous emissions from agricultural soil? A global meta-analysis. *Environmental Research* 202: 111789.

Shopify. (2024). https://biochartoday.com/2025/01/24/shopifys-investment-in-biochar-and-carbon-removal-grows-in-2024/

Siddiqa, A. (2021). Determinants of unemployment in selected developing countries: a panel data analysis. *Journal of Economic Impact* 3 (1): 19–26.

Singh, J. (2025). Google strikes World's largest biochar carbon removal deal with Indian startup varaha. https://techcrunch.com/2025/01/16/google-strikes-worlds-largest-biochar-carbon-removal-deal-with-indian-startup-varaha/ (retrieved 05 April 2025).

Sirohi, R., Tarafdar, A., de Souza Vandenberghe, L.P. et al. (2025). *Sustainable Technologies for Food Waste Management*. CRC Press.

Solomon, N. and Isaac Kow, T. (2023). Potential ecological risk assessment of heavy metals associated with abattoir liquid waste: a narrative and systematic review. *Heliyon* 9 (8): e17359. https://doi.org/10.1016/j.heliyon.2023.e17359.

Sughosh, P., Savitha, G., Girisha, S. et al. (2021). Collection of municipal solid waste by clustering of dynamic nodes for vehicle path optimization. *International Conference on Civil Engineering Trends and Challenges for Sustainability*.

Sujeeun, L. and Thomas, S.C. (2022). Biochar rescues native trees in the biodiversity hotspot of Mauritius. *Forests* 13 (2): 277.

THE 17 GOALS | Sustainable Development. (2024). https://sdgs.un.org/goals (retrieved 25 September 2024).

The European Biochar Certificate (EBC). (2024). https://www.european-biochar.org/en/ct/2-EBC-and-WBC-guidelines-documents (retrieved 10 January 2025).

United Nations Climate Change Convention. (2023). U.S. Biochar Coalition Response to UNFCCC Article 6 Call for Comments Issue. https://unfccc.int/sites/default/files/resource/US_BioChar_Coalition.pdf

United Nations. (2022). Biochar and the UN Sustainable Development Goals, Issue. https://sdgs.un.org/sites/default/files/2022-05/USA.pdf

Upadhyay, V., Choudhary, K.K., and Agrawal, S.B. (2024). Use of biochar as a sustainable agronomic tool, its limitations and impact on environment: a review. *Discover Agriculture* 2 (1): 20.

USBI. (2024). United States Biochar Initiative (USBI) Conference Sacramento, California., Issue. https://biochar-us.org/sites/default/files/presentations/USBI-NABC24-Chaiwe-CaDev-Biochar-Investment-Opportunities-In-Africa.pdf

Van Nguyen, T.T., Phan, A.N., Nguyen, T.-A. et al. (2022). Valorization of agriculture waste biomass as biochar: as first-rate biosorbent for remediation of contaminated soil. *Chemosphere* 307: 135834.

Verde, S.F. and Chiaramonti, D. (2021). *The Biochar System in the EU: The Pieces are Falling into Place, but Key Policy Questions Remain.*

Vinti, G., Bauza, V., Clasen, T. et al. (2021). Municipal solid waste management and adverse health outcomes: a systematic review. *International Journal of Environmental Research and Public Health* 18 (8): 4331.

Wang, J., Xiong, Z., and Kuzyakov, Y. (2016). Biochar stability in soil: meta-analysis of decomposition and priming effects. *GCB Bioenergy* 8 (3): 512–523.

Wang, X., Chang, V.W.-C., Li, Z. et al. (2021). Co-pyrolysis of sewage sludge and organic fractions of municipal solid waste: synergistic effects on biochar properties and the environmental risk of heavy metals. *Journal of Hazardous Materials* 412: 125200.

Weijuan, L., Xinxin, Z., Yingxian, W. et al. (2024). Research progress of biochar modification technology and its application in environmental remediation. *Biomass and Bioenergy* 184: 107178. https://doi.org/10.1016/j.biombioe.2024.107178.

World Economic Forum. (2023). https://www.weforum.org/stories/2023/11/biochar-carbon-removals-jack-of-all-trades-for-immediate-climate-action/ (retrieved 18 March 2025).

Xia, L., Cao, L., Yang, Y. et al. (2023). Integrated biochar solutions can achieve carbon-neutral staple crop production. *Nature Food* 4 (3): 236–246.

Xiao, Y., Zhou, G., Qiu, X. et al. (2024). Biodiversity of network modules drives ecosystem functioning in biochar-amended paddy soil. *Frontiers in Microbiology* 15: 1341251.

Yining, G., Pan, W., Paramsothy, J. et al. (2022). Biochar as a potential strategy for remediation of contaminated mining soils: mechanisms, applications, and future perspectives. *Journal of Environmental Management* 313: 114973. https://doi.org/10.1016/j.jenvman.2022.114973.

You, S., Li, W., Zhang, W. et al. (2022). Energy, economic, and environmental impacts of sustainable biochar systems in rural China. *Critical Reviews in Environmental Science and Technology* 52 (7): 1063–1091.

Yuan, H.-M., Xue, W., Roiloa, S. et al. (2024). Increasing biochar diversity promotes the impacts of plant diversity on remediating cadmium in soil. *Journal of Plant Ecology* 17 (6): rtae068.

Zhao, J., Qiu, Y., Yi, F. et al. (2024). Biochar dose-dependent impacts on soil bacterial and fungal diversity across the globe. *The Science of the Total Environment* 930: 172509. https://doi.org/10.1016/j.scitotenv.2024.172509.

Zheng, X., Xu, W., Dong, J. et al. (2022). The effects of biochar and its applications in the microbial remediation of contaminated soil: a review. *Journal of Hazardous Materials* 438: 129557.

Zilberman, D., Laird, D., Rainey, C. et al. (2023). Biochar supply-chain and challenges to commercialization. *GCB Bioenergy* 15 (1): 7–23.

18

Environmental Stability of Biochar in Natural Systems

Shishu Zhu[1], Lanfang Han[2], and Ke Sun[3]*

[1] *The Key Lab of Pollution Control and Ecosystem Restoration in Industry Clusters, Ministry of Education, School of Environment and Energy, South China University of Technology, Guangzhou, China*
[2] *Key Laboratory for City Cluster Environmental Safety and Green Development of the Ministry of Education, School of Ecology, Environment and Resources, Guangdong University of Technology, Guangzhou, China*
[3] *State Key Laboratory of Water Simulation, School of Environment, Beijing Normal University, Beijing, China*

*Corresponding author: zhushishu@scut.edu.cn

18.1 Environmental Reactivity of Biochar

Biochar as one of pyrogenic materials produced through incomplete combustion or pyrolysis of biomass has garnered much attention worldwide owing to its positive role in carbon sequestration, environmental remediation, and soil enhancement (McBeath and Smernik 2009; Knicker et al. 2008). Particularly, its physicochemical properties such as porosity, abundant functional groups, and redox characteristics govern its reactivity toward environmental matrices including minerals, microorganisms, and contaminants (Lal 2004; Nguyen et al. 2007). In this section, the chemical behavior of biochar in the soils and sediments is systematically reviewed, and its interactions with microorganisms are discussed in combination with advanced researches. The environmental reactivity of BC is generally determined by its multi-scale compositions, primarily correlating to the pyrolysis temperature. For instance, the biochar prepared at relatively low temperature biochar (< 500 °C) retains oxygenated functional groups (e.g., carboxyl and hydroxyl groups), enabling cation exchange and metal complexation (Joseph et al. 2010). As pyrolysis temperature increases, the total content of oxygen atom declines along with decomposition of oxygen-containing functional groups (Joseph et al. 2010; Oh and Seo 2016), subsequently leading to the weakness of cation exchange capacity (CEC) of biochar. The biochar also serves as the typical electron shuttles due to its electron donating/accepting ability, which refers to the maximum amounts of electrons that can be provided or obtained (Tian et al. 2021). The functional groups on surface of biochar such as quinone/phenolic groups and aromatic rings are redox-active and capable of reversibly gaining or losing electrons, thus determining the measured electron-accepting capacity (EAC) and electron-donating capacity (EDC) of biochar (Jing et al. 2020). Especially, studies have observed that the EAC of biochar increased significantly at elevated pyrolysis temperature and reached a maximum value at ~400–500 °C, probably attributed to evolution of radical-like oxidizing groups at varied pyrolysis temperatures (Jing et al. 2020; Prévoteau et al. 2016). In comparison, the EDC of biochar decreased with pyrolysis temperature due to the significant decay of alcohols, phenols, and other reducing groups (Saquing et al. 2016). Li et al. (2020) also proved that the type of biomass significantly impacted electron transfer capacity of resultant biochar using the partial least-squares modeling. Notably, it is found that specific surface area (S_{BET}) played a role in regulating the redox properties of biochar, since it is one of the most independent variables for the EDC and EAC. The presence of a porous structure or reduction of particle size can enhance the contact between the biochar and the soluble exogenous redox components to affect its redox reactivity (Keiluweit et al. 2010). In addition, π–π-electron system and graphitized region in biochar have certain electrical conductivity and play a key role in biochar-mediated chemical reactions (Xu et al. 2013). In addition, ubiquitous persistent free radicals (PFRs) in biochar, including semiquinones and carbon-centered radicals, will form via electron transfer between phenolic compounds and transition metals during pyrolysis (Figure 18.1). The PFRs in biochar are redox-active and can give electrons to O_2 and H_2O_2 to generate reactive oxygen species, which are of importance on redox reactions of contaminants in soils and sediments (Zhen et al. 2021; Wu et al. 2021). Zhu et al. (2020) have also proved that PFRs in biochar could significantly improve the adsorption and transformation of heavy metals, while carbon-centered PFRs on the surface of biochar reduce Fe(III) to Fe(II) and triggered Fenton reactions that mineralizes alachlor (Qin et al. 2017). Transition metals in biochar largely contribute to the formation of PFRs. For example, metal compositions in biomass feedstocks are thought to be more important than phenolic compounds in the formation of oxygen-centered PFRs, because the concentration of oxygen-centered PFRs after removal of metals was lower than that after extraction of phenolic compounds (Huang et al. 2019).

The special reactivity of biochar determines its environmental impact and stability. The complexation and redox abilities of biochar directly result in the immobilization and detoxification of heavy metals such as As, Cr, and Hg (Zhang et al. 2018; Dong et al. 2014). In addition, some active organic or mineral substances minerals involved in biochar may also act as catalysts for reaction with

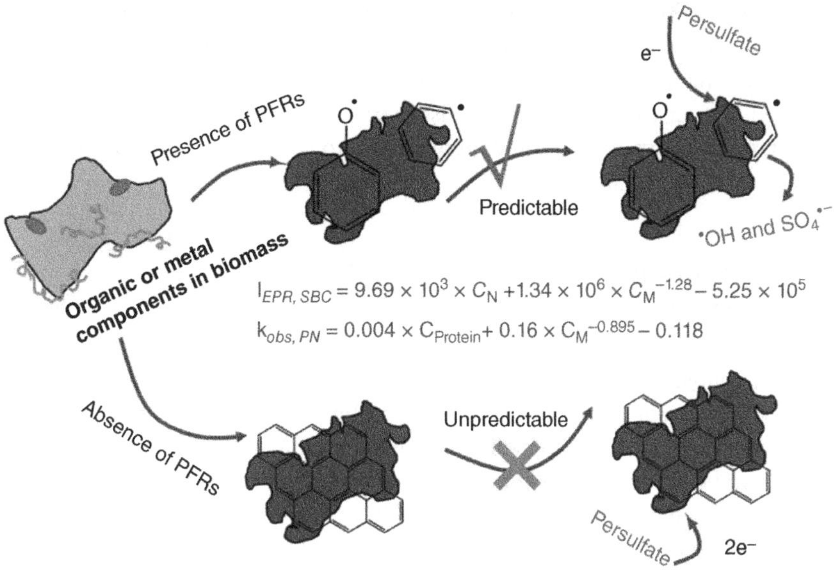

Figure 18.1 Proposed mechanisms of PFR formation and peroxide activation by biochar. *Source:* Wu et al. 2021/with permission of American Chemical Society.

pollutants including explosives, pesticides, and antibiotics (Joseph et al. 2010; Oh and Seo 2016). For example, biochar can accelerate the reductive reaction of dithiothreitol and its metabolites in the presence of sulfides (Ding and Xu 2016). Functional groups on surface of biochar can also mediate electron transfer between zero-valent iron and sulfate, accelerating the formation of sulfate radicals and promoting the degradation of trichloroethylene (Yan et al. 2015). Such reactivity of biochar as electron shuttles can also mediate the reversible electron transfer between microorganisms and extracellular electron acceptors (Saquing et al. 2016; Kappler et al. 2014). Endogenous redox mediators such as melanin, flavin and phenazine are mainly secreted by microorganisms themselves, while naturally occurring or artificially synthesized redox substances include humus, sulfur-containing compounds and coordination substances (Kappler et al. 2014). It has been found that wood biochar can act as an "electron shuttle" to promote the dissimilar metal-reducing bacterium *Shewanella oneidensis* MR-1 to reduce amorphous Fe(III) oxides in soil (Kappler et al. 2014). Some studies have further found that biochar can accelerate the reduction of high-crystallinity hematite Fe(III) mediated by MR-1, during which released semi-quinone radicals play an important role (Xu et al. 2016). Besides, electrons provided by biochar also mediates microbial decomposition of electrophilic pollutants (Yu et al. 2015). Studies have found that straw biochar can mediate the electron transfer from the microorganisms to pentachlorophenol to enhance dichlorination process, in which the conductivity of biochar contributed to the enhanced oxidation of acetic acid by microbials and electron transfer to nitrates (Saquing et al. 2016). The redox reactivity of biochar can also accelerate interspecies electron transfer within microbial community and promote the growth and reproduction of microorganisms (Chen et al. 2014).

18.2 Aggregation and Transport Behaviors of Biochar Colloids

Biochar colloids with particle size < 0.45 μm can generate during biomass pyrolysis, or form through natural weathering, decomposition, artificial fragmentation, or aggregation of dissolved biochar in environmental matrices. As one of most active components, colloidal fraction significantly determines the environmental impact and stability of biochar, and its behaviors are profoundly influenced by specific surface area, surface functional groups, and specific colloidal properties.

18.2.1 Colloidal Stability and Aggregation

The colloidal stability of biochar is fundamentally regulated by a delicate balance between colloidal aggregation and dispersion mechanisms, as elucidated by the Derjaguin–Landau–Verwey–Overbeek (DLVO) theory. This framework posits out that particle interactions are dominated by two opposing forces: long-range van der Waals attraction force driving particle aggregation, and electrostatic repulsion force from overlapping electrical double layers (EDLs) maintaining colloidal stability. The van der Waals force, arising from instantaneous dipole-induced dipole interactions, is particularly significant at short separation distances (< 5 nm) and favor for particle coalescence. Conversely, the electrostatic repulsion force between similarly charged surfaces creates an energy barrier to prevent particle approach, in which the magnitude of this repulsion is intrinsically linked to the surface charge density of biochar colloids (Derjaguin and Landau 1993). The surface charge characteristics of biochar colloids are thus related to their stability and principally determined by their oxygen-containing functional groups and aromatic π-electron systems, also undergoing protonation/deprotonation reactions in response to solution chemistry. The pH-dependent ionization of these functional groups controls zeta potential with typical isoelectric points between pH 2 and 4 for most biochar. Solution parameters including ionic strength (IS), electrolyte valence, and dissolved organic matter (DOM) further modulate biochar colloidal behavior through charge screening effects. In addition, some non-DLVO forces (e.g., hydration repulsion and steric interactions) serve as additional stabilization mechanisms in biochar colloid systems (van Oss 1995; van Oss et al. 1986).

Understanding these colloidal interactions is critical for stability of biochar in environmental remediation application. Wang et al. (2019) demonstrated that aging processes (e.g., nitric acid oxidation) significantly increase contents of carboxyl and phenolic groups on biochar colloids, elevating negative charge density and thereby enhancing colloidal stability in solution through intensified electrostatic repulsion. Concurrently, aging process improved hydrophilicity and surface roughness of biochar colloids, further promoting their dispersion in aquatic environments. Wang et al. (2013) revealed that low-temperature pyrolyzed biochar (350 °C) possessed lower surface negative charge compared to that prepared at high temperature (550 °C), resulting in more stable colloidal systems. Bisiaux et al. (2011) observed rapid attenuation of biochar colloid pulses originating from wildfires and urban runoff in fluvial systems, as evidenced by unexpected particle aggregation and/or fragmentation. The elevated aggregation rates may be partially attributed to wildfire-derived tar fractions, whose non-polar constituents likely mediate charge shielding effects. This charge neutralization mechanism reduces interparticle electrostatic repulsion among wildfire-associated biochar colloids, thereby enhancing their heteroaggregation. The preparation method also influences colloidal stability, for example, ultrasonically derived biochar colloids exhibit higher oxygen content and reduced aromaticity, leading to superior aqueous dispersibility (Liu et al. 2018). The aggregation behavior of biochar colloids is also regulated by solution chemistry and interfacial interactions. Monovalent ions (e.g., Na^+) can compress the EDL to weaken electrostatic repulsion. However, due to strong hydration forces between biochar colloids in Na^+-containing solutions, they still remain stable even under extremely high ISs (800 mM NaCl). In contrast, divalent ions (e.g., Ca^{2+}) significantly promote biochar colloid aggregation through charge neutralization and bridging effects, primarily attributed to the formation of coordination bonds between Ca^{2+} and carboxyl groups (Xu et al. 2017a). Lian et al. (2019) demonstrated that biochar colloid aggregation is influenced by environmental minerals. In that case, biochar colloids could form 1–2 μm aggregates with goethite via "face-to-face" or "edge-to-face" configurations, during which aggregation kinetics were governed by synergistic effect of mineral surface charge and solution pH. Under acidic conditions (pH 5.3), the enhanced positive surface charge of goethite strengthened electrostatic attraction with negatively charged biochar colloids and promoted heteroaggregation. At neutral pH (7.4), aggregation of biochar colloids occurred predominantly through hydrogen bonding and Lewis acid–base interactions. Furthermore, environmental organic compounds and microorganisms may modify both the conformation and surface charge of biochar colloids via bridging complexation, thereby influencing their aggregation behaviors (Lian and Xing 2024; Yang et al. 2022a).

18.2.2 Deposition and Transport in Porous Media

The transport behavior of biochar colloids in porous media plays a pivotal role in governing their environmental fate and stability in natural systems. As introduced into soil-sediment matrices, the migration of these submicron biochar particles is governed by multiscale interfacial processes. At the macroscopic scale, hydrodynamic conditions (e.g., flow velocity and pore geometry) dominate physical sedimentation and mechanical retention mechanisms. Stokes sedimentation and inertial collision effects drive colloid entrapment at pore throats. At the mesoscale, surface filtration operates through size-exclusion mechanisms, where retention efficiency increases significantly when colloid diameter exceeds one-third of the pore throat aperture (based on classical pore throat theory). At the microscopic scale, diffusion-controlled interfacial attachment follows colloid deposition kinetics, whose rates are modulated by dynamic equilibrium between Brownian motion intensity and local hydrodynamic shear forces (Yao et al. 1971). In addition, the binding of biochar colloids to porous media surfaces is influenced by electrostatic interactions and non-DLVO forces (e.g., steric hindrance, hydrophobic effects, hydrogen bonding, and cation/anion bridging) (Wang et al. 2020).

Long-term aging process of biochar colloids markedly alters particle size and surface charge density, thereby modulating their transport within the porous media. For example, Wang et al. (2019) demonstrated that in spite of enhanced hydrophilicity to reduce hydrophobic adsorption, increased surface roughness of aged biochar colloids promoted their deposition via multi-point contact. Owing to decomposition of negatively charged oxygen-containing groups, elevated pyrolysis temperatures could weaken electrostatic repulsion with negatively charged sand grains and enhancing retention ability of biochar colloids (Wang et al. 2013). In addition, redox transformations of biochar colloids affect their retention. For example, reductive transformation decreased oxygen-containing groups along with increasing hydrophobicity and aggregation to impede mobility, whereas oxidation enhances hydrophilicity and suppresses aggregation to facilitate transport (Sigmund et al. 2018). Biochar colloid transport is further regulated by solution chemistry and soil composition. Despite sharing negative surface charges with quartz sand, biochar colloids exhibit ~30% increase in deposition rate under high IS (e.g., 50 mM NaCl) due to EDL compression, whereas aged or low temperature-pyrolyzed biochar colloids retain relatively high mobility under such conditions (Wang et al. 2019). Elevated environmental pH also enhances transport by increasing zeta potential (Yang et al. 2020). Transport of biochar colloids can be improved significantly in coarse-textured soils and unsaturated conditions due to modified pore structures, interaction forces, and preferential flow pathways (Li et al. 2024). In comparison, soil iron oxides inhibit transport via surface coating that masks negative charges. For instance, under iron oxyhydroxide-coated sand conditions (surface coverage of 0.4), biochar colloids pyrolyzed at 350 °C exhibited 49% reduction in maximum relative effluent concentration, decreasing from $C/C_0 = 0.79$ to 0.40 and the breakthrough occurred later (Wang et al. 2013). DOM emerges as a critical regulator of biochar colloid mobility, driven by its ubiquitous occurrence in environmental matrices and strong interactions with biochar interfaces through ligand exchange and π–π interactions (Yang et al. 2019). The sorption of DOM can render the zeta potential of biochar colloids more negative or competitively desorb mineral-bound biochar colloids and thereby affects biochar colloids transport behavior (Wang et al. 2013; Yang et al. 2019). Cao et al. (2022) revealed that humic acid (HA) enhances biochar colloids transport through synergistic mechanisms involving enhancing negatively charged surfaces and steric effects under neutral pH. However, HA and IS commonly show antagonistic effects on biochar colloids mobility, where the presence of one factor significantly attenuates the influence of the other. Notably, biochar colloids act as carriers for hydrophobic pollutants (e.g., polycyclic aromatic hydrocarbons) or heavy metals to induce co-transport behaviors (Swaren et al. 2022). Zhou et al. (2023) found that Cd^{2+} could promote the deposition of biochar colloids on quartz sand by increasing IS, further elevating additional sorption sites for Cd^{2+} during co-transport scenarios. Phenanthrene can infiltrate hetero-aggregates formed between biochar colloids and mineral phases (e.g., goethite), expanding nanopore structures and enhancing its adsorption. This dual behavior, i.e., pore restructuring and increased pollutant binding, prompt biochar colloid-phenanthrene complexes to migrate farther in low-IS environment but aggregate and co-deposit in high-IS condition (Lian et al. 2019). Such mechanisms highlight how biochar colloids act as "mobile platforms" for hydrophobic pollutants, governing their spatial redistribution. Moreover, π–π interactions between biochar colloids and non-polar aromatics (e.g., naphthalene and phenanthrene) may lead to the ordered assembly of colloidal aggregates (Xiao et al. 2017).

18.3 Biochar Carbon Stability

Biochar's global deployment serves as a sustainable carbon sequestration strategy, which can mitigate up to 1.8 Gt CO_2—C equivalent annually (offsetting ~12% of anthropogenic emissions) and maintain ecosystem integrity and agricultural productivity (Mašek et al. 2013). While inert fraction can endure in soils for centuries or even millennia, the role of biochar as a climate mitigation tool critically

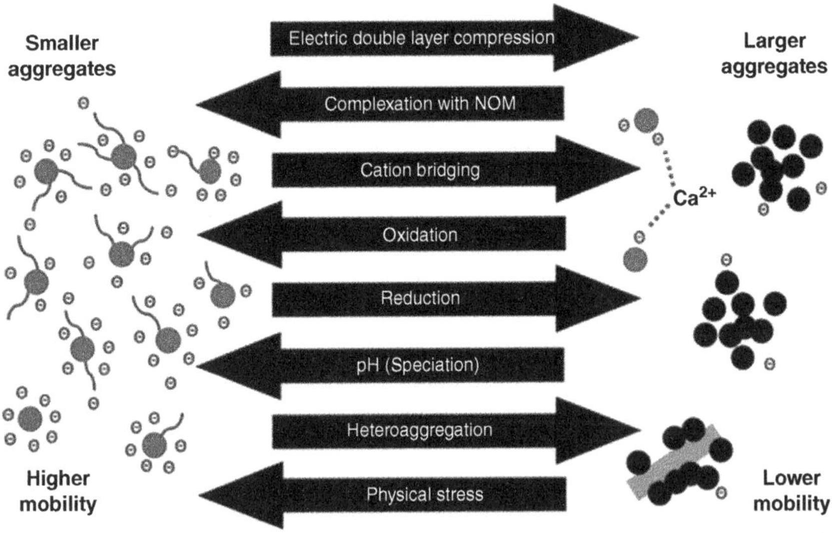

Figure 18.2 Transformation processes affecting environmental behavior of biochar colloids. *Source:* Sigmund et al. 2018/Royal Society of Chemistry/CC BY 3.0.

hinges on its carbon stability (Wang et al. 2016). Such stability also determines the role of biochar in regulating the pollutant dynamics and biogeochemical processes (Figure 18.2). In this part, we will comprehensively present the basic definition, assessment methods, and related mechanisms of carbon stability in biochar, alongside promising strategies to optimize its stability. The carbon stability of biochar refers to its resistance to abiotic (e.g., physical disintegration, chemical oxidation, and leaching) and biotic (e.g., microbial metabolism) decomposition processes in soil and aquatic systems. This stability is not intrinsic property but a dynamic outcome of biochar's physicochemical characteristics upon interactions with soil constituents, typically categorizing into two concepts: (1) recalcitrance—the inherent biochemical resistance relying on biomass feedstocks and pyrolysis conditions, and (2) persistence—the actual longevity of biochar, mainly mediated by soil redox processes, mineral interactions, and microbial activities. It can be quantified by metrics such as mean residence time (MRT) generally ranging from decades to millennia (Wang et al. 2016; Kuzyakov et al. 2014; Gómez et al. 2016).

18.3.1 Assessment Methodologies for Biochar Stability

The H/C and O/C ratios are commonly recognized indicators describing the carbon stability of biochar, as standardized by the International Biochar Initiative (IBI).

A lower O/C molar ratio (< 0.2) indicates advanced carbonization and high oxidative resistance, largely correlating to long-term stability (> 1000 years of half-life), whereas a higher O/C molar ratio (> 0.6) reflects less stable biochar with a half-life below 100 years. Similarly, the lower H/C molar ratio (< 0.7) demonstrated an increased aromaticity with strong resistance toward degradation, though reliability varies with feedstock and pretreatment (e.g., acid-washing to remove inorganic hydrogen) (Leng et al. 2019; Crombie et al. 2013; Ghysels et al. 2022). However, for biochar with high ash content, total carbon quantification may be compromised by inorganic carbon interference (e.g., carbonates), thereby limiting practical applicability of this method. Besides, the H/C ratio is preferred over O/C ratio in IBI because H content can be directly measured experimentally, whereas oxygen content is calculated by mass difference (O=100−C−H−N−S−ash) and may be overestimated (Yang et al. 2018a, b).

Solid-state ^{13}C nuclear magnetic resonance (NMR) is particularly effective for probing smaller aromatic clusters in low-temperature biochar, which exhibits limited crystallinity due to incomplete carbonization. The aromaticity and aromatic condensation degree of biochar, as quantified by the NMR, exhibited significant negative correlations to H/C and O/C molar ratios as well as its labile C content. Conversely, the NMR-measured aromatic degree showed a positive correlation to the MRT of biochar in soil environments. Notably, the variability in NMR instrumentation may introduce inconsistencies in carbon structure determination of biochar. Comparative analyses across different NMR equipment reveal significantly discrepant results, emphasizing the requirement for standardized reference materials of biochar to minimize inter-equipment variability. Although aromatic condensation measurements are of potential applicability, their reliability needs further validation through systematic testing and inter-laboratory comparative studies. However, the NMR is also proven to be difficult to resolve nano-scale heterogeneity of biochar–mineral complexes that are prevalent in natural soils (Lian and Xing 2024; Leng et al. 2019; Ghysels et al. 2022). Based on oxidation resistance of biochar relative to graphite, an R_{50} index can be attained by thermogravimetric analysis (TGA) to assess the stability of biochar. In that case, biochar produced at higher pyrolysis temperature generally exhibit increased value of R_{50}, with those having R_{50} values above 0.5 being considered stable. Nevertheless, the presence of iron-containing minerals in biochar may promote thermal decomposition, potentially reducing the accuracy of TGA assessments (Crombie et al. 2013; Yang et al. 2018b; Chen et al. 2016). The benzene polycarboxylic acids (BPCAs) have been established as standard molecular markers over the past two decades for quantifying black carbon in charcoal, soils, and sediments. For this quantitative method, complex polycyclic aromatics involved in biochar are chemically digested and depolymerized into individual BPCAs, which are labile to analysis via gas or liquid chromatography.

The quantification of BPCAs serves as molecular indicators of biochar stability through two key parameters: (1) the "BPCA pattern," defined by the relative abundance of carboxylated single-ring aromatic monomers (B3CA, B4CA, B5CA, and B6CA, denoting three to six carboxylic groups, respectively), reflecting the size of original polycyclic aromatic structures; (2) the ratio of total BPCA to organic carbon (T-BPCA/TOC), dominantly correlating to the molecular aromaticity. In addition, the proportion of highly carboxylated monomers (e.g., B6CA/total BPCA) also represents the degree of aromatic condensation, corresponding to the aromatic cluster scale in biochar (Wang et al. 2016; Kuzyakov et al. 2014; Zhu et al. 2024).

Near-edge X-ray absorption fine structure (NEXAFS) spectroscopy can assess the aromaticity of biochar via spectral deconvolution, which quantifies distributions of varied functional group as percentages of total carbon. In general, the characteristic absorption peak near ~285.0 eV can be unambiguously attributed to C=C bonds. Bulk NEXAFS analysis quantifies that aromatic carbon accounts for ~40% of total carbon content in biochar, further applied to evaluate their aromaticity and aromatic condensation indices. However, it is noted that the aromatic condensation metrics quantified by NEXAFS analysis are usually inconsistent with those by solid-state ^{13}C NMR. The observed discrepancies may stem from the intrinsic nature of NEXAFS resonances, which signify collective electronic environment of carbon species rather than discrete functional group assignments, thereby limiting the reliability of simplified spectral indices for structural interpretation (Wiedemeier et al. 2015; Heymann et al. 2011). X-ray diffraction (XRD) analysis enables the determination of three-dimensional crystal parameters of carbonaceous materials such as biochar, including crystallite size, aromaticity, and interlayer spacing, with particular advantages for highly crystalline biochar. The XRD spectra also facilitate the evaluation of aromaticity and aromatic condensation degree in biochar. Specifically, the low-angle diffraction peaks in the XRD profile can be deconvoluted into two distinct bands corresponding to aromatic and saturated structural components, respectively. The resultant difference in intensity between these bands provides a quantitative basis for assessing aromaticity in coal-derived chars (Leng et al. 2019; Lu et al. 2000). X-ray photon spectroscopy (XPS) spectral analysis can semi-quantitatively characterize surface carbon speciation, such as the ratio of unoxidized carbon to oxidized carbon species and the relative distribution of oxygen-containing functional groups. In particular, the binding energy assignments of C1s and O1s enable functional group differentiation. The temporal variation of O/C molar ratio and oxygenated carbon moieties detected in XPS analysis largely correlates to biochar's short-term stability, while quantitative assessment of aromatic C—H groups through C1s spectral deconvolution is expected to be reliable stability indicator. Relatively, rigorous correction for inorganic oxygen contribution must be implemented prior to O/C ratio interpretation, as its corrected values may lead to significant discrepancies in biochar stability (Yang et al. 2016). Fourier Transform Infrared Spectroscopy (FTIR) primarily serves as a qualitative method and is a more prevalent analytical technique for functional group analysis compared to XPS analysis. Its application in biochar stability assessment relies fundamentally on characterization of carbon-containing chemical functional groups. Specifically, this analytical approach focuses on tracking the depletion of functional groups and aliphatic C—H moieties, contrasted against the progressive accumulation of aromatic C—H and C=C structures. While the FTIR data combined with XPS analysis is qualitative, qualitative determination of oxygen-containing functional groups can be additionally achieved with supplement of Boehm titration. Recent advancements have implemented the FTIR approach to spatially resolve carbon speciation (e.g., aromatic carbon) within entire micro-aggregates in carbon-deficient soils with biochar amendment, promoting semi-qualitative characterization of functional groups. This microscopic technique enables visual mapping of functional group distributions within aggregate matrices, resulting in diagnostic fingerprints for principal molecular architectures present in biochar-amended soil systems (Hernandez-Soriano et al. 2016).

Oxidation resistance assessment refers to an analytical protocol, involving the selective oxidative removal of labile carbon fractions from biochar using chemical reagents, allowing quantification of the residual stable carbon content. An accelerated aging method (also called Edinburgh stability tool) capable of simulating oxidative degradation of biochar in soil has been developed, during which the accelerated oxidation is performed by adding 5% H_2O_2 to biochar sample under heating at 80 °C for 2 days. The biochar carbon stability (Æ) can be evaluated by determining the mass percentage of residual stable carbon relative to the original total carbon content of pristine biochar according to Eq. (18.1).

$$\text{Æ} = \frac{B_r \times B_rC \times 100\%}{B_t \times B_tC} \tag{18.1}$$

where B_t and B_r are the residual and total biochar's mass; B_tC and B_rC are the C content in residual and total biochar's mass. The carbon stability of biochar derived from various feedstocks quantified by this method is predicted to range from 60 to 80% (w/w) (Cross and Sohi 2013). Further systematic validation should be developed and optimized to increase the accuracy by comparing with other measurements.

The oxidation using $K_2Cr_2O_7$ as agents enables chemical stability assessment of biochar without the thermal pretreatment, extensively functioning as a recognized protocol for BC quantification in soil and sediment matrices (Knicker et al. 2007). The consumption of $K_2Cr_2O_7$ can be determined spectrophotometrically and the mass of oxidized carbon can be calculated by Eq. (18.2). Subsequent multiplication by the total carbon content of biochar allows derivation of the proportion of oxidizable carbon (labile carbon fraction).

$$2K_2Cr_2O_7 + 3C + 8H_2SO_4 = 2K_2SO_4 + 2Cr_2(SO_4)_3 + 3CO_2\uparrow + 8H_2O \quad (18.2)$$

The carbon loss is considered to be related to aromatic and oxygen-containing functional groups (analyzed by infrared method) as well as ratios of O/C and H/C (from elemental analysis). Besides, $K_2Cr_2O_7$ agent has been found to display stronger oxidizing ability toward biochar than commonly used H_2O_2. However, it is worth mentioning that the sample pool size is too small in these studies and their conclusions need more verifications (Chen et al. 2016; Xu et al. 2017b).

18.3.2 Intrinsic Mechanisms of Carbon Stability of Biochar

Biochar samples attained under different conditions include distinct chemical compositions and structures (e.g., aromaticity and condensation degree), further exhibiting varying abilities to resist mineralization. Most current studies indicate that aromatic C plays a dominant role in determining the carbon stability of biochar, while condensed aromatics are more stable than those amorphous aromatic hydrocarbons (Kuzyakov et al. 2014; Chen et al. 2016). Especially, the high-temperature biochar is equipped with polycyclic aromatic hydrocarbon skeleton (i.e., graphitized microcrystalline) featuring a high resistance to microbial decomposition (McCarthy et al. 2008). In addition, pore filling is considered an important protection mechanism for organic carbon in soil, during which biochar can further enhance the adsorption of OC due to its abundant porosity. Therein, the adsorbed OC localized into micropores (< 2 nm) and mesopores (2–50 nm) of biochar significantly limits its contact with microorganisms (McCarthy et al. 2008). The biochar will release a large amount of dissolved OC, directly elevating the content of total dissolved OC of amended soil and altering molecular composition of soil OC due to physical blending. The low-molecular-weight organic substances generated during biomass pyrolysis usually show relatively high solubility (Han et al. 2018), primarily considered as the main form of labile C (Lin et al. 2012). Such unstable OC from biochar may lower the stability of total soil organic carbon (SOC) components by promoting microbial growth (i.e., positive priming effect) (Magill and Aber 2000; Weng et al. 2018). Nevertheless, a negative priming effect of biochar has also been reported, partly attributed to the possible toxicity of biochar to microorganisms (Wang et al. 2016). This can be also ascribed to a fact that the biochar and its derived dissolved OC are of lower microbial decomposition rate and higher protective effect (Bamminger et al. 2014). In addition, it has been recently reported that some aromatic components can maintain stronger stability during soil redox processes, during which aliphatic hydrocarbons and others rich in carboxyl groups are more prone to formation of carbon-centered radicals and further abiotic humification for carbon sequestration (Fu et al. 2024).

Biochar will not only be altered by interacting with soil components but also regulate the composition of SOC by multiple pathways during a long-term application. Although microbial oxidation causes changes in property and structure of biochar, most of the OC components in biochar are difficult to be mineralized as CO_2 over a long period of time. This means that the biochar can largely participate in biogeochemical processes in soil to further affect the SOC stability. Specifically, it is widely manifested as follows: (1) inhibition of microbial activity by biochar, i.e., biochar adsorbs extracellular enzymes in soil or releases phenolic substances, slowing down the SOC decomposition rate (antibacterial effect of lignin-derived polyphenols); (2) protection of microbial residues by biochar, i.e., residues of the death oligotrophic microorganisms colonized on surface of biochar can be protected and form a stable carbon pool (e.g., accumulation of amino sugars) (Weng et al. 2022); (3) humification process induced by biochar, i.e., the addition of biochar, can directly/indirectly affect microbial activity as well as abundance of bacteria and fungi, further mediating the humification of labile OC and influencing the carbon stability (Huang and Hardie 2009). Furthermore, the stabilization of SOC by biochar is also mediated by other factors: (1) synergistic enhancement by biochar combined with mineral constitutes, i.e., metal ions such as Ca^{2+} and Fe^{3+} on surface of biochar reassemble stable complexes with soil minerals (e.g., iron oxide-coated particles); (2) the influence of biochar on soil aggregates and the transformation of OC by PFRs. For example, biochar promotes the formation of macro-aggregates typically stabilizing SOC through physical protection, polymerization, and humification processes (Wu et al. 2021; Sarker et al. 2018).

The formation conditions of biochar are the most critical factors affecting the stability of biochar. Low-temperature biochar contains more aliphatic carbon and soluble components (such as cellulose and hemicellulose derivatives), which are easily decomposed by microorganisms. High temperature (> 500 °C) increases the degree of aromatization and reduces surface functional groups (for example, the half-life of biochar at 600 °C can reach the millennium level), thus leading to a change in the stability mode of biochar. In addition, the DOC content derived from high-temperature biochar is lower, accompanied by a higher degree of decarboxylation and dehydration, a lower content of functional groups, and lower water solubility, and thus accompanied by more stable organic carbon (Keiluweit et al. 2010). The type of raw material also affects carbon stability by influencing the structural formation trend. Lignin is more likely to form a stable aromatic structure during pyrolysis. Straw contains more cellulose and ash. Low-temperature biochar is prone to dissolve DOM and accelerate oxidative degradation. However, at high temperature (500 °C), minerals in the ash (such as Fe and Al oxides) can form stable complexes with biochar, enhancing the antioxidant capacity (Han et al. 2018; Xiao and Chen 2017).

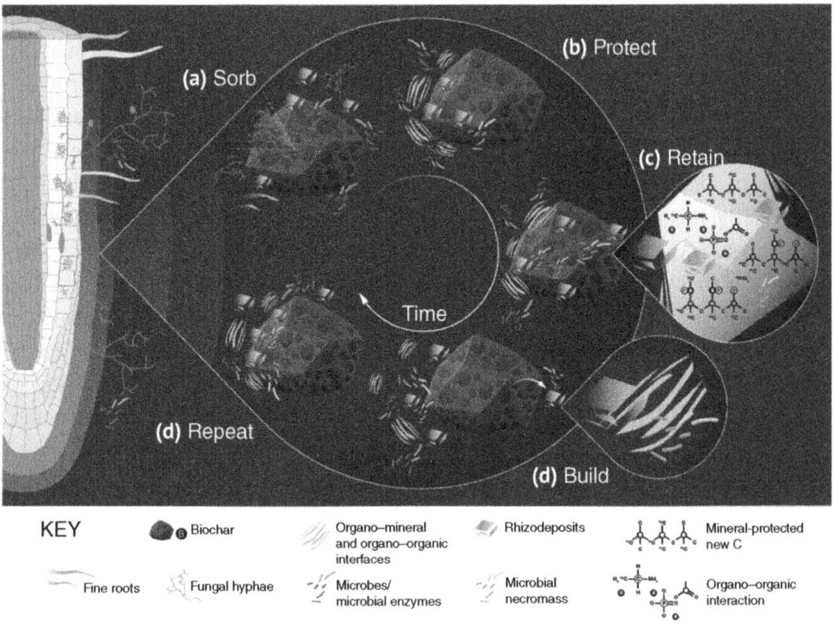

Figure 18.3 Carbon stabilization mechanisms for biochar (Weng et al. 2022). *Source:* Qin et al. 2017/Springer Nature/CC BY 4.0.

The main factors affecting the carbon stability of biochar include the following (Figure 18.3):

1. Soil Mineral Composition: Clay minerals such as montmorillonite and kaolinite fix biochar particles through surface adsorption and pore filling. For example, the organic–mineral complexes formed by clay minerals and biochar can reduce the probability of microbial contact, making the carbon retention rate of biochar in clay more than 30% higher than that in sandy soil. In addition, iron oxides (such as hematite and goethite) and aluminum oxides combine with the carboxyl groups on the surface of biochar through coordination bonds or electrostatic interactions to form stable complexes.
2. pH and Temperature: Microbial activity is relatively strong in soils with high pH, and the mineralization rate of biochar is high. However, biochar with a high ash content (such as biochar from livestock and poultry manure) can buffer the pH of acidic soils by releasing Ca^{2+} and Mg^{2+}, indirectly enhancing its own stability (Fang et al. 2015; Luo et al. 2011). For every 10 °C increase in temperature, the mineralization rate of biochar increases by 1.5–2 times, and the half-life of biochar in a hot and humid environment (such as a tropical rainforest) can be shortened to less than 50 years.

3. Redox Conditions: The anaerobic conditions in waterlogged soils inhibit the chemical oxidation and microbial decomposition of biochar (for example, paddy fields are more conducive to carbon preservation than dry lands, and the preservation of biochar in Amazonian dark earth (Terra Preta) on a millennium scale benefits from long-term waterlogging). Alternate wetting and drying not only lead to the disintegration of the microporous structure but also promote the generation of free radicals and reduce the SOC content (Zhao et al. 2023).

The optimization of biochar properties can be divided into precursor regulation and post-treatment activation. Precursor regulation involves co-pyrolyzing biomass with transition metal salts, inorganic inert compounds, and organic/inorganic nitrogen sources (such as sludge) to shorten the graphitization temperature range, promote the dehydrogenation and polycondensation reactions of biochar precursors, and improve the aromatization degree and chemical stability of biochar (Zhang et al. 2024; Zhu et al. 2018). Post-treatment activation modifies biochar with inorganic reagents after carbonization to change the functional groups and physical structure of biochar, strengthening its structural and chemical stability. For example, (1) acid/alkali activation regulates the carboxyl functional groups of biochar to improve oxygen tolerance; (2) steam activation optimizes the pore channels to reduce the risk of micropore collapse; (3) low-temperature hydrothermal treatment enhances the hydrophobicity and antioxidant properties of biochar (Liu et al. 2023). During the application process, considering the diversity of coexisting substances in actual soils, the stability of biochar can be strengthened by adding specific compounds to act together with biochar. Common methods include microbial regulation, humus addition, and directional combination of soil minerals (Luo et al. 2023). Combining specific microorganisms (such as urease-producing bacteria) with biochar can form mineral–organic complexes (such as calcium carbonate coating layers) to inhibit the oxidation and decomposition of carbon. The introduction of HA or compost reduces the exposure of active sites on the biochar surface through complexation, delaying the carbon mineralization rate. In addition, blending biochar with soil clay minerals (such as montmorillonite) can form stable complexes through electrostatic adsorption and mineral encapsulation, further stabilizing biochar-derived OC and SOC (Liu et al. 2023).

18.4 Perspectives

Biochar is one of the most active components in natural sediments, for it contains various groups with high chemical reactivity. Although increasing studies have noticed the unique functions of biochar in reacting with pollutants and microbials in both natural processes and engineered applications, more attention is needed on its long-term interaction with minerals in soil, such as the adsorption mechanism and the influence on the generation of PFRs. Especially, future research could prioritize leveraging machine learning to gain a more comprehensive understanding of biochar's chemical behavior, such as conducting life cycle assessments to predict its long-term behavior under diverse environmental scenarios.

Current methods for evaluating biochar stability—including elemental ratios (H/C, O/C), thermal oxidation indices (R_{50} index), and isotopic tracing—are often inconsistent and inaccurate. For instance, H/C ratio fails to address mineral-associated carbon or heterogeneous biochar–soil interactions, thereby overestimating the stability of biochar in clay-rich soils (Yang et al. 2018a; Yang et al. 2016; Adhikari et al. 2024). Even advanced techniques like solid-state ^{13}C NMR struggle to distinguish biochar-derived aromatic carbon from native soil organic matter, particularly in long-term field studies (Kuzyakov et al. 2014). Furthermore, oxidation resistance assessments (e.g., H_2O_2 oxidation) may fail to replicate the natural degradation pathways that biochar undergoes, as they neglect microbial and enzymatic processes (Leng et al. 2019; Kim et al. 2021).

The stability of biochar exhibits pronounced context specificity, being modulated by soil characteristics, climates, and agricultural managements. For example, accelerated mineralization kinetics in sandy soils, which demonstrates 2–3 folds enhancement relative to clay-rich systems, have been mechanistically linked to diminished physical protection and amplified microbial metabolic activities (Yang et al. 2022b). Besides, current studies predominantly emphasize controlled laboratory simulations, neglecting field-relevant variables like tillage-induced aggregate disruption, extreme hydrologic fluctuations (drought-flood cycles), and freeze–thaw dynamics (Zhu et al. 2024; Wang et al. 2022).

Emerging challenges from climate change (including intensified freeze–thaw cycles, prolonged droughts, extreme heatwaves, and erratic precipitation) is reshaping the environmental behavior of biochar and poses new threats to biochar carbon stability in soils. Research on the carbon sequestration stability of biochar must transcend simplistic analyses of inherent physicochemical attributes (e.g., aromaticity, H/C ratio, and pyrolysis temperature) and adopt holistic evaluation frameworks that integrate multifactorial interactions between endogenous biochar properties and extrinsic environmental drivers (e.g., soil mineral adsorption, climate-moisture coupling, and microbial community dynamics), thereby establishing predictive models for dynamic stability under heterogeneous environmental stressors (Zhu et al. 2024; Adhikari et al. 2024; Kim et al. 2021).

Machine learning models trained on multi-omics datasets could be prioritized to evaluate interactions between biochar and soil components, as well as their impact on biochar carbon stability. For example, biochar stability can be predicted by integrating relevant evaluation indices (e.g., R_{50} index, H/C and O/C ratios), soil parameters (e.g., pH, mineral content, and soil aggregates), and climate change projections (Yang et al. 2018b). The prediction results could further optimize biochar production conditions to enhance carbon sequestration potential.

References

Adhikari, S., Moon, E., Paz-Ferreiro, J. et al. (2024). Comparative analysis of biochar carbon stability methods and implications for carbon credits. *Science of the Total Environment* 914: 169607.

Bamminger, C., Marschner, B., and Jüschke, E. (2014). An incubation study on the stability and biological effects of pyrogenic and hydrothermal biochar in two soils. *European Journal of Soil Science* 65 (1): 72–82.

Bisiaux, M.M., Edwards, R., Heyvaert, A.C. et al. (2011). Stormwater and fire as sources of black carbon nanoparticles to Lake Tahoe. *Environmental Science & Technology* 45 (6): 2065–2071.

Cao, G., Qiao, J., Ai, J. et al. (2022). Systematic research on the transport of ball-milled biochar in saturated porous media: effect of humic acid, ionic strength, and cation types. *Nanomaterials* 12 (6): 988.

Chen, D., Yu, X., Song, C. et al. (2016). Effect of pyrolysis temperature on the chemical oxidation stability of bamboo biochar. *Bioresource Technology* 218: 1303–1306.

Chen, S., Rotaru, A.-E., Shrestha, P.M. et al. (2014). Promoting interspecies electron transfer with biochar. *Scientific Reports* 4 (1): 5019.

Crombie, K., Mašek, O., Sohi, S.P. et al. (2013). The effect of pyrolysis conditions on biochar stability as determined by three methods. *GCB Bioenergy* 5 (2): 122–131.

Cross, A. and Sohi, S.P. (2013). A method for screening the relative long-term stability of biochar. *GCB Bioenergy* 5 (2): 215–220.

Derjaguin, B. and Landau, L. (1993). Theory of the stability of strongly charged lyophobic sols and of the adhesion of strongly charged particles in solutions of electrolytes. *Progress in Surface Science* 43 (1–4): 30–59.

Ding, K. and Xu, W. (2016). Black carbon facilitated dechlorination of DDT and its metabolites by sulfide. *Environmental Science & Technology* 50 (23): 12976–12983.

Dong, X., Ma, L.Q., Gress, J. et al. (2014). Enhanced Cr(VI) reduction and As(III) oxidation in ice phase: important role of dissolved organic matter from biochar. *Journal of Hazardous Materials* 267: 62–70.

Fang, Y., Singh, B., and Singh, B.P. (2015). Effect of temperature on biochar priming effects and its stability in soils. *Soil Biology and Biochemistry* 80: 136–145.

Fu, H., Zheng, W., Duan, W. et al. (2024). Overlooked roles and transformation of carbon-centered radicals produced from natural organic matter in a thermally activated persulfate system. *Environmental Science & Technology* 58 (33): 14949–14960.

Ghysels, S., Rathnayake, D., Maziarka, P. et al. (2022). Biochar stability scores from analytical pyrolysis (Py-GC-MS). *Journal of Analytical and Applied Pyrolysis* 161: 105412.

Gómez, N., Rosas, J.G., Singh, S. et al. (2016). Development of a gained stability index for describing biochar stability: relation of high recalcitrance index (R50) with accelerated ageing tests. *Journal of Analytical and Applied Pyrolysis* 120: 37–44.

Han, L., Ro, K.S., Wang, Y. et al. (2018). Oxidation resistance of biochars as a function of feedstock and pyrolysis condition. *Science of the Total Environment* 616-617: 335–344.

Hernandez-Soriano, M.C., Kerré, B., Kopittke, P.M. et al. (2016). Biochar affects carbon composition and stability in soil: a combined spectroscopy-microscopy study. *Scientific Reports* 6 (1): 25127.

Heymann, K., Lehmann, J., Solomon, D. et al. (2011). C 1s K-edge near edge X-ray absorption fine structure (NEXAFS) spectroscopy for characterizing functional group chemistry of black carbon. *Organic Geochemistry* 42 (9): 1055–1064.

Huang, D., Luo, H., Zhang, C. et al. (2019). Nonnegligible role of biomass types and its compositions on the formation of persistent free radicals in biochar: insight into the influences on Fenton-like process. *Chemical Engineering Journal* 361: 353–363.

Huang, P.M. and Hardie, A.G. (2009). Formation mechanisms of humic substances in the environment. *Biophysico-Chemical Processes Involving Natural Nonliving Organic Matter in Environmental Systems* 41–109. Wiley.

Jing, Y., Zhang, Y., Han, I. et al. (2020). Effects of different straw biochars on soil organic carbon, nitrogen, available phosphorus, and enzyme activity in paddy soil. *Scientific Reports* 10 (1): 8837.

Joseph, S., Camps Arbestain, M., Lin, Y. et al. (2010). An investigation into the reactions of biochar in soil. *Australian Journal of Soil Research* 48: 501–515.

Kappler, A., Wuestner, M.L., Ruecker, A. et al. (2014). Biochar as an electron shuttle between bacteria and Fe(III) minerals. *Environmental Science & Technology Letters* 1 (8): 339–344.

Keiluweit, M., Nico, P.S., Johnson, M.G. et al. (2010). Dynamic molecular structure of plant biomass-derived black carbon (biochar). *Environmental Science & Technology* 44 (4): 1247–1253.

Kim, H.-B., Kim, J.-G., Kim, T. et al. (2021). Interaction of biochar stability and abiotic aging: influences of pyrolysis reaction medium and temperature. *Chemical Engineering Journal* 411: 128441.

Knicker, H., Hilscher, A., González-Vila, F.J. et al. (2008). A new conceptual model for the structural properties of char produced during vegetation fires. *Organic Geochemistry* 39 (8): 935–939.

Knicker, H., Müller, P., and Hilscher, A. (2007). How useful is chemical oxidation with dichromate for the determination of "black carbon" in fire-affected soils? *Geoderma* 142 (1): 178–196.

Kuzyakov, Y., Bogomolova, I., and Glaser, B. (2014). Biochar stability in soil: decomposition during eight years and transformation as assessed by compound-specific 14C analysis. *Soil Biology and Biochemistry* 70: 229–236.

Lal, R. (2004). Soil carbon sequestration impacts on global climate change and food security. *Science* 304 (5677): 1623–1627.

Leng, L., Huang, H., Li, H. et al. (2019). Biochar stability assessment methods: a review. *Science of the Total Environment* 647: 210–222.

Li, P., Yan, M., Li, M. et al. (2024). Migration rules and mechanisms of nano-biochar in soil columns under various transport conditions. *Nanomaterials* 14 (12): 1035.

Li, S., Shao, L., Zhang, H. et al. (2020). Quantifying the contributions of surface area and redox-active moieties to electron exchange capacities of biochar. *Journal of Hazardous Materials* 394: 122541.

Lian, F. and Xing, B. (2024). From bulk to nano: formation, features, and functions of nano-black carbon in biogeochemical processes. *Environmental Science & Technology* 58 (36): 15910–15925.

Lian, F., Yu, W., Wang, Z. et al. (2019). New insights into black carbon nanoparticle-induced dispersibility of goethite colloids and configuration-dependent sorption for phenanthrene. *Environmental Science & Technology* 53 (2): 661–670.

Lin, Y., Munroe, P., Joseph, S. et al. (2012). Water extractable organic carbon in untreated and chemical treated biochars. *Chemosphere* 87 (2): 151–157.

Liu, G., Zheng, H., Jiang, Z. et al. (2018). Formation and physicochemical characteristics of nano biochar: insight into chemical and colloidal stability. *Environmental Science & Technology* 52 (18): 10369–10379.

Liu, Q., Meki, K., Zheng, H. et al. (2023). Biochar application in remediating salt-affected soil to achieve carbon neutrality and abate climate change. *Biochar* 5 (1): 45.

Lu, L., Sahajwalla, V., and Harris, D. (2000). Characterization of chars prepared from various pulverized coals at different temperatures using drop-tube furnace. 2000, 4: 869—876; *Energy & Fuels* 14 (5): 1125–1125.

Luo, L., Wang, J., Lv, J. et al. (2023). Carbon sequestration strategies in soil using biochar: advances, challenges, and opportunities. *Environmental Science & Technology* 57 (31): 11357–11372.

Luo, Y., Durenkamp, M., De Nobili, M. et al. (2011). Short term soil priming effects and the mineralisation of biochar following its incorporation to soils of different pH. *Soil Biology and Biochemistry* 43 (11): 2304–2314.

Magill, A.H. and Aber, J.D. (2000). Variation in soil net mineralization rates with dissolved organic carbon additions. *Soil Biology and Biochemistry* 32 (5): 597–601.

Mašek, O., Brownsort, P., Cross, A. et al. (2013). Influence of production conditions on the yield and environmental stability of biochar. *Fuel* 103: 151–155.

McBeath, A.V. and Smernik, R.J. (2009). Variation in the degree of aromatic condensation of chars. *Organic Geochemistry* 40 (12): 1161–1168.

McCarthy, J.F., Ilavsky, J., Jastrow, J.D. et al. (2008). Protection of organic carbon in soil microaggregates via restructuring of aggregate porosity and filling of pores with accumulating organic matter. *Geochimica et Cosmochimica Acta* 72 (19): 4725–4744.

Nguyen, T.H., Cho, H.-H., Poster, D.L. et al. (2007). Evidence for a pore-filling mechanism in the adsorption of aromatic hydrocarbons to a natural wood char. *Environmental Science & Technology* 41 (4): 1212–1217.

Oh, S.-Y. and Seo, Y.-D. (2016). Polymer/biomass-derived biochar for use as a sorbent and electron transfer mediator in environmental applications. *Bioresource Technology* 218: 77–83.

Prévoteau, A., Ronsse, F., Cid, I. et al. (2016). The electron donating capacity of biochar is dramatically underestimated. *Scientific Reports* 6 (1): 32870.

Qin, Y., Zhang, L., and An, T. (2017). Hydrothermal Carbon-mediated Fenton-like reaction mechanism in the degradation of alachlor: direct electron transfer from hydrothermal carbon to Fe(III). *ACS Applied Materials & Interfaces* 9 (20): 17115–17124.

Saquing, J.M., Yu, Y.-H., and Chiu, P.C. (2016). Wood-derived black carbon (biochar) as a microbial electron donor and acceptor. *Environmental Science & Technology Letters* 3 (2): 62–66.

Sarker, T.C., Incerti, G., Spaccini, R. et al. (2018). Linking organic matter chemistry with soil aggregate stability: insight from 13C NMR spectroscopy. *Soil Biology and Biochemistry* 117: 175–184.

Sigmund, G., Jiang, C.J., Hofmann, T. et al. (2018). Environmental transformation of natural and engineered carbon nanoparticles and implications for the fate of organic contaminants. *Environmental Science: Nano* 5 (11): 2500–2518.

Swaren, L., Safari, S., Konhauser, K.O. et al. (2022). Pyrolyzed biomass-derived nanoparticles: a review of surface chemistry, contaminant mobility, and future research avenues to fill the gaps. *Biochar* 4 (1): 33.

Tian, R., Dong, H., Chen, J. et al. (2021). Electrochemical behaviors of biochar materials during pollutant removal in wastewater: a review. *Chemical Engineering Journal* 425: 130585.

van Oss, C.J. (1995). Hydrophobicity of biosurfaces—origin, quantitative determination and interaction energies. *Colloids and Surfaces B: Biointerfaces* 5 (3): 91–110.

van Oss, C.J., Good, R.J., and Chaudhury, M.K. (1986). The role of van der Waals forces and hydrogen bonds in "hydrophobic interactions" between biopolymers and low energy surfaces. *Journal of Colloid and Interface Science* 111 (2): 378–390.

Wang, C.Z., Wang, R.Y., Huo, Z.L. et al. (2020). Colloid transport through soil and other porous media under transient flow conditions-a review. *Wiley Interdisciplinary Reviews-Water* 7 (4): e1439.

Wang, D., Zhang, W., Hao, X. et al. (2013). Transport of biochar particles in saturated granular media: effects of pyrolysis temperature and particle size. *Environmental Science & Technology* 47 (2): 821–828.

Wang, H., Nan, Q., Waqas, M. et al. (2022). Stability of biochar in mineral soils: assessment methods, influencing factors and potential problems. *Science of the Total Environment* 806: 150789.

Wang, J., Xiong, Z., and Kuzyakov, Y. (2016). Biochar stability in soil: meta-analysis of decomposition and priming effects. *GCB Bioenergy* 8 (3): 512–523.

Wang, Y., Zhang, W., Shang, J. et al. (2019). Chemical aging changed aggregation kinetics and transport of biochar colloids. *Environmental Science & Technology* 53 (14): 8136–8146.

Weng, Z., Van Zwieten, L., Singh, B.P. et al. (2018). The accumulation of rhizodeposits in organo-mineral fractions promoted biochar-induced negative priming of native soil organic carbon in Ferralsol. *Soil Biology and Biochemistry* 118: 91–96.

Weng, Z., Van Zwieten, L., Tavakkoli, E. et al. (2022). Microspectroscopic visualization of how biochar lifts the soil organic carbon ceiling. *Nature Communications* 13 (1): 5177.

Wiedemeier, D.B., Abiven, S., Hockaday, W.C. et al. (2015). Aromaticity and degree of aromatic condensation of char. *Organic Geochemistry* 78: 135–143.

Wu, W., Zhu, S., Huang, X. et al. (2021). Determination of instinct components of biomass on the generation of persistent free radicals (PFRs) as critical redox sites in pyrogenic chars for persulfate activation. *Environmental Science & Technology* 55 (11): 7690–7701.

Xiao, X. and Chen, B. (2017). A direct observation of the fine aromatic clusters and molecular structures of biochars. *Environmental Science & Technology* 51 (10): 5473–5482.

Xiao, X., Chen, B., Zhu, L. et al. (2017). Sugar cane-converted graphene-like material for the superhigh adsorption of organic pollutants from water via coassembly mechanisms. *Environmental Science & Technology* 51 (21): 12644–12652.

Xu, F., Wei, C., Zeng, Q. et al. (2017a). Aggregation behavior of dissolved black carbon: implications for vertical mass flux and fractionation in aquatic systems. *Environmental Science & Technology* 51 (23): 13723–13732.

Xu, S., Adhikari, D., Huang, R. et al. (2016). Biochar-facilitated microbial reduction of hematite. *Environmental Science & Technology* 50 (5): 2389–2395.

Xu, W., Pignatello, J.J., and Mitch, W.A. (2013). Role of black carbon electrical conductivity in mediating hexahydro-1,3,5-trinitro-1,3,5-triazine (RDX) transformation on carbon surfaces by sulfides. *Environmental Science & Technology* 47 (13): 7129–7136.

Xu, X., Hu, X., Ding, Z. et al. (2017b). Effects of copyrolysis of sludge with calcium carbonate and calcium hydrogen phosphate on chemical stability of carbon and release of toxic elements in the resultant biochars. *Chemosphere* 189: 76–85.

Yan, J., Han, L., Gao, W. et al. (2015). Biochar supported nanoscale zerovalent iron composite used as persulfate activator for removing trichloroethylene. *Bioresource Technology* 175: 269–274.

Yang, F., Xu, Z., Yu, L. et al. (2018a). Kaolinite enhances the stability of the dissolvable and undissolvable fractions of biochar via different mechanisms. *Environmental Science & Technology* 52 (15): 8321–8329.

Yang, F., Zhao, L., Gao, B. et al. (2016). The interfacial behavior between biochar and soil minerals and its effect on biochar stability. *Environmental Science & Technology* 50 (5): 2264–2271.

Yang, W., Bradford, S.A., Wang, Y. et al. (2019). Transport of biochar colloids in saturated porous media in the presence of humic substances or proteins. *Environmental Pollution* 246: 855–863.

Yang, W., Li, B., and Shang, J. (2022a). Aggregation kinetics of biochar nanoparticles in aqueous environment: interplays of anion type and bovine serum albumin. *The Science of the Total Environment* 833: 155148.

Yang, W., Shang, J.Y., Li, B.G. et al. (2020). Surface and colloid properties of biochar and implications for transport in porous media. *Critical Reviews in Environmental Science and Technology* 50 (23): 2484–2522.

Yang, Y., Sun, K., Han, L. et al. (2018b). Effect of minerals on the stability of biochar. *Chemosphere* 204: 310–317.

Yang, Y., Sun, K., Han, L. et al. (2022b). Biochar stability and impact on soil organic carbon mineralization depend on biochar processing, aging and soil clay content. *Soil Biology and Biochemistry* 169: 108657.

Yao, K.-M., Habibian, M.T., and O'Melia, C.R. (1971). Water and waste water filtration. Concepts and applications. *Environmental Science & Technology* 5 (11): 1105–1112.

Yu, L., Yuan, Y., Tang, J. et al. (2015). Biochar as an electron shuttle for reductive dechlorination of pentachlorophenol by *Geobacter sulfurreducens*. *Scientific Reports* 5 (1): 16221.

Zhang, X., Fu, W., Yin, Y. et al. (2018). Adsorption-reduction removal of Cr(VI) by tobacco petiole pyrolytic biochar: batch experiment, kinetic and mechanism studies. *Bioresource Technology* 268: 149–157.

Zhang, Y., Liu, L., Huang, G. et al. (2024). Enhancing humification and microbial interactions during co-composting of pig manure and wine grape pomace: the role of biochar and Fe_2O_3. *Bioresource Technology* 393: 130120.

Zhao, G., Tan, M., Wu, B. et al. (2023). Redox oscillations activate thermodynamically stable iron minerals for enhanced reactive oxygen species production. *Environmental Science & Technology* 57 (23): 8628–8637.

Zhen, Y., Zhu, S., Sun, Z. et al. (2021). Identifying the persistent free radicals (PFRs) formed as crucial metastable intermediates during Peroxymonosulfate (PMS) activation by N-doped carbonaceous materials. *Environmental Science & Technology* 55 (13): 9293–9304.

Zhou, D., Bao, X., Chen, Z. et al. (2023). Experimental and numerical investigations of biochar-facilitated Cd^{2+} transport in saturated porous media: role of solution pH and ionic strength. *Biochar* 5 (1): 10.

Zhu, L., Chen, N., Zhang, X. et al. (2024). Freeze–thaw cycle events enable the deep disintegration of biochar: release of dissolved black carbon and its structural-dependent carbon sequestration capacity. *Environmental Science & Technology* 58 (47): 20979–20989.

Zhu, S., Huang, X., Ma, F. et al. (2018). Catalytic removal of aqueous contaminants on N-doped graphitic biochars: inherent roles of adsorption and nonradical mechanisms. *Environmental Science & Technology* 52 (15): 8649–8658.

Zhu, S., Huang, X., Yang, X. et al. (2020). Enhanced transformation of Cr(VI) by heterocyclic-N within nitrogen-doped biochar: impact of surface modulatory persistent free radicals (PFRs). *Environmental Science & Technology* 54 (13): 8123–8132.

19

Risk Assessment of Biochar in Soil and Aquatic Ecosystem

Xiaochen Huang[1], Peng Xie[2], Huijun Li[1], and Shih-Hsin Ho[2]*

[1] *State Key Lab of Biocontrol, Guangdong Provincial Key Laboratory of Plant Stress Biology, School of Agriculture and Biotechnology, Shenzhen Campus of Sun Yat-sen University, Shenzhen, P. R. China*
[2] *State Key Laboratory of Urban Water Resource and Environment, School of environment, Harbin Institute of Technology, Harbin, P. R. China*

*Corresponding author: huangxch55@mail.sysu.edu.cn

Although biochar can exhibit superior performance in environment remediation, its application would inevitably result in its diffusion into aquatic or soil environments via multiple processes, such as weathering, decomposition, infiltration, and runoff. The continuous release of biochar would cause potential ecological risks, which cannot be overlooked and requires to be thoroughly understood prior to the large-scale engineered applications of biochar. In this chapter, negative effects and potential mechanisms of biochar on soil and aquatic organisms are introduced in detail. Possible methods for risk avoiding of biochar are also mentioned.

19.1 Negative Impacts of Biochar on Soil Ecosystem

Biochar is widely used in the application of soil modification agents, and it can be diffused into the environment through weathering, decomposition, leaching, and runoff during application. This diffusion can negatively affect plants, animals, and microorganisms, either directly or indirectly, thereby posing potential risks to soil ecosystem (Figure 19.1).

Waste-derived Biochar for Sustainable Rural Revitalization, First Edition. Edited by Yuqing Sun and Daniel C.W. Tsang.
Published 2026 by John Wiley & Sons, Inc.

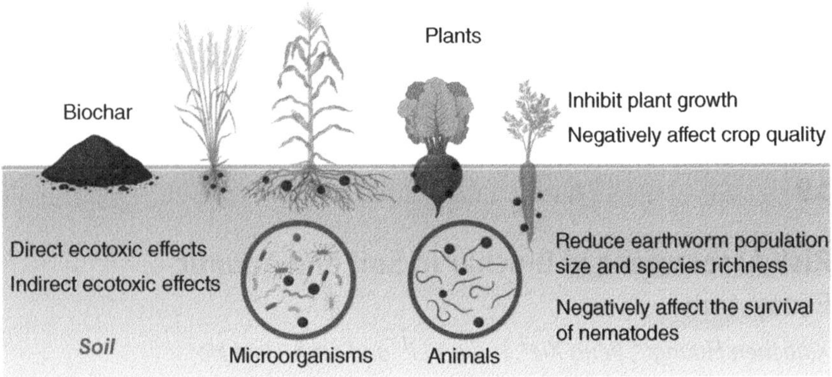

Figure 19.1 Negative impacts of biochar on soil ecosystem.

19.1.1 Terrestrial Plants

As a soil additive, the positive effects of biochar on plant growth are well known. However, adverse effects were also found from biochar with high levels of pollutants (e.g., heavy metals and polycyclic aromatic hydrocarbons [PAHs]). For example, Cu in biochar negatively affected both seed germination and root elongation in cucumber, watercress, and sorghum (Visioli et al. 2016). Krzyszczak-Turczyn et al. (2024) observed an increase in the concentration of PAHs' derivatives in biochar-modified soil in a long-term experiment. As persistent organic pollutants in biochar, PAHs can adversely affect microbial nitrogen-cycling processes, thereby impeding plant assimilation of bioavailable nitrogen through disruption of soil nitrogen transformation (Laghari et al. 2015). Furthermore, environmentally persistent free radicals (EPFRs) derived from biochar demonstrate phytotoxic effects that significantly inhibit seed germination and plant growth (Lian and Xing 2017). This phytotoxicity is primarily attributed to oxidative damage caused by EPFRs-generated reactive oxygen species (ROS), which induce peroxidative degradation of phospholipid bilayers and compromise plasma membrane integrity in plant cells (Alfei and Pandoli 2024; Odinga et al. 2020).

Particle size is a key factor in regulating phytotoxicity of biochar, with smaller particles generally exhibiting higher specific surface areas that enhance interfacial interactions with plants and amplify growth inhibitory responses (Huang et al. 2020). Zhang et al. (2020) found that nano-biochar (<100 nm) collected from high-temperature (700 °C) biochar and lignin-rich feedstock had an inhibiting effect on reed that dramatically decreased shoot length and biomass. Inhibitory effects of nanobiochar were caused not only by phenolic compounds on its surface but also by the blocking effect on epidermal openings, resulting in a reduced transfer of nutrients and water.

19.1.2 Soil Animals

Biochar addition to soil can also pose a potential threat to soil animals. Earthworms are important bioindicators for assessing soil health, as they are often the first to sense changes in the soil environment when biochar is applied. Llovet et al. (2021) found that earthworms (macrofauna) appeared to respond negatively to increased biochar rate applications. Briones et al. (2020) reported that a 50 t/ha biochar addition had a negative effect on soil earthworms, significantly reducing their population sizes and species richness. Nematodes are another important species of soil organisms. High concentrations of biochar also negatively affect the survival of nematodes in the soil. For instance, at a 50 t/ha biochar addition rate, nematodes were disturbed and had a reduced presence (Llovet et al. 2021). Lieke et al. (2018) demonstrated that biochar treatments at high concentrations (1000 and 2000 mg C/L) significantly affect the locomotion behavior of *Caenorhabditis elegans* by decreasing the number of body bends as well as the relative movement. However, the exposure dose of rice straw-derived biochar (500 °C) below 500 mg C/L had a hormetic effect on model organism *C. elegans*. It is observed a 7.8 and 8.9% elevation of body bends number and relative movement speed (all p values <0.05) at the dose of 250 mg C/L. Further experiments confirmed the abundance of free radicals in biochar, suggesting that oxidative damage induced by EPFRs may be a potential reason for the evaluated neurotoxic effect of biochar.

In addition, the adsorption capacity of biochar for agricultural pesticides establishes novel contaminant transfer pathways in soil ecosystems (Thamiris Ferreira and Dias Ferreira 2024). Soil organisms, such as earthworms and nematodes, can indirectly ingest pesticides by the casual predation of biochar particles. Pesticides may be released in the organisms' gut, eventually causing biotoxicity.

19.1.3 Soil Microorganisms

Microorganisms are widespread in the environment and serve as critical bioindicators for evaluating and analyzing the ecotoxicological risks associated with biochar applications. Biochar addition to soil can have a direct or indirect adverse impact on soil microorganisms. Direct ecotoxic effects of biochar on microbial viability primarily arise from its released heavy metals and organic pollutants, mediated through multiple exposure pathways (trophic transfer and dermal contact) (Liu et al. 2019). For example, the growth of bacterial and actinomycetes populations may be inhibited by Cd in biochar, while organic pollutants, for instance, PAHs, carboxylic acids, furans, phenols, benzene, and ketones, may inhibit microbial activity at high concentrations (Dong et al. 2025). PAHs produced unintentionally during biochar pyrolysis have mutagenic effects on *Salmonella*/microsomes (Anjum et al. 2014). Moreover, EPFRs in biochar can cause electron transfer between the biochar surface and specific cells, thereby altering the microbial community structure (Xiang et al. 2021).

Indirect ecotoxic effects of biochar on soil microorganisms are mediated by changes in the environment, such as pH, or other factors related to the ecological tolerance range of the exposed species. For example, Fan et al. (2020) found changes in soil bacterial community composition after 6 years of biochar amendment and noted that biochar affected the soil microbial community indirectly by altering the soil properties in the long term. Biochar can also sequester microbial quorum-sensing signals, disrupting intercellular communication and suppressing key gene expression and transmission of related processes (Ren et al. 2021). In short, the influence of biochar on the structure of the microbial community depends on the type of biochar as well as complex and changeable mechanisms.

19.2 Negative Impacts of Biochar on Aquatic Ecosystem

In addition to soil environment, biochar can be diffused into aquatic environment via multiple processes, such as weathering, decomposition, infiltration, and runoff (Lian and Xing 2017). The long term presence of biochar in water would induce aquatic ecosystem risks. As for biochar's risk assessment, previous work generally refers to the methods of evaluating the ecotoxicity of nanomaterials. Phytoplankton can serve as a principal test organism due to their pivotal role in aquatic ecosystems and high sensitivity to environmental perturbations. As primary producers, phytoplankton form the foundation of aquatic food webs, driving global biogeochemical cycles through photosynthesis, oxygen production, and carbon sequestration. The selection of phytoplankton as model organisms in nanotoxicology studies is predicated on their rapid proliferation rates, standardized culturing protocols, and capacity to reveal sublethal effects at molecular and physiological levels. The negative impacts of biochar on phytoplankton are introduced in Figure 19.2.

Previous studies have confirmed the negative effects of biochar on phytoplankton were also in a species-dependent manner. For example, Zhang et al. separately selected freshwater alga *Scenedesmus obliquus* and the luminescence of *Photobacterium phosphoreum* to assess the toxicity of pine needle-derived biochar (300–600 °C), finding that biochar could cause higher inhibition growth rate to *S. obliquus* than *P. phosphoreum* under the same exposure condition, with 72 h-EC_{50} ranging from 60.6 to 133.6 mg/L (Zhang et al. 2019). Similar phenomenon was also determined by other previous work, which also found alga *S. obliquus* was more sensitive to carbon quantum dots than zebrafish and zooplankton (Yao et al. 2018). It is suggested that the organisms with higher tropical levels generally

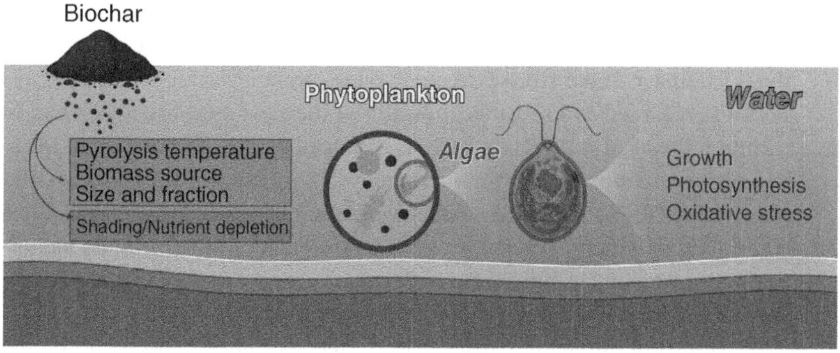

Figure 19.2 Negative impacts of biochar on aquatic ecosystem.

had more tolerance to biochar in the environment. Huang et al. (2021, 2023) selected *Desmodesmus* sp. and *Chlorella vulgaris* as model microalgae to evaluate the effects of biochar in aquatic environment. It is found that (1) pyrolysis temperature, (2) biomass resource, (3) particle size, and (4) fractions of biochar could significantly impact microalgal growth, photosynthesis, and oxidative stress.

1. **Pyrolysis temperature:** Increasing pyrolysis temperature from 400 to 600 °C could significantly alleviate the 96-h algal growth inhibition, due to the change of physicochemical properties of biochar. That is, elevated pyrolysis temperature could increase carbon content and the D/G peak intensity ratios (I_D/I_G) as well as reduce persistent free radicals (PFRs) intensity. The correlation between PFRs intensity and growth inhibition confirmed the negative role of PFRs in risks of biochar (Huang et al. 2020). Furthermore, the effects of nanosized biochar on freshwater algae *C. vulgaris* were also regulated by pyrolysis temperature. Nanosized biochar prepared at 700 °C induced the highest 72-h growth inhibition ratio than those prepared at 350 and 500 °C (Huang et al. 2023). In addition to the induced oxidative stress, algal fatty acid metabolism was disrupted, including saturated fatty acids (C16 : 0 and C18 : 0), mono-unsaturated fatty acids (C18 : 1 n-9 and C16 : 1 n-7), and poly-unsaturated fatty acids (C18 : 2 n-6 and C18 : 3 n-3). These changes would lead to the disruption of cell membrane permeability (Huang et al. 2023).
2. **Biomass source:** A large amount of waste biomass can be prepared to biochar, exhibiting different physicochemical properties and biological effects. When three biomass sources (rice husk, wood sawdust, and corn straw) were selected to prepare biochar and compare their biotoxicity, wood sawdust-derived nano-biochar (nano-WBC) caused the most severe growth inhibition of *C. vulgaris* with a 72 h-EC50 of 148.8 mg/L than rice husk (386.8 mg/L) and corn straw (556.9 mg/L) (Huang et al. 2021). In general,

malondialdehyde (MDA), ROS, superoxide dismutase (SOD), catalase (CAT), and mitochondrial membrane potential (MMP) could be selected to assess the cell damage under stresses. Among these physiological indexes, excessive ROS was regarded as a sensitive indicator of the algal response to biochar. In addition, the algal cell membrane was damaged via the observation of ultrastructure by scanning electron microscopy (SEM) and transmission electron microscopy (TEM) as well as the determination of intact-injured cells ratios using flow cytometry (Huang et al. 2021). Notably, three biochars significantly adsorbed micronutrients (B, Co, Cu, Fe, and Zn) from algal medium through oxygen-containing functional groups (C=O, C—O, and —OH), with nano-WBC showing particularly strong adsorption. This could be the reason for the strongest inhibition of wood sawdust-derived biochar to algae. The nutrient depletion caused substantial algal growth inhibition, where nano-WBC induced 27.0% adsorption-related inhibition (40.9% of total inhibition), surpassing rice husk-derived nano-biochar (nano-RBC; 11.7%) and corn straw-derived nano-biochar (nano-CBC; 9.7%). While macronutrients (K, P) increased due to biochar release, critical micronutrient deficiencies persisted. Strong negative correlations emerged between algal growth and residual Fe ($r = -0.985$) and Zn ($r = -0.967$) concentrations, indicating the adsorption of nutrients (especially Fe and Zn) on biochar could be a key factor in their distinct biological effects on algae. In line with the contribution of nutrient adsorption to toxicity, graphene-family nanomaterials, particularly graphene oxide (GO) at 50 mg/L, demonstrated 28.7% algal growth inhibition attributed to nutrient adsorption effects (Zhao et al. 2017).

3. **Size effects:** In recent years, nano-biochar has been prepared to apply environmental remediation due to the superior specific surface area and reactivity compared to bulk biochar. Nano-biochar colloids are likely formed with a size commonly less than 100 nm, which was much smaller than bulk biochar and can be stably present in aquatic environment (Liu et al. 2018; Zhang et al. 2019; Wang et al. 2013; Wang et al. 2019). Considering bulk biochar has demonstrated inhibitory effects on certain organisms, the unique physicochemical properties of nano-biochar suggest potentially more severe ecological impacts. Reduced particle size would facilitate its interaction with organisms, resulting in greater growth inhibition (Prodana et al. 2019). Indeed, previous work reported that nano-biochar induced stronger negative effects (reduction of 96 h-cell number) on algae than its bulk counterpart, likely due to its enhanced nutrient adsorption capabilities and environmental persistence (Huang et al. 2021).
4. **Contribution of different biochar fractions:** Upon environmental release, biochar undergoes natural partitioning into dissolved and undissolved fractions. The dissolved biochar (DBC) could be generally extracted using water

for certain hours. Similar to pristine biochar (PBC), the biological effects of DBC also depend on biomass source and pyrolysis temperature. For example, cellulose-derived DBC pyrolyzed at 300 °C exhibited stronger microalgal growth inhibition compared to those generated at higher pyrolysis temperatures. Conversely, aqueous extracts from 300 °C lignin-derived biochar showed no significant growth inhibitory effects on *Synechococcus* sp. (Smith et al. 2016). Wang et al. (2017) found biochar derived from rice husk, sawdust, and *Acorus calamus* could extract DBCs with distinct chemical profiles, subsequently causing differential biological responses across test organisms. The inhibitory bio-effects were found on *Pseudomonas aeruginosa*, *Triticum* spp., and *C. elegans*. However, microalgae residue-derived DBC could significantly accelerate growth of *Desmodesmus* sp. The release of polysaccharides from DBC may be a main reason for the stimulatory effects, which served as carbon sources for cell growth (Huang et al. 2020). On the other hand, the undissolved fractions that also name water-unextractable fraction of biochar (WBC) exert growth inhibition, while these inhibitory effects were lower than PBC due to the reduced potential toxic elements, PAHs and EPFRs, in WBC.

Notably, as for phytoplankton, shading effect is mainly considered as an indirect factor contributing to the nanotoxicity. The dark coloration of biochar in aquatic environments can reduce light penetration, inhibiting photosynthetic efficiency in microalgae. A significant positive correlation ($p < 0.05$) was observed between biochar's optical density at 680 nm (chlorophyll-a absorption peak) and elevated biochar exposure concentrations, suggesting light limitation caused by particulate shading was identified as a primary driver of microalgal growth suppression. This phenomenon is consistent with observations from other carbonaceous materials such as graphene-family nanomaterials (Zhao et al. 2017; Long et al. 2012).

19.3 Combined Effects of Biochar and Pollutants to Organisms

Biochar has emerged as a promising tool for water remediation, particularly in mitigating pollution through the adsorption of heavy metals, organic contaminants, and excess nutrients. However, its potential ecological risks to aquatic organisms remain a subject of growing concern. Microalgae and fish are particularly susceptible to changes in water chemistry and sediment dynamics induced by biochar amendments. Microalgae is one of the main organisms for evaluating the toxicity of biochar to aquatic organisms (Xiang et al. 2021). The introduction of biochar into aquatic systems can influence algal communities through nutrient modulation, shading effects, and potential toxicity (Zhang et al. 2019).

Additionally, EPFRs in biochar have been shown to produce hydroxyl radicals in aquatic environments, potentially triggering ROS formation in aquatic plant cells and leading to cellular and organ damage (Odinga et al. 2020). Another critical concern is the release of biochar-derived contaminants, including PAHs and heavy metals, which may exert toxic effects on microalgal cells by disrupting membrane integrity and metabolic processes (Laranjeiro et al. 2024). The potential risks to fish stem from the physicochemical properties of biochar. Biochar particles suspended in the water column can cause mechanical irritation or clogging of gills, leading to respiratory stress and impaired oxygen exchange (Jovanović and Palić 2012). Fine biochar particles, particularly those in the nanoscale range, may also penetrate epithelial barriers, triggering inflammatory responses and disrupting ion regulation (Zhao et al. 2011).

Moreover, the adsorption properties of biochar can influence the bioavailability of essential nutrients and trace elements, potentially leading to deficiencies that affect fish growth, immune function, and reproductive success. The ingestion of biochar particles, whether directly from the water column or indirectly through contaminated prey, raises concerns regarding bioaccumulation and gastrointestinal effects (Johnston et al. 2010). Fish that consume biochar-laden sediments or filter-feed in biochar-amended water may experience digestive disturbances, including alterations in gut microbiota composition and nutrient absorption efficiency. Additionally, biochar can act as a carrier for adsorbed contaminants, such as heavy metals and hydrophobic organic compounds, which may desorb in the gastrointestinal tract due to pH and enzymatic conditions, leading to systemic toxicity (Ma and Lin 2013). Studies have shown that exposure to biochar-associated contaminants can induce oxidative stress, impair liver function, and disrupt endocrine signaling pathways in fish, potentially affecting their development, behavior, and reproductive success (Ma and Lin 2013). Moreover, biochar-mediated changes in sediment composition and benthic habitat structure may have cascading effects on fish populations by altering food web dynamics and habitat availability (Canesi et al. 2012). The modification of sediment characteristics, such as increased porosity and altered microbial community structure, can influence the abundance and diversity of benthic invertebrates, a key food source for many fish species (Xiang et al. 2021; Wang et al. 2020). Changes in prey availability and nutritional quality may, in turn, affect fish growth rates, foraging efficiency, and overall population dynamics. While biochar is often regarded as an environmentally friendly material with potential benefits for pollution mitigation, its ecological risks to fish remain insufficiently understood, particularly in long-term exposure scenarios and complex natural ecosystems. Further research is needed to assess the chronic effects of biochar exposure on fish physiology, behavior, and population dynamics, taking into account factors such as biochar composition, particle size distribution, and interactions with co-occurring contaminants. Addressing these knowledge gaps will be essential for ensuring that biochar applications in aquatic systems are implemented with minimal ecological disruption, allowing for the development of guidelines that optimize its benefits while mitigating potential risks to fish and other aquatic organisms.

In addition, biochar is widely recognized for its benefits in improving soil quality and promoting plant growth. However, its potential risks to crops under specific conditions cannot be ignored. One concern is the presence of hazardous organic and inorganic compounds, such as PAHs and heavy metals, which may leach from biochar into the soil and accumulate in plants, potentially posing health risks to humans and animals (Lehmann et al. 2011; Rombola et al. 2019). During pyrolysis, biochar can retain these toxic substances, leading to elevated PAH concentrations in biochar-amended soils (Gelardi et al. 2019). Studies have shown that crops grown in such soils can accumulate PAHs, negatively impacting their growth and increasing health risks through consumption (Wang et al. 2018). Heavy metals in biochar, such as Cu and Zn, can also adversely affect plant germination and root elongation, particularly at higher biochar application rates (Visioli et al. 2016). Root growth appears to be more sensitive to metal toxicity than seed germination, with significant inhibition observed at biochar application rates exceeding 10% (w/w). The presence of EPFRs in biochar has also been linked to reduced seed germination and plant survival (Lian and Xing 2017). These radicals can generate ROS, which destabilize cell membranes and induce oxidative stress, leading to impaired root and shoot growth (Odinga et al. 2020). Additionally, the accumulation of volatile organic compounds (VOCs) and low molecular weight organic molecules (LOMs) on biochar surfaces further exacerbates its phytotoxicity, suppressing plant growth (Joseph et al. 2014; Buss and Mašek 2014). Moreover, biochar's strong adsorption capacity can reduce the availability of essential nutrients, such as nitrogen and phosphorus, hindering plant growth (Kim et al. 2015; Rajkovich et al. 2012). It may also immobilize plant hormones, disrupting growth regulation (Lehmann et al. 2011; Zhu et al. 2017). Furthermore, biochar can alter the mobility of toxic elements in soil, potentially increasing the bioavailability of heavy metals like arsenic and cadmium under specific environmental conditions (Rinklebe et al. 2020; El-Naggar et al. 2019a).

19.4 Potential Measures for Risk Avoidance

In summary, while biochar offers notable benefits to environment, its potential risks to aquatic and/or soil ecosystems stem from (1) the presence of contaminants such as PAHs, heavy metals, EPFRs, and VOCs, which can inhibit plant germination and growth; (2) the toxic effects of biochar nanoparticles, mainly depending on biomass sources and pyrolysis temperatures; and (3) biochar's ability to adsorb essential nutrients, plant hormones, and toxic pollutants. Careful selection of biochar feedstock, pyrolysis conditions, and application rates is crucial to minimizing these risks and optimizing its benefits for further large-scale application. Toxicity determinants primarily originate from two key parameters: biomass source and pyrolysis conditions. Selection protocols should prioritize raw materials with minimal contaminants, particularly favoring plant-derived biomass that inherently contains reduced PAH precursors (Quilliam et al. 2013).

Production optimization should emphasize slow pyrolysis protocols (0.1–2 °C/s), as empirical evidence demonstrates their ecological superiority over rapid pyrolysis methods (>50 °C/s). Slow-pyrolyzed biochar exhibits triple advantages: (1) decreased concentrations of phytotoxic compounds, (2) diminished nutrient immobilization capacity in soil matrices, and (3) enhanced mineralization potential (El-Naggar et al. 2019b; Bruun et al. 2012). The thermal regulation also proves to be a particularly crucial factor. That is, low-temperature processing (<500 °C) yields biochar with 40–60% lower PAH content compared to high-temperature counterparts (>700 °C), while extended pyrolysis duration (≥2 h) further reduces residual contaminants through progressive thermal decomposition (Hale et al. 2012). These combined parameters critically determine the biochar's ecotoxicological profile, which should be carefully considered for large application of biochar in environmental remediation and agriculture.

References

Alfei, S. and Pandoli, O.G. (2024). Biochar-derived persistent free radicals: a plethora of environmental applications in a light and shadows scenario. *Toxics* 12 (4): 245.

Anjum, R., Krakat, N., Toufiq Reza, M. et al. (2014). Assessment of mutagenic potential of pyrolysis biochars by Ames *Salmonella*/mammalian-microsomal mutagenicity test. *Ecotoxicology and Environmental Safety* 107: 306–312.

Briones, M.J.I., Panzacchi, P., Davies, C.A. et al. (2020). Contrasting responses of macro- and meso-fauna to biochar additions in a bioenergy cropping system. *Soil Biology and Biochemistry* 145: 107803.

Bruun, E.W., Ambus, P., Egsgaard, H. et al. (2012). Effects of slow and fast pyrolysis biochar on soil C and N turnover dynamics. *Soil Biology and Biochemistry* 46: 73–79.

Buss, W. and Mašek, O. (2014). Mobile organic compounds in biochar-a potential source of contamination–phytotoxic effects on cress seed (*Lepidium sativum*) germination. *Journal of Environmental Management* 137: 111–119.

Canesi, L., Ciacci, C., Fabbri, R. et al. (2012). Bivalve molluscs as a unique target group for nanoparticle toxicity. *Marine Environmental Research* 76: 16–21.

Dong, M., Jiang, M., He, L. et al. (2025). Challenges in safe environmental applications of biochar: identifying risks and unintended consequence. *Biochar* 7: 12.

El-Naggar, A., El-Naggar, A.H., Shaheen, S.M. et al. (2019a). Biochar composition-dependent impacts on soil nutrient release, carbon mineralization, and potential environmental risk: a review. *Journal of Environmental Management* 241: 458–467.

El-Naggar, A., Shaheen, S.M., Hseu, Z.-Y. et al. (2019b). Release dynamics of As, Co, and Mo in a biochar treated soil under pre-definite redox conditions. *The Science of the Total Environment* 657: 686–695.

Fan, S., Zuo, J., and Dong, H. (2020). Changes in soil properties and bacterial community composition with biochar amendment after six years. *Agronomy* 10 (5): 746.

Gelardi, D.L., Li, C., and Parikh, S.J. (2019). An emerging environmental concern: biochar induced dust emissions and their potentially toxic properties. *The Science of the Total Environment* 678: 813–820.

Hale, S.E., Lehmann, J., Rutherford, D. et al. (2012). Quantifying the total and bioavailable polycyclic aromatic hydrocarbons and dioxins in biochars. *Environmental Science & Technology* 46 (5): 2830–2838.

Huang, X., Zhu, S., Zhang, H. et al. (2021). Biochar nanoparticles induced distinct biological effects on freshwater algae via oxidative stress, membrane damage, and nutrient depletion. *ACS Sustainable Chemistry & Engineering* 9 (32): 10761–10770. https://doi.org/10.1021/acssuschemeng.1c01814.

Huang, X., Zhang, C., Zhu, S. et al. (2020). Effects of biochar on microalgal growth: difference between dissolved and undissolved fractions. *ACS Sustainable Chemistry & Engineering* 8 (24): 9156–9164.

Huang, Y., Lv, J., Liu, S. et al. (2023). Physicochemical properties of nanosized biochar regulated by heat treatment temperature dictates algal responses: from the perspective of fatty acid metabolism. *Journal of Hazardous Materials* 444: 130342.

Johnston, B.D., Scown, T.M., Moger, J. et al. (2010). Bioavailability of nanoscale metal oxides TiO_2, CeO_2, and ZnO to fish. *Environmental Science & Technology* 44 (3): 1144–1151.

Joseph, S., Graber, E.R., Chia, C. et al. (2014). Shifting paradigms: development of high-efficiency biochar fertilizers based on nano-structures and soluble components. *Carbon Management* 4: 323–343.

Jovanović, B. and Palić, D. (2012). Immunotoxicology of non-functionalized engineered nanoparticles in aquatic organisms with special emphasis on fish-Review of current knowledge, gap identification, and call for further research. *Aquatic Toxicology* 118–119: 141–151.

Kim, H.S., Kim, K.R., Kim, H.J. et al. (2015). Effect of biochar on heavy metal immobilization and uptake by lettuce (*Lactuca sativa* L.) in agricultural soil. *Environmental Earth Sciences* 74: 1249–1259.

Krzyszczak-Turczyn, A., Dybowski, M.P., Kończak, M. et al. (2024). Increased concentration of PAH derivatives in biochar-amended soil observed in a long-term experiment. *Journal of Hazardous Materials* 465: 133159.

Laghari, M., Mirjat, M.S., Hu, Z. et al. (2015). Effects of biochar application rate on sandy desert soil properties and sorghum growth. *Catena* 135: 313–320.

Laranjeiro, F., Rotander, A., López-Ibáez, S. et al. (2024). Comparative assessment of the acute toxicity of commercial bio-based polymer leachates on marine plankton. *Science of The Total Environment* 946: 174403.

Lehmann, J., Rillig, M.C., Thies, J. et al. (2011). Biochar effects on soil biota–a review. *Soil Biology & Biochemistry* 43: 1812–1836.

Lian, F. and Xing, B. (2017). Black carbon (biochar) in water/soil environments: molecular structure, sorption, stability, and potential risk. *Environmental Science & Technology* 51 (23): 13517–13532.

Lieke, T., Zhang, X., Steinberg, C.E.W. et al. (2018). Overlooked risks of biochars: persistent free radicals trigger neurotoxicity in *Caenorhabditis elegans*. *Environmental Science & Technology* 52 (14): 7981–7987.

Liu, G., Zheng, H., Jiang, Z. et al. (2018). Formation and physicochemical characteristics of nano biochar: insight into chemical and colloidal stability. *Environmental Science & Technology* 52 (18): 10369–10379.

Liu, J., Yin, M., Zhang, W. et al. (2019). Response of microbial communities and interactions to thallium in contaminated sediments near a pyrite mining area. *Environmental Pollution* 248: 916–928.

Llovet, A., Mattana, S., Chin-Pampillo, J. et al. (2021). Long-term effects of gasification biochar application on soil functions in a Mediterranean agroecosystem: higher addition rates sequester more carbon but pose a risk to soil faunal communities. *Science of The Total Environment* 801: 149580.

Long, Z., Ji, J., Yang, K. et al. (2012). Systematic and quantitative investigation of the mechanism of carbon nanotubes' toxicity toward algae. *Environmental Science & Technology* 46 (15): 8458–8466.

Ma, S. and Lin, D.H. (2013). The biophysicochemical interactions at the interfaces between nanoparticles and aquatic organisms: adsorption and internalization. *Environmental Science: Processes & Impacts* 15 (1): 145–160.

Odinga, E.S., Waigi, M.G., Gudda, F.O. et al. (2020). Occurrence, formation, environmental fate and risks of environmentally persistent free radicals in biochars. *Environment International* 134: 105172.

Prodana, M., Silva, C., Gravato, C. et al. (2019). Influence of biochar particle size on biota responses. *Ecotoxicology and Environmental Safety* 174: 120–128.

Quilliam, R.S., Rangecroft, S., Emmett, B.A. et al. (2013). Is biochar a source or sink for polycyclic aromatic hydrocarbon (PAH) compounds in agricultural soils? *GCB Bioenergy* 5 (2): 96–103.

Rajkovich, S., Enders, A., Hanley, K. et al. (2012). Corn growth and nitrogen nutrition after additions of biochars with varying properties to a temperate soil. *Biology and Fertility of Soils* 48: 271–284.

Ren, X., Tang, J., Wang, L. et al. (2021). Combined effects of microplastics and biochar on the removal of polycyclic aromatic hydrocarbons and phthalate esters and its potential microbial ecological mechanism. *Frontiers in Microbiology* 12: 647766.

Rinklebe, J., Shaheen, S.M., El-Naggar, A. et al. (2020). Redox-induced mobilization of Ag, Sb, Sn, and Tl in the dissolved, colloidal and solid phase of a biochar-treated and un-treated mining soil. *Environment International* 140: 105754.

Rombola, A.G., Fabbri, D., Baronti, S. et al. (2019). Changes in the pattern of polycyclic aromatic hydrocarbons in soil treated with biochar from a multiyear field experiment. *Chemosphere* 219: 662–670.

Smith, C.R., Hatcher, P.G., Kumar, S. et al. (2016). Investigation into the sources of biochar water-soluble organic compounds and their potential toxicity on aquatic microorganisms. *ACS Sustainable Chemistry & Engineering* 4 (5): 2550–2558.

Thamiris Ferreira, D.S. and Dias Ferreira, G.M. (2024). Biochars as adsorbents of pesticides: laboratory-scale performances and real-world contexts, challenges, and prospects. *ACS ES&T Water* 4 (10): 4264–4282.

Visioli, G., Conti, F.D., Menta, C. et al. (2016). Assessing biochar ecotoxicology for soil amendment by root phytotoxicity bioassays. *Environmental Monitoring and Assessment* 188 (3): 166.

Wang, D., Felice, M.L., and Scow, K.M. (2020). Impacts and interactions of biochar and biosolids on agricultural soil microbial communities during dry and wet-dry cycles. *Applied Soil Ecology* 152: 103570.

Wang, D., Zhang, W., Hao, X., and Zhou, D. (2013). Transport of biochar particles in saturated granular media: effects of pyrolysis temperature and particle size. *Environmental Science & Technology* 47 (2): 821–828.

Wang, J., Xia, K., Waigi, M.G. et al. (2018). Application of biochar to soils may result in plant contamination and human cancer risk due to exposure of polycyclic aromatic hydrocarbons. *Environment International* 121: 169–177.

Wang, Y.Y., Jing, X.R., Li, L.L. et al. (2017). Biotoxicity evaluations of three typical biochars using a simulated system of fast pyrolytic biochar extracts on organisms of three kingdoms. *ACS Sustainable Chemistry & Engineering* 5 (1): 481–488.

Wang, Y., Zhang, W., Shang, J. et al. (2019). Chemical aging changed aggregation kinetics and transport of biochar colloids. *Environmental Science & Technology* 53: 8136–8146.

Xiang, L., Liu, S., Ye, S. et al. (2021). Potential hazards of biochar: the negative environmental impacts of biochar applications. *Journal of Hazardous Materials* 420: 126611.

Yao, K., Lv, X., Zheng, G. et al. (2018). Effects of carbon quantum dots on aquatic environments: comparison of toxicity to organisms at different trophic levels. *Environmental Science & Technology* 52 (24): 14445–14451.

Zhang, K., Wang, Y., Mao, J. et al. (2020). Effects of biochar nanoparticles on seed germination and seedling growth. *Environmental Pollution* 256: 113409.

Zhang, Y., Yang, R., Si, X. et al. (2019). The adverse effect of biochar to aquatic algae-the role of free radicals. *Environmental Pollution* 248: 429–437.

Zhao, J., Cao X., Wang, Z. et al. (2017). Mechanistic understanding toward the toxicity of graphene-family materials to freshwater algae. *Water Research* 111: 18–27. https://doi.org/10.1016/j.watres.2016.12.037.

Zhao, J., Wang, Z.Y., Liu, X.Y. et al. (2011). Distribution of CuO nanoparticles in juvenile carp (*Cyprinus carpio*) and their potential toxicity. *Journal of Hazardous Materials* 197: 304–310.

Zhu, X., Chen, B., Zhu, L. et al. (2017). Effects and mechanisms of biochar-microbe interactions in soil improvement and pollution remediation: a review. *Environmental Pollution* 227: 98–115.

Index

Note: Page numbers in *italics* and **bold** refers to figures and tables respectively.

X

Z

Printed and bound by CPI Group (UK) Ltd, Croydon, CR0 4YY
17/09/2026
14964329-0008